Marine electronic navigation

Marine electronic navigation

S. F. Appleyard C ENG MIERE MRIN
R. S. Linford MSC, BSC (HONS), C ENG MIEE
P. J. Yarwood MSC, C ENG, MIERE

Incorporating 'The gyro-compass'
by the late G. A. A. Grant MA FRIN FRGS
and the late J. Klinkert FRIN FRMETS

Routledge & Kegan Paul · London and New York

First published 1980
Revised edition published in 1988 by
Routledge & Kegan Paul Ltd
11 New Fetter Lane, London EC4P 4EE

Published in the USA by
Routledge & Kegan Paul Inc.
in association with Methuen Inc.
29 West 35th Street, New York, NY 10001

Set in Times 11/13
by Hope Services, Abingdon
and printed in Great Britain
by Butler & Tanner Ltd
Frome and London

Library of Congress Cataloging in Publication Data
Appleyard, S. F.
Marine electronic navigation / S.F. Appleyard, R. S. Linford, P.J. Yarwood; incorporating 'The gyro-compass' by the late G.A.A. Grant and the late J. Klinkert.—Rev. ed.
p. cm.
Includes index.
ISBN 0–7102–1271–2
1. Electronics in navigation. I. Linford, R.S. II. Yarwood, P.J. III. Grant, G.A.A. Gryro-compass.
VK560.A66 1987
623.89'3—dc 19

British Library CIP Data also available
ISBN 0–7102–1271–2

To our long suffering families

Contents

Preface to the second edition

Over the past seven years *Marine electronic navigation* has become established as a reference text for both nautical students and for all who have a professional involvement with marine electronic navigation systems. In response to demand, this second edition has been substantially enlarged to include all of the electronic systems now encountered by navigation/communication personnel. To achieve this I sought co-authors with both theoretical and practical knowledge of their respective topics – Dick Linford and Peter Yarwood have these qualities. The section on the gyro-compass is extracted from *The ship's compass*; there can have been no greater authorities on their subject than J. Klinkert and G. A. A. Grant.

On reading the final text I am pleased to see that my original objectives have been maintained. Again I give no excuse for including basic theory which in some cases will require a little effort in digesting, but the benefits of not operating equipment as a black-box, but with real understanding, ultimately makes the effort worthwhile. Again I say that it is least important to consider specific manufacturers' hardware since this is always well covered in their manuals; it is always most important to highlight the fundamental deficiencies of each system since the consequences of believing them to be perfect can ultimately be a serious mistake.

Acknowledgements are given by all authors to the following companies, organizations and individuals; US Defense Mapping Agency Hydrographic/Topographic Center, Washington D.C. 20315, Racal-Decca Marine Navigation Ltd, Racal Marine Radar Ltd, Marconi International Marine Co Ltd, Mr J. B. Hacking of Marconi Communication Systems Ltd and David Green of Marconi Research for valuable comment on the section on communictions.

Finally my gratitude to Mrs Klinkert for her permission to use the work of her late husband.

S. F. Appleyard

1 Radiation and propagation

An understanding of the behaviour of radio waves during propagation from their source to point of reception is fundamental to an understanding of all radio navigation systems. It is particularly relevant to an appreciation of their limitations, since propagational effects are invariably the major source of error. (A more detailed treatment of this topic is included in chapter 19.)

1.1 Electromagnetic waves

Radio waves form a specific part of the total spectrum of electromagnetic radiation of which light is also a part. An electromagnetic wave can be considered as an oscillating electric force travelling through space, and inseparably accompanied by an oscillating magnetic force in a plane at right angles to it (Figure 1.1). The plane of the electric field in space provides the basis of defining the wave's

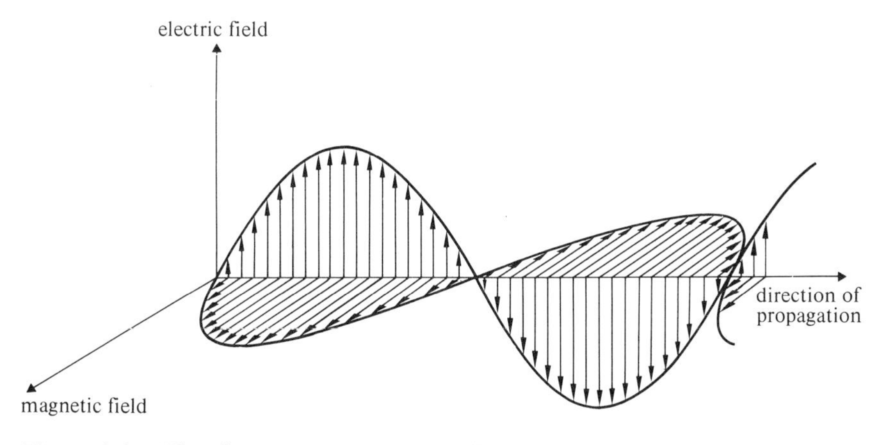

Figure 1.1 The electromagnetic wave, illustrating the relationship between electric and magnetic fields, and direction of propagation.

polarization. A wave with a vertical electric field is said to be vertically polarized.

The relationship of radio waves with the rest of the spectrum of electromagnetic radiation is illustrated in Figure 1.2. Radio waves are usually specified in terms of their frequency (f), which is related to wavelength (λ) by the expression:

$$f = \frac{c}{\lambda} \qquad \textbf{[1.1]}$$

where c is the velocity of electromagnetic radiation in a vacuum (free space) and has been determined as being 299, 792 km/s, although the approximation of 3×10^5 km/s is often used in equation **[1.1]**.

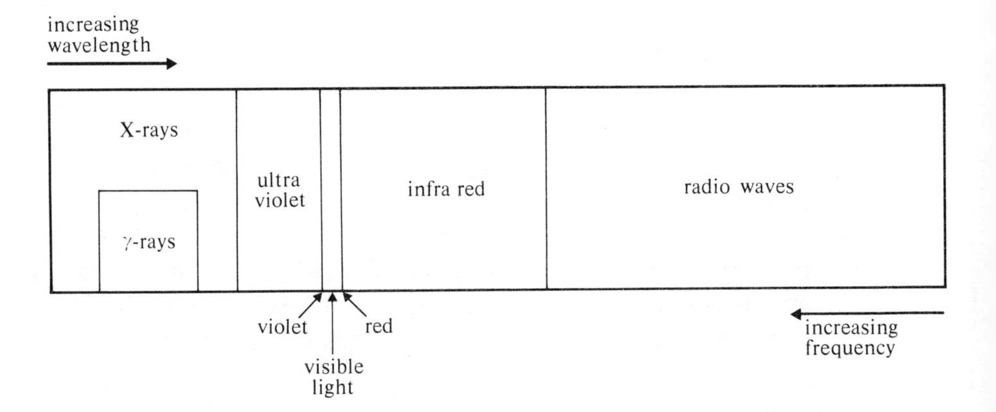

Figure 1.2 The spectrum of electromagnetic radiation.

The unit of frequency is the Hertz (Hz), and the radio wave part of the electromagnetic spectrum extends from 3×10^3 Hz to 30×10^9 Hz, although these are not rigidly defined limits. Since very large numbers are involved the following prefixes are assigned:

10^3Hz = 1 kilohertz (kHz)
10^6Hz = 1 Megahertz (MHz)
10^9Hz = 1 Gigahertz (GHz)

Figure 1.3 illustrates the position of the different radio navigation aids in the radio wave spectrum and relates them with other sources of radio waves.

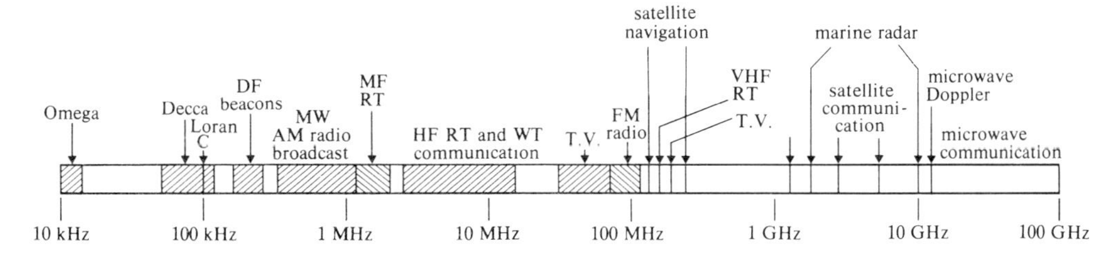

Figure 1.3 The radio wave spectrum.

1.2 Propagation

The behaviour of radio waves and the influences which affect them during their passage from transmitter to receiver are dependent upon the frequency of the wave. All radio waves within a given frequency band will have the same propagational characteristics irrespective of their use, and so the following descriptions of propagation apply equally to other radio signals on the same or adjacent frequencies.

When energy is radiated from an omnidirectional transmitting antenna some energy will travel away from the earth, and some will travel away from the antenna remaining (initially) parallel with the ground. In explaining the mechanisms of propagation these two directions are considered separately, and are termed 'skywave' and 'groundwave' respctively. The relative importance of the skywaves and groundwaves depends upon many factors, which include the frequency of the transmission, the time of day, and the distance between transmitter and receiver.

1.2.1 *Groundwaves*

The groundwave can be subdivided into two components, the 'space wave' and the 'surface wave'. The space wave can be further divided into the 'direct wave' and the 'ground reflected wave'. These latter two waves illustrated by Figure 1.4 are of little significance in the various radio navigation systems described in this book, since their range is short and in many cases the two waves cancel at the receiver.

Of more significance is the surface wave, since in this case the earth's surface and the lower atmosphere influence the wave in such

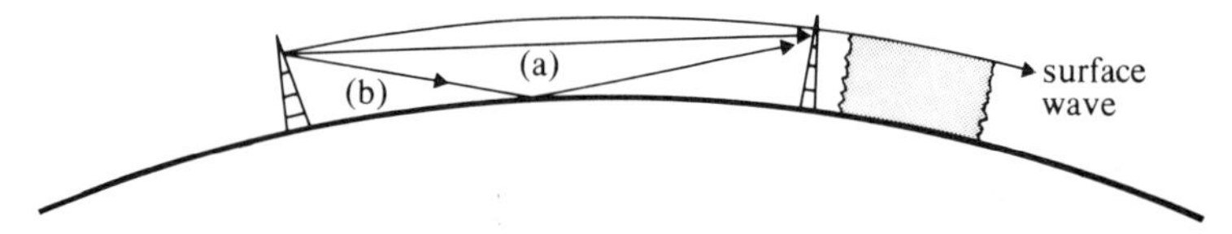

Figure 1.4 *The two components of groundwave propagation, space wave and surface wave. Space wave has two components,* (a) *direct wave, and (b) ground reflected wave.*

a way as to cause it to follow the curvature of the earth. Since energy is transferred from this wave to the ground, the distance over which the wave can propagate depends upon the frequency of the transmission, and the properties of the ground over which the wave passes. The distance over which a surface wave can travel before suffering unacceptable attenuation varies from only hundreds of feet to many thousands of miles.

At low and medium frequencies, horizontally polarized surface waves suffer much greater attenuation than vertically polarized surface waves. In this frequency range therefore, antennas are designed to transmit and receive vertically polarized waves.

Since the space wave does not play any significant part at the frequencies of the radio navigation aids described in this book, the general expression groundwave is used throughout to mean surface wave.

1.2.2 *Skywaves*

It may be thought that waves travelling away from the surface of the earth would be lost into space and thus play no further part. This is by no means always the case since around the earth is the ionosphere, a belt of ionized gases which extends from approximately thirty miles to several hundreds of miles from the earth's surface. The effect of the ionosphere is to cause waves of certain frequencies to refract, and ultimately reflect back to the surface.

1.3 The ionosphere

During day-time the densest region of ionization exists between altitudes from sixty to six hundred miles. Throughout this region there are several layers in which the ionization density is at a maximum, known as the D-, E- and F-layers (Figure 1.5a). During the day, the F-layer splits into two layers and these are designated F1 and F2. The density of ionization of these layers depends upon many factors including time of day, season, latitude, and the phase of the eleven-year sunspot cycle. During night-time, all layers of the ionosphere slowly de-ionize. In particular the D-region quickly disappears in the absence of the sun and quickly ionizes shortly after the following sunrise (Figure 1.5b).

During both day- and night-time the ionosphere has the effect of refracting the radio waves which pass through it. The amount by which the waves are refracted is dependent upon the density of the ionization of each of the layers, and on the frequency of the radio waves. In general, as the frequency decreases the amount of refraction increases, until the point is reached where the wave is

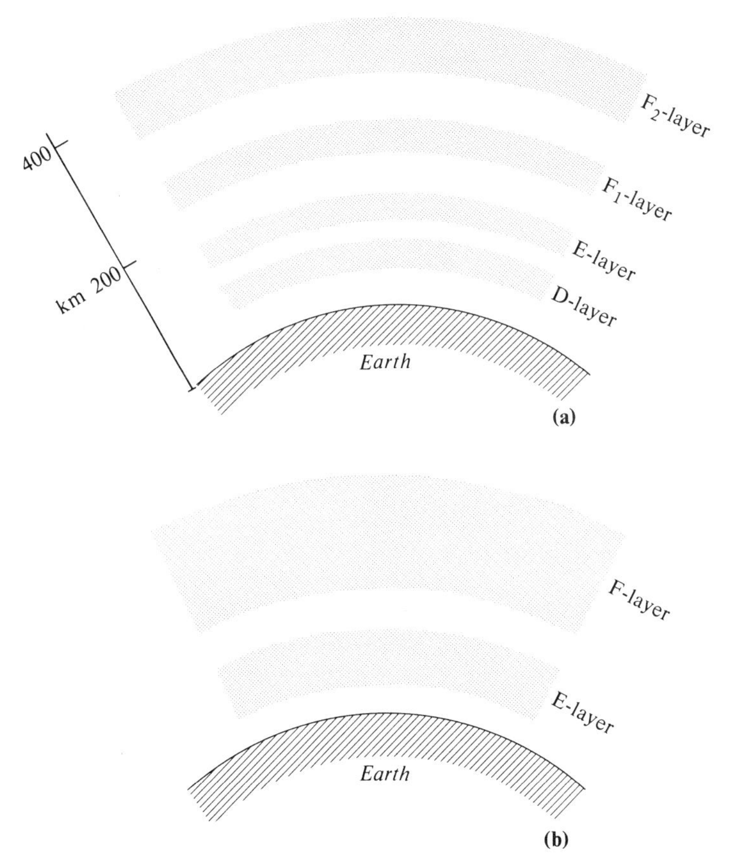

Figure 1.5 The ionosphere: (a) *day-time;* (b) *night-time.*

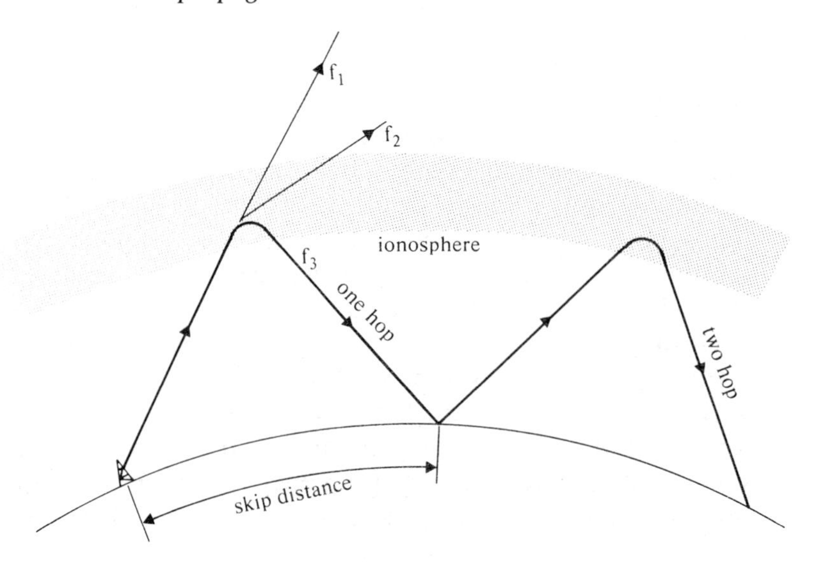

Figure 1.6 Ionospheric refraction and reflection of radio waves, frequency $f_1 > f_2$ *and* $f_2 > f_3$.

actually reflected back from the ionosphere (Figure 1.6). Since the density of the ionosphere varies daily and seasonally, a radio wave of a given frequency may be reflected at some times and not at others.

The skywave returning to earth provides signal reception at a distant point from the transmitter, termed the skip distance (Figure 1.6). Skywaves can undergo two or more reflections. When radio signals are used for communication, the presence of the reflected skywave is of great value since it makes communication over many thousands of miles possible, far beyond the range of the groundwave signal. Communication frequencies are therefore often chosen to make optimum use of skywave signals.

The opposite to this is generally the case for radio navigation systems, since these rely upon a precise knowledge of the propagation times of the signal from transmitter to receiver. For communication purposes accurate ionospheric predictions can be made relating to the presence of skywave reflections for given frequencies, but it is the precise time delays caused by the ionosphere which are difficult to predict with precision. In certain cases skywaves are used to give extended operation, Loran C being one example, but the positional accuracies are considerably reduced from those normally achieved with groundwaves. Usually the presence of a skywave is a matter of nuisance, and in extreme cases the skywave interferes with the groundwave to cause a system to become unusable.

Transmissions from Loran C (100 kHz), Decca (70–130 kHz) and direction-finding (DF) beacons (up to 350 kHz) behave in a similar manner. During day-time the ionized D-region attenuates the skywave both before and after it is reflected by the E-region. The skip distance falls within the groundwave, but the skywave has been attenuated sufficiently to prevent serious interference with the groundwave.

At night the D-region de-ionizes and the attenuation of the skywave is now less. Reflections occur from both the E- and F-layers, with some signals returning within the groundwave and some beyond. The precise effect of these skywaves on the performance of each of the radio navigation skywaves is discussed under the heading of the particular system.

1.4 Very low frequency (VLF) propagation

The ray method of describing and analysing skywave propagation becomes cumbersome at very low frequencies (<30 kHz), since for distances beyond about 1000 kilometres from the transmitter the signal has suffered many successive ionospheric reflections. A more convenient method of describing VLF propagation is to use waveguide theory, one wall of the waveguide being the surface of the earth and the opposite wall being the lower region of the ionosphere. The effective height of the waveguide varies daily (and seasonally), between approximately seventy kilometres by day and ninety kilometres by night. At the frequencies of the Omega VLF navigation system (10.2–13.6 kHz), this represents some two to four wavelengths.

The normal waveguide theory of two component waves travelling obliquely to the walls of the guide, fully applies in the case of the earth-ionosphere waveguide. A wave which makes the double passage across the guide and back, and undergoes reflections at two walls must be self consistent. This is only possible for certain discrete directions of the wave normals, and so leads to a series of discrete modes. In a loss-free system there is no change of amplitude on reflection, and so the condition for self consistency is that the total change of phase in the double passage across the guide and back (with two reflections) is an integral multiple of 2π radians.

In a conventional waveguide the walls are clearly defined, reflecting waves incident upon their surface. Waves can also be reflected from a wall whose boundary is not clearly defined, but with a refractive index which varies continuously across the boundary. Such is the case with the ionosphere, although at Omega frequencies the wave has totally reflected within the D-layer. The layer can be considered as having a discrete boundary at a specific reflecting height, and with an equivalent reflection coefficient.

When the earth ionosphere waveguide is excited by a radiating antenna, many modes exist within the first 500 kilometres from the antenna, consequently within this distance the mode theory becomes cumbersome. Fortunately the high-order modes are attenuated more than the low-order and so beyond 500 kilometres it is necessary only to consider the first- and second-order modes.

Omega transmissions are launched from vertically polarized antennas which create transverse-magnetic propagation modes within the waveguide (Figure 1.7). At a distance of 1000 kilometres from the antenna only the TM1 and TM2 modes are large enough to be of interest, beyond this the amplitude of the TM2 mode falls below that of the TM1, and then only this latter mode remains significant.

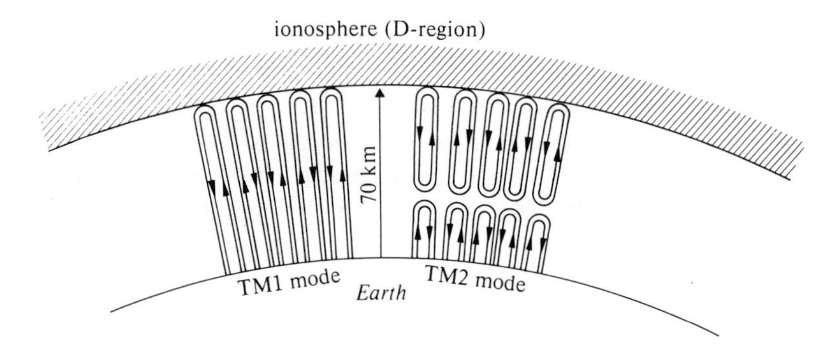

Figure 1.7 VLF propagation within the earth-ionosphere waveguide, illustrating the two primary modes TM1 and TM2.

1.4.1 *Phase and group velocity*

Figure 1.8 is a plot of the transverse field along the longitudinal axis of a waveguide. The diagram is drawn for one instant of time and for a guide transmitting an unmodulated sinusoidal wave. If Figure 1.8

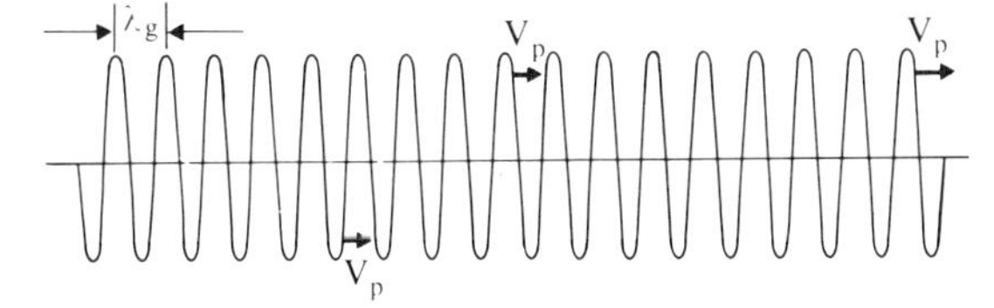

Figure 1.8 Unmodulated sinusoidal wave in a waveguide.

is redrawn for several successive instants of time, a changing field pattern is obtained and the wave appears to move along the X-axis with a velocity V_p called the 'phase velocity'.

The value of V_p may be found from the fact that the wave in Figure 1.8 moves forward a distance equal to λ_g in one cycle. Thus, in one unit of time the wave moves forward f cycles, or a distance $f.\lambda_g$, where f is the frequency of the wave. That is:

$$V_p = f.\lambda_g.$$

In a similar way, the velocity of any electromagnetic wave in free space is:

$$c = f.\lambda_a.$$

Dividing these two equations and multiplying by c yields:

$$V_p = \frac{\lambda_g}{\lambda_a}.c.$$

Thus the phase velocity in the guide is greater than the velocity of light in the same ratio that the wavelength in the guide is greater than the wavelength in free space. This high value of V_p is possible despite the physical principle that signals cannot be transmitted with a velocity greater than c. The wave in Figure 1.8 is unmodulated and therefore does not transmit any intelligence.

The actual phase velocity of Omega signals within the earth-ionosphere waveguide is dependent upon the height of the ionosphere and the ground conductivity, and can vary between 0.9992c and 1.003.5c. The nominal phase velocity which is used in computing Omega hyperbolae is 1.0026c, which is 300,574 km/s.

If a waveform is modulated or is in the form of a pulse, the corresponding velocity of the waveform is termed the group velocity, and cannot be greater than c.

2 Measurement of time

The accurate measurement of time has always been fundamental to navigation, and the history of man's continuous endeavour to improve his abilities in this field is well documented.

Progress has ultimately led to the measurement of time using fundamental properties of the atom, and this has resulted in a redefinition of the second as being the duration of 9,192,631,770 periods of the radiation corresponding to the transition between two hyperfine levels of the ground state of the cesium atom 133.

The navigator however is still primarily concerned with time related to the daily rotation of the earth. The relationship between this and atomic time is summarized by Figure 2.1.

The measurement of time in electronic navigation is more usually concerned with the lapse of time between the occurrence of two events, such as the time between the transmitted and received pulses of an echosounder, or the time between reception of master and slave pulses in a Loran C system.

The principal requirements of an electronic navigation system are therefore, (1) an integral source of time which is precise and stable over the measurement interval, and (2) a means of measuring the time lapse.

The source of time is usually an oscillator, the frequency of which (hence time, since $t = 1/f$) is maintained constant within the required stability tolerance. One method of maintaining a stable oscillation frequency is to utilize the natural resonances of quartz crystal. By maintaining the temperature of a crystal oscillator constant, frequency stabilities of better than 1×10^{-9}/day are achievable. Such an oscillator could then be used to measure the lapse of time in the manner illustrated by Figure 2.2. In this case the oscillator is set precisely to a frequency of 1 MHz; the period of the oscillation is thus one microsecond ($t = 1/f$). The oscillator's sine wave output voltage is converted to pulses which are then fed to a pulse counting circuit. The pulse counting commences at the start of event A (eschosounder transmission, etc.) and terminates at the start of event B (returned echo, etc.). The number of pulses counted

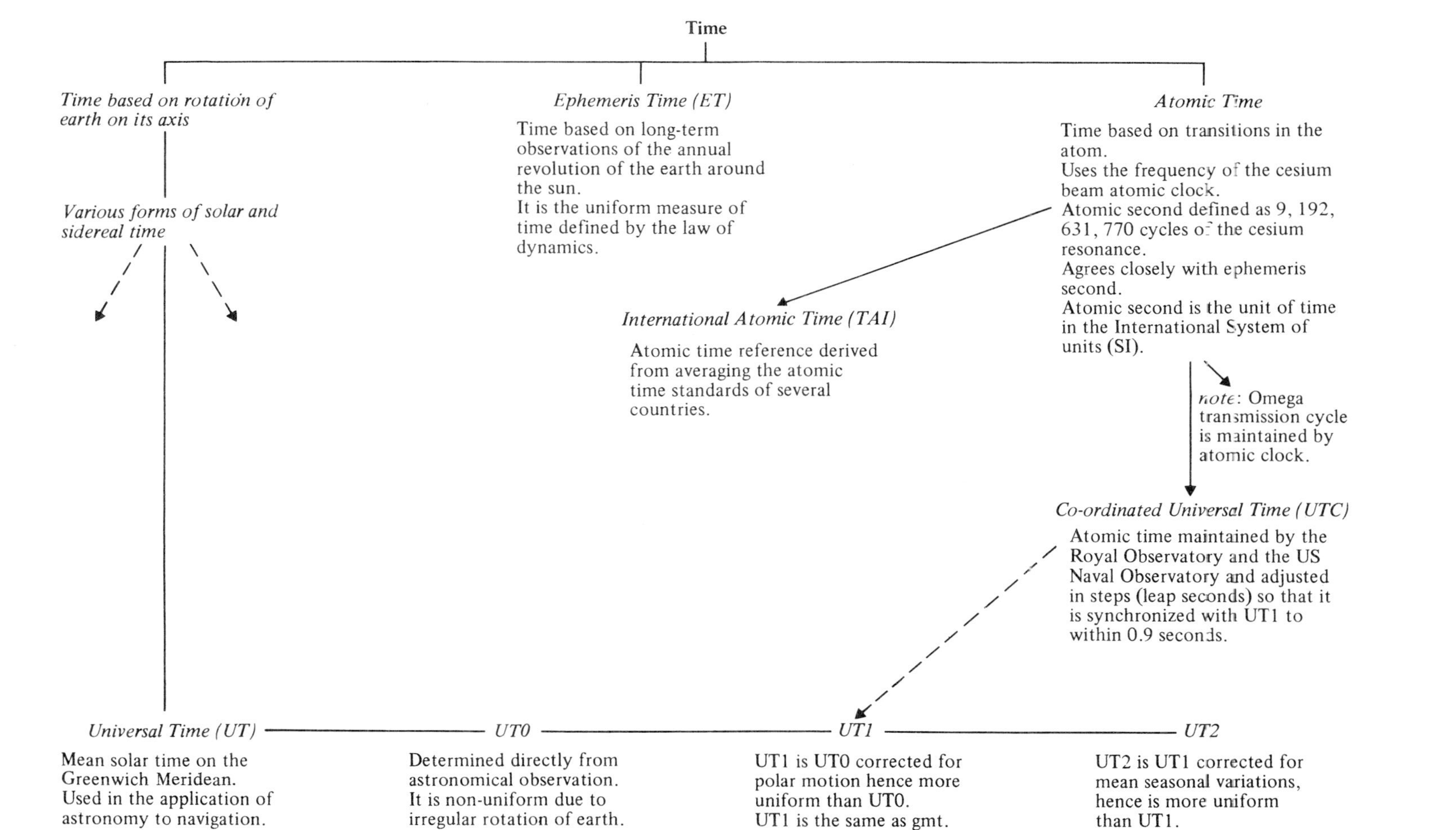

Time
Time based on rotation of earth on its axis
Ephemeris Time (ET)
Time based on long-term observations of the annual revolution of the earth around the sun.
It is the uniform measure of time defined by the law of dynamics.
Atomic Time
Time based on transitions in the atom.
Uses the frequency of the cesium beam atomic clock.
Atomic second defined as 9, 192, 631, 770 cycles of the cesium resonance.
Agrees closely with ephemeris second.
Atomic second is the unit of time in the International System of units (SI).
Various forms of solar and sidereal time
International Atomic Time (TAI)
Atomic time reference derived from averaging the atomic time standards of several countries.
note: Omega transmission cycle is maintained by atomic clock.
Co-ordinated Universal Time (UTC)
Atomic time maintained by the Royal Observatory and the US Naval Observatory and adjusted in steps (leap seconds) so that it is synchronized with UT1 to within 0.9 seconds.
Universal Time (UT)
UT0
UT1
UT2
Mean solar time on the Greenwich Meridean.
Used in the application of astronomy to navigation.
Determined directly from astronomical observation.
It is non-uniform due to irregular rotation of earth.
UT1 is UT0 corrected for polar motion hence more uniform than UT0.
UT1 is the same as gmt.
UT2 is UT1 corrected for mean seasonal variations, hence is more uniform than UT1.

Figure 2.1 The inter-relationship of the different methods of measuring and defining time.

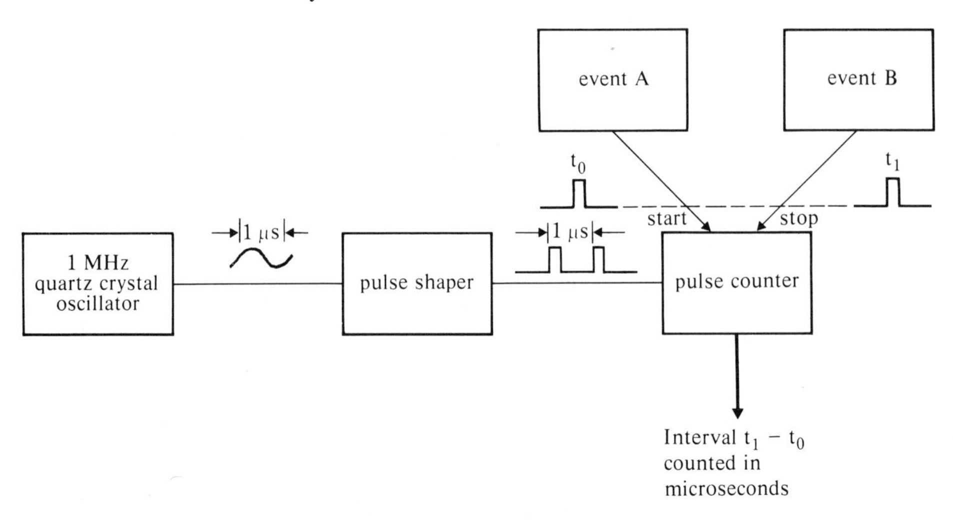

Figure 2.2 Simplified illustration of measuring time lapse in electronic navigation systems.

represents the time lapse in microseconds; a greater degree of timing resolution is achieved by using a higher frequency of oscillation.

One timing requirement which cannot be adequately met by a crystal oscillator is that associated with range-range navigation. For example, in the case of range-range operation of Omega, the phase of the reference oscillator of the receiver must remain precisely synchronized with the transmitted signal, so that any subsequent change in phase of the reference with respect to the received signal is due only to movement of the vessel.

To maintain a timing accuracy of this order over an operational period of several weeks, requires a stability which is better than that of quartz crystal oscillators. It therefore becomes necessary to use an atomic frequency standard, of which there are two basic types commercially available, cesium and rubidium. To describe their basic operation, the same simplified block diagram can be used for both the cesium and the rubidium standards (Figure 2.3). The output frequency f_0 (typically 1–10 MHz) of a low-noise quartz crystal oscillator is multiplied and synthesized to the atomic resonance frequency (+6834 MHz for rubidium and +9192 MHz for cesium). The signal is frequency modulated to sweep through the atomic resonance frequency, causing the beam intensity in the cesium tube, or transmitted light through the rubidium cell, to vary.

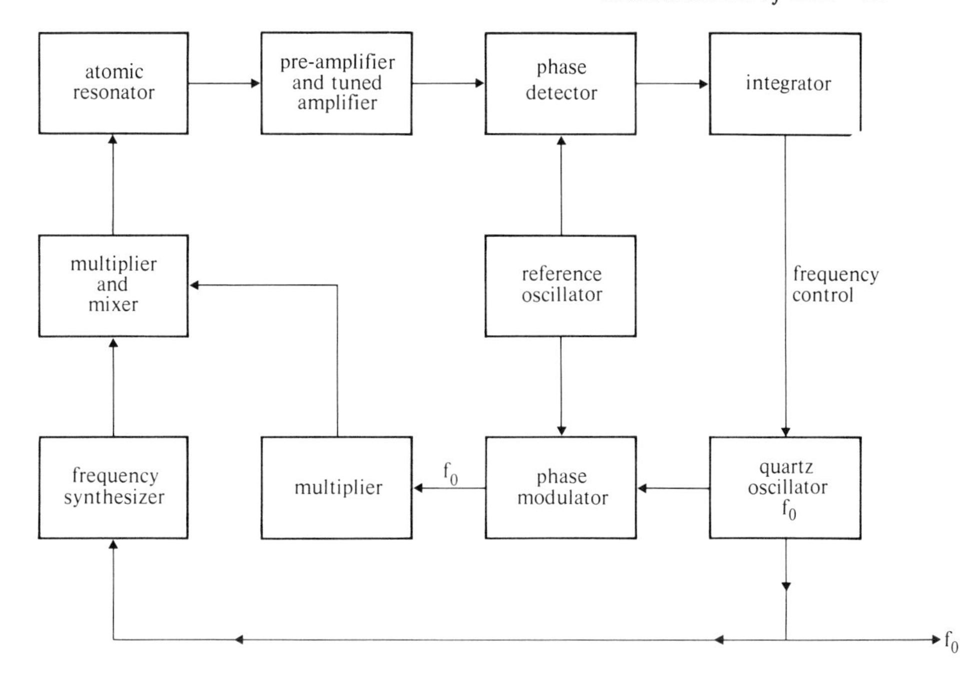

Figure 2.3 Simplified illustration of an atomic frequency standard (courtesy Hewlett Packard Ltd).

The output signal is amplified and through a phase detector controls the frequency f_0 of the quartz crystal oscillator.

The invariant resonance frequency of the cesium atoms passing through a microwave cavity, maintains the output frequency of the cesium standard constant to extremely high accuracy. The cesium beam standard is a primary standard, meaning that it does not require any other reference for calibration. An accuracy of $\pm 1 \times 10^{-11}$ is typical for a commercially available standard, and with a stability of $\pm 3 \times 10^{-12}$ over the life of the beam tube.

The rubidium standard is similar to the cesium beam standard in that an atomic resonant element prevents drift of a quartz oscillator through a frequency lock loop. In this case however, the performance of the rubidium gas cell is dependent upon gas mixture and gas pressure in the cell. It therefore requires calibration and is subject to a small degree of drift, although this is typically only $\pm 1 \times 10^{-11}$/month, which is at least one hundred times better than the best quartz crystal standard.

Some navigation systems include absolute time as an inherent part of their transmission. Commercial Transit satellite navigation receivers can determine absolute time from the satellite transmissions to better than 25 μs. The newer Navstar system is likely to permit the determination of time to better than 1 μs.

This ability to determine time to a high degree of accuracy on board ship could lead to new schemes, such as a collision avoidance system based on time transmission. If all vessels transmitted a precisely synchronized pulse – the time from transmission to the first received emission – would give the distance to the nearest vessel with the same system.

3 Error

A simple account is given here of the terminology and the mathematical expressions used in describing navigational errors. Reference to specific error causes is made only as a means of illustration, since these are discussed fully under each particular system heading.

Within the context of this explanation, error is defined as the difference between the value indicated by the navigation instrument and the true value which corresponds to the actual position or position line (LOP). The resultant error of a navigation system will usually have contributions from the transmission system, the propagation path, and the receiver. In the case of the particular receivers which numerically display position information, it can generally be assumed that there is no error contribution due to inaccuracies in reading the instrument's display.

In general errors can be described as being either 'systematic' or 'random'. Systematic errors are basically those errors which follow a law, the simplest type being constant errors such as those caused by the misalignment of the bearing pointer of the DF receiver, or the oscillator off-set error of a satellite navigation receiver. More complex examples of systematic error are the quadrantal error encountered in DF installations, and the diurnal propagation variations which influence Omega lines of position. Systematic errors can be compensated for, the effectiveness of the compensation depending ultimately upon how well the law governing the error is understood. In contrast, random errors are governed by the laws of probability and so cannot be either predicted or compensated. Random errors occur because of short-term effects, such as short-term variations in the ionosphere.

3.1 One-dimensional error distribution

Although a random error cannot be predicted in isolation, a large number of successive random errors of a given system will exhibit certain characteristics, such as a law governing the distribution of the errors, and a numerical representation of their spread. This is readily illustrated by the following example.

The displayed readings of one position line were taken from an Omega receiver at ten-second intervals. After applying the diurnal propagation correction prevailing at that time, the results obtained were as tabulated below. The correct value for the LOP was determined by computation as being 981.23.

Omega LOP (L)		*Occurrence* (N)	
L_1	981.20	N_1	1
L_2	981.21	N_2	2
L_3	981.22	N_3	5
L_4	981.23	N_4	11
L_5	981.24	N_5	20
L_6	981.25	N_6	30
L_7	981.26	N_7	37
L_8	981.27	N_8	40
L_9	981.28	N_9	38
L_{10}	981.29	N_{10}	29
L_{11}	981.30	N_{11}	19
L_{12}	981.31	N_{12}	12
L_{13}	981.32	N_{13}	4
L_{14}	981.33	N_{14}	2
			250 total readings (N)

Using these results a mean value (μ) can be calculated for the LOP:

$$\mu = \frac{L_1N_1 + L_2N_2 + \ldots}{N_1 + N_2 + \ldots}$$

This yields a mean of 981.27 (Figure 3.1).
The difference of four centilanes between this mean value and the correct value is due to uncompensated systematic error.

A single number can now be determined which describes the amount of spread about the mean. This is termed the 'root mean square' (RMS) error and is given by:

$$\text{RMS error} = \sqrt{\frac{\Sigma.d^2}{N}}$$

where d is the deviation from the mean.

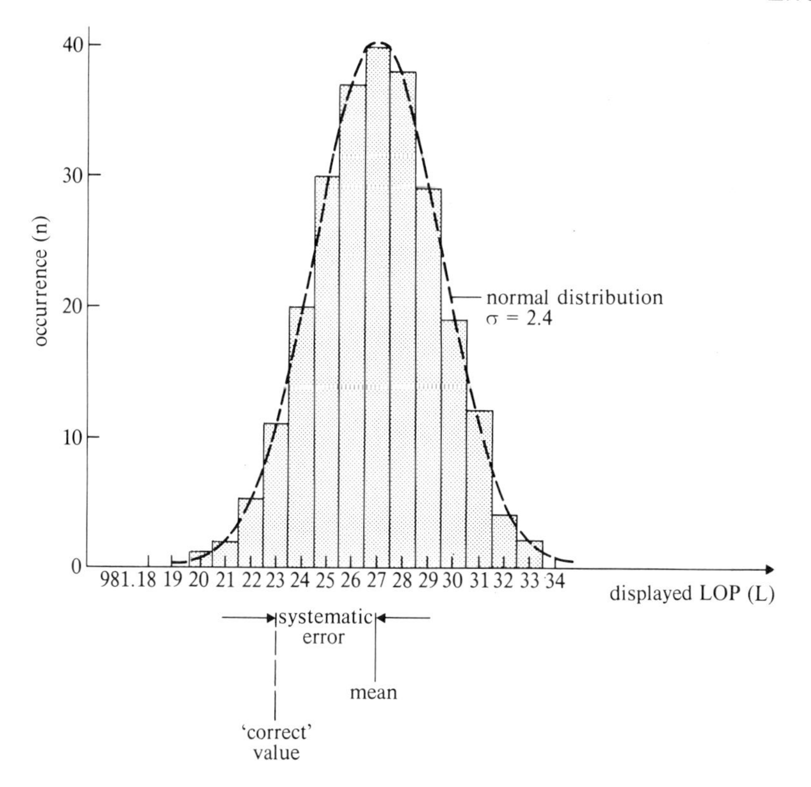

Figure 3.1 *Histogram of the distribution of Omega readings shown in the example on p.16.*

In the case of simple one-dimensional errors which are being considered here, RMS error is more commonly termed 'standard deviation' and denoted by the Greek letter sigma (σ).

For the example quoted:

$$\text{RMS error (or } \sigma) = \sqrt{\frac{1.(7^2) + 2.(6^2) + 5.(5^2) + 11.(4^2)}{250}}$$

This yields $\sigma = 2.4$.

Numerically, sigma corresponds to 68.27 per cent of the distribution. That is, if a large number of random measurements are made then 68.27 per cent will fall within one sigma (1σ) of the mean.

In general random variations of this type follow a Normal (Gaussian) distribution given by:

$$f(x). = \frac{1}{\sigma.\sqrt{2\pi}} .e^{-\frac{(x-\mu)^2}{2\sigma^2}}$$

where μ is the mean of the distribution.

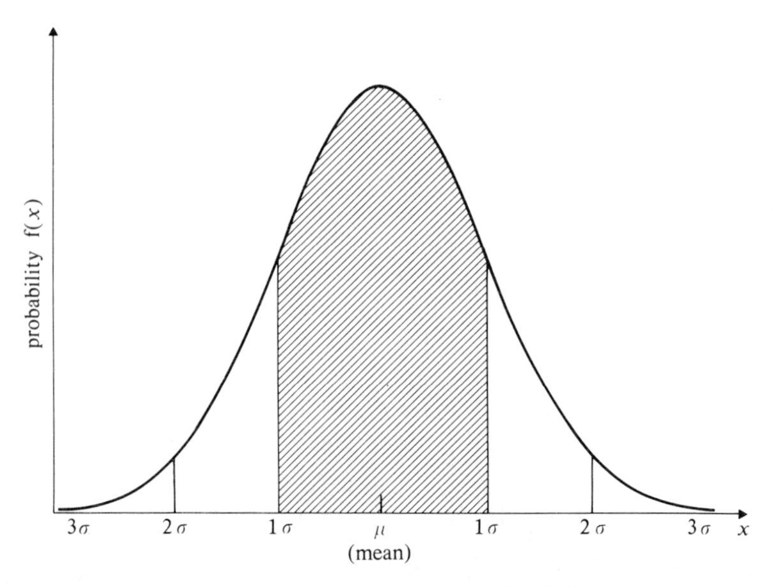

Figure 3.2 The Normal distribution; the shaded area represents 68.27 per cent of the samples.

In the Normal distribution 95.45 per cent of the sample falls within 2σ of the mean, and 99.73 per cent falls within 3σ. A navigation system's performance is often stated in terms of its 2σ error, meaning that there is a 95 per cent probability of the accuracy of a position line being better than the figure quoted.

The term 'probable error' is also used, being the error which represents 50 per cent of the distribution. That is, no more than half of the errors in the sample are greater in value than the probable error. Numerically the probable error is equal to 0.6745σ.

Not all random errors which are met in navigation and navigation systems follow precisely a normal distribution; however they are usually assumed to be normally distributed, at least up to the 2σ error points.

3.2 Two-dimensional error distribution

Discussion so far has related only to the one-dimensional errors of individual position lines. Two-dimensional errors of an actual position fix require a more complex analysis and definition.

Considering first the special case of orthogonal position lines having equal standard deviations and zero means. The resulting error distribution is described as being 'radial normal' or circular normal' and has an equation of the form:

$$f_{(R)} = 1 - e^{-\frac{R^2}{2\sigma^2}}$$

Because the distribution is now radial σ has a different numerical value, 1σ corresponds to 39.35 per cent probability, and 2σ to 86.47 per cent. For circular normal distributions the error is often specified by the 50 per cent probability radius which is termed the 'circular probable error' (CPE or CEP).

Orthogonal position lines with equal standard deviations rarely occur in practice, a more usual situation is that of Figure 3.3 in which the lines of position cross at an angle θ, and each position line has an independent error probability. In this general case the resulting error distribution is elliptical. An elliptical error distribution is not particularly meaningful to a navigator, and so this has led to the use of 'circles of equivalent probability', the radii of which are calculated so that the probability of a position falling within the

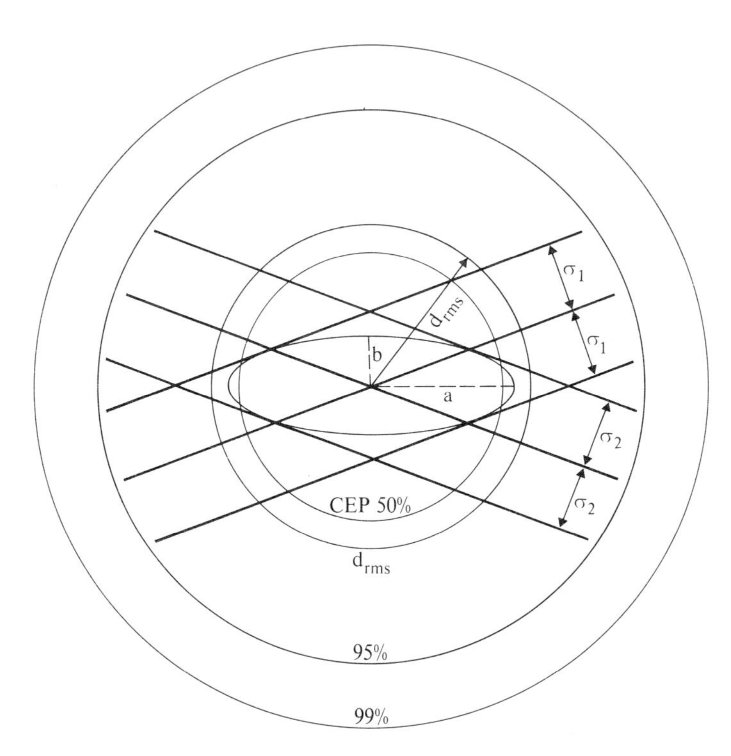

Figure 3.3 The error ellipse and circle of equivalent probability.

circle, is the same as (or related by a factor to) the probability of it falling within the ellipse.

There are various methods of converting error ellipses to circles of equivalent probability; some are rigorous, requiring the use of tables, and others make approximations such as equation **[3.1]** which gives an approximate value for the 50 per cent probability radius:

$$\text{CEP} \simeq 0.62\,\sigma_a + 0.56\,\sigma_b \qquad [3.1]$$

Multiplying the radius of the CEP circle by 2.08 gives the radius of the 95 per cent probability circle, and by 2.58 gives the 99 per cent radius.

In the case of one-dimensional errors the standard deviation and the root mean square error are numerically equal; this is not so in the case of two-dimensional errors. In this latter case the terms radial error, root mean square error, and d*rms* have the same meaning, which is the square root of the sum of the squares of the one sigma error components along the major and minor axes of the probability ellipse. Similarly, using 2σ values yields the 2d*rms* probability.

The numerical values of probability associated with d*rms* and 2d*rms* are not fixed quantities, but vary as a function of the eccentricity of the ellipse. Typically d*rms* can vary from being 63.21 per cent probability when $\sigma_a = \sigma_b$, to 68.2 per cent when σ_b is ten times greater than σ_a. Over this same range the probability associated with 2d*rms* will vary from 98.2 to 95.4 per cent.

4 Radio direction finding

Radio direction finding was the earliest of all radio navigation aids, its development following only a few years after the first use of radio for communication between ship and shore in 1898. Direction finding remains unique as the only radio navigation system providing the capability of determining the relative bearing of other vessels from their radio transmissions. The DF receiver can therefore play an important role in distress situations, and for this reason the DF receiver will almost certainly continue to be specified as mandatory by SOLAS (International Convention for the Safety of Life at Sea), for vessels over 1600 tons. Direction finding also continues to be a useful general radio navigation aid, providing a position fixing capability throughout many parts of the world, including areas not yet within the coverage of other systems.

In common with other radio navigation aids, a knowledge of the factors which affect system performance is essential. In general the most significant factors are those which influence the propagation of the radio waves, and this is true for radio direction finding, but in this case the influence of the ship itself on performance is also very significant.

Irrespective of the differences in outward appearance of the many types of DF receivers, the basis of their operation is always fundamentally the same: each utilizes an antenna exhibiting directional properties. A study of direction finding should therefore commence with a study of DF antenna operation.

4.1 DF antennas

4.1.1 The polar diagram

To describe the directional characteristics of antennas, use is made of the polar diagram. This is a technique of graphically presenting the response of an antenna to signals arriving from different directions. In the normal operation of marine DF receivers, the radio signal propagates only as a groundwave and so only the polar diagram relating to azimuth need be considered.

The simplest example of a polar diagram is that associated with a vertical, open wire antenna. This type of aerial is omnidirectional, meaning that it does not exhibit directional properties and so its diagram is simply a circle (Figure 4.1).

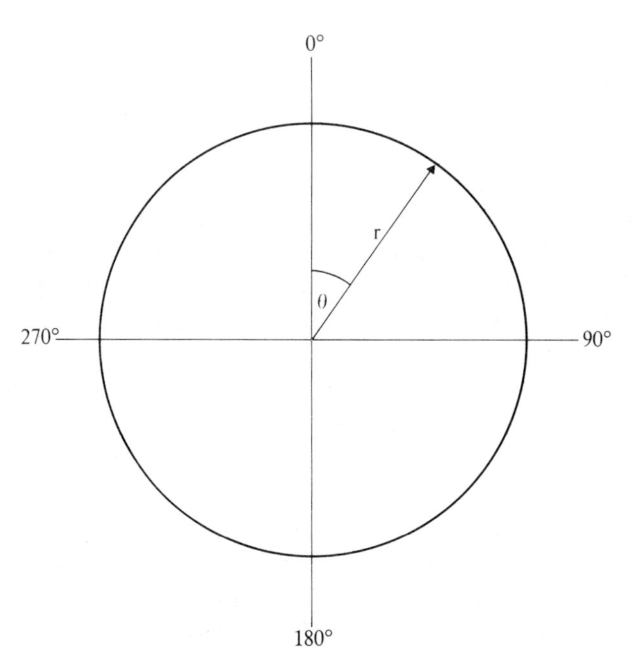

Figure 4.1 Polar diagram of an omnidirectional antenna.

It is often convenient to consider the simple polar responses as mathematical expressions. In general terms the expression will be of the form $r = f(\theta)$, where r is proportional to the voltage induced in the antenna for a given angular displacement θ.

The expression for the simple case of the omnidirectional antenna is:

$$r = C(\text{constant}) \qquad \textbf{[4.1]}$$

4.1.2 *The loop antenna*

Many different types of antenna exhibit directional characteristics and in theory they could each be used as the basis of a DF system.

The antenna type which in practice is used almost without exception in low-frequency marine DF is the simple loop, since it has the ideal polar response.

The loop antenna is simply a closed conducting loop the shape of which is unimportant, and so circular, triangular, and rectangular constructions are all equally common. The shape most suited to describing the theory of the loop antenna is the rectangle, since other shapes can be simply resolved into the horizontal and vertical components of the rectangular loop.

Since the waveform impinging on the antenna is normally vertically polarized, there will be voltage induced in the vertical elements only. The horizontal elements can therefore be considered as simply completing the electrical circuit.

In describing the operation of the loop antenna, two orientations are considered: (1) with the plane of the loop perpendicular to the direction of propagation of the wave (Figure 4.2); and, (2) with the plane of the loop parallel to the direction of propagation of the wave (Figure 4.3).

In the first case, the distance of the two vertical elements from the

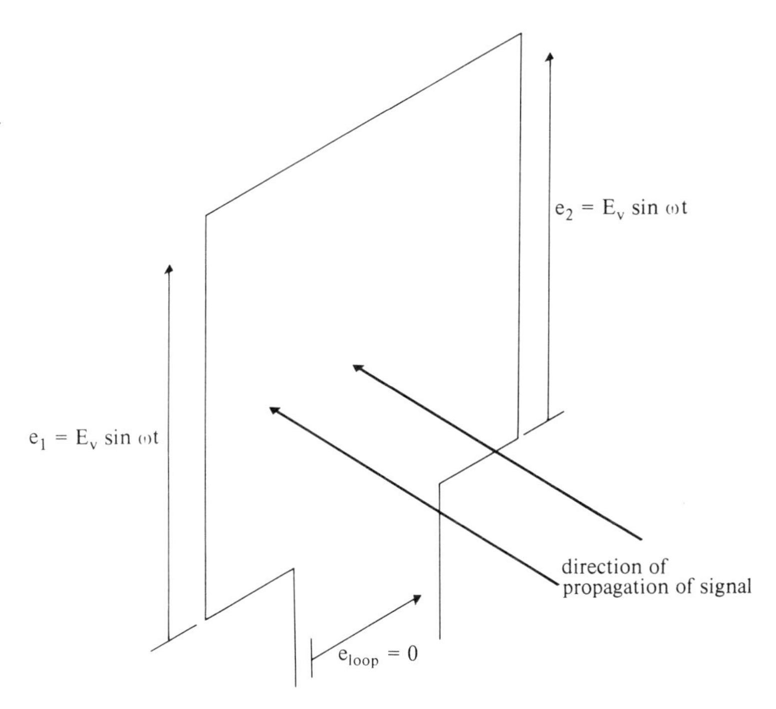

Figure 4.2 Induced voltages with loop normal to direction of propagation.

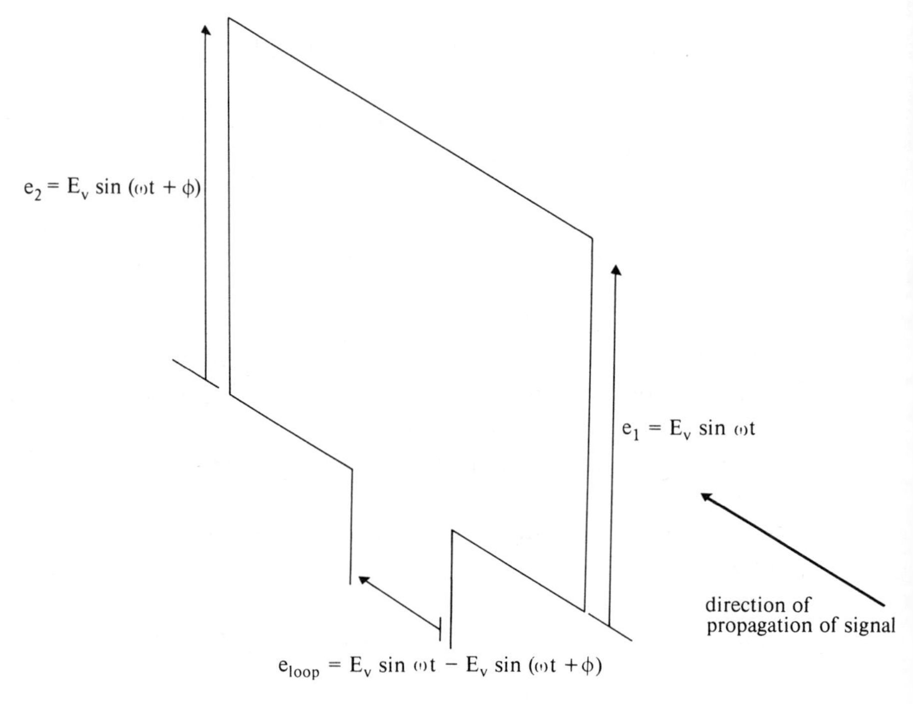

Figure 4.3 Induced voltages with loop parallel to direction of propagation.

transmitter is the same (Figure 4.2), and the voltages induced in each are therefore equal to amplitude and phase. Since no voltage is induced into the horizontal elements, the sum of the voltages around the loop is zero, and consequently there will be no circulating loop current.

Considering now the second case when the plane of the loop is in line with the direction of propagation of the incident wave (Figure 4.3). Since the distance between transmitter and antenna is always considerably greater than the dimensions of the loop, the voltage induced into the two vertical elements will always have the same maximum value E_v. The separation in the vertical elements now causes a difference in the instantaneous amplitude of the induced voltages e_1 and e_2. This is illustrated in Figure 4.4; voltage e_2 is now removed from voltage e_1 by the phase angle ϕ.

This can be expressed as:

$$e_1 = E_v \sin \omega t$$

and $e_2 = E_v \sin(\omega t + \phi)$

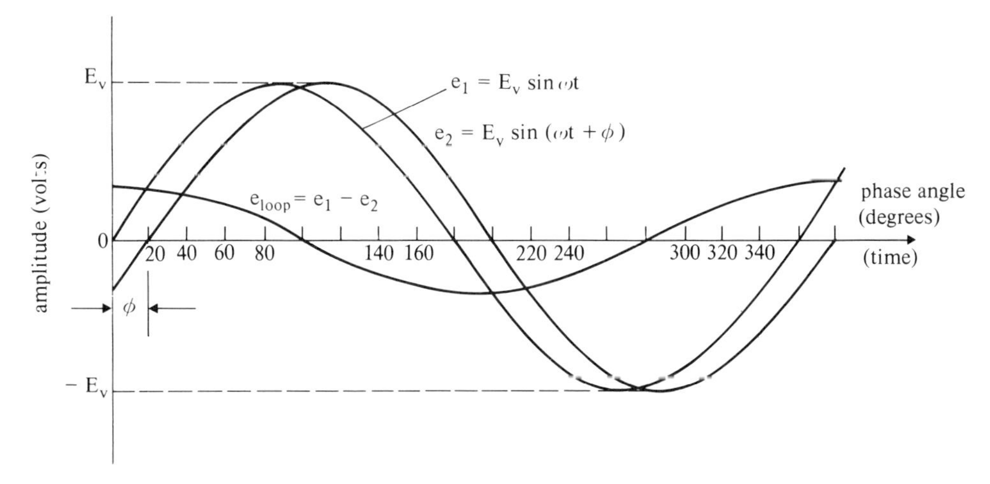

Figure 4.4 Induced loop voltages e_1 *and* e_2 *and resultant voltage* e_{loop}.

Summing the induced voltages gives:

$$\begin{aligned} e_{loop} &= (e_1) + (-e_2) \\ &= E_v \sin \omega t + (-E_v \sin (\omega t + \phi) \\ &= E_v \sin \omega t - E_v \sin (\omega t + \phi) \end{aligned}$$

expanding $E_v \sin (\omega t + \phi)$ to $E_v (\sin \omega t \cos \phi + \cos \omega t \sin \phi)$ and making $\cos \phi = 1$:

$$e_{loop} = E_v \cos \omega t \sin \phi \qquad \textbf{[4.2]}$$

This expression shows two facts relating to the loop antenna:

- The phase of the resultant voltage, e_{loop}, is now in quadrature ($\cos \omega t$) with the voltage induced in a single vertical element ($\sin \omega t$).
- The maximum loop voltage is $E_v \sin \phi$. At low and medium frequencies ϕ is always very small and so the resultant loop voltage will be much less than the voltage induced in a vertical element.

The peak amplitude of the loop voltage depends upon the phase angle ϕ, this in turn is a function of the separation of the vertical elements, measured along the direction of propagation.

From Figure 4.5, $\phi \propto d \cos \theta$. The peak loop terminal voltage for a given angle θ is therefore:

$$E_{loop} = E \sin \phi . \cos \theta .$$

Plotting this for all values of θ from 0 to 360 degrees gives the polar diagram of the loop aerial (Figure 4.6). From this can be seen the directional properties necessary for basic direction finding.

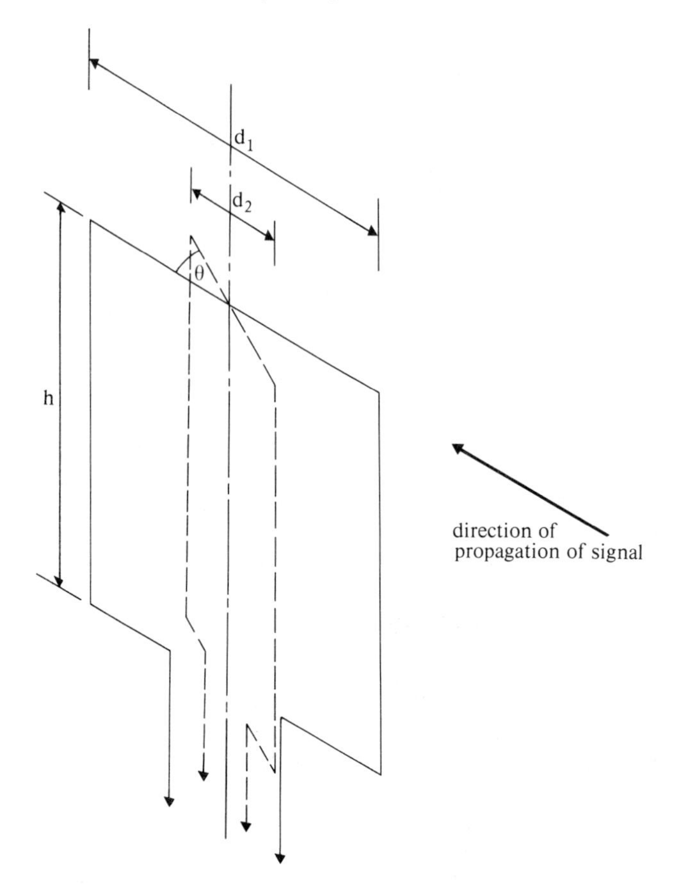

Figure 4.5 *Separation of vertical elements* (d), *is a function of rotational angle* θ.

There are two angular positions ($\theta = 0°$ and 180°) of maximum loop voltage and two positions of zero loop voltage ($\theta = 90°$ and 270°).

A basic direction finder is constructed simply by coupling a loop antenna capable of rotation to a radio receiver. The direction of a given transmission can be determined (after tuning the receiver to the appropriate frequency) by rotating the loop until the amplitude of the signal is either at a maximum or a minimum. From the angular position of the loop, the direction of the transmission is thus determined. The positions of signal minima are used in preference to the positions of maxima since the rate of change of loop voltage with rotational angle θ is considerably greater in the region of the minima. The positions of zero loop voltage can therefore be found more precisely than positions of maximum loop voltage.

4.1.3 The loop antenna – construction

The shape of the basic DF loop can take any form since all shapes can be reduced to vertical and horizontal components.

The expression for loop voltage was shown to be:

$$e_{\text{loop}} = E_{\text{v}} \cos \omega t \sin \phi$$

and since at low and medium frequencies the phase angle is always very small:

$$\sin \phi \simeq \phi.$$

The phase angle ϕ is directly proportional to the horizontal loop dimension d and the induced voltage E_{v} is directly proportional to

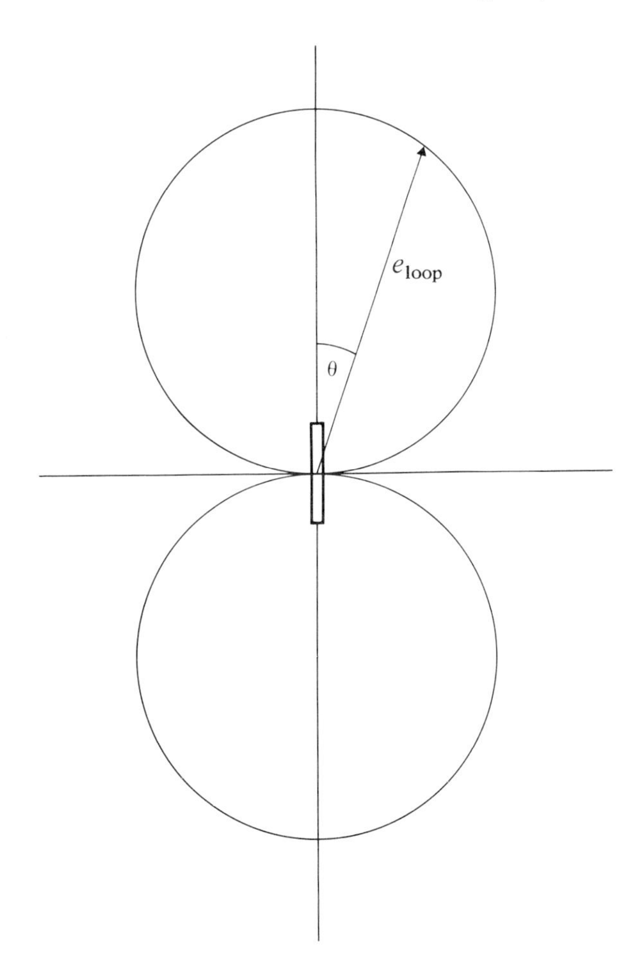

Figure 4.6 Polar diagram of loop antenna.

the length of the vertical components (h) of the loop (for the normal case of a vertically polarized wave).

Therefore $e_{loop} \propto$ d.h

$\propto$ area enclosed by the loop.

The length of the vertical component is directly proportional to the number of 'turns' of conductor within the loop. However, since the inductance of the loop is proportional to the square of the number of turns, any increase in induced voltage obtained from increasing the turns is lost due to now requiring a reduced transformer step up ratio when coupling the loop to the receiver. In summary therefore the effectiveness of a DF loop antenna is proportional to its enclosed area but is independent of the number of turns of the conductor.

For correctly defined zeros to occur in the loop polar diagram, the voltages induced in the vertical sections must be equal in amplitude. An inequality in amplitude has the effect of producing an asymmetrical polar diagram and consequently incorrectly positioned zeros (Figure 4.7). One possible cause of this type of problem is an unbalanced capacitive coupling between the loop vertical elements and nearby metallic objects. A simple remedy exists to ensure that this never occurs, and that is to place the loop inside an earthed metallic tube or 'shroud'. This presents a fixed and uniform capacitance to earth from all parts of the loop (Figure 4.9), but the

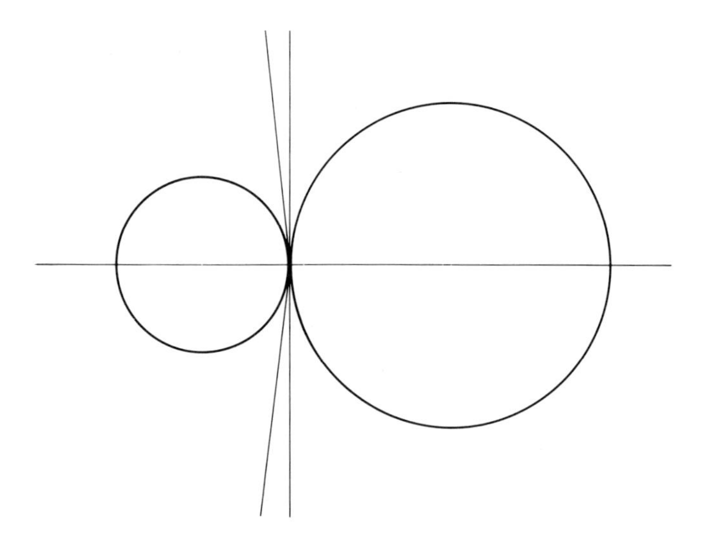

Figure 4.7 Effect of unbalanced loop voltages (unequal amplidude) on antenna polar diagram.

shroud must not form a complete turn otherwise it would screen the loop from the incident waves.

The unbalance may be such that the induced voltages are equal in amplitude but unequal in phase. In this situation the null points are correctly positioned, but the loop voltage does not fall to zero (Figure 4.8). The effect is reduced by the use of the loop shroud, but can be further reduced by introducing an equal and opposite voltage to the residual. This voltage is coupled from the sense antenna via a differential capacitor which is usually called the 'zero clearing' or 'zero sharpening' control.

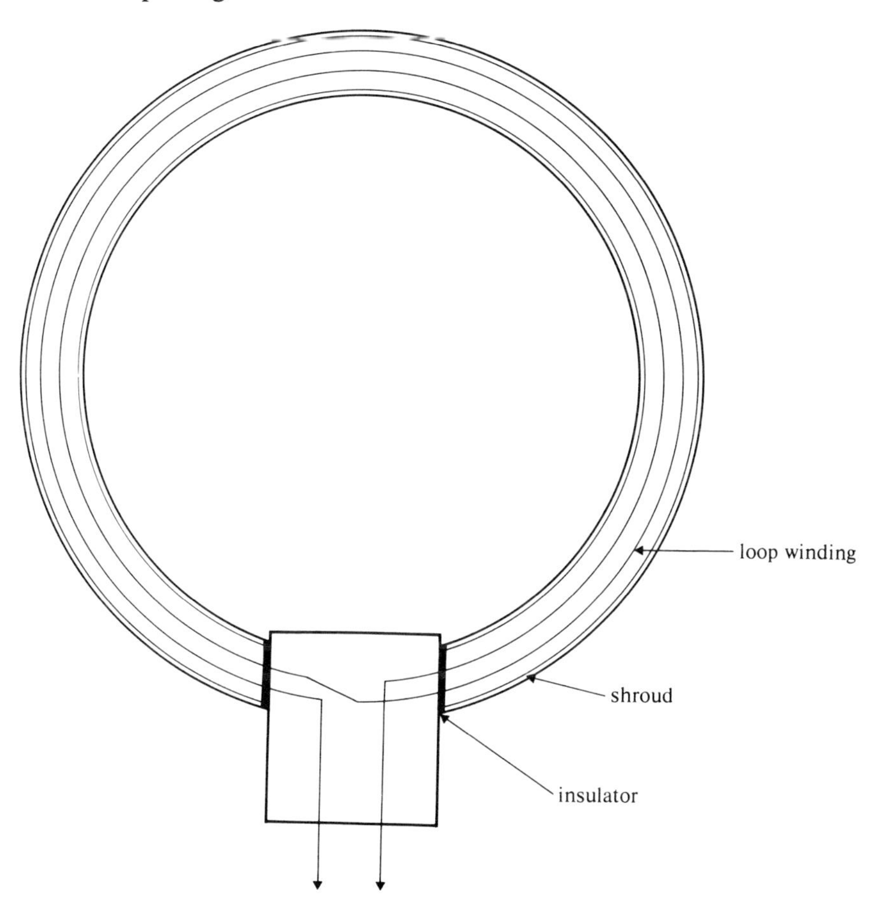

Figure 4.8 Construction of loop antenna.

4.2 Bearing ambiguity

The figure-of-eight polar diagram of the loop antenna provides the ideal basis for a marine DF system, since the well defined positions

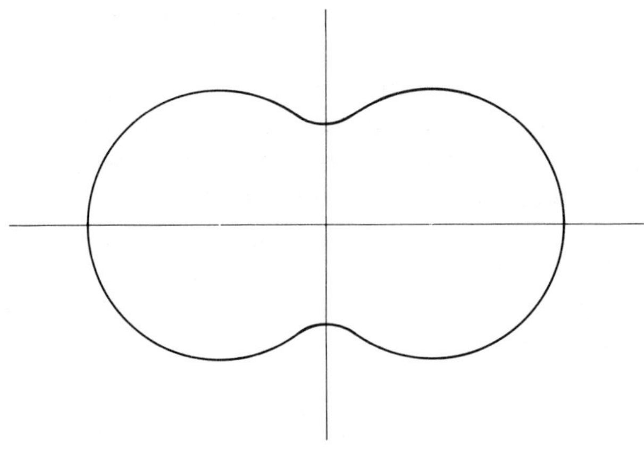

Figure 4.9 Effect of unbalanced loop voltages (unequal phase) on antenna polar diagram.

of zero loop voltage can be made to correspond with the relative direction of the transmission. There is one basic difficulty with this polar diagram, and this is caused by the fact that there are two orientations of the antenna which produce the zero loop voltage condition. Since these are 180 degrees of rotation apart, there will always be two possible relative bearings. In reality this does not usually present a problem, since of the two bearings the correct one is readily obvious, but in the small number of situations when the bearing direction is totally unknown a means is required for resolving this 'sense' ambiguity. For this reason an omnidirectional polar response is added to the loop's figure-of-eight response, resulting in a polar diagram which has only one position of zero voltage (Figure 4.10).

The omnidirectional response is obtained from a simple vertical, 'sense' antenna, but before it can be summed with the loop voltage, the phase of one of the voltages must be changed by ninety degrees to bring it into phase with the other. This is necessary since the previously developed expression for loop terminal voltage:

$$e_{\text{loop}} = E_{\text{v}} \cos \text{t}\theta t \sin \phi$$

showed that the loop terminal voltage was ninety degrees removed in phase from the voltage induced in a simple vertical conductor.

The resultant then of adding the loop terminal voltage:

$$E \sin \phi . \cos \theta$$

to the vertical sense voltage C (equation [**4.1**]) after making $C = E \sin \phi$ is:

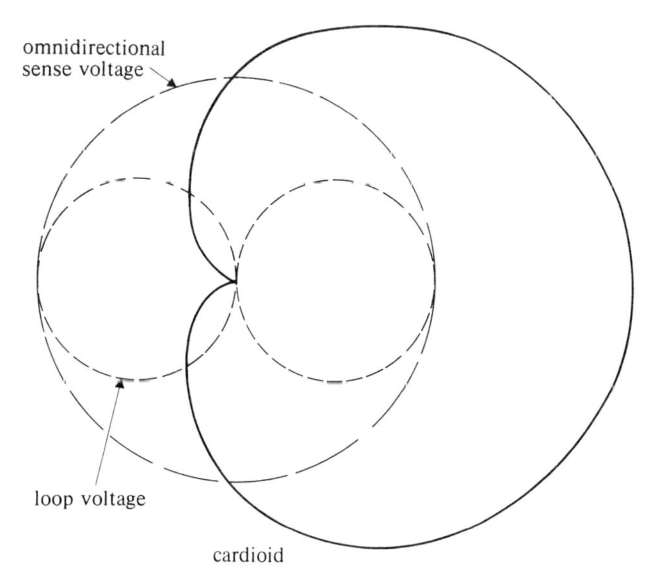

Figure 4.10 *Addition of loop and omnidirectional polar diagrams produce cardioid polar diagram for sense determination.*

$$E \sin \phi + E \sin \phi \cos \theta = E \sin \phi(1 + \cos \theta)$$

which is the expression for the cardioid polar diagram.

4.2.1 *Procedure for sense determination*

The method of determining the sense of a received signal usually follows the same general procedure for all rotating loop DF receivers:

- The loop is slowly rotated and by observing the signal amplitude, the null point is found (Figure 4.11a). (There will be a second null on the reciprocal bearing to this.)
- If sense ambiguity exists, the sense antenna is switched into circuit to produce the cardioid polar diagram (Figure 4.11b).

Since the null position of the cardioid is physically ninety degrees removed from the two nulls of the loop antenna, a 'sense' pointer (ninety degrees removed from the bearing pointer) indicates the position of the sense null.

The loop is rotated so that the sense pointer is set first at one possible bearing (Figure 4.11b), and then at the other (Figure

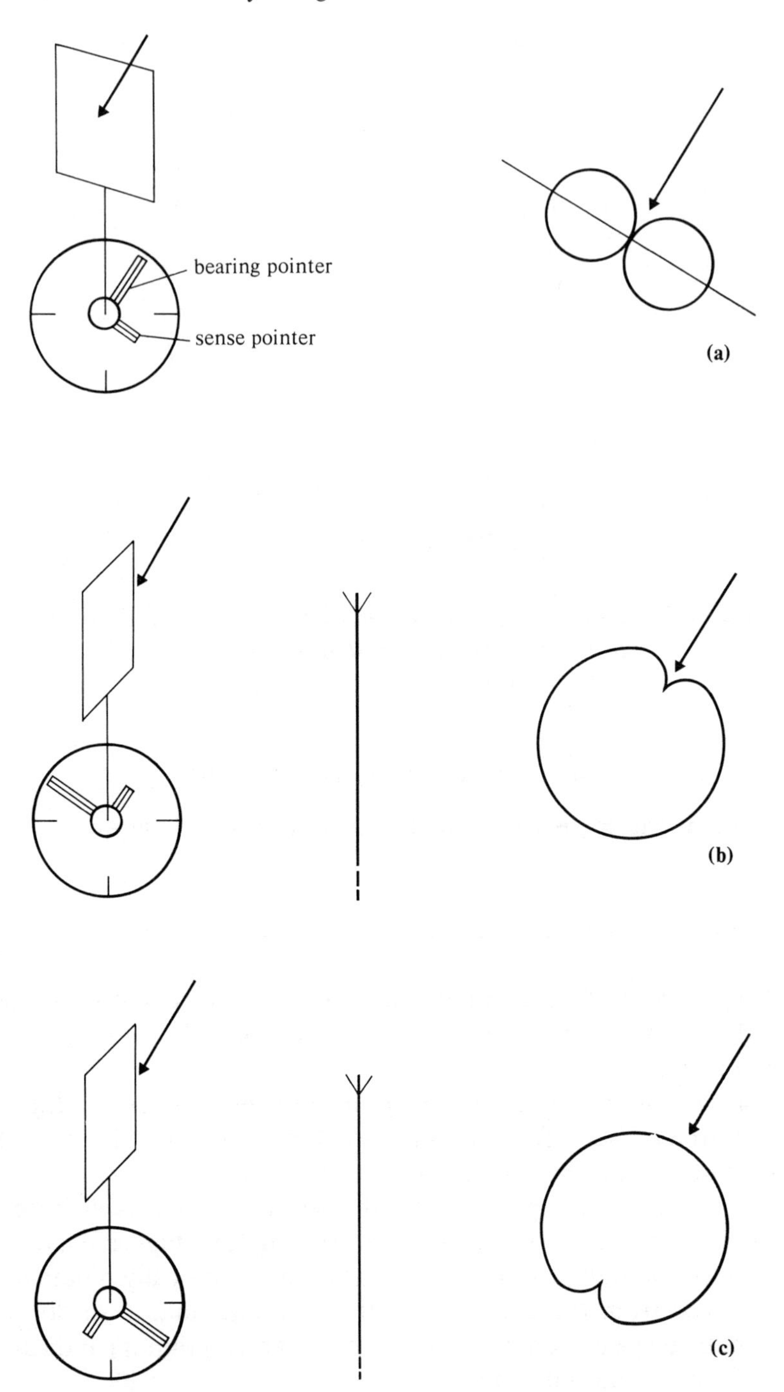

Figure 4.11 *Procedure for sense determination:* (a) *taking bearing;* (b) *correct sense;* (c) *incorrect sense.*

4.11c). Of the two possible bearings the one which produces the least signal is the correct bearing.

The accuracy of the cardioid null depends on the maximum loop voltage and the vertical sense voltage being of equal amplitude. If the sense voltage is greater than the loop voltage, the null point becomes ill-defined (Figure 4.12). If the sense antenna voltage is smaller than the loop voltage, two null points exist, neither at the position of the correct null (Figure 4.13). This dependence on amplitude equality makes the cardioid unsuitable for use directly to determine bearing, but when used to solve the problem of bearing ambiguity by comparing two signal amplitudes (Figure 4.11, b and c), the amplitude balance is no longer critical.

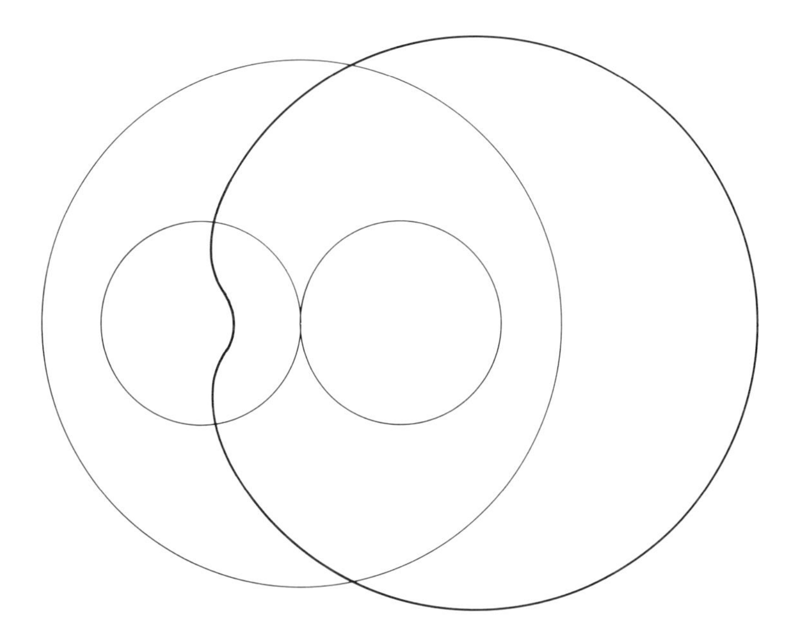

Figure 4.12 Effect on cardioid diagram when omnidirectional antenna voltage is greater than maximum loop voltage.

4.3 The rotating loop DF receiver

The circuit functions of a basic rotating loop DF receiver are shown by Figure 4.14. The loop antenna terminal voltage is smaller than the vertical antenna voltage, and thus requires amplification before

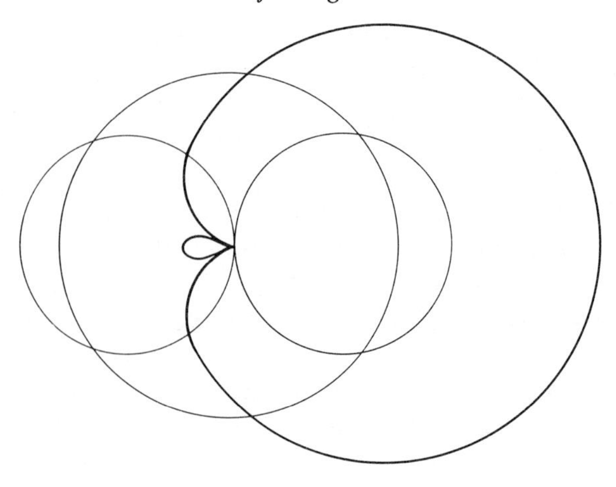

Figure 4.13 Effect on cardioid diagram when omnidirectional antenna voltage is less than maximum loop voltage.

the two voltages can be combined. The variable attenuation through which the vertical antenna signal is fed permits adjustment to be made to precisely balance the loop and sense voltages.

A phase shift circuit changes the phase of the sense signal by ninety degrees to bring it into phase with the loop voltage. The addition of the two voltages usually occurs within an inductance having three separate windings: two for the application of the loop

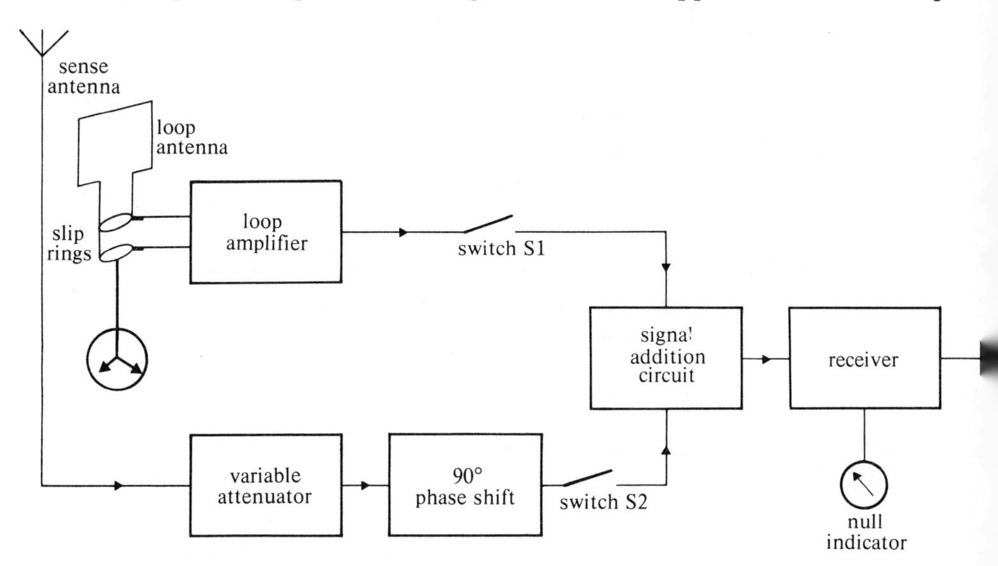

Figure 4.14 Schematic diagram of basic rotating loop DF receiver.

and sense voltages; and the third, an output winding from which the summation is derived.

Switches S1 and S2 provide the three modes of operation:

- S1 closed and S2 open circuit, gives the figure-of-eight loop polar diagram for basic direction finding.
- S1 closed and S2 closed, gives the cardioid polar diagram for sense determination.
- S1 open and S2 closed, gives the omnidirectional polar diagram of the vertical antenna, which can be used for normal receiver operation.

4.4 The Bellini-Tosi crossed loop

The basic limitation of the simple rotating loop DF receiver is the requirement for a mechanical coupling between the loop antenna and the bearing pointer. For the bearing pointer to be conveniently sited near to the receiver usually restricts the siting of the loop to a less than optimum location. The development by Bellini and Tosi in 1909 of a DF antenna based on two fixed loops set at right angles (Figure 4.15) overcame this problem, and today it still provides the

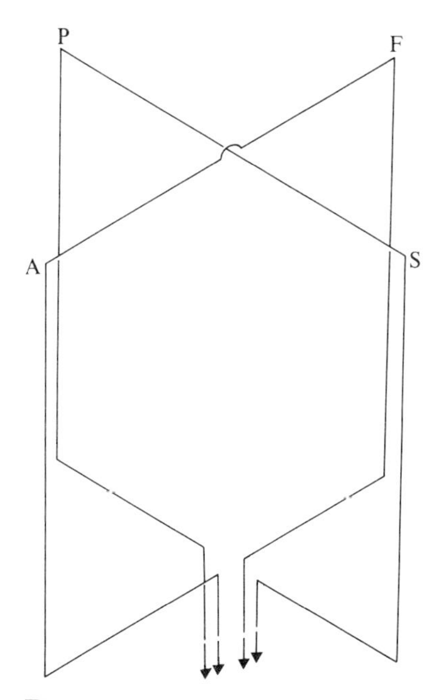

Figure 4.15 Basic Bellini-Tosi crossed loop antenna.

basis of most marine DF receivers. All the previous theory relating to the single loop applies equally to the two individual loops of the Bellini-Tosi crossed loop antenna.

In a marine DF installation the antenna is positioned such that one loop lies along the vessel's centre line, the other then being athwartships (Figure 4.16). The two loops are subsequently referred to as the fore-aft (FA) loop and port-starboard (PS) loop respectively.

Figure 4.16 Plan view of ship showing orientation of crossed loop antenna.

A received signal on a relative bearing θ will produce a loop terminal voltage $E \cos \theta$ in the FA loop and $E \sin \theta$ in the PS loop (Figure 4.17). The angle θ is now represented by two voltage amplitudes, and can therefore be coupled to the receiver by four interconnecting wires, thus removing the mechanical constraints of the rotating loop.

4.4.1 *The goniometer*

The fundamental difference between the various types of DF receivers in current use is the means by which the relative bearing angle θ is obtained from the two loop voltages $E \cos \theta$ and $E \sin \theta$. The original device developed for the purpose and still used in many DF receivers is the 'radiogoniometer' (usually abbreviated to goniometer).

The goniometer consists of two coils set in the same relative planes as the two loops of the antenna. A third coil is free to rotate within the magnetic field produced by the currents through the coils (Figure 4.18). The amplitude of the magnetic field is thus proportional to the two loop voltages: the magnetic field of FA goniometer coil is

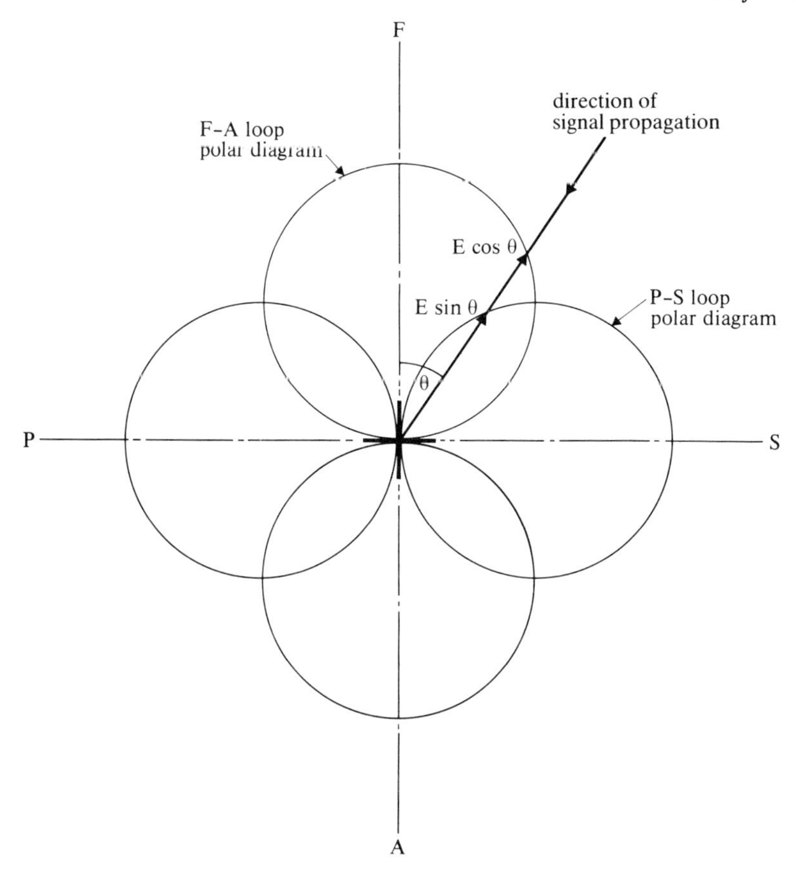

Figure 4.17 Polar diagram of crossed loop DF antenna.

proportional to $E \cos \theta$, and the magnetic field of PS goniometer coil is proportional to $E \sin \theta$.

The voltage induced into the rotating coil by each of these magnetic fields is a function of the rotor angle ϕ. The voltage induced by the FA coil is proportional to $\cos \phi$, and by the PS coil to $\sin \phi$.

The resultant voltage induced into the rotating coil is proportional to:

$E \cos \theta \cos \phi - E \sin \theta \sin \phi$
and when $E \cos \theta \cos \phi - E \sin \theta \sin \phi$
that is, $\cot \theta = \tan \phi$
the induced rotor voltage is zero.
This is true when $\theta = \phi + 90 + n\,180$
where $n = 0$ or any integer.

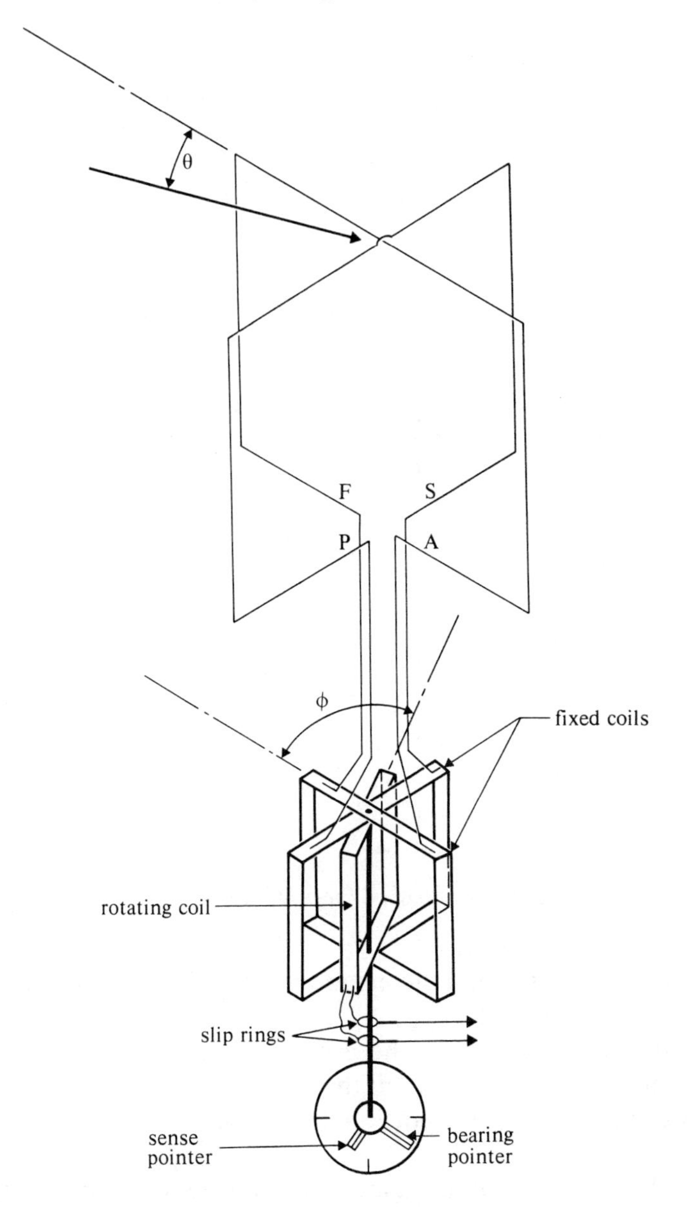

Figure 4.18 Diagrammatic representation of crossed loop antenna and goniometer.

The rotating coil of the goniometer behaves exactly as the rotating loop, and therefore it can be considered as having the same figure-of-eight polar diagram. When the goniometer is incorporated into a DF receiver, a bearing indicator is attached to its rotor which indicates the relative DF bearing when the rotor is set to the signal null position.

Since the goniometer has the same polar diagram as the simple loop, it will also suffer the same bearing ambiguity problem. The problem is solved in the same way – by adding an omnidirectional sense voltage. The circuits of a typical crossed loop with goniometer type of DF receiver are illustrated schematically by Figure 4.19. Apart from the goniometer and antenna, it is identical in other respects to the rotating loop receiver of Figure 4.14.

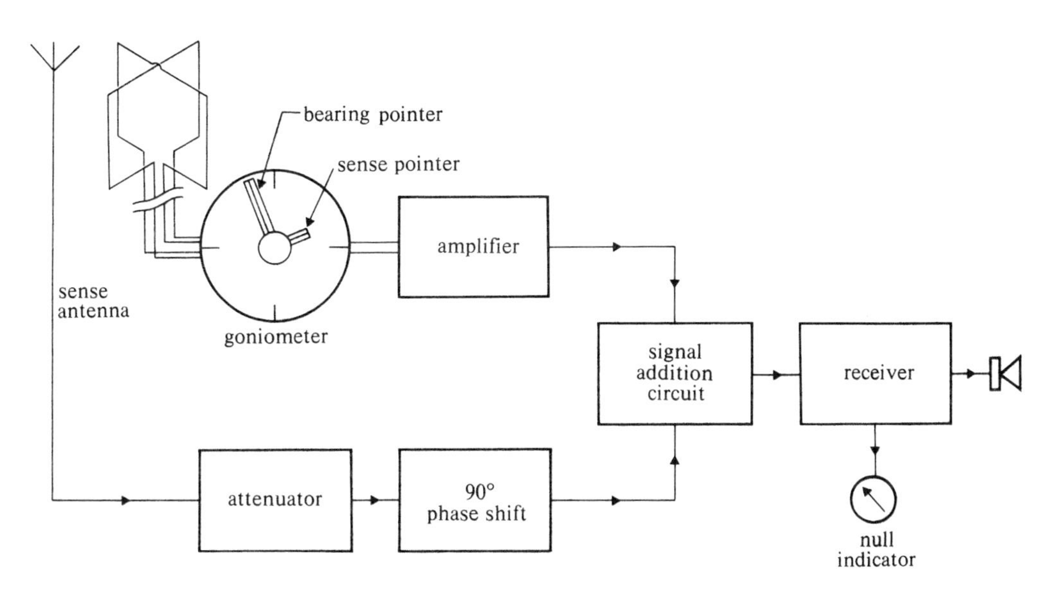

Figure 4.19 Schematic diagram of basic crossed loop DF receiver.

4.5 Automatic direction finding receivers

The rotating and crossed loop receivers of Figure 4.14 and Figure 4.19 are both termed 'manual' DF receivers, referring to the fact that in operation it is necessary to turn by hand either the antenna or the goniometer, to arrive at the bearing of a given signal. There have been many schemes devised by which the bearing is indicated automatically once the receiver has been tuned to the appropriate signal.

4.5.1 The spinning goniometer

In this system the crossed loop antenna and the goniometer are conventional, except that the goniometer is coupled to a motor which rotates at high speed. Figure 4.20 illustrates the output waveform from the goniometer when the antenna receives a continuous transmission. The position of the null points on the X-axis defines the bearing of the transmission, and the position of this null in terms of angular displacement of the goniometer is determined by mechanically coupling an a.c. generator to the motor. This produces a reference voltage to which the position of the null can be related. The basic circuit arrangement of the spinning goniometer is shown in Figure 4.21. This type of receiver has the same bearing ambiguity problem as the manual types of receiver, and the solution is basically the same – the addition of an omnidirectional sense voltage. From this is produced a modulation envelope with the number of minima halved, and since this is fundamentally the cardioid the minima are displaced in time in relation to the original bearing minima (Figure 4.22).

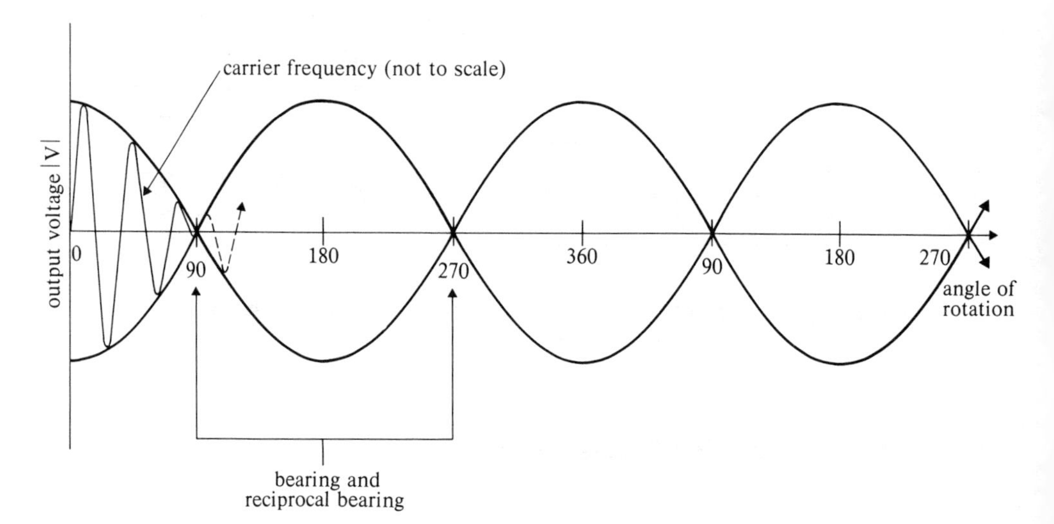

Figure 4.20 Output waveform from spinning goniometer.

4.5.2 Electronic goniometer

By modulating the output signal of each loop, the effect of the spinning goniometer can be produced by pure electronic means.

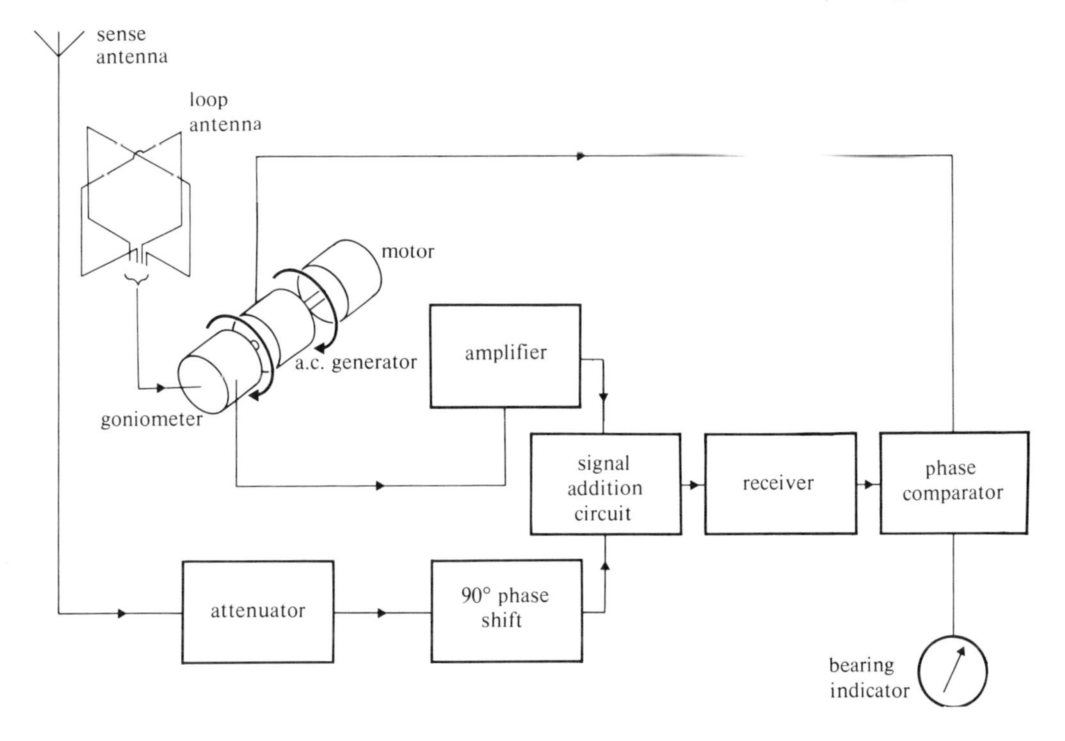

Figure 4.21 Schematic diagram of spinning goniometer DF receiver.

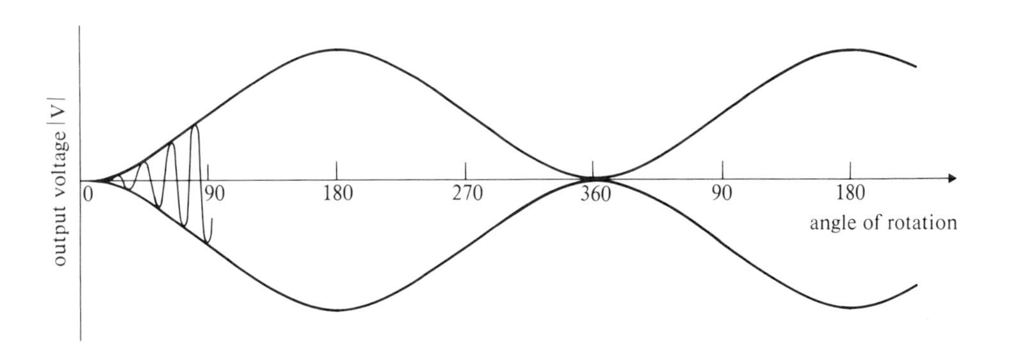

Figure 4.22 Spinning goniometer DF receiver; modulation envelope for sense operation.

The FA loop voltage, $E \cos \theta$, is modulated by a voltage $V \sin \omega t$ and the PS loop voltage $E \sin \theta$ is modulated with $V \cos \omega t$ (Figure 4.23).

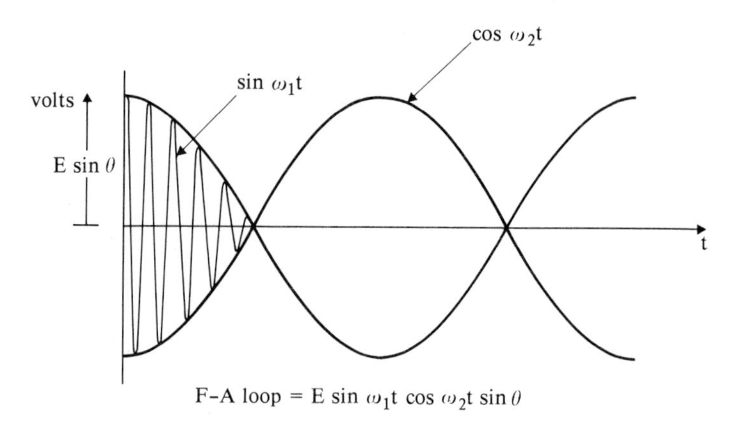

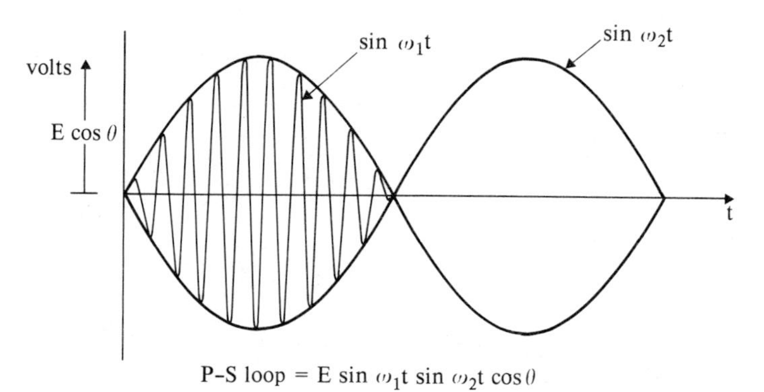

Figure 4.23 Loop voltages modulated by an electronic goniometer circuit.

This results in:
FA loop voltage $= V \cos \theta \sin \omega t$
PS loop voltage $= V \sin \theta \cos \omega t$.
At two points during each cycle of the modulation, the amplitude of the modulated loop voltages are equal: when
$V \cos \theta \sin \omega t = V \sin \theta \cos \omega t$
that is when $\omega t = \theta$.
By detecting the points at which the two loop voltages are equal, and relating these to the phase of the modulation voltage, the bearing angle θ is determined.

The amplitudes of the modulated loop voltages are twice equal during each cycle of the modulation, due to the presence of the reciprocal bearing. The ambiguity caused by the reciprocal bearing is resolved by adding the voltage derived from an omnidirectional sense antenna as previously explained.

4.5.3 Servo-operated goniometer

One group of automatic DF receivers base their operation on the principle of a servo motor driving the goniometer (Fig. 4.24). The goniometer and receiver form part of a servo loop which causes the servo motor to drive the goniometer rotor to the position of signal null. For a system of this type to truly give the bearing automatically, the possibility of arriving at the reciprocal bearing must be removed. An omnidirectional sense antenna is used, but not in this case to produce a cardioid polar diagram, but to ensure that the error voltage is generated with the correct phase for the servo motor to always drive the goniometer rotor towards the correct null. It is possible for the situation to arise whereby the goniometer rotor is already in the position of the reciprocal bearing null on tuning to a

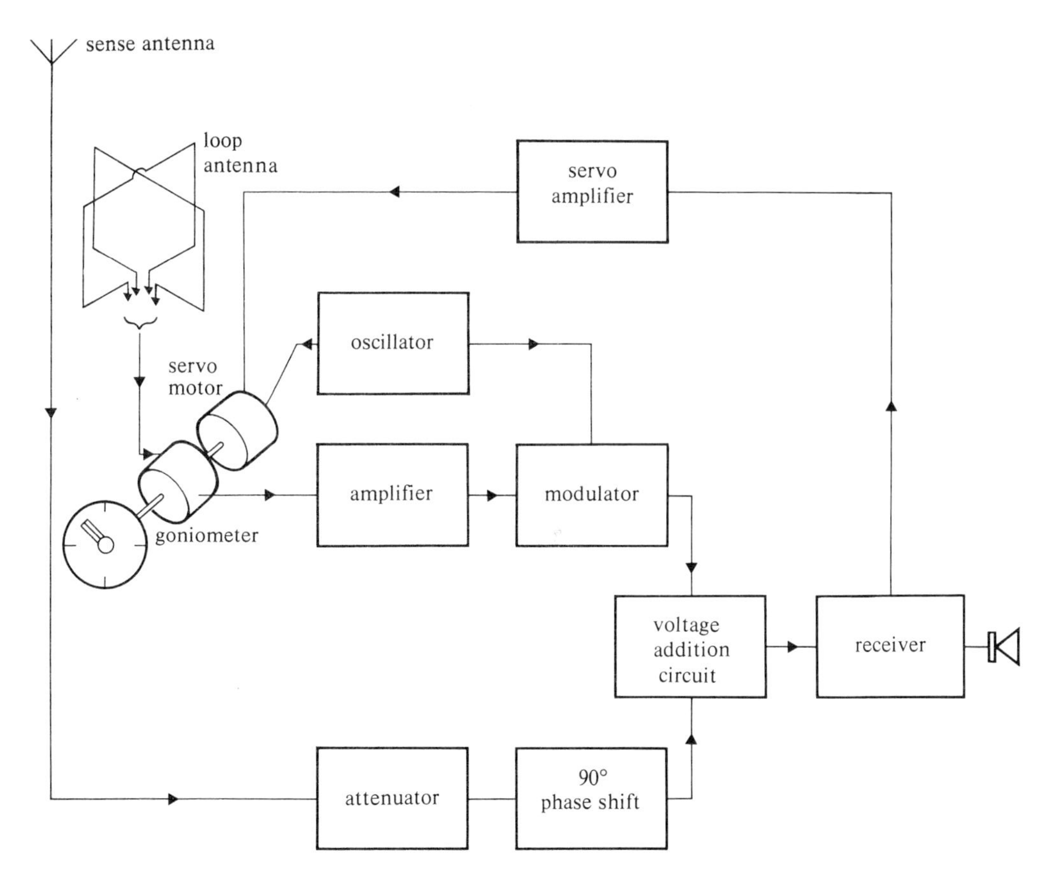

Figure 4.24 Schematic diagram of servo-operated automatic DF receiver.

given signal. Then in theory there would be no error voltage and the goniometer would remain at the reciprocal bearing. But in practice receiver noise is sufficient to prevent it remaining statically on this false null, and once the goniometer rotor moves slightly off the null an error voltage is produced which drives it to the correct bearing.

4.5.4 *Cathode ray tube DF receivers*

A cathode ray tube (CRT) can be utilized to resolve the relative bearing angle from the two loop voltages of a Bellini-Tosi crossed loop antenna. A DF receiver which makes use of a CRT to display bearing information, has the voltage from one loop of the antenna coupled to the X-deflection circuit, and the other loop coupled to the Y deflection circuit of the cathode ray tube.

If the voltage from the FA loop causes Y deflection and the voltage from the PS loop causes X deflection, the spot will be instantaneously deflected to a position P (Figure 4.25a).

$$e \tan \phi = \frac{e \sin \theta}{e \cos \theta}$$

and since $\dfrac{e \sin \theta}{e \cos \theta} = e \tan \theta$

therefore $\tan \theta = \tan \phi$

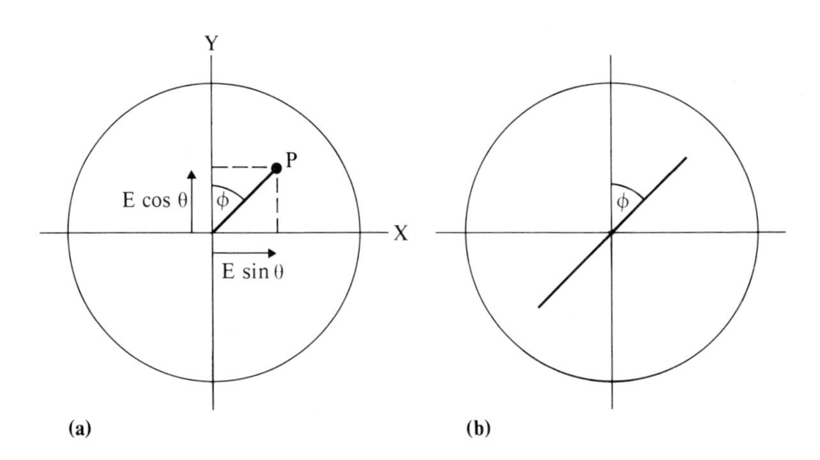

Figure 4.25 CRT display DF receiver; (a) *point P corresponds to the maximum amplitude of the loop signals;* (b) *bearing line produced by sinusoidal loop signals.*

and $\theta = \phi$

where θ = relative bearing angle

and e = instantaneous loop voltage

thus the angle ϕ is equal to the relative bearing angle θ.

Since the signal amplitude is sinusoidal ($E \sin \omega t$) the spot will be deflected along a line maintaining the angle ϕ (Figure 4.25b).

A basic CRT type of DF receiver is illustrated by Figure 4.26. It is fundamentally necessary for the two receiver channels to have identical gain and phase responses to accurately maintain the ratio of the amplitudes of the two loop voltages.

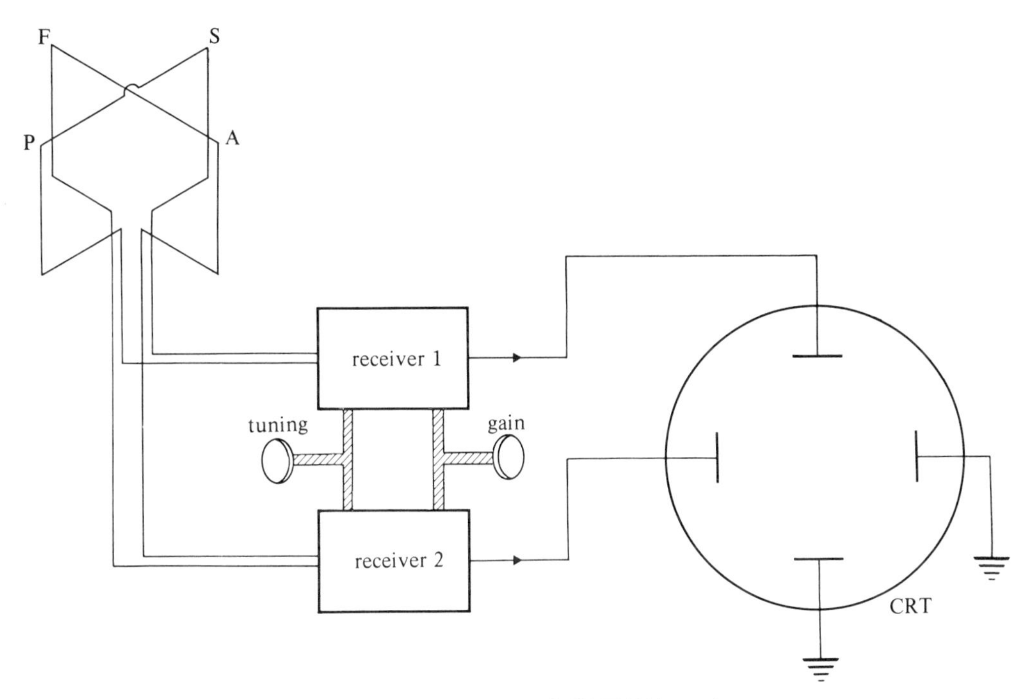

Figure 4.26 Schematic representation of CRT DF receiver.

For example: if the difference in receiver gains is not to cause a bearing error of greater than ±0.5 degrees at mid-quadrant, that is when $\theta = 45°$, $\phi \ngtr 45.5°$,

$$\frac{E \sin \theta \times \text{gain of receiver 1}}{E \cos \theta \times \text{gain of receiver 2}} \ngtr \tan 45.5 \ (\ngtr 1.0176)$$

but $\theta = 45°$

and $\tan 45° = 1.000$.

If the error of the displayed bearing ϕ is not to exceed 0.5 degrees the difference in gains of the two receiver channels must not exceed 1.76 per cent. This equality in gain must be held throughout the tuning range of the receivers.

An inequality in the phase responses of the receiver channels will cause the deflecting voltages applied to the CRT to have differing phases. The effect of this is to turn the single deflection line into an elliptical trace (Figure 4.27c).

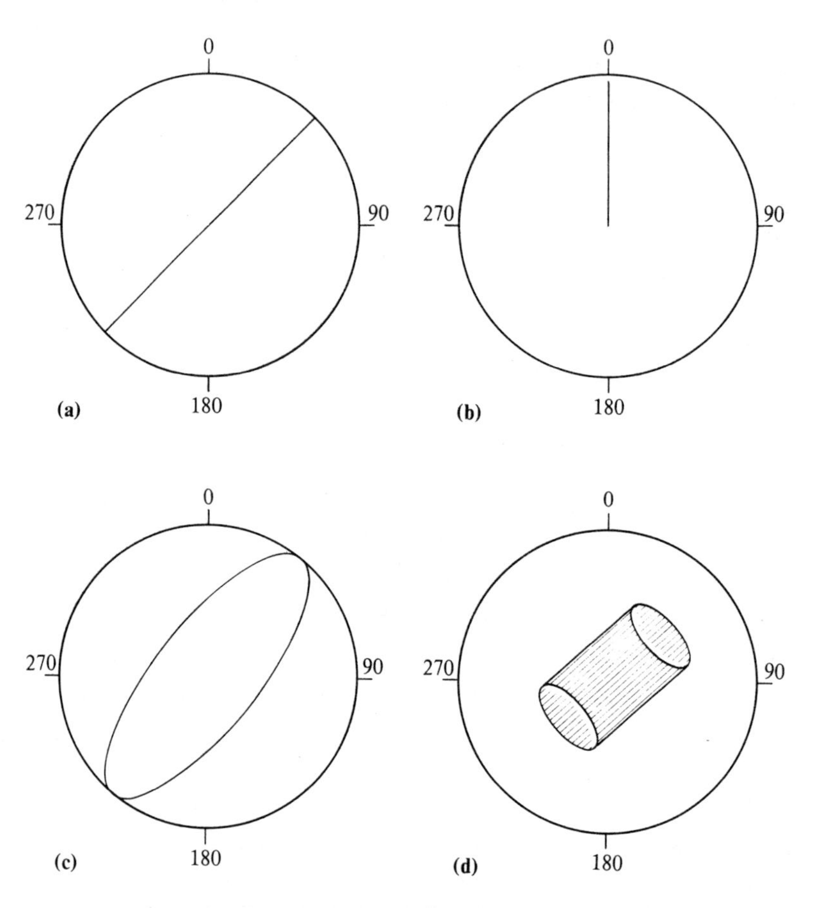

Figure 4.27 (a) *normal trace of CRT DF receiver;* (b) *display for sense determination;* (c) *effect of loop signals with differing phases;* (d) *effect of interfering signal.*

An elliptical trace occurs not only due to phase differences caused within the receiver, but also when the waves impinging on the two loops induce voltages which differ in phase. This can occur when the wave deviates from the normal vertical polarization, to become horizontally or elliptically polarized. The CRT type of display is useful in its ability to indicate abnormal propagation conditions. By observing the angle made by the major axis of the elliptical trace, a relatively accurate bearing can still be determined.

This type of DF receiver suffers from sense ambiguity, but the problem is resolved using an omnidirectional sense antenna. It is usual to add the sense antenna voltage to only one of the two loops, and the resulting trace then resolves the ambiguity by showing the signal source to be either ahead or astern (or to port or starboard). Figure 4.27b is a sense indication showing that the correct bearing of Figure 4.27a is ahead, that is 045 not 225 degrees.

Figure 4.27d illustrates the nature of the displayed trace when an interfering signal is on the same frequency or close to the frequency of the wanted signal. Bearings can be taken of both the wanted and interfering signals by determining the relative angles of the axes of the parallelogram.

4.6 Direction finding: operation

Direction finding receivers are usually designed and manufactured to a standard which will, under ideal circumstances, enable the measurement of a radio bearing to a high degree of accuracy. The actual accuracies achieved with ship-borne DF systems become a function of the many influences outside the equipment itself. In order to achieve the optimum performance from a DF receiver it is necessary to be fully aware of all possible errors and to understand their source and their potential magnitude.

Basically all DF bearing errors occur because the observed direction of propagation of a received wave does not coincide with the actual bearing line between transmitter and the DF antenna. This deviation of the radio wave can occur along the path from the transmitter to the ship, or it can be caused by the vessel itself. There are several causes of error falling into each of these two groups.

4.6.1 Half-convergency error

Since radio waves follow great circle paths, plotting a straight bearing line onto a chart of Mercator projection will cause an error; Figure 4.28 shows that the error increases with longitudinal

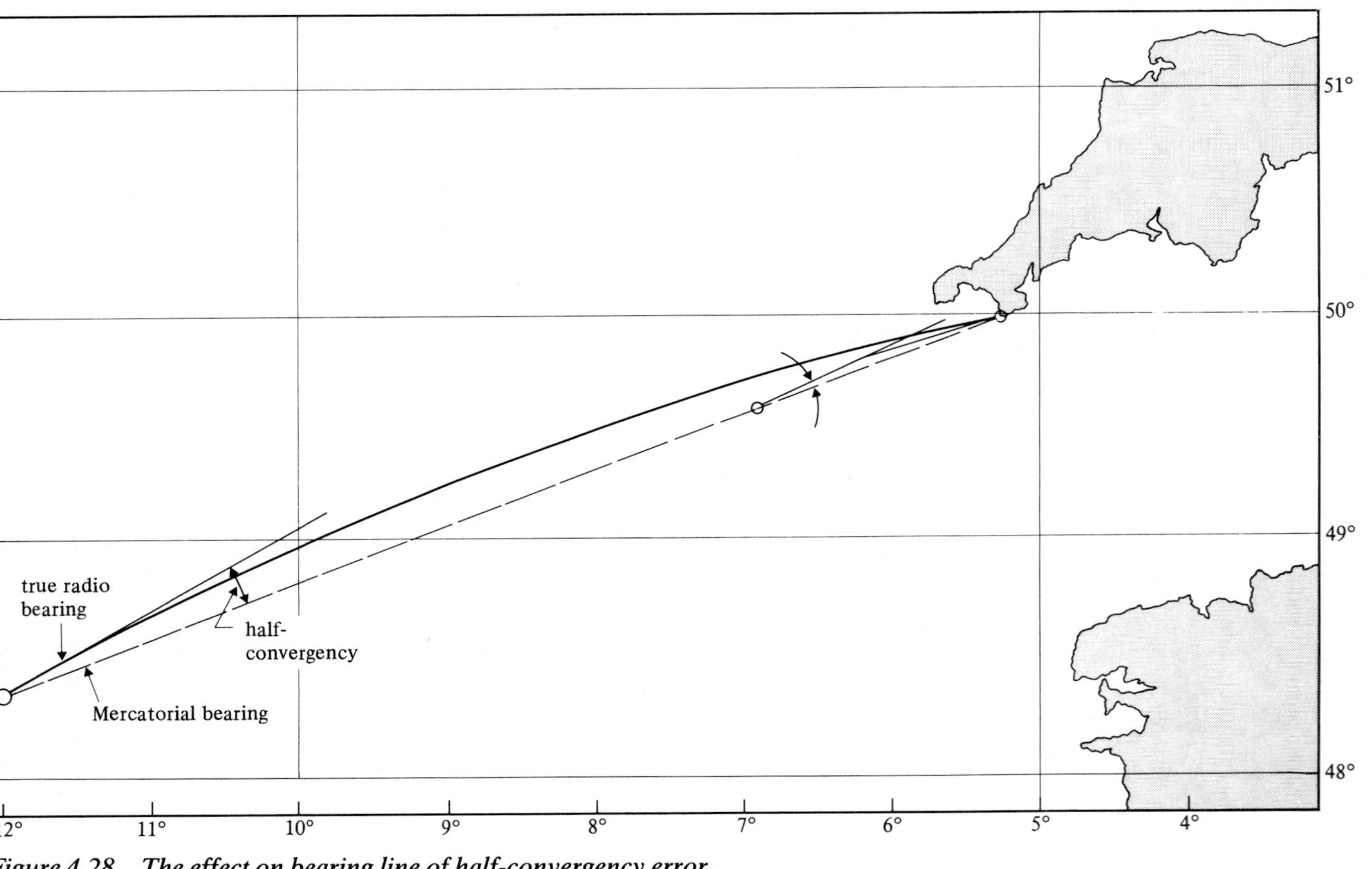

Figure 4.28 The effect on bearing line of half-convergency error.

separation between transmitting beacon and ship. This effect can be corrected by applying half the convergency correction from the approximate formula: convergency in minutes = difference of longitude in minutes × sine of the mean latitude. Taking the example of Figure 4.28:

Beacon latitude = 49.9°N. longitude = 5.2°W.
Vessel latitude = 48.3°N. longitude = 12.0°W.

Mean latitude = 49.1°N. *Difference in longitude* = 6.8° = 408′

Convergency angle = 408 × sin 49.1 = 308′ = 5.13°.
Half-convergency angle = 2.57°.

The true radio bearing of the beacon was 067°; the Mercatorial bearing is therefore 069.6°. The half-convergency angle is always added as a correction to the observed radio bearing in the direction towards the equator.

4.6.2 *Night effect*

Normally DF bearings are taken from signals which have propagated as stable, vertically polarized groundwaves, but under certain conditions of distance and time, less stable skywave signals are received. The effect of these skywaves, which usually vary in amplitude, phase, and polarization, is to cause bearing errors and signal fading. In general the skywave effects are at their worst during sunrise and sunset, when the amplitude of the reflected wave approximately equals the amplitude of the groundwave. The effect of the two signals on the DF receiver's performance can change rapidly as the skywave moves in and out of phase with the groundwave. At other times during the hours of darkness, the skywaves predominate and the situation becomes slightly more stable.

In general, reliance should not be placed on DF bearings if the possibility of skywave interference exists, and at night this could be at any distance beyond twenty miles from the beacon. Bearings should never be taken within one hour of either sunrise or sunset.

4.6.3 Coastal refraction

The phenonemon of refraction of electromagnetic waves, due to a change of velocity occurring when the waves pass from one medium to another, is well known. An example is the observed bending of a light beam as it passes through an air-glass or air-water boundary. Since the propagation velocity of a groundwave is dependent on the conductivity of the ground over which it passes, a bending of the wave occurs if there is an abrupt change in ground conductivity. Such a change occurs at a coastline and gives rise to the effect of 'coastal refraction' (Figure 4.29). Bearings should be avoided which make acute angles with the coastline, since refraction effects may give rise to errors in excess of five degrees.

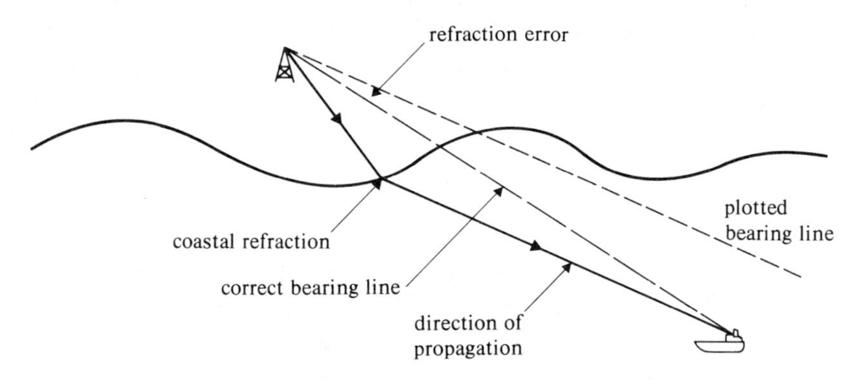

Figure 4.29 The effect on bearing line of coastal refraction.

4.6.4 Vessel effects

In the planning and installation of shore based DF installations, care is taken to ensure that an ideal location is chosen for the siting of the antenna. Ideal sites, totally free of spurious influences which may affect the symmetry of the electromagnetic field, do not exist within the immediate area around a ship. To achieve the optimum performance from a shipborne DF system it is necessary to be aware of the mechanisms by which the wave can be influenced by the ship's structure and thus lead to bearing errors. Some sources of error can be removed; others which cannot can often be compensated for within the DF receiver.

To measure the total influence of the vessel on the DF performance, it is always necessary to calibrate the DF installation.

A 'correction curve' is plotted from the results of the calibration, and from this the causes of the errors can usually be determined and the necessary corrective measures taken.

4.6.5 *Quadrantal error*

Quadrantal error is so termed because the error is at a maximum for bearings in each of the quadrants, and falls to zero for bearings at 00, 090, 180 and 270 degrees, resulting in the characteristic quadrantal error curve (Figure 4.30). The error is caused by circulating currents induced into the ship's hull, which in turn produce a secondary field. This re-radiated magnetic field is in phase with the external field which caused it, and so it therefore adds to the external field causing it to deviate from its correct orientation. The effect is to pull the relative bearing into the fore-aft line (Figure 4.31). In the case of relative bearings which are precisely ahead or astern, the secondary field will be large, but since it has the same orientation as the primary field, there will be no resultant error. For relative bearings of 090 and 270 degrees, orientation of the wave relative to the hull will be such that there will be no circulating currents induced into the hull. There will consequently be no secondary field, and so no error for these relative bearings.

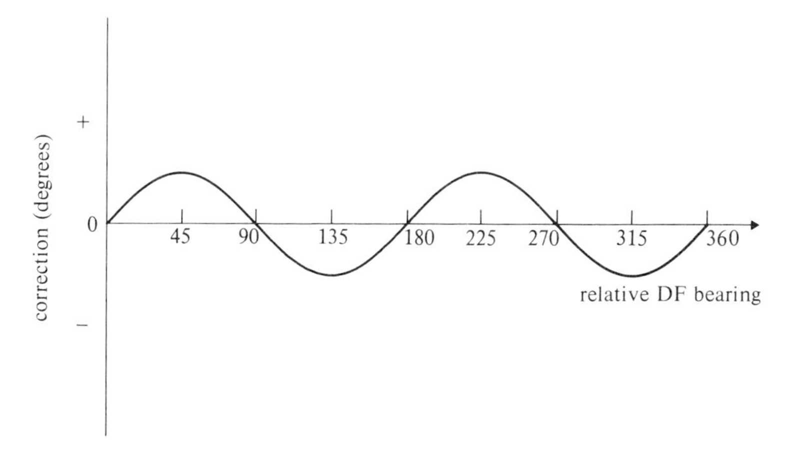

Figure 4.30 Typical correction curve for quadrantal error.

The magnitude of the secondary field is dependent on the vessel's beamwidth and free-board. Fortunately a change in free-board appears to have only a small effect on the secondary field, which can

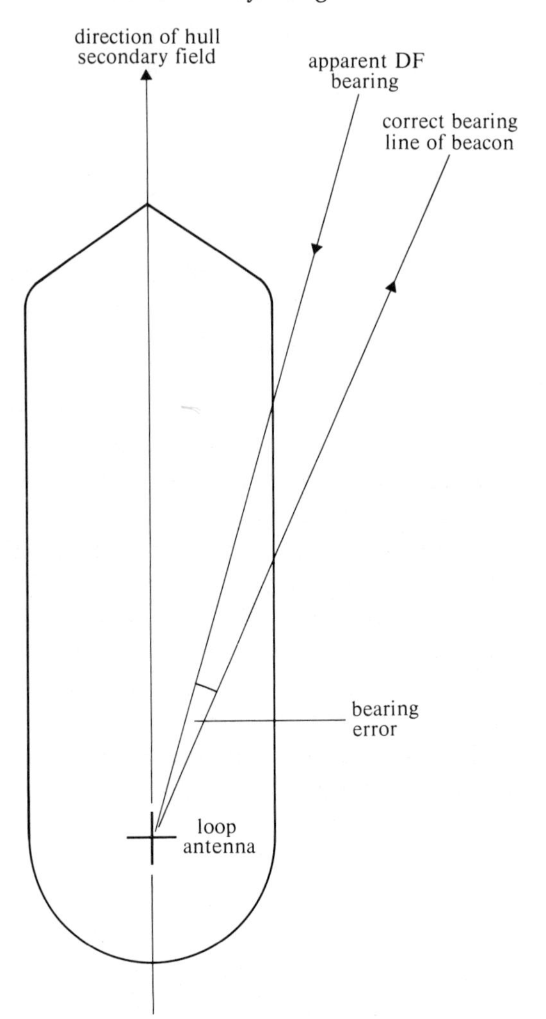

Figure 4.31 Effect on bearing line of hull secondary field.

be neglected in practice. The beamwidth is of course a fixed quantity and so for a given antenna location the quadrantal error effect remains constant. The effect of the secondary field in causing bearing errors reduces with increasing height of the DF antenna above the hull (Figure 4.32).

In the case where the DF antenna is sited on or above a large bridge superstructure, thc secondary field from this may have a greater effect than the field from the hull (Figure 4.33).

Mathematical analyses of the secondary field produced by the hull and bridge superstructures have been developed, but are extremely

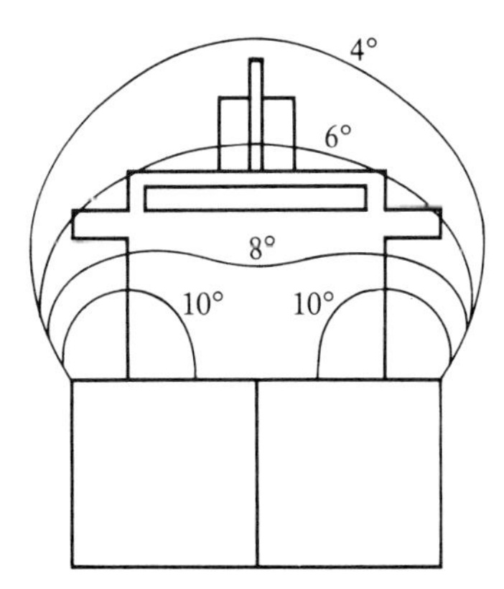

Figure 4.32 Typical contours of maximum quadrantal error for hull secondary field.

complex and are little used in predicting the magnitude of errors for given loop sites. On present-day vessels the DF antenna has to compete for the prime locations with radar, VHF and satellite navigation antennas, and in many cases the DF has a less than ideal antenna site.

The precise form of the quadrantal error which produces the curve shown in Figure 4.30 is given by the expression:

$$\tan \phi = C \tan \theta$$

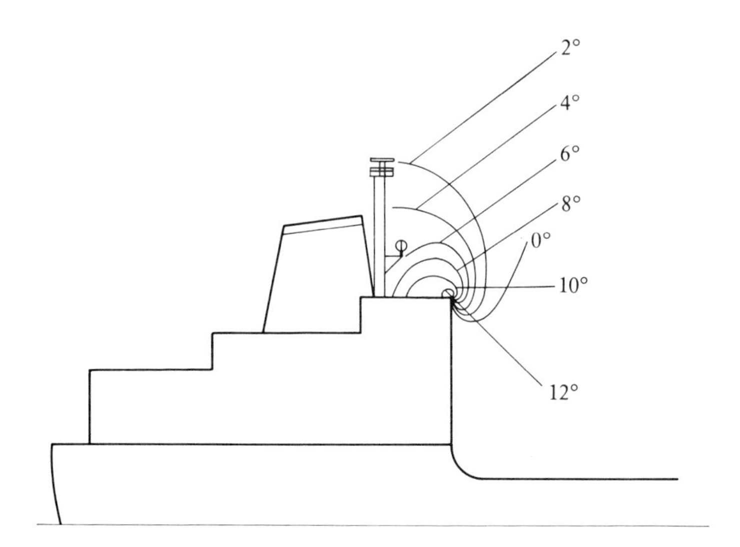

Figure 4.33 Typical contours of maximum quadrantal error for secondary field from superstructure.

where θ = actual relative bearing

ϕ = DF relative bearing

C = constant for a given vessel.

From this expression a family of quadrantal error curves can be drawn showing the points of maximum error (Figure 4.34).

The secondary fields from the hull and superstructure are in general independent of frequency, and so a quadrantal error correction curve produced from calibration results will hold true over a range of frequencies.

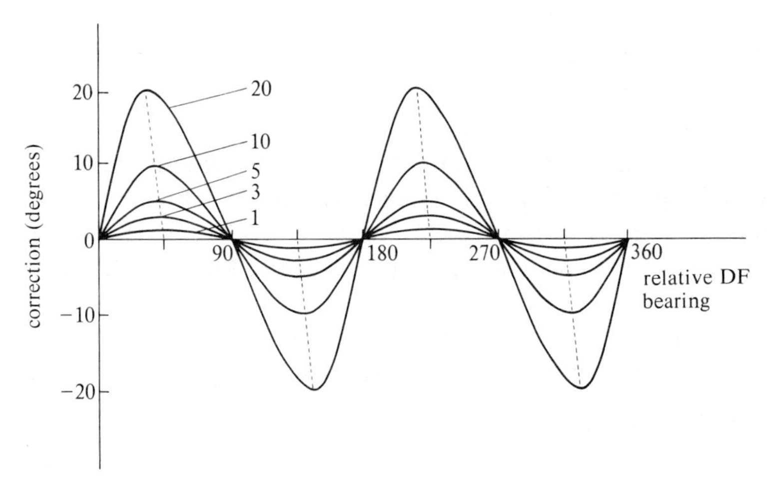

Figure 4.34 Family of quadrantal error correction curves.

It is relatively simple to compensate for quadrantal error in a crossed loop DF system by placing a shunting impedance across the terminals of one loop, thereby reducing its terminal voltage. Taking as an example the case where the secondary field has the effect of pulling the relative DF bearing closer to the centreline (Figure 4.31): the FA loop antenna terminals are shunted to reduce the effective pickpup in that loop, by an amount sufficient to restore the bearing to its correct position.

In the absence of quadrantal error, the shunt impedance would itself produce a curve of the form: $\tan \phi = C \tan \theta$; in this case C is constant for a given shunting effect.

By placing a variable shunt impedance across the appropriate loop of the antenna and adjusting its effect, it should always be possible to totally eliminate pure quadrantal error.

A practical shunting impedance is usually a complex arrangement

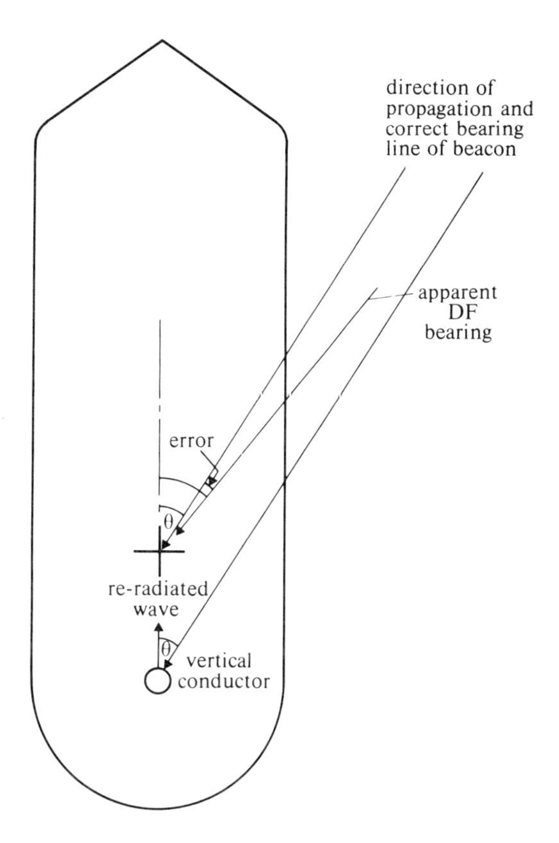

Figure 4.35 Effect on bearing line of vertical re-radiator.

of capacitors and inductors, designed to give a constant amount of correction over a range of frequencies.

4.6.6 *Semicircular error*

As with quadrantal error, semicircular error is named after the shape of the correction curve which results from the presence of the error (Figure 4.36a). The curve shows two points of maximum error which occur 180 degrees apart. The cause of this type of error is a re-radiated signal from a vertical conductor close to the DF antenna. In the practical case of a shipboard DF installation an offending vertical conductor could be a mast, funnel, halyard, stay or antenna. The re-radiation is at its most intense when the length of the vertical conductor relates to the wavelength of the primary field

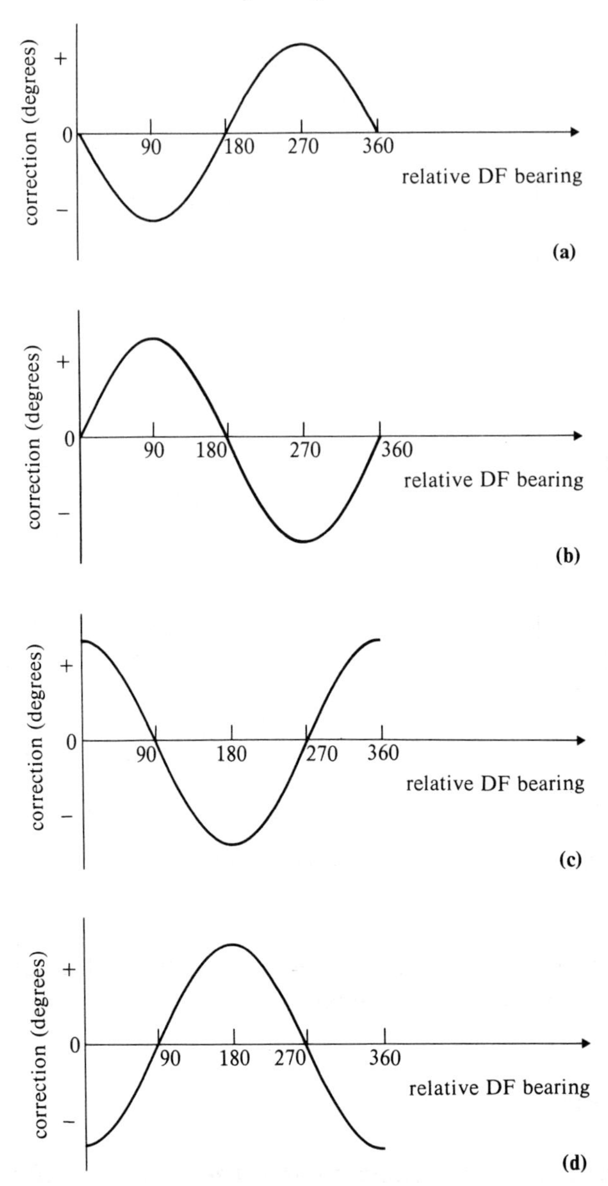

Figure 4.36 *Correction for semicircular error when re-radiator is* (a) *directly astern of loop;* (b) *directly ahead of loop;* (c) *port side of loop;* (d) *starboard side of loop.*

in the ratio λ/4, 3λ/4, 5λ/4, etc. (λ = wavelength of signal to which the DF receiver is tuned).

The correction curve of Figure 4.36a occurs when the re-radiated energy from the vertical conductor arrives at the loop antenna in-phase with the direct wave. The induced loop voltages are then a result of both waves, and so the bearing angle (ϕ) determined by the goniometer deviates from the actual bearing (θ) of the direct wave. The two points of zero error occur when the beacon, re-radiator, and DF antenna are in a straight line.

The correction curve for semicircular error always takes the form of Figure 4.36a, but the positions of maximum and minimum error depend on the position of the re-radiator relative to the DF antenna. The four diagrams of Figure 4.36 show the basic shape of the correction curves associated with the four cardinal positions of the re-radiator relative to the DF antenna.

4.6.7 *Re-radiation (out-of-phase)*

It is possible for the physical dimensions of the vertical conductor (or its separation from the DF antenna) to be such that the re-radiation is in phase-quadrature with the primary field incident on the DF antenna. In this case the re-radiation does not combine with the primary field, and so the bearing accuracy is not affected. The effect instead, is to 'blur' the null, since at the null position of the goniometer there still exists the induced voltage from the re-radiated field. This residual voltage can be eliminated by deliberately introducing an antiphase voltage derived from the sense antenna, and adjusting its amplitude using the 'zero sharpen' control.

4.6.8 *Re-radiation: general*

As already mentioned the magnitude of the secondary field depends on the length of the vertical conductor and is at a maximum when its length is equal to one-quarter wavelength ($\lambda/4$); smaller peaks occur at $3\lambda/4$, $5\lambda/4$, etc. The precise lengths for resonance can vary slightly from this, depending on the cross-section dimensions of the vertical conductor.

In the MF beacon band, wavelengths are typically of the order of one thousand metres and thus it is very unlikely that vertical conductors will exist which approach the quarter wave resonant length. Consequently secondary fields at these frequencies will always be very small.

At frequencies of the order of two MHz the quarter wave resonant length is approximately thirty-seven metres, and consequently secondary fields from stays, halyards etc., are liable to be considerably more severe. The performance of a DF receiver on the 2182 kHz distress frequency is frequently affected in this way. A large amount of re-radiation results in a correction curve of the form shown in Figure 4.37. The bearings corresponding with the re-entrant parts of the curve are unusable since there is more than one possible relative bearing for an indicated DF bearing.

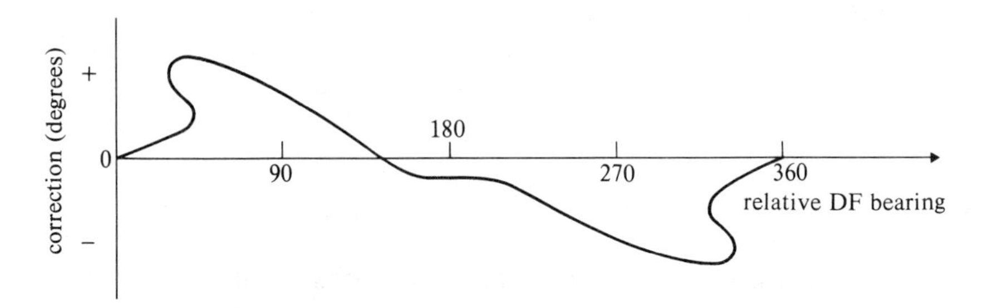

Figure 4.37 Typical calibration curve produced when maximum value of re-radiation becomes excessive.

In the extreme case the re-radiation reaches a magnitude such that the DF receiver permanently indicates the relative bearing of the re-radiator, irrespective of the actual direction of the incident signal.

In the 300 kHz MF beacon band, the error caused by re-radiation from a vertical conductor never reaches the extremes of these latter cases. Nevertheless semicircular error, no matter how small, is unacceptable since the error does not remain constant for all frequencies, and the correction curve produced from a calibration will not be applicable at frequencies other than that of the calibration. The correction may not even be applicable at this one frequency at all times, since the re-radiating vertical conductor may contain mechanical joints whose electrical properties vary, causing a variation in the intensity of the re-radiation.

The only acceptable remedies to the problems associated with re-radiation from a vertical conductor are either to inhibit it at source, or to move the DF antenna away from its influence. The method of inhibiting the re-radiation depends on the physical properties of the vertical conductor. Clearly little can be done with a mast or funnel, but steel halyards can be replaced with rope. Stays can be broken at intervals to reduce their conducting length and consequently raise their resonant frequency to a point well above the operational DF frequencies. Adjacent receiving and transmitting antennas can be particularly troublesome, even though the physical length of the antenna may not approach one-quarter wavelength of the direction finding frequency. This is by virtue of their connection to a tuned circuit within the transmitter or receiver. The remedy is thus to disconnect all antennas from their associated equipments.

As an example of the required separation between a DF antenna and a vertical conductor, to remove the DF from the effect of the re-radiation, the worst case is considered. This is a vertical conductor, with a length equivalent to one-quarter wavelength of the DF received frequency. In this case the secondary field at deck level reduces with distance from the base of the conductor, such that the field has reduced to the level of the primary field at a distance of one-quarter wavelength. At a distance of five times the vertical height of the re-radiator, the secondary field has reduced to one-fifth of the primary field.

4.6.9 *Field alignment error*

Any physically large horizontal structure will contribute to the total quadrantal error, which more usually originates from circulating currents in the hull or bridge structure. Commonly encountered large horizontal structures on a vessel are stowed derricks and deck cranes. They are often stowed at an angle to the centre line and so the field associated with any circulating currents will be misaligned with the hull field. The loop antenna may be sited sufficiently close to the horizontal structure to cause a distortion of the DF correction curve. This is illustrated by two actual cases (Figures 4.38 and 4.39).

4.6.10 *Loop alignment error*

In the initial installation of a crossed loop direction finder the antenna is aligned with the centreline of the ship. Any deviation from this would introduce a constant error, which if no other errors were present would result in the correction curve of Figure 4.40. This error could be compensated for by adjustment in alignment of the bearing pointer of the DF receiver, but this is not a sound practice since the receiver cannot then be easily changed for servicing purposes. The error should always be corrected by alignment of the antenna.

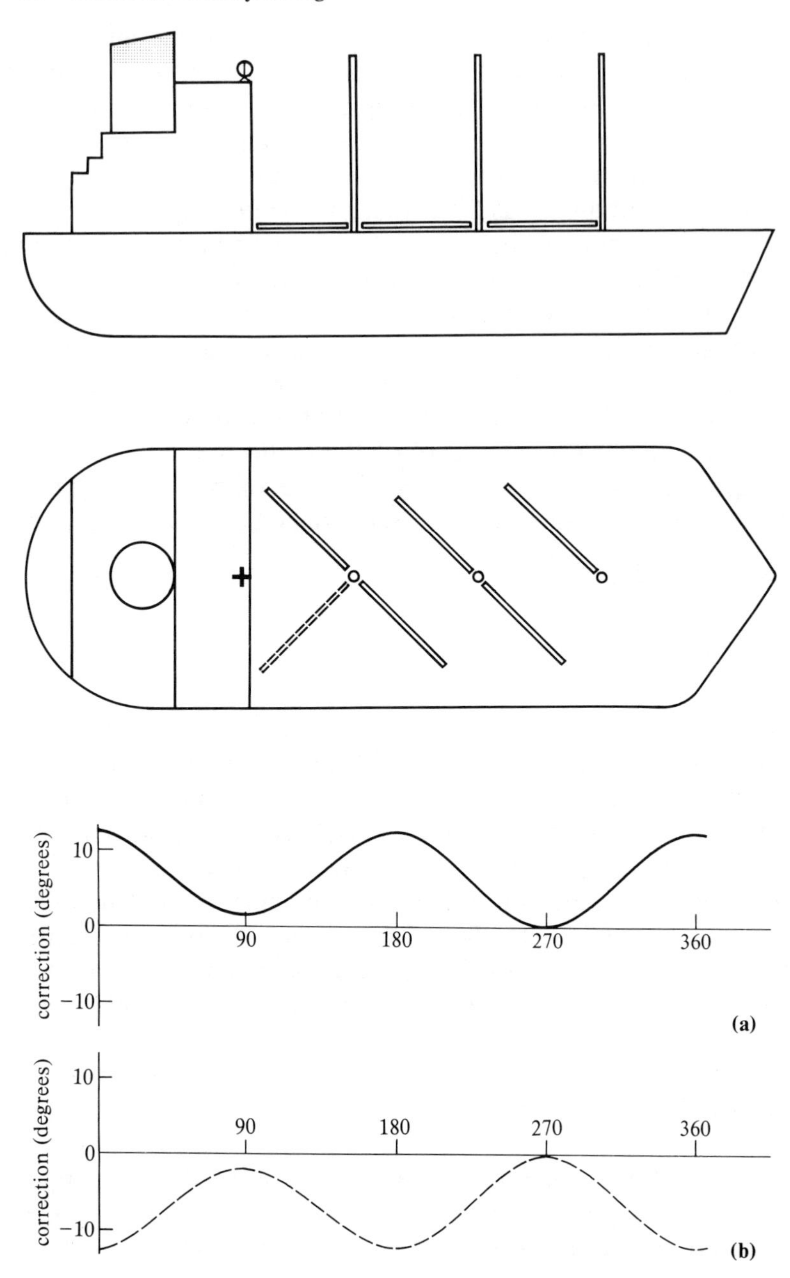

Figure 4.38 *Example of field misalignment due to stowed derricks. Distortion of the correction curve* (a) *is due mainly to the aftmost derrick as shown by the second calibration curve;* (b) *resulting from the stowage of this derrick on the starboard side.*

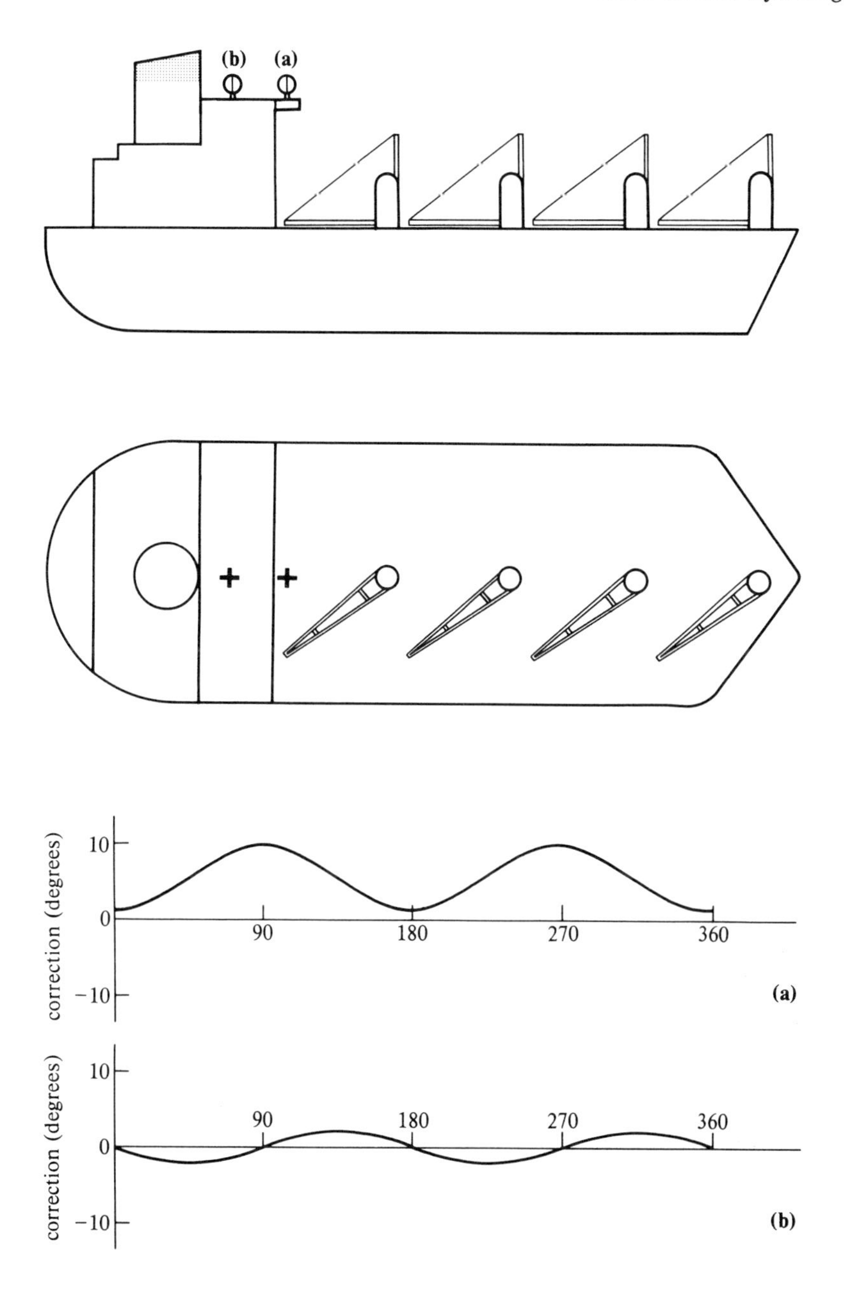

Figure 4.39 *Example of field misalignment due to stowed cranes. In this case the effect was minimized by re-siting the loop antenna in position* (b).

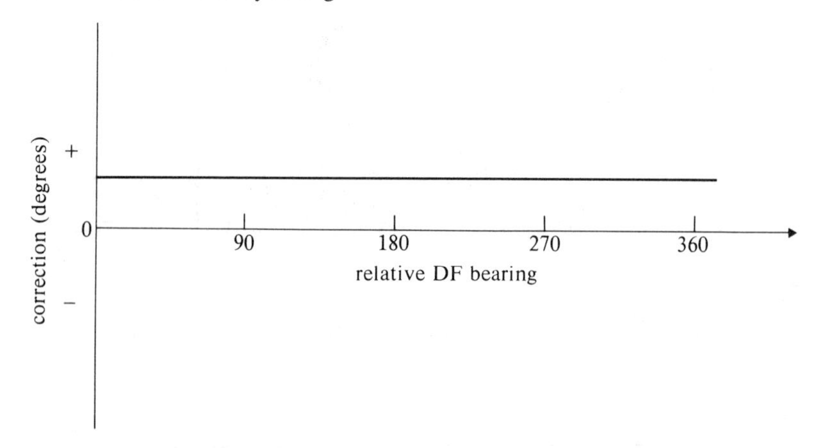

Figure 4.40 *Correction curve associated with loop alignment error.*

4.6.11 *Deck containers*

Most of the basic DF research, both theoretical and practical, has been carried out in connection with warships. The theory is then applied to merchant vessels and is generally found to hold true. An error source which has not been investigated in this way, because there is no naval counterpart, is that of deck containers. Observations have shown that deck containers can cause an error, although no rigorous and conclusive tests are known to have been conducted. The worst case is obviously when a container vessel is initially calibrated without deck containers, and subsequently carries a full deck cargo of perhaps three tiers. In this case, deviations from the initial calibration of up to five degrees have been reported. It is prudent therefore, that container ships should be calibrated at least twice to cover the extreme situations, and so determine the potential magnitude of the problem. Frequent check bearings should be taken in other load conditions, to establish the validity of the curves for intermediate loads.

4.6.12 *Composite errors*

It is convenient, for the purpose of achieving a simple explanation, to show the calibration curve which would result from each of the error sources in isolation. In reality two or more errors usually occur simultaneously, and the resulting calibration curve is a composite of these. Typical examples are quadrantal error with semicircular error (Figure 4.41) and quadrantal error with loop alignment error (Figure 4.42).

In these situations each calibration curve should be broken down into the basic error types and each error then either eliminated or corrected.

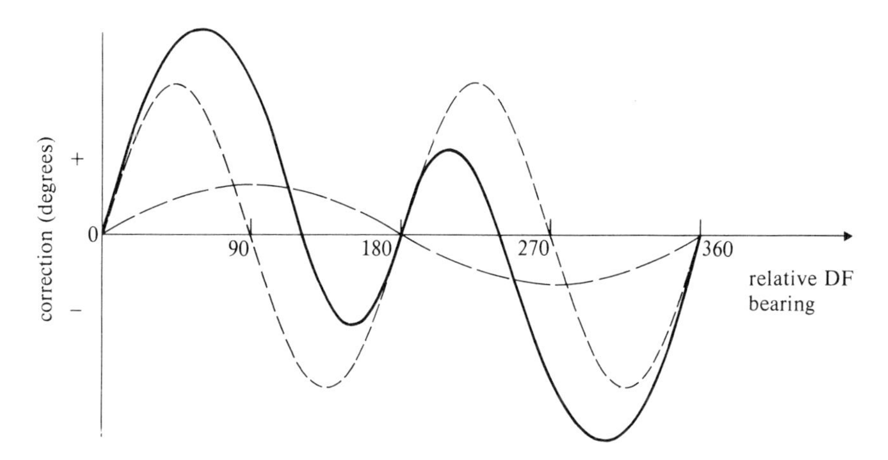

Figure 4.41 *A correction curve resulting from quadrantal error and semicircular error.*

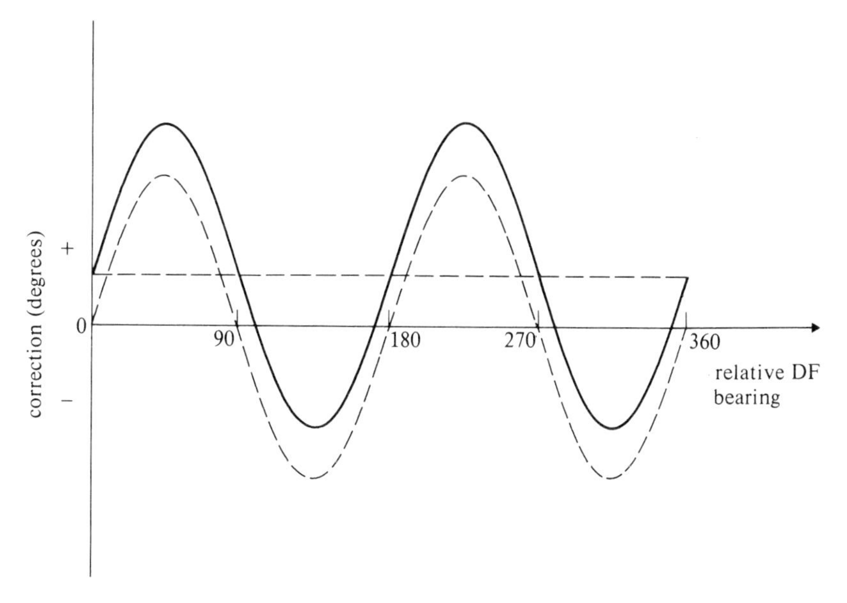

Figure 4.42 *A correction curve resulting from quadrantal error and loop alignment error.*

4.7 Calibration

The purpose of calibrating a marine DF installaton is to measure the effect that the ship has on the accuracy of the DF bearings. It is not a calibration of the DF equipment, since a correctly functioning instrument will have more than adequate inherent accuracy. Apart from determining the effect of the vessel, a satisfactorily completed calibration confirms the correct operation of the DF installation. It is always necessary to calibrate immediately after installing a DF system, and subsequently if there are significant structural changes, particularly in the vicinity of the DF antenna.

The calibration of a marine DF installation is basically a four-stage process: (1) preliminary check of the system for correct DF operation; elimination of potential sources of re-radiation where practicable; (2) the first calibration swing; plotting and analysing of correction curve; (3) elimination and correction of errors; (4) second calibration swing and plotting of final correction curve. These four stages can now be considered in detail.

(1) Interfering with a ship's normal trading in order to carry out a calibration is costly, and so it is obviously prudent to confirm correct functioning of the DF equipment immediately prior to the calibration. Other pre-calibration checks are: determine that there are no obvious sources of re-radiation close to the DF antenna; check that the vessel is in seagoing trim with derricks and cranes stowed; check that the communication antennas are isolated. Time spent on these preliminary checks increases the probability that the calibration will be carried out quickly and successfully.

Another preliminary check is to test the communication link between the DF receiver and the pelorus with which the visual bearings are taken. The usual method is a gong or bell sited near to the pelorus and activated by a switch at the DF receiver. In this case the visual observer notes the relative visual bearing at the time of signalling by the DF operator, although it is often more convenient for the visual observer to signal at regular intervals of visual bearing, and for the DF operator to note the corresponding DF bearing.

(2) The calibration should take place during daylight hours, avoiding sunrise and sunset by at least two hours. The vessel should be in a position of good visual contact with a continuously transmitting DF beacon, but not closer than half a mile. The beacon should be transmitting on a frequency within the marine beacon band of 285–315 kHz.

The ship should be swung through a full 360 degrees at a rate of about six degrees per minute. The relative DF bearing should be noted at each interval of five degrees, with the visual observer noting the corresponding relative visual bearing on receipt of the appropriate signal from the DF operator.

(3) From these results an interim calibration curve is drawn and the errors analysed into the basic error types. The following action should be taken for each of the errors identified.

Semicircular error: a calibration curve which shows the presence of semicircular error indicates a totally unsatisfactory situation, since the error is likely to vary with frequency or time. It is essential that the cause of the error is identified and its effect eliminated.

Quadrantal error: if a correction facility for compensation of quadrantal error exists within the DF receiver, then the appropriate amount of correction should be applied. If there is no compensation facility it is acceptable for this error to appear in the final correction curve.

Field alignment error: the distortion of the normal shape of the quadrantal error curve means that the quadrantal error correction will not be totally effective. The optimum amount of compensation is a matter of judgment, and the residual error is acceptable since it is not variable with frequency or time.

Loop alignment error: this error is removed by turning the loop antenna through an appropriate number of degrees. The correct adjustment is most easily carried out by holding the vessel in a position with the calibrating beacon visually dead ahead, and then slowly turning the antenna until the DF bearing is also ahead. The antenna is then firmly secured in this position.

(4) After eliminating and compensating for the errors shown up by the first swing, the vessel is turned for a second time and the procedure of recording visual and DF bearings is repeated. From these results, the final calibration curve is drawn and is displayed near to the DF receiver. It is then used in the routine operation of the direction finder to apply the appropriate amount of correction.

A typical calibration table is shown in Figure 4.43. Referring to this, there are two points to particularly note. The long axis corresponds to relative DF bearing *not* visual bearing, and the short axis to degrees of correction *not* error. Correction being: relative visual bearing – relative DF bearing.

As soon as possible after the calibration, several check bearings should be taken in each quadrant to confirm the accuracy of the calibration curve. Check bearings should continue to be taken

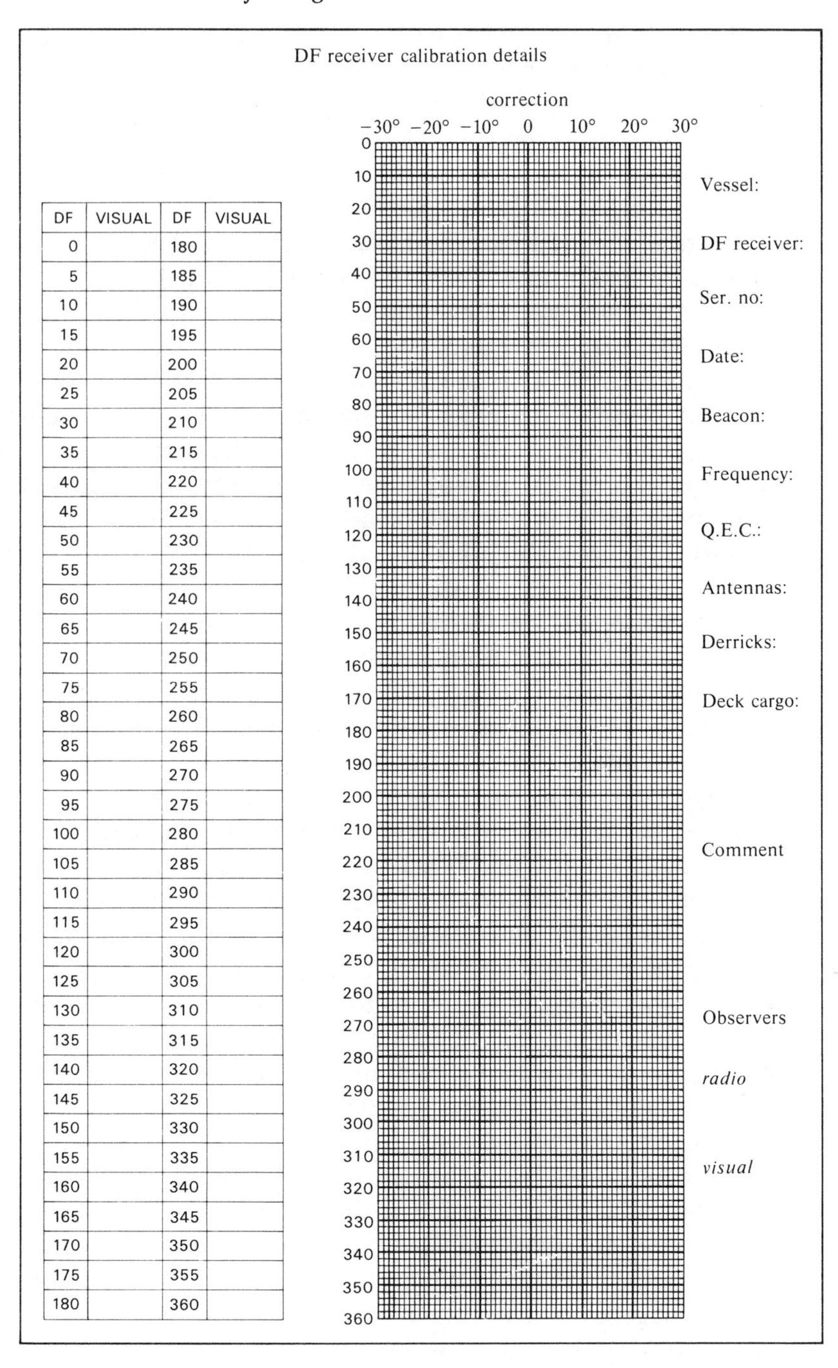

DF receiver calibration details

DF	VISUAL	DF	VISUAL
0		180	
5		185	
10		190	
15		195	
20		200	
25		205	
30		210	
35		215	
40		220	
45		225	
50		230	
55		235	
60		240	
65		245	
70		250	
75		255	
80		260	
85		265	
90		270	
95		275	
100		280	
105		285	
110		290	
115		295	
120		300	
125		305	
130		310	
135		315	
140		320	
145		325	
150		330	
155		335	
160		340	
165		345	
170		350	
175		355	
180		360	

Vessel:

DF receiver:

Ser. no:

Date:

Beacon:

Frequency:

Q.E.C.:

Antennas:

Derricks:

Deck cargo:

Comment

Observers

radio

visual

Figure 4.43 A typical format for display DF calibration results.

regularly to prove the correction operation of the DF installation, and the continued validity of the calibration curve.

4.7.1 *Shortened calibration procedure*

To ensure a satisfactory calibration with the minimum of residual error, the four-stage procedure as previously outlined should be fully implemented. However, if there is not sufficient time available for two complete swings, the first (stage 2) can be omitted, but with the possibility of a greater residual error in the final curve.

The first swing and curve are replaced by the following shortened procedure. The vessel is first held in position with the calibrating beacon dead ahead, to check loop alignment. The vessel is then turned to give relative visual bearings of first 045 and then 315, the corresponding DF bearings are noted.

From these two bearings the required amount of quadrantal error compensation is determined. For example:

Visual bearing	*DF bearing*	*Correction*
045	039	+6
315	319	−4

The average magnitude of the error is five degrees and its effect is to pull the bearing towards the centreline. In this case therefore, five degrees of quadrantal error correction is required across the fore-aft loop. A difference in magnitude of the errors at 045 and 315 degrees could be caused by the presence of a semicircular error, which will be seen as a residual error in the final calibration curve.

4.7.2 *Calibration with a portable transmitter*

It is often convenient to calibrate the DF installation while the vessel is at anchor, moving the calibration transmitter around the vessel by launch. This procedure is satisfactory provided that the launch maintains a distance from the calibrating vessel of at least half a mile, and that there are no other objects within half a mile of either vessel.

4.8 DF beacons

Over 1600 beacons are available worldwide for maritime navigation by direction finder. Details of all the beacons are listed in a number of publications, including *The Admiralty List of Radio Signals* vol. 2, and *The US Radio Navigational Aids* publications. The majority of these beacons are specially set up and maintained for the maritime service by national administrations, although a proportion of those listed are aircraft beacons conveniently sited near to the coast. While these latter beacons can be used by ships because of their convenient location, they may be more liable to coastal refraction effects since their siting is not specifically optimized for seaward transmission paths.

The aero beacons and some of the marine beacons transmit continuously, but other marine beacons transmit cyclically in a group. A typical group is illustrated by Figure 4.44: each beacon transmits

Group transmission frequency 310.3 kHz
Modulation A2 752 Hz

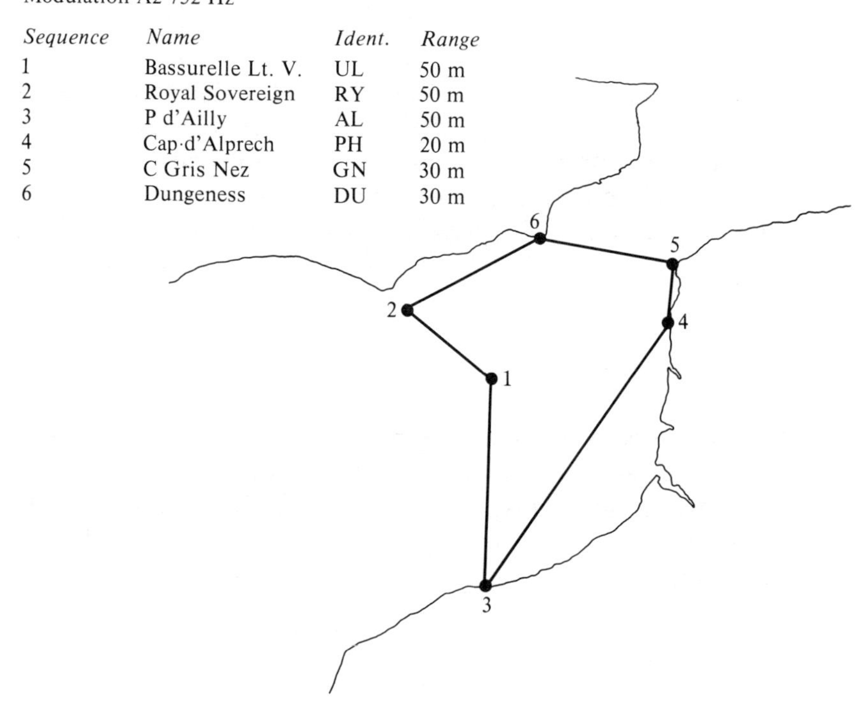

Sequence	*Name*	*Ident.*	*Range*
1	Bassurelle Lt. V.	UL	50 m
2	Royal Sovereign	RY	50 m
3	P d'Ailly	AL	50 m
4	Cap·d'Alprech	PH	20 m
5	C Gris Nez	GN	30 m
6	Dungeness	DU	30 m

Figure 4.44 Typical DF beacon group.

for one minute, a full cycle therefore occupies a six-minute period and always commences on the hour and at H + 06, H + 12 etc.

In groupings of less than six beacons, some beacons in the group may transmit more than one time slot to make up the six-minute sequence. In some areas the groupings are based on other than a six-minute cycle. Japan is one such area; here the beacons are usually operated in groups of two or three, each transmitting for a five-minute period.

Beacons transmit their identifying call sign in morse code followed by a continuous transmission, during this time the DF bearing is taken.

The publications which list the DF beacons usually indicate a range in nautical miles. This is the distance at which the field strength of the groundwave falls to a level beyond which the performance of the DF receiver becomes unacceptable.

The transmission frequencies of beacon groups are chosen so that adjacent groups have different frequencies, and groups with the same frequency are separated by distances sufficient to minimize interference. Nevertheless interference can occur at night-time due to the skywave transmission paths, and for this reason the radiated power of many beacons is reduced at night.

4.9 QTG service

Many of the medium frequency coast radio stations will transmit a signal for DF purposes when so requested. The request takes the form of a message to the coast station which includes the abbreviation QTG, and the required transmission frequency. The frequency of 410 kHz is usually used for this purpose.

5 Consol

The Consol navigation system originates from the German 'Sonne' system which was developed during the Second World War. The system has limited use today, providing low-accuracy long-range navigation. Only a brief description of Consol is given here, since its value to navigators of merchant vessels is minimal.

Fundamentally Consol falls within the classification of a hyperbolic system, since a position line is determined from the difference in arrival times of synchronized transmissions. The similarity with more conventional hyperbolic navigation aids extends little further than this. Consol signals which result in a line of position are radiated from three antennas, closely spaced, and all part of the same transmitter system. A position line is obtained using only a simple radio receiver, since the time difference is determined from a sequence of dots and dashes.

5.1 The system

The radiating antennas of a transmitting system are equally spaced in a line, two and a half to three kilometres apart. At the frequencies of Consol beacons this represents approximately three wavelengths. Since the baseline distance is very short the hyperbolae approximate to great circle tracks passing through the centre transmitting antenna.

The two outer antennas A_1 and A_2 (Figure 5.1) radiate signals which are of equal amplitude but of opposite phase. The phases of the signal from the outer antennas can be varied with respect to that of the centre antenna A_0, the amplitude of which is four times greater than the amplitude of A_1 and A_2. Consider first the situation depicted by Figure 5.1, in which the phase of A_0 lags A_1 by ninety degrees, and leads A_2 by ninety degrees. For this situation hyperbolic lines can be drawn representing every point at which A_1 and A_2 signals are received in antiphase, and hence mutually cancel.

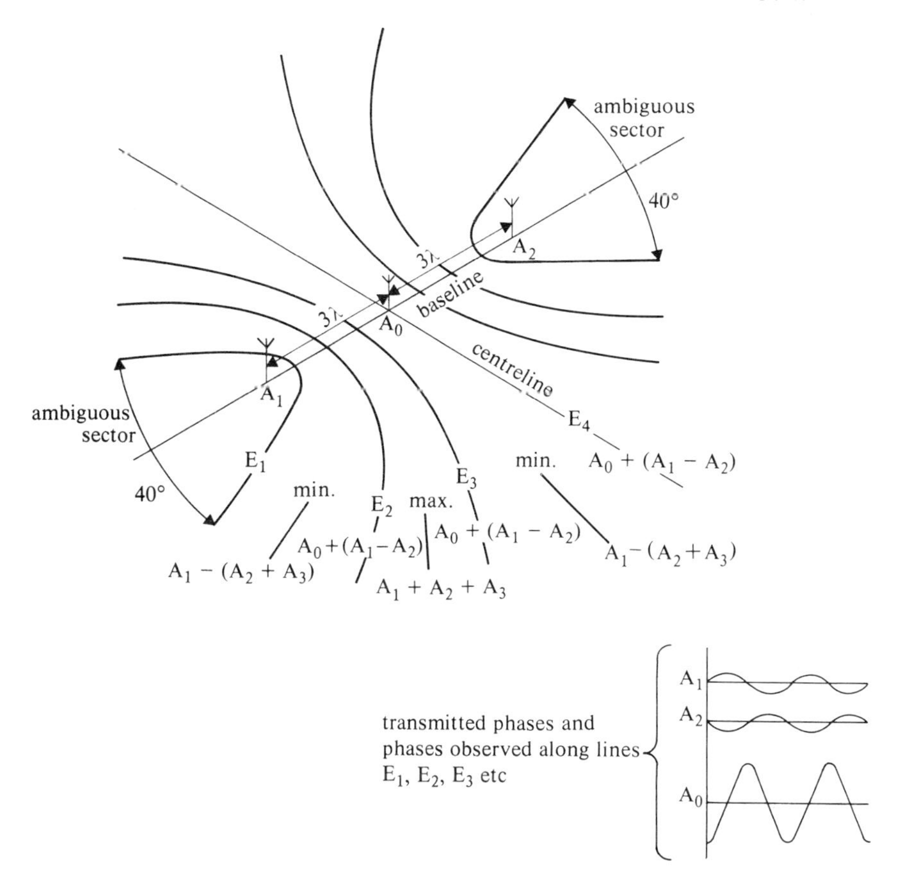

Figure 5.1 *Along the hyperbolic lines only the* A_0 *signal is heard as* A_1 *and* A_2 *are in antiphase and hence cancel. In the area of maximum signal* A_0, A_1 *and* A_2 *are all in phase. In the area of minimum signal* A_1 *and* A_2 *are in phase, but in antiphase with* A_0.

Along the hyperbolic lines therefore, only the A_0 signal is received. In between these lines the signal will reach a maximum when all signals are in-phase, and a minimum when the A_1 and A_2 signals have the same phase but are in antiphase with the A_0 signal. Since the hyperbolic lines approximate to straight lines centred on the A_0 antenna, the space between adjacent lines is usually termed a sector rather than a lane.

Having shown how the hyperbolic lines are derived, the mechanism which allows a position line to be obtained using a simple radio receiver, can be described.

If the phases of the A_1 and A_2 radiated signals are suddenly switched through 180 degrees, all areas in which the signal was previously at a minimum now become areas of maximum signal, and, conversely, the maximum signal becomes a minimum signal. By switching the phase for typically one-eighth of a second in every half second, a user in a weak area hears a series of dots, and a user in a maximum signal area hears a series of dashes. Along the hyperbolic lines there will be neither dots nor dashes, since the A_1 and A_2 signals cancel leaving only the steady A_0 signal, which is termed the equisignal.

If now the phase of the signals radiated by the A_1 and A_2 antennas are slowly changed by equal and opposite amounts, the positions of the equisignal lines will move. In effect the whole hyperbolic lattice will rotate, with the two halves of the pattern rotating in opposite directions (Figure 5.2). This creates a means whereby the user can determine a position line, since it is related to the time at which the equisignal is heard. Before the arrival of the equisignal dots (or dashes) will be heard, and after the equisignal these will becomes dashes (or dots). During each thirty second pattern rotation period,

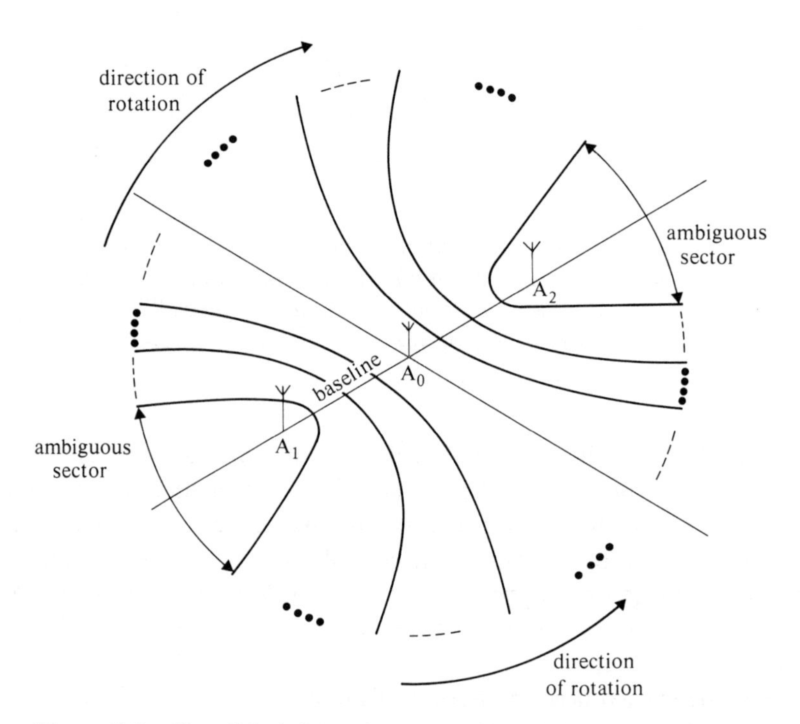

Figure 5.2 Simplified Consol pattern, illustrating directions of pattern rotation. Observed transmission sequence commences with dots and dashes in alternate sectors.

a total of sixty dots and sixty dashes are transmitted, and it is the number of dots (or dashes) which are received before the equisignal which defines the user's position line.

5.2 Using Consol

The occurrence of the equisignal is not always a clearly defined moment in time; under circumstances of poor signal conditions several dots and dashes immediately before and after the equisignal, may not be heard. To compensate for this effect the normal procedure when using Consol is to count both dots and dashes, and since it is known that the total should be sixty, any shortfall is assumed to be due to the loss of an equal number of dots and dashes. For example:

actual count	= 16 dashes, 38 dots
	$16 + 38 = 54$
lost count	$= 60 - 54 = 6$
	= 3 dashes and 3 dots
corrected count	= 19 dashes and 41 dots
	19 dashes thus defines the LOP.

A position line can now be plotted onto an overprinted chart (Figure 5.3), or by using the tables such as those published in the *Admiralty List of Radio Signals* vol. 5, plotting can be directly onto the navigation chart.

A basic ambiguity exists within the Consol system between alternate sectors; the user must therefore know his position to within one sector.

Almost any receiver capable of receiving continuous wave (A1) signals and covering the appropriate frequencies, can be used for receiving Consol signals. All Consol stations are within the frequency range 257–363 kHz.

Consol stations commence their transmission sequence with a callsign, followed by a continuous tone from which a DF bearing can be taken to assist in resolving sector ambiguity. Finally there is the thirty second period of pattern rotation.

There are Consol stations situated in Norway, France, Spain and the USSR. In the USA there is a variant of Consol, called Consolan, which is operationally the same as Consol.

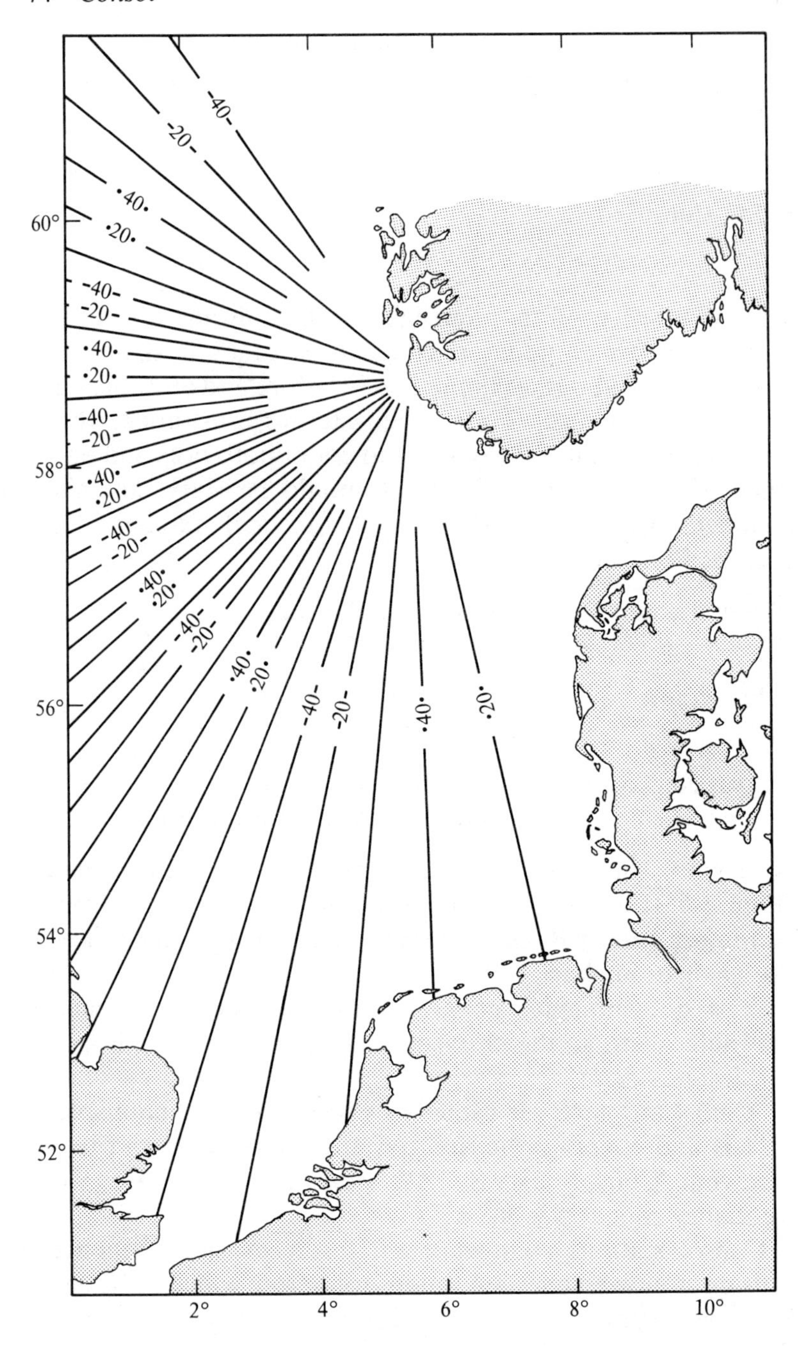

Figure 5.3 One half of the pattern of the Stavanger Consol beacon.

The operational range of Consol is typically from a minimum of thirty nautical miles to a maximum of 1500 nautical miles.

The accuracy of a position line is dependent upon the distance of the user from the centreline. The accuracy is at its best on the centreline, which in the day-time an angular error of better than 0.3 degrees can be expected, degrading during night-time to 0.7 degrees. Off the centreline the accuracy progressively decreases as the sector becomes wider, reducing to 1.2 degrees during day-time and 2.7 degrees during night-time.

Users of the Consol system should always be aware that these angular errors can represent very large positional errors at the longer ranges of operation.

6 Hyperbolic navigation

6.1 Basic principles

Most electronic navigation systems use the precisely known velocity at which electromagnetic energy (radio waves) travel between their source and the point of reception, to determine distance and ultimately position. The precise method of using propagation velocity to determine position differs for the various navigation systems. An understanding of the basic principles involved is essential for an understanding of overall system operation.

Several navigation systems, including Loran, Omega, and Decca Navigator, utilize the same basic principle, and are collectively known as 'hyperbolic' navigation systems. This principle is illustrated by Figure 6.1. Simultaneous transmissions occur from points A and B, and at the point of reception the difference in their time of arrival is determined.

For an observer on the centreline, the transmissions will be received simultaneously (Figure 6.1a), but at all points other than on the centreline there will be a difference in the arrival time. The locus of points corresponding to a given time difference is a hyperbola (Figure 6.1b); a family of hyperbolae can be drawn representing discrete time differences (Figure 6.2), and hyperbolae for other intervals determined by interpolation.

These hyperbolae can be overprinted on to navigation charts so that after measuring the time difference, a position line can be plotted directly on to the chart. The second position line required to produce the position fix is obtained from a second pair of transmitting stations, one of which can be one of the original pair (Figure 6.3).

The inherent advantage of the hyperbolic system is the need to make only a measurement of time lapse. This is in contrast with the method known as circular (or range-range) navigation, in which the actual time of signal propagation from transmitter to receiver is measured. In this case the loci of points of constant propagation

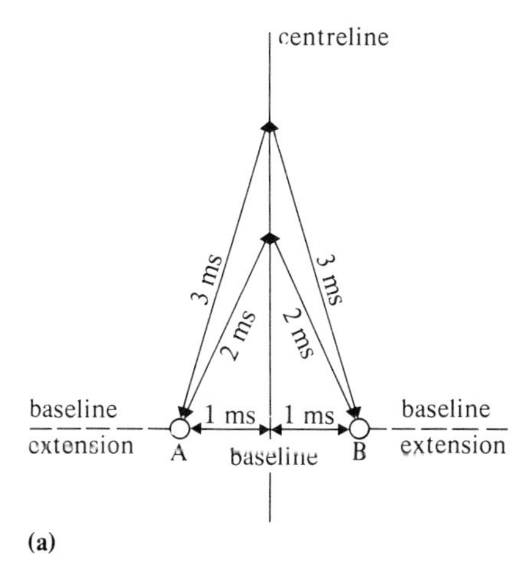

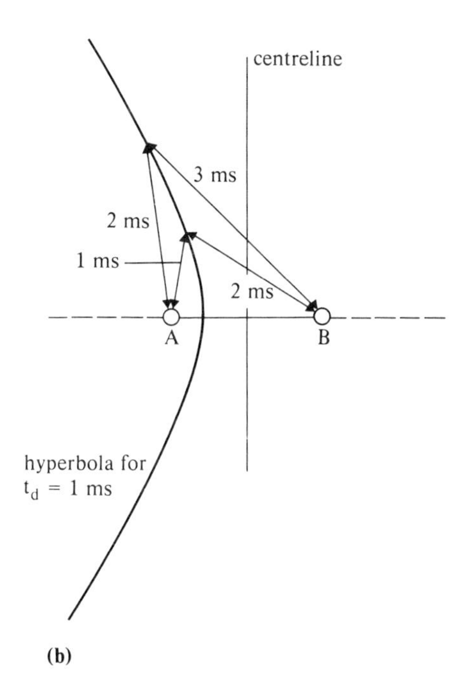

Figure 6.1 (a) *The locus of zero time difference;*
(b) *A locus of constant time difference.*

time are circles, centred on the transmitting stations. In circular navigation systems the receiver clock must be accurately synchronized with the transmission for it to be possible to determine the propagation time. To maintain this synchronization over a long

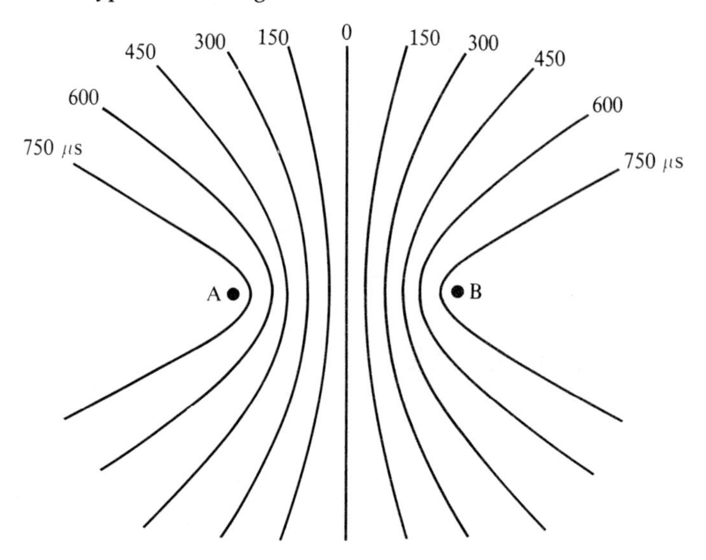

Figure 6.2 A set of hyperbolae for specific time differences.

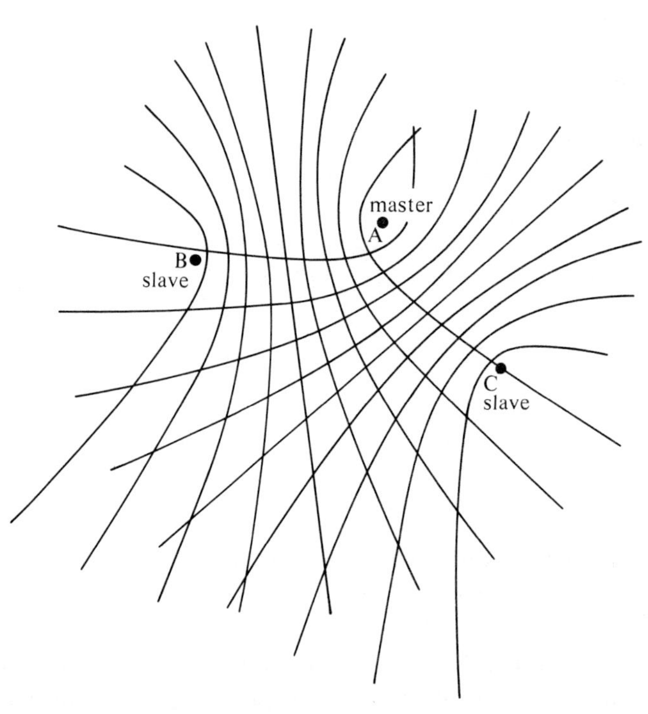

Figure 6.3 Two hyperbolic patterns can be obtained from three transmitters.

period requires a high degree of reference oscillator stability, usually necessitating the use of an atomic frequency standard.

6.2 Transmitter synchronization

The reference oscillator of a hyperbolic navigation receiver need not have the same high degree of long-term stability since synchronization with the transmissions is not necessary. However it is necessary for mutual synchronization to be accurately maintained by the pairs of transmitting stations. In the past this has been achieved by one station monitoring the transmissions of the other, and then automatically adjusting its own transmission timing to maintain precise synchronization. With the inherent stability of atomic frequency standards, the need for constant monitoring and regular adjustment has now been removed.

If in practice both transmissions occur simultaneously, the resulting hyperbolic pattern contains a fundamental ambiguity, since the two halves of the pattern represent the same range of time differences, the only difference being the order in which the transmissions are received. This ambiguity could be difficult to resolve near to the centreline, and so measures are taken to ensure that it does not occur. This can be simply achieved by arranging that one transmission occurs at a fixed interval before the other, the interval being longer than the maximum time difference. Subtracting the fixed delay from the measured time delay at the receiver gives the actual propagation delay.

In hyperbolic navigation systems the station which radiates first is usually termed the master, and it is this which acts as the common station to produce two or more hyperbolic patterns. It is the slave transmission timing which is adjusted when necessary to maintain the precise time relationship.

6.3 The time difference measurement

The signals radiated from the transmitters must be of a form which will allow the time difference measurement to be made with optimum accuracy.

Fundamentally there are two methods by which the time

difference can be determined. One is the method employed in the Loran A system, in which the transmission takes the form of a short pulse of energy, and the time difference is measured from the beginning of pulse A to the beginning of pulse B (Figure 6.4b). The second method is that utilized in the Omega and Decca Navigator systems; here it is the phases of the transmitted carrier frequencies which are compared, to derive the time difference (Figure 6.4a).

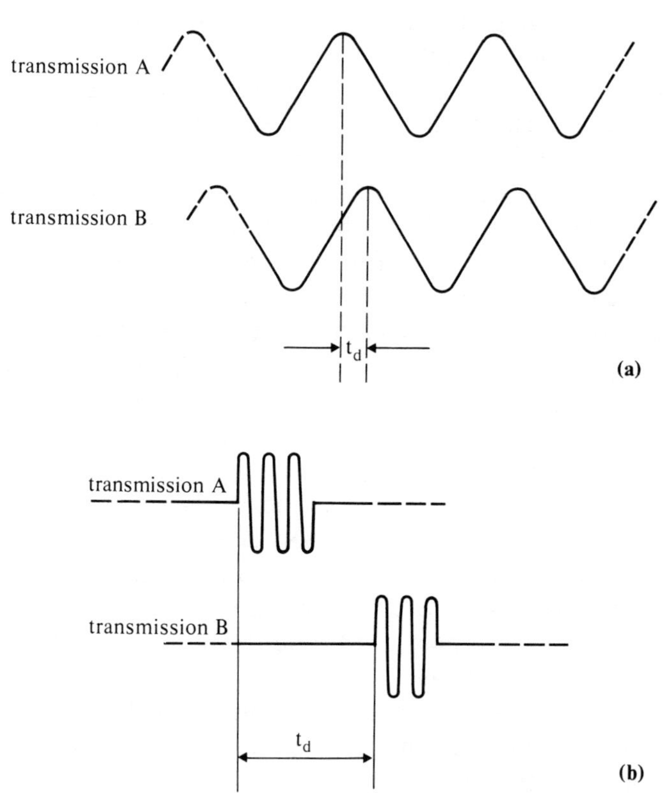

Figure 6.4 *Measurement of time difference* (a) *by phase comparison;* (b) *by difference in pulse arrival times.*

6.4 Lane width

A time difference of given accuracy has the least error as a position line when the user is on the baseline, since here the lane width is at its narrowest. Away from the baseline the lanes expand, and the position line accuracy reduces by a corresponding amount. The lane expansion factor E, is given by:

$$E = \text{cosec}\ \alpha/2$$

where α is the angle between the two stations, as seen from the observer (Figure 6.5).

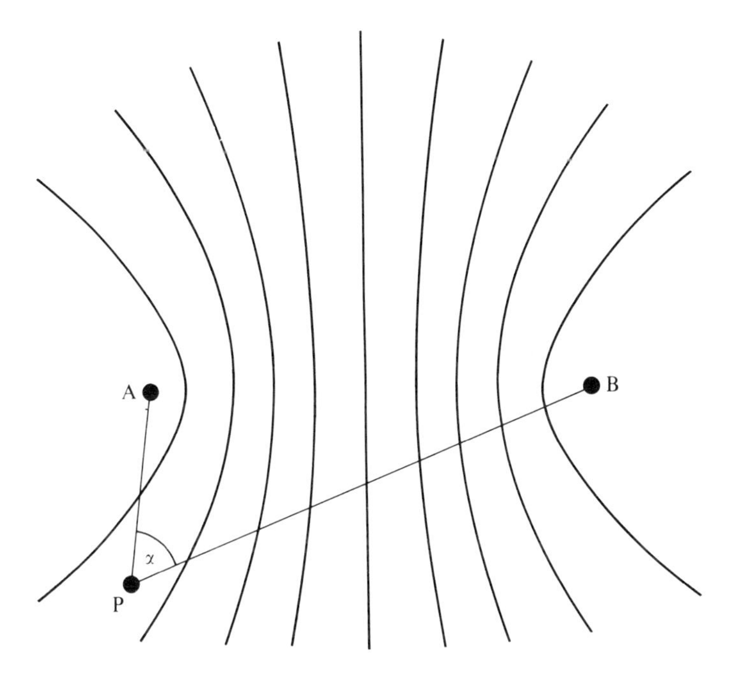

Figure 6.5 Illustration of lane expansion factor.

The Omega system with its very long baselines benefits from the effect of spherical excess. On a spherical surface the sum of the three angles of a triangle is not exactly 180 degrees as on a plane, but may be considerably more. When the length of a hyperbolic baseline approaches or exceeds the radius of the earth the effect is very beneficial, and the lane expansion is very much less than it would otherwise be on a plane surface.

6.5 General

The various hyperbolic navigation systems differ in many aspects, including transmission frequency, baseline distance, and method of time difference measurement. Each is chosen to optimize the system's performance in a manner best suited to its intended navigational role. These and many other factors are fully described under each of the specific system headings.

6.6 Range-range navigation

Range-range (or circular) navigation differs from hyperbolic navigation in that instead of measuring time difference, a measure is made of actual signal propagation time between transmitter and receiver. The resulting loci are therefore circles centred on the signal source and with radii corresponding to the distances which are equivalent to the measured propagation time (Figure 6.6). There are several advantages to be gained by using range-range in preference to hyperbolic navigation. Only two stations are required to determine position and there is no dilution of accuracy through hyperbolic lane expansion. It is also an easier technique for processing several signals (such as received from four or five Omega stations), since

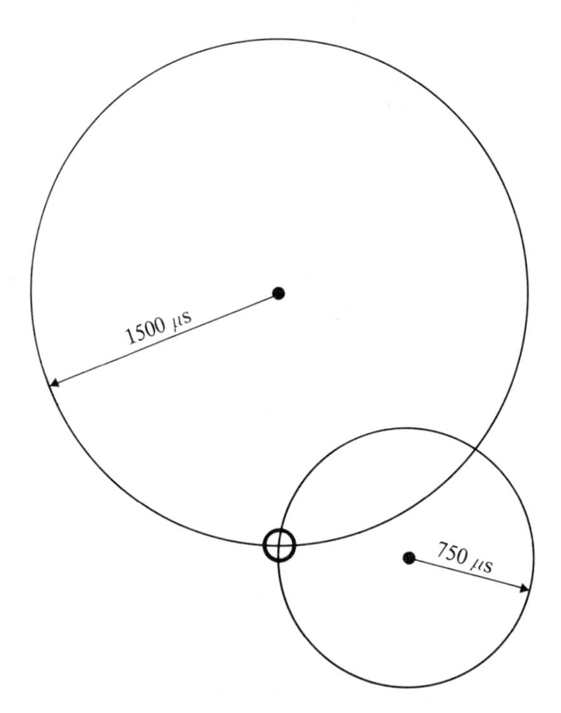

Figure 6.6 The circular loci of range-range navigation.

each can be dealt with in isolation, rather than having to consider the many possible combinations of paired stations.

The most severe limitation of range-range navigation is the requirement to maintain absolute time with a high degree of precision. The clock stability necessary for the duration of an ocean passage would demand the use of a cesium frequency standard. For short coastal voyages a more moderate rubidium standard could possibly be used, but in most cases both of these would be prohibitive in terms of their cost.

An alternative approach is that of pseudo range-range, in which the timing of received signals is measured with respect to an unknown reference. Provided that three sets of measurements are made the equations can be solved for the three unknowns, latitude, longitude and time. This method requires the use of a minimum of three transmitting stations, but allows the use of a low-cost quartz crystal reference oscillator.

7 Loran A

Loran A is the forerunner of Loran C, developed during the Second World War in the United States and implemented and operated by the US Coast Guard. By the end of the war there were seventy transmitting stations and 75,000 user equipments. Although the more recent Loran C system is more accurate than Loran A, the latter system has remained operational due to its intensive use. In 1970 the US Government published its National Plan for Navigation which approved Loran C as the preferred navigation system for its coastal confluence zone. As a result of this, a programme of expansion of the Loran C system and of phasing out of Loran A has been implemented. Because of this planned obsolescence only a brief description of the Loran A system is given here.

7.1 System description

Loran A is a basic hyperbolic navigation system in which the time difference is that observed between individual pulses from two transmitters. The transmitting stations operate in master-slave configurations on one of the frequencies 1850 kHz, 1900 kHz, or 1950 kHz, radiating pulses which are approximately forty microseconds long. A master station which controls two slaves is 'double pulsed', meaning that it transmits entirely separate pulses for each slave.

Adjacent pairs may use the same transmission frequency, but are differentiated at the receiver by their pulse repetition intervals (PRI) – the time between successive pulses. A chain is therefore specified by its frequency and its pulse repetition interval. In order to simplify this method of designation, from the need to specify the actual frequency and pulse repetition interval, a coded designation is used:

Frequency code

1 1950 kHz
2 1850 kHz
3 1900 kHz

Pulse repetition interval code

	S	*L*	*H*	
0	50 000	40 000	30 000	(*microseconds*)
1	49 900	39 900	29 900	
2	49 800	38 800	29 800	
3	49 700	39 700	29 700	
4	49 600	39 600	29 600	
5	49 500	39 500	29 500	
6	49 400	39 400	29 400	
7	49 300	39 300	29 300	

A rate of 1H5 for example, means a frequency of 1950 kHz and pulse repetition interval of 29,500 microseconds.

7.2 Propagation

Transmitting stations which make up a master-slave pair are generally located 200 to 400 miles apart. During day-time propagation is almost entirely due to groundwave, and at night-time it is due to both groundwave and skywave. During night-time there may be received pulses due to various modes of skywave, including one and two hop E-layer and one and two hop F-layer. By making use of the first skywave pulse (the one hop E-layer), operational range of Loran A can be extended from about 700 miles for groundwave up to approximately 1500 miles. Multiple E-layer reflections and all F-layer reflections are too variable for reliable navigation.

7.3 The Loran A receiver

Most Loran A receivers utilize a cathode ray tube (CRT) to visually display the master and slave pulses. The spot is made to deflect across the tube face at a precisely controlled rate. If the rate coincides with the pulse repetition interval of a given chain, it will cause the pulses from that chain to appear stationary on the tube face, and pulses from other chains (with different PRIs) will appear to be moving across the tube. It is usual to split the total spot deflection into two halves, making a top trace and a bottom trace (Figure 7.1). It is also usual for 'pedestals' to be generated on the two traces to facilitate alignment of the master and slave pulses. The pedestal part of the sweep is expanded in time to simplify the operation of locating the Loran pulses on the pedestals. A typical Loran A receiver is illustrated in Figure 7.2. Selection of the 'rate'

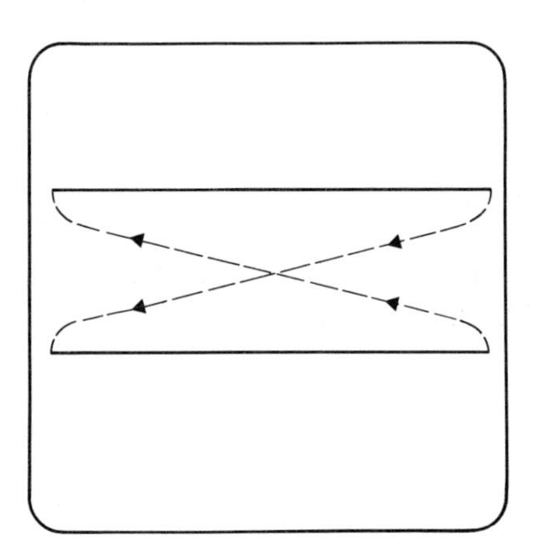

Figure 7.1 The two traces of a Loran A CRT display. The spot is not seen while traversing the dashed sections.

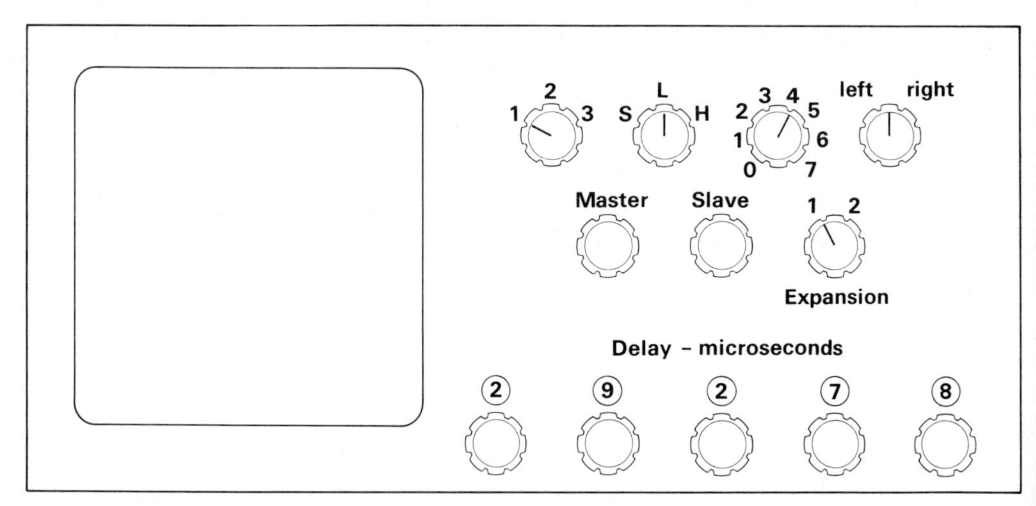

Figure 7.2 A typical Loran A receiver.

for a given chain adjusts the receiver tuning to the appropriate frequency, and sets the CRT time base to be synchronized with the pulse repetition interval. The left-right control is then adjusted to make the master pulse coincident with the top trace pedestal (Figure 7.3). The delay switches are then adjusted to make the slave pulse coincident with the pedestal of the bottom trace. At this point the trace can be expanded, making the top of the pedestal the full width of the CRT (Figure 7.4). Fine adjustment of the delay control is then made, to achieve precise alignment between the two pulses. The time delay between master and slave pulses can be read directly from the delay switches, and this reading is then used to plot a

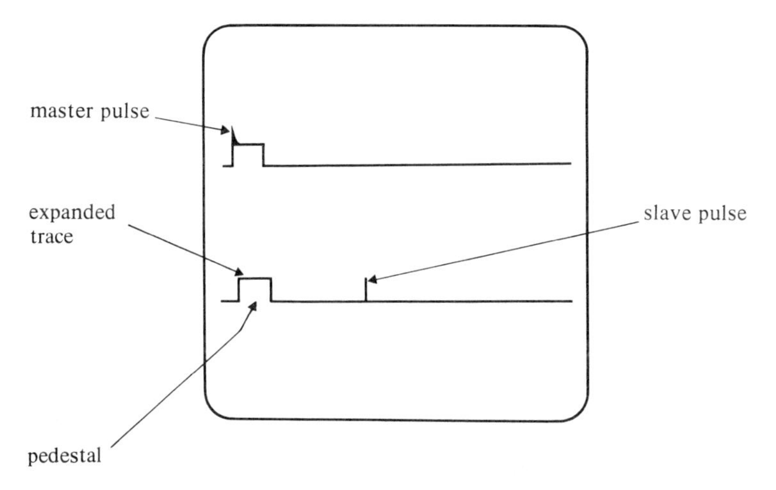

Figure 7.3 *A typical Loran A display showing pedestals and master and slave pulses.*

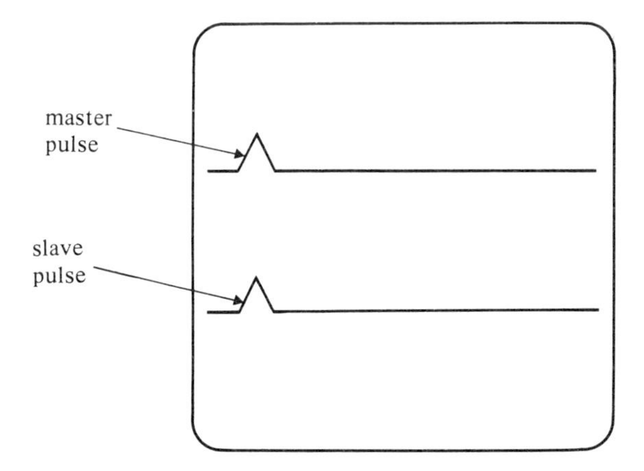

Figure 7.4 *A typical expanded Loran A display.*

position line onto an overprinted Loran A chart. A second position line can be determined by repeating the procedure, using a second master-slave pair.

In the areas which are permanently within groundwave coverage, using a Loran A receiver is basically simple, requiring little operator skill. At greater distances from the transmitter a number of skywave signals may be received, including one and two hop E-layer and one and two hop F-layer, thus making it more difficult to determine the correct pulse to use. Additional confusion may arise due to the skywave pulse splitting into two. Of the skywave pulses, only one hop E-layer can be used for navigation, and then only after first applying a skywave correction.

8 Loran C

Loran C is a low-frequency, pulsed, hyperbolic navigation system, managed and operated by the US Coastguard. It is available for use by any vessel in the coverage area carrying the appropriate receiving apparatus.

The first Loran C chain came into operation in 1957 and since that time has developed into the major navigation system which it is today (Figure 8.1). It has been selected by the United States government as the sponsored radio navigation system for civil marine use in the US coastal and confluence zone. Consequently the most complete areas of coverage of Loran C are around the US coastline, although many other areas of the world are also well covered.

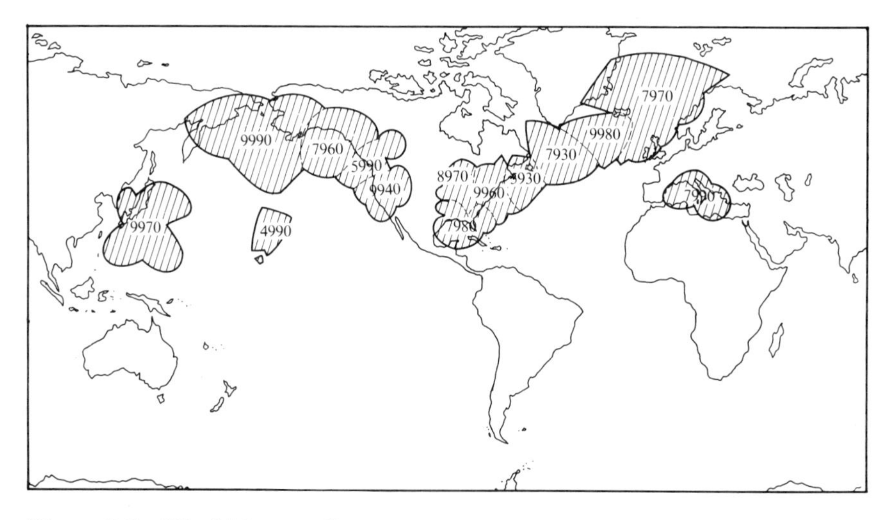

Figure 8.1 World Loran C coverage.

The propagation distances at the transmission frequency of 100 kHz, and the stability of the propagated waves, give Loran C a range adequate for ocean navigation and an accuracy potential adequate for coastal navigation.

8.1 Chain configuration

The hyperbolic line of position is derived from the difference in arrival time of pulses from two transmitting stations. The system is organized such that one station is the master and always transmits first. The second station, the slave, is synchronized with the master and transmits at a precise interval after the master transmission. This interval, known as the secondary coding delay, ensures that even in situations when a vessel's location is close to the slave transmitter, the master pulses will still arrive before the slave pulses.

A Loran C chain is comprised of at least two slaves, and more usually either three or four. These are designated W, X, Y and Z, and each has its own coding delay, so that the slaves are always received in the same sequence throughout the coverage area (Figure 8.2).

All transmitters of all chains radiate at the same frequency of 100 kHz, so to prevent interference between chains, the repetition rate of the pulse groups of each chain is made unique. This is referred to as the group repetition interval (GRI), and for a given chain is selected to be of sufficient length to contain the master and slave transmissions.

On charts and other Loran C publications, a chain is designated by its group repetition interval in microseconds, but omitting the last zero. Taking as an example the Mediterranean chain (Figure 8.3); the GRI of this chain is 70,900 μs (microseconds), and the chain is therefore designated as 7990. The secondary transmitting stations X, Y, and Z have coding delays of 11,000, 29,000 and 47,000 μs respectively. The relevant section of Loran C chart (Figure 8.4) shows the hyperbolic lines resulting from time differences between master and slave X (7990X), and master and slave Y (7990Y). Consider the particular line of position 7990X, 12,500 μs; this time difference is a composite of the secondary coding delay of 11,000 μs and the time difference due to the actual propagation times. But for all practical purposes it can be considered that 12,500 μs and the other charted time differences are the actual propagated time differences.

In the case of some adjacent chains, one transmitting station may be common to both chains. For example in the North Atlantic chain (9980), the master station of this chain is also the secondary Y of the Norwegian chain (7970). Slave X of 7930 chain is Slave W of chain 9980 (Figure 8.3).

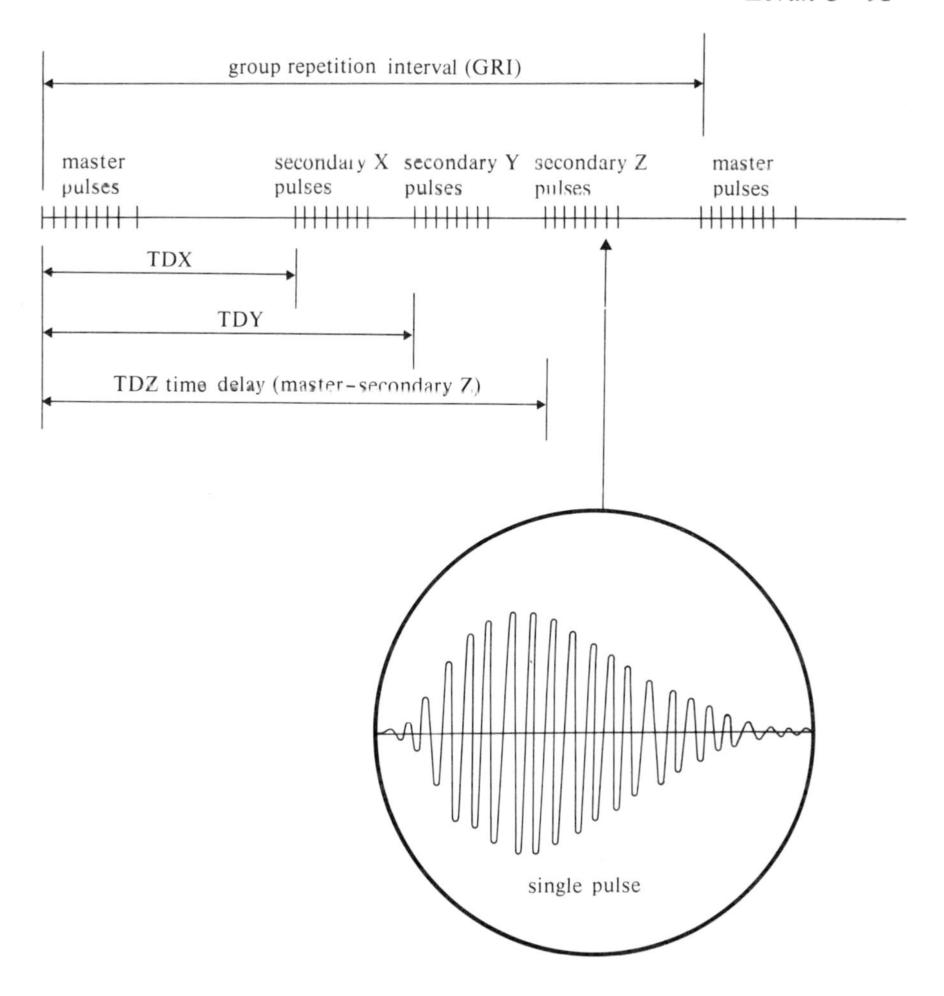

Figure 8.2 Sequence of received Loran C pulses.

Older charts and older Loran C receivers may show chains designated not by their group repetition interval, but by a designation such as SS1 and SL1. The conversion from the old to the new designations is given in Figure 8.5.

Each chain has one or two system monitoring stations; in the case of the Mediterranean chain they are on the islands of Rhodes and Sardinia. The function of these stations is to aid the system in maintaining proper synchronization.

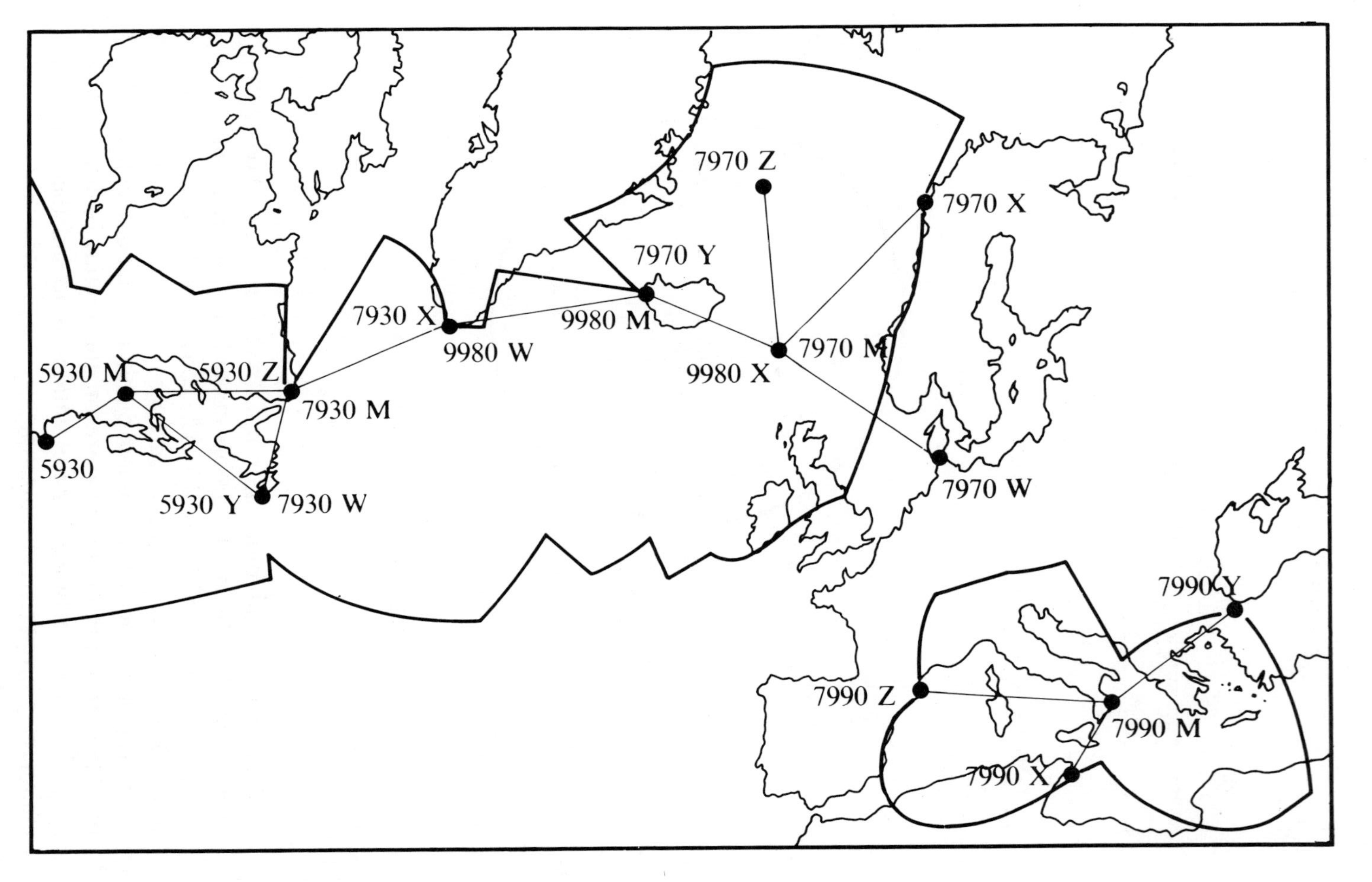

Figure 8.3 North Atlantic and Mediterranean Loran C chains (1984).

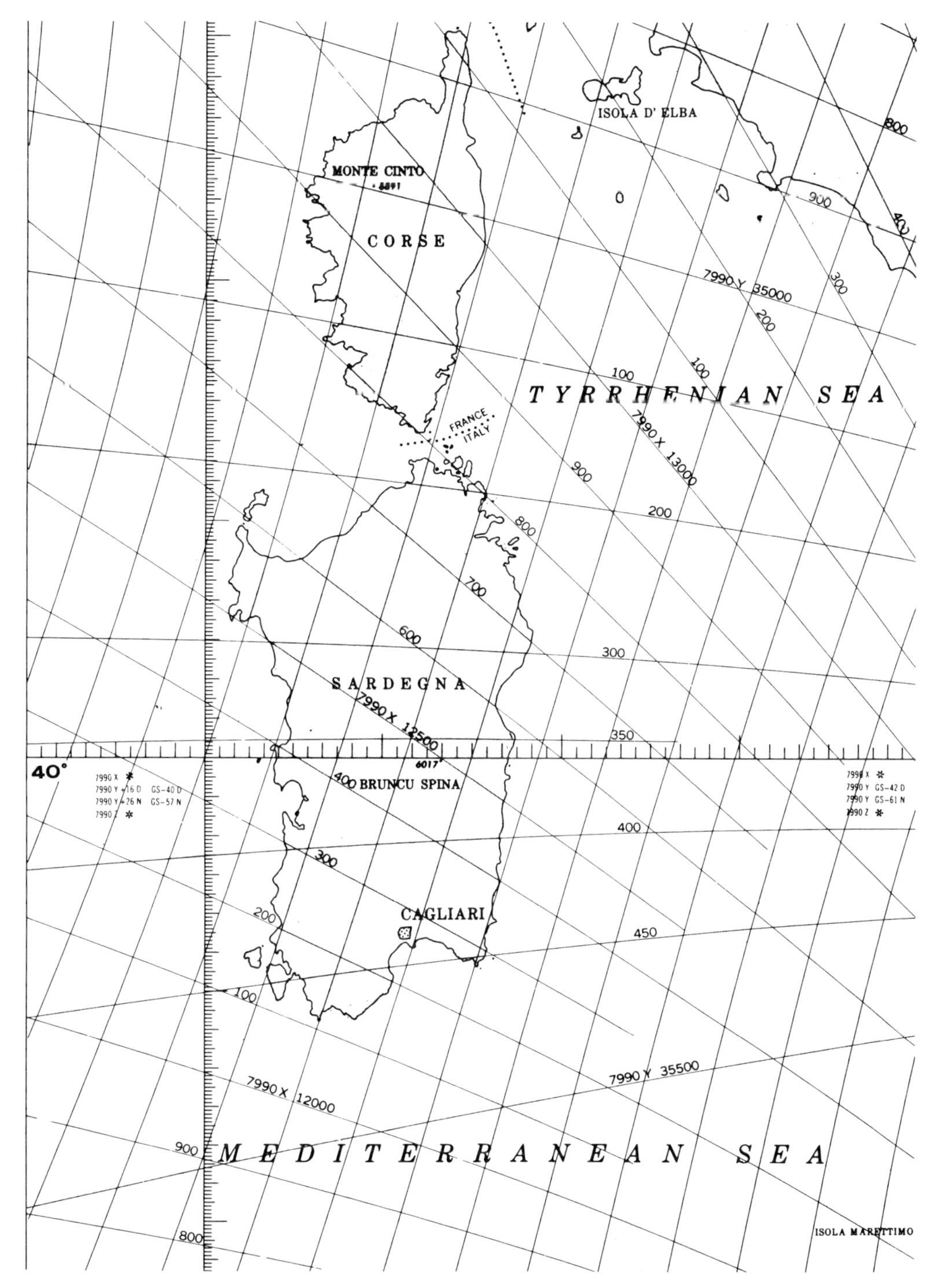

Figure 8.4 *Part of Mediterranean Loran C chart (courtesy US Defence Mapping Agency Hydrographic/Topographic Center).*

Old designation	*New designation*
SS1	9990
SS3	9970
SS7	9930
SL1	7990
SL3	7970
SL7	7930
SH1	5990
SH3	5970
SH7	5930
S1	4990
S3	4970
S7	4930

These designations specify the group repetition interval (GRI) e.g. 9990 = 99,900 microseconds.

Figure 8.5 Conversion from old to new Loran C chain designations.

8.2 Pulse format

To achieve a simple time difference measurement, it is only necessary for the master and slave stations to transmit one pulse each. In reality a multiple pulse group is radiated, since this provides more energy at the receiver without increasing the peak transmitter power. The master pulse group consists of eight pulses spaced 1000 microseconds apart, and a ninth pulse 2000 microseconds after the eighth (Figure 8.6). The slave transmissions simply consist of the first eight pulses.

Unlike Loran A which uses only the basic pulse envelope for the time difference measurement, Loran C makes use of the fundamental (100 kHz) carrier to achieve a more precise time difference. This is termed cycle matching. To obtain the maximum accuracy, measurements must take place during the first few cycles of carrier, since the latter part of the pulse may be contaminated by a delayed skywave signal. It is therefore desirable that the pulse rises quickly to make optimum use of these first few cycles, but a compromise is necessary

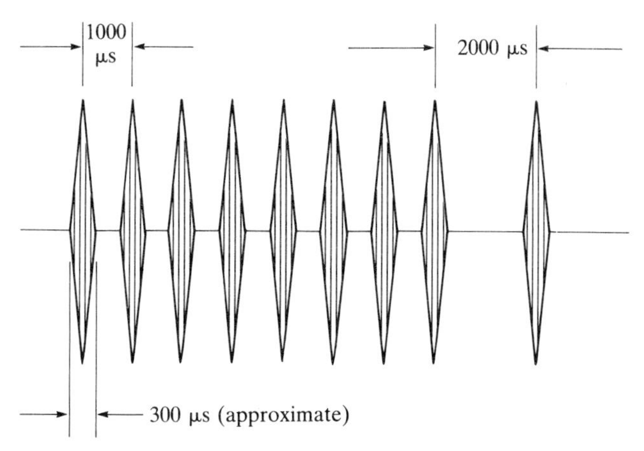

Figure 8.6 Master station pulse group; slave stations transmit first eight pulses only.

since a fast rising pulse occupies a wide bandwidth. The allocated frequency band for Loran C is 90–110 kHz, and the radiated pulse shape is therefore chosen to restrict over 99 per cent of the transmitted energy to this 20 kHz band (Figure 8.7).

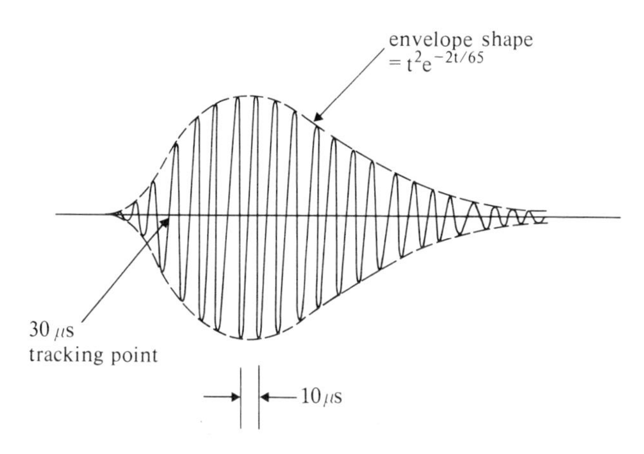

Figure 8.7 A single Loran C pulse.

8.2.1 *Phase coding*

It is possible for a skywave to be delayed on the groundwave by as much as 1,000 microseconds and so overlapping the groundwave of the next pulse. To prevent long delay skywaves from affecting the time difference measurement, the phase of the 100 kHz carrier is

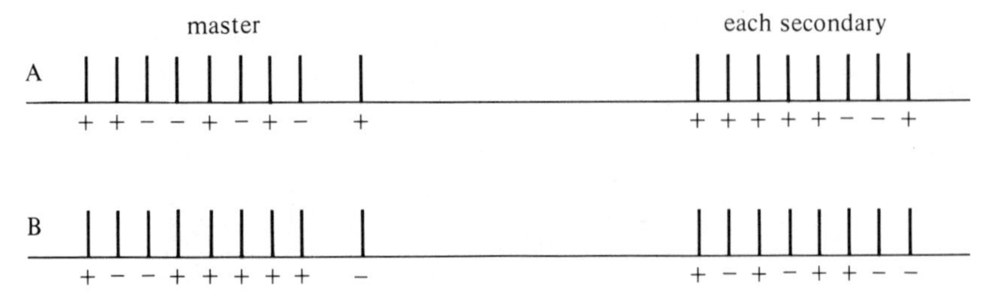

Each transmitter alternates the A and B sequences

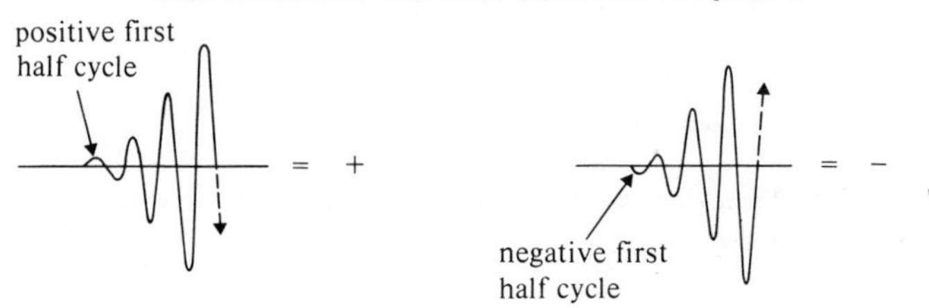

Figure 8.8 Phase coding of Loran C pulse groups.

changed in successive pulses of a group (Figure 8.8). The different phase codes for the master and secondary signals also allow automatic receivers to use the code for master and secondary station identification.

8.2.2 Blink

The ninth pulse in the master group can be switched on and off to warn users that there is an error in the transmissions of a given station. The slave station of the unusable pair also 'blinks', by turning its first two pulses off and then on, for approximately one-quarter second every four seconds. The code format of the master station blinks is given in Figure 8.9. Modern commercial Loran C receivers generally do not have the capability to react to blink warnings.

8.3 Propagation

8.3.1 Groundwave

The high position-fixing accuracy of the Loran C system is achieved

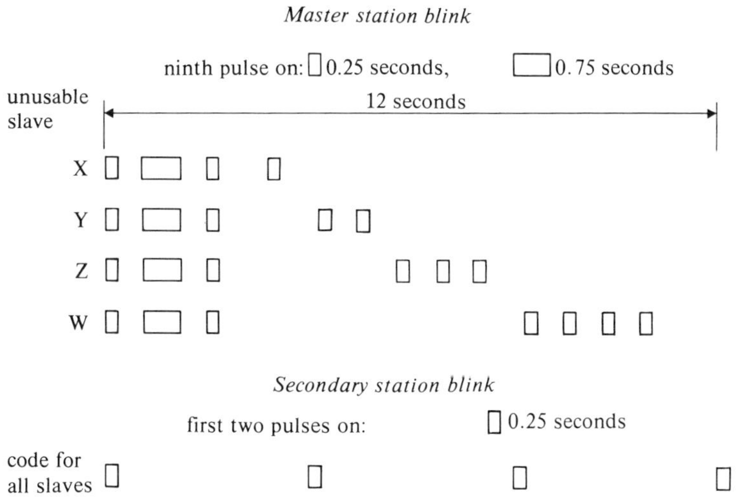

Master station blink

ninth pulse on: 0.25 seconds, 0.75 seconds

unusable slave

12 seconds

X

Y

Z

W

Secondary station blink

first two pulses on: 0.25 seconds

code for all slaves

Figure 8.9 The master station blink code.

only when groundwave signals are received. These propagate out to typically 1,000 miles from the transmitting station. Absolute position accuracy depends upon the accuracy with which the hyperbolic time difference lattices are calculated, which in turn is a function of the actual propagation velocity. This velocity varies with ground conductivity from $v = 0.9997\ c$ (c = free space velocity electromagnetic radiation) over water, to $v = 0.9977\ c$ over low conductivity ground, which has the effect of reducing velocity by absorption. In the process of overprinting the time difference lattices onto charts, a total seawater path ($v = 0.9997\ c$) is usually assumed. This correction from the free space electromagnetic velocity to that over seawater is termed the 'secondary phase correction'. When the signal path is partially over land (with varying conductivity), and partially over water, the prediction of mean propagation velocity becomes far more complex. The additional correction for the overland path is termed the 'additional secondary phase correction' and is derived by analytical and empirical means. It is taken into account in the overprinting of time difference lattices on some coastal charts.

8.3.2 Skywave

The 100 kHz Loran C signals will be reflected by the lower regions of the ionosphere (D- and E-layers), with an amplitude which is a function of ionization density and distance from the transmitting station. In general, during the day-time the amplitude of the skywave will be significantly lower than the amplitude of the groundwave up to a distance of 200–300 miles from the transmitting station. During night-time the skywave signals may be equal in amplitude to the groundwave signals at a distance of only 100 miles from the transmitting station, but they should not present a problem since they will be delayed on the groundwave well past the thirty microsecond tracking point (Figure 8.10).

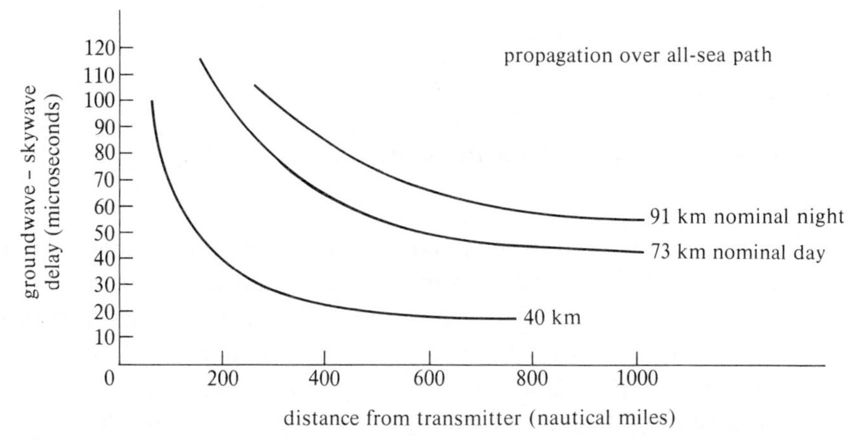

Figure 8.10 Relationship of groundwave – skywave time delay and ionospheric height.

The delay between the groundwave and skywave pulses reduces with distance from the transmitter, so that with a low reflecting height of the ionosphere it is possible for the skywave to arrive before the thirty microsecond tracking point of the groundwave pulse, at distances greater than 500 miles from the transmitter. This may cause a troublesome situation to arise within the Loran C receiver depending upon the relative amplitudes of the skywave and groundwave signals. At distances beyond the normal groundwave coverage, continued use of Loran C must rely entirely on skywave signals, but in these areas the overprinted time difference hyperbolae are still based on groundwave propagation velocities. Therefore it becomes necessary to compensate for the time delays caused by the longer propagation paths of skywaves. A basic correction is often printed onto Loran C charts in the approximate area to which it applies. Considering the example of the section of chart shown in

Figure 8.4, the correction to apply to the 7990Y time difference if master and slave skywave signals are received during daytime is +16 microseconds. During night-time a correction of +26 microseconds should be applied. In this particular area it is possible that the master pulses are received as a groundwave propagated signal, and the slave received as a skywave. The appropriate correction for this situation is indicated by GS, and in the example the correction to apply is −40 during day-time, and −57 during night-time. The opposite case of a skywave propagated master and a groundwave slave would have the correction preceded by SG.

Knowing when to apply the appropriate correction depends on the user's experience to judge when groundwave or skywave propagation is likely to be dominant. For any given Loran C transmitting station there will be an area which is clearly always covered by a dominant groundwave, and there will be a distant area where only skywave prevails. There will also be intermediate areas in which the received pulses could be either predominantly skywave or predominantly groundwave, depending upon ionospheric conditions.

Skywave corrections which are more accurate than those printed on Loran C charts are given in the Loran C tables published by the US Defense Mapping Agency Hydrographic/Topographic Center. Taking as an example Publication 221 (1007) Pair 7970 Z (master and slave Z): these tables give the corrections to apply when receiving skywaves from both master and slave (Figure 8.11), or one skywave and one groundwave. As a guide to the user, an illustration such as Figure 8.12 is given at the front of each volume of Loran C tables; this shows the area covered by the corrections. These boundaries should not be taken as the rigorous limits of the skywave and groundwave signals from the two transmitters.

8.4 Charts

Except for the type of Loran C receiver which automatically computes a position as a latitude and longitude directly from the time difference hyperbolae, users of conventional Loran C receivers require a means of converting time differences to absolute positions.

One such means is to use the published Loran C lattice tables (Publication 221), (Figure 8.13). The full procedure for using these tables, with examples, is given at the beginning of each volume.

7970-W **SKYWAVE CORRECTION** FS

DAYTIME (h = 73 km.)

Longitude—29°W to 15°W

LATITUDE NORTH	29	28	27	26	25	24	23	22	21	20	19	18	17	16	15	LATITUDE NORTH
89																89
88																88
87																87
86																86
85																85
84																84
83																83
82	0.0	0.0	0.0	0.0	0.0	0.0	0.0	0.0	0.0	0.0	0.0	0.0	0.0	0.0	0.0	82
81	0.1	0.1	0.1	0.1	0.1	0.1	0.1	0.2	0.2	0.2	0.2	0.2	0.3	0.3	0.3	81
80	0.3	0.3	0.4	0.4	0.4	0.4	0.4	0.4	0.4	0.4	0.4	0.4	0.5	0.5	0.5	80
			ADD					**ADD**					**ADD**			
79	0.5	0.5	0.6	0.6	0.6	0.6	0.6	0.7	0.7	0.7	0.7	0.7	0.8	0.8	0.8	79
78	0.8	0.8	0.8	0.8	0.8	0.9	1.0	1.0	1.0	1.0	1.0	1.1	1.1	1.1	1.2	78
77	1.0	1.0	1.0	1.1	1.2	1.2	1.2	1.3	1.4	1.4	1.4	1.5	1.5	1.5	1.6	77
76	1.2	1.3	1.4	1.5	1.6	1.6	1.6	1.7	1.8	1.9	2.0	2.0	2.0	2.1	2.2	76
75	1.6	1.7	1.8	1.9	2.0	2.1	2.2	2.3	2.4	2.5	2.6	2.6	2.7	2.8	2.8	75
74	2.0	2.2	2.3	2.4	2.6	2.7	2.8	2.9	3.0	3.2	3.3	3.4	3.6	3.7	3.8	74
73	2.6	2.7	2.8	3.0	3.2	3.4	3.6	3.8	4.0	4.1	4.3	4.5	4.6	4.8	5.0	73
72	3.0	3.3	3.5	3.7	4.0	4.2	4.5	4.8	5.0	5.3	5.6	5.8	6.0	6.3	6.4	72
71	3.6	3.9	4.2	4.5	4.8	5.2	5.6	5.9	6.3	6.7	7.0	7.4	7.7	8.0	8.3	71
70	4.2	4.6	5.0	5.4	5.8	6.3	6.8	7.3	7.8	8.3	8.8	9.3	9.8	10	11	70
			ADD					**ADD**					**ADD**			
69	4.8	5.3	5.8	6.4	7.0	7.6	8.2	8.9	9.6	10	11	12	12	13	14	69
68	5.4	6.0	6.7	7.4	8.2	8.9	9.7	10	11	12	14	15	16	17	18	68
67	6.0	6.7	7.5	8.3	9.2	10	11	12	14	15	16	18	20	21	23	67
66	6.4	7.2	8.2	9.1	10	11	13	14	16	17	20	22	24	27	30	66
65	6.8	7.6	8.6	9.6	11	12	14	15	18	20	22	25	29	32	37	65
64	6.8	7.7	8.8	9.8	11	13	14	16	19	21	25	28	33	38		64
63	6.8	7.6	8.7	9.8	11	13	15	16	19	22	26	30	35			63
62	6.4	7.3	8.4	9.4	11	12	14	16	19	21	25	29	34			62
61	6.0	6.8	7.8	8.7	10	11	13	15	17	20	23	26	31	36		61
60	5.4	6.2	7.0	7.9	9.0	10	12	13	15	17	20	23	27	31	36	60
			ADD					**ADD**					**ADD**			
59	4.8	5.5	6.2	7.0	7.8	8.9	10	11	13	15	17	19	22	25	28	59
58	4.2	4.8	5.4	6.0	6.7	7.6	8.5	9.6	11	12	14	15	17	19	21	58
57	3.6	4.1	4.6	5.1	5.7	6.3	7.0	7.9	8.8	9.7	11	12	13	14	16	57
56	3.0	3.4	3.9	4.2	4.7	5.2	5.7	6.3	7.0	7.7	8.4	9.2	10	11	12	56
55	2.5	2.8	3.2	3.5	3.8	4.2	4.6	5.0	5.5	6.0	6.5	7.0	7.5	8.0	8.4	55
54	2.0	2.3	2.6	2.9	3.1	3.4	3.6	3.9	4.3	4.6	4.9	5.2	5.5	5.8	6.0	54
53	1.6	1.8	2.1	2.3	2.5	2.7	2.9	3.1	3.3	3.5	3.7	3.9	4.0	4.1	4.2	53
52	1.3	1.4	1.6	1.8	2.0	2.1	2.3	2.4	2.5	2.6	2.7	2.8	2.9	2.9	2.9	52
51	1.0	1.1	1.3	1.4	1.6	1.7	1.8	1.8	1.9	2.0	2.0	2.0	2.0	2.0	1.9	51
50	0.7	0.8	0.9	1.1	1.2	1.4	1.4	1.4	1.4	1.5	1.5	1.5	1.4	1.3	1.2	50
			ADD					**ADD**					**ADD**			
49	0.5	0.6	0.7	0.8	0.9	1.0	1.1	1.1	1.1	1.1	1.1	1.0	1.0	0.9	0.8	49
48	0.3	0.4	0.4	0.5	0.6	0.7	0.8	0.9	0.9	0.8	0.8	0.7	0.6	0.6	0.4	48
47	0.1	0.2	0.2	0.3	0.4	0.5	0.5	0.6	0.7	0.6	0.6	0.5	0.4	0.3	0.2	47
46	0.0	0.0	0.0	0.1	0.2	0.2	0.3	0.4	0.4	0.5	0.4	0.4	0.3	0.2	0.1	46
45				0.0	0.0	0.0	0.1	0.2	0.2	0.3	0.3	0.2	0.2	0.1	0.0	45
44							0.0	0.0	0.0	0.0	0.1	0.1	0.1	0.0	0.1	44
43											0.0	0.0	0.0	0.1	0.2	43
42														0.0	0.0	42
41																41
40																40
													SUBTRACT			
39																39
38																38
37																37
36																36
35																35
34																34
33																33
32																32
31																31
30																30
	29	28	27	26	25	24	23	22	21	20	19	18	17	16	15	

Longitude—29°W to 15°W

DAYTIME

Figure 8.11 Example of published skywave corrections (courtesy US Defense Mapping Agency Hydrographic/Topographic Center).

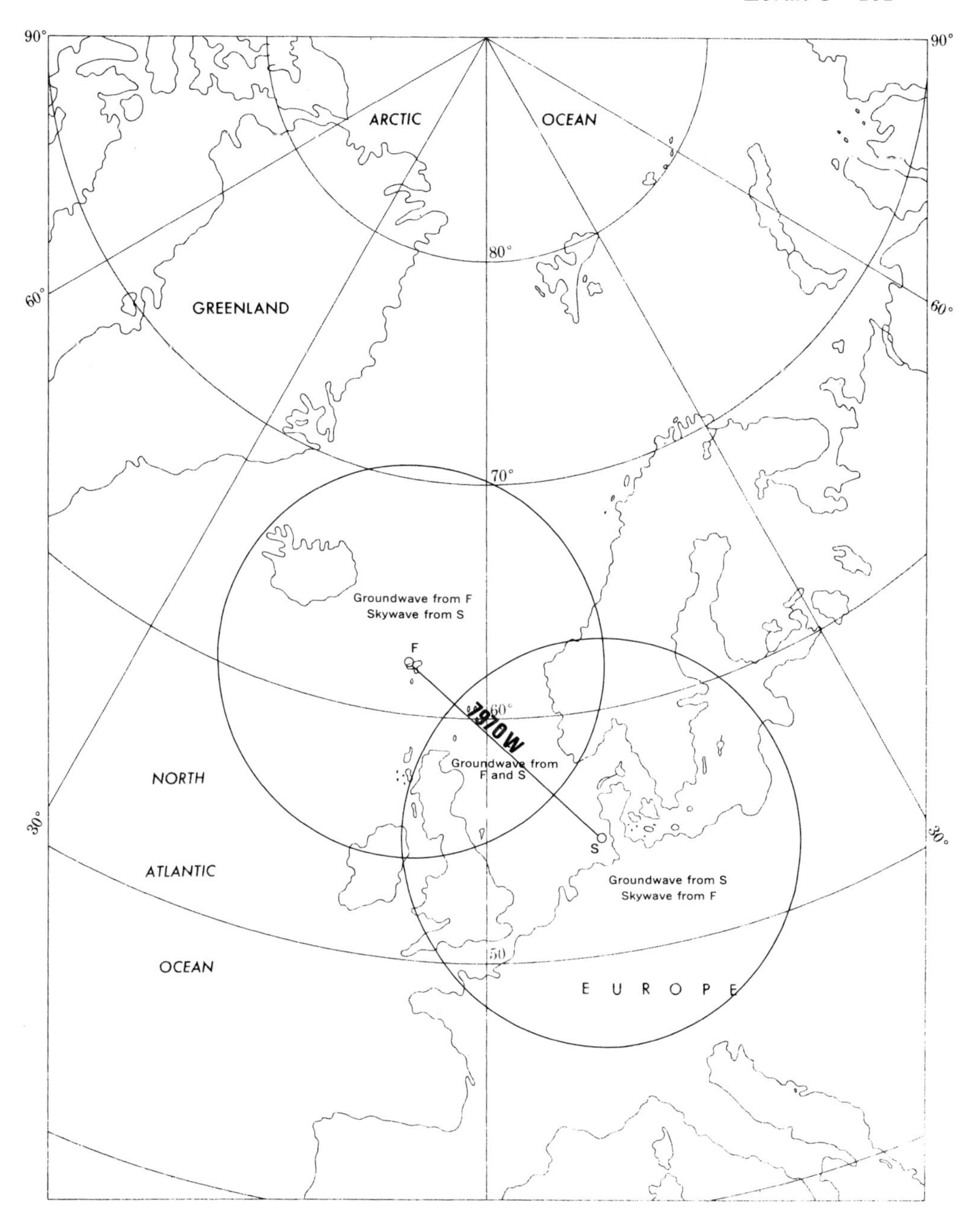

Figure 8.12 Example of illustration included in each volume of skywave corrections to show the limits of the published corrections (courtesy US Defense Mapping Agency Hydrographic/Topographic Center).

7970-W FS 133

T Lat ° ′	27650 ° ′	Δ	27660 ° ′	Δ	27670 ° ′	Δ	27680 ° ′	Δ	27690 ° ′	Δ	T Long ° ′
	57 41.1N	28	57 43.9N	28	57 46.7N	28	57 49.5N	28	57 52.3N	28	30 E
	57 33.5N	29	57 36.5N	29	57 39.4N	28	57 42.3N	29	57 45.2N	29	31 E
	57 25.3N	30	57 28.4N	30	57 31.5N	30	57 34.5N	29	57 37.5N	29	32 E
	57 16.5N	31	57 19.7N	31	57 22.9N	31	57 26.0N	31	57 29.2N	31	33 E
	57 07.2N	32	57 10.4N	32	57 13.7N	32	57 17.0N	32	57 2C.2N	32	34 E
	56 57.1N	34	57 00.5N	34	57 04.0N	33	57 07.3N	33	57 10.7N	33	35 E
	56 46.5N	35	56 50.0N	35	56 53.5N	35	56 57.1N	34	57 00.5N	34	36 E
	56 35.2N	36	56 38.9N	36	56 42.5N	36	56 46.1N	35	56 49.7N	35	37 E
	56 23.3N	37	56 27.0N	37	56 30.8N	37	56 34.6N	37	56 38.3N	37	38 E
	56 10.7N	39	56 14.6N	38	56 18.5N	38	56 22.4N	38	56 26.2N	38	39 E
	55 57.5N	40	56 01.5N	40	56 05.5N	40	56 09.5N	40	56 13.5N	40	40 E
	55 43.5N	41	55 47.6N	41	55 51.8N	41	55 55.9N	41	56 0C.1N	41	41 E
	55 28.8N	43	55 33.1N	43	55 37.4N	42	55 41.7N	42	55 46.CN	42	42 E
	55 13.4N	44	55 17.9N	44	55 22.4N	44	55 26.8N	43	55 31.2N	43	43 E
	54 57.3N	46	55 02.0N	46	55 06.6N	45	55 11.1N	45	55 15.7N	45	44 E
	54 40.5N	48	54 45.3N	47	54 50.0N	47	54 54.7N	47	54 59.4N	47	45 E
	54 22.9N	49	54 27.8N	49	54 32.7N	48	54 37.6N	48	54 42.4N	48	46 E
	54 04.5N	50	54 09.6N	50	54 14.6N	50	54 19.7N	50	54 24.7N	49	47 E
	53 45.2N	52	53 50.5N	52	53 55.8N	52	54 00.9N	51	54 0E.1N	51	48 E
	53 25.2N	54	53 30.6N	54	53 36.0N	54	53 41.4N	53	53 4E.7N	53	49 E
	53 04.3N	56	53 09.9N	55	53 15.4N	55	53 21.0N	55	53 26.5N	55	50 E
	52 42.4N	58	52 48.2N	57	52 54.0N	57	52 59.7N	57	53 05.4N	56	51 E
	52 19.8N	59	52 25.7N	59	52 31.6N	59	52 37.5N	59	52 43.4N	58	52 E
	51 56.1N	62	52 02.3N	61	52 08.4N	60	52 14.5N	60	52 2C.6N	60	53 E
	51 31.4N	63	51 37.8N	63	51 44.2N	63	51 50.4N	62	51 56.7N	62	54 E
	51 05.8N	65	51 12.4N	65	51 19.0N	65	51 25.4N	64	51 31.9N	64	55 E
	50 39.2N	67	50 46.0N	67	50 52.7N	67	50 59.4N	66	51 06.1N	66	56 E
	50 11.5N	70	50 18.5N	69	50 25.4N	69	50 32.3N	69	50 39.3N	68	57 E
	49 42.6N	72	49 49.9N	72	49 57.1N	71	50 04.2N	71	50 11.3N	70	58 E
	49 12.7N	74	49 20.2N	74	49 27.6N	73	49 35.0N	73	49 42.3N	73	59 E
	48 41.6N	77	48 49.3N	76	48 56.9N	76	49 04.6N	75	49 12.1N	75	60 E
	48 09.3N	79	48 17.2N	79	48 25.1N	78	48 33.0N	78	48 40.8N	77	61 E
	47 35.7N	82	47 43.9N	81	47 52.0N	81	48 00.2N	80	48 C8.2N	80	62 E
	47 00.8N	85	47 09.3N	84	47 17.7N	83	47 26.1N	83	47 34.4N	82	63 E
T	27650		27660		27670		27680		27690		T

Figure 8.13 Example of Loran C lattice table (courtesy US Defense Mapping Agency Hydrographic/Topographic Center).

The more commonly used conversion procedure is that of plotting directly onto special Loran C charts, overprinted with the time difference hyperbolae. The position of the hyperbolae are usually computed for a constant propagation velocity, equivalent to an all-sea path with a nominal surface conductivity of 5 mhos/metre. The US Defense Mapping Agency Hydrographic/Topographic Center publish the 7800 series of charts which cover all Loran C areas at a scale of one inch to thirty nautical miles. Additionally this and other charting authorities publish larger scale charts, usually for their own coastal regions or other specific areas such as fishing grounds. To improve the accuracy of these charts, some include the effect of over-land propagation (additional secondary phase correction) on the overprinted time difference hyperbolae.

8.5 Accuracy

The accuracy of Loran C positions are affected by the factors of lane expansion and angle of cut which are common to all hyperbolic navigation systems. The sources of error specifically pertaining to the Loran C system are described in the following sections, with error magnitudes expressed as time difference errors in microseconds. An error of one microsecond on the baseline produces a position line error of 145 metres.

8.5.1 Synchronization error

There is a small error caused by discrepancies in the precise timing intervals between master and slave transmissions. The monitoring stations coupled with sophisticated control techniques maintain this error to within ±0.1 microseconds. If the error exceeds ±0.2 microseconds the blink warning comes into play, although this is an extremely rare occurrence.

8.5.2 Groundwave propagation error

A position plotted onto a Loran C chart which has had no allowance made for additional secondary phase correction will be in error, unless the total propagation path is over water (with a conductivity of 5 mhos/m). The error increases with the amount of land over which the signal has propagated. Although it is possible that this error could be as large as ten microseconds, it does not usually exceed three microseconds in marine navigation.

8.5.3 *Envelope to cycle discrepancy (ECD)*

The medium over which the groundwave propagates can have the effect of distorting the pulse shape by propagating some parts of the pulse's frequency spectrum more readily than others. The effect of this is to upset the precise relationship between the phase of the 100 kHz carrier and the pulse shape. The magnitude of a resulting error in the measured time difference is difficult to predict, since it is a function of receiver performance and its susceptibility to this effect. An error of one to two microseconds may arise, or in the worst case the receiver may jump to an incorrect tracking point, introducing a ten microsecond error. The US Coast Guard intentionally distorts the transmitted ECD where necessary, to limit the received ECD error where large overland paths affect the propagated ECD.

8.5.4 *Receiver errors*

Loran C receivers will introduce processing errors in a number of different ways but the total error is usually less than 0.1 microseconds, except for one potentially large error, that of incorrect third cycle identification. The methods by which Loran C receivers achieve identification of the third cycle tracking point vary in their sophistication and fallibility. In severe noise conditions, or due to skywave contamination, some Loran C receivers may track the wrong cycle and thus display time differences which are in error by ten, twenty or even thirty microseconds.

8.5.5 *Skywave errors*

The time difference hyperbolae of Loran C charts and lattice tables are based on groundwave propagation times. The additional time delay of the skywave transmission path can to some extent be compensated for by the published skywave corrections. These though, are a crude form of correction since they assume that the reflective height of the ionosphere is seventy-three kilometres during day-time and ninety-one kilometres during night-time. In reality the effective reflecting height of the ionosphere varies daily, seasonally, and throughout the eleven-year sunspot cycle. The two specific values of ionospheric height used in computing the skywave corrections can therefore only ever be approximations. Even after applying the skywave correction, errors can exist of thirty microseconds and more.

8.5.6 *Overall accuracy*

A description of error sources such as those given here tends to accentuate the weaknesses of a system, and so to restore perspective it should be said that Loran C has a good reputation for accuracy and reliability.

8.5.7 *Repeatable accuracy*

The Loran C system is seen at its most accurate when the repeatable accuracy of groundwave signals with good chain geometry is considered. Earth surface parameters affect the accuracy with which time differences are translated into geographical co-ordinates, but since these parameters are relatively stable, they have much less effect on the accuracy with which a user can return to previously obtained Loran C time difference co-ordinates. For each Loran C chain the US Defense Mapping Agency Hydrographic/Topographic Center publish reliability diagrams. These show the 500 ft, 750 ft, and 1500 ft accuracy contours (2 d*rms* and 1:10 signal-to-noise ratio). In practice repeatable accuracies of 50–300 ft are regularly achieved within the areas of good chain geometry.

8.5.8 *Relative accuracy*

For the same reason that Loran C has a high repeatable accuracy, it also has a good rendezvous capability. Two vessels each equipped with Loran C receivers can accurately rendezvous, since as they come closer together the propagation paths become nearer to being the same, and they will then both experience the same groundwave and skywave errors.

8.5.9 *Absolute accuracy*

1 Groundwave
When a position derived in Loran C time difference co-ordinates is

translated into absolute geographical co-ordinates, the accuracy is affected by the groundwave propagation error. If a nearly total sea-water path exists, or if the published charts have a note indicating ASF (additional secondary phase factor) corrections are based on actual measurements, then absolute position accuracies of better than 0.25 nautical miles (2 d*rms*) are usually achieved within the stated groundwave coverage area.

2 Skywave
The variation in propagation times of skywave signals can add a large contribution to the error of fixes derived from those time differences which include skywave signals. Such a fix could be from one to more than ten nautical miles in error.

8.6 Loran C receivers

Marine Loran C receivers vary from simple modified Loran A receivers which display the master and slave pulses on a cathode ray tube, to at the other extreme, computer/microprocessor receivers which display positions directly as a latitude and longitude.

8.6.1 The Loran A/C receiver

The basic Loran A receiver utilizes a cathode ray tube to display the received pulses. The time base is synchronized to the group repetition interval of the selected chain, and the calibrated time base is used to measure the time differences from the leading edge of a master pulse, to the leading edges of the corresponding slave pulses. Such a receiver can be converted for use with the Loran C system by adding an additional receiver circuit tuned to 100 kHz, and by providing the facility of synchronizing the time base to the Loran C GRI rates. The accuracy of most receivers of this type is fundamentally limited, since the user is required to physically identify an arbitrary point on the front edge of the pulse, rather than use the more precise timing provided by the 100 kHz carrier frequency within the pulses. Advanced versions of Loran A/C receivers provide a time base expansion which allows the operator to see the carrier frequency, and thus carry out cycle matching, but the accuracy of this method falls short of that achieved by digital processing, and the risk of whole cycle error is high.

Since receivers of this type rely on visual identification of a point on just one pulse, no benefit is gained from the fact that each Loran C station transmits eight pulses to provide (if processed correctly) an improvement in signal/noise ratio.

8.7 Automatic Loran C receivers

Within the group of so-called automatic Loran C receivers, there are many variations in the particular functions which are made automatic in their operation. In general there are two which are essential for a receiver to fall into this category:

- the automatic location and tracking of the selected master and slave stations;
- automatic measurement and display of master-slave time differences.

Additional functions which can be automated are:

- chain selection,
- selection of best slave stations,
- indication of lost signal,
- indication of 'blink' warning,
- indication of envelope-to-cycle discrepancy,
- attenuation of interfering signals,
- conversion of time difference LOP to a position displayed as a latitude and longitude.

The desirability and effectiveness of making some or all of these functions automatic is open to differing views. Hence the existence of equipments which cover the spectrum, from totally manual to fully automatic receivers. The following sections describe some of the more important aspects of Loran C receiver operation.

8.7.1 Interference filters

It is essential that Loran C pulses are received and amplified with an absolute minimum of distortion of the pulse envelope shape. To achieve this, Loran C receivers have wide bandwidths (typically 20–30 kHz), but this in turn leads to the problem of adjacent channel interference affecting signal acquisition and time difference measurements. To overcome this problem most receivers incorporate adjustable interference filters, which attenuate interfering signals on the frequency to which the filters are set. Many receivers have two or more filters adjustable from front panel controls, and with these equipments it is left to the user to regularly check for the presence of interference, and to adjust the filters to minimize its effect.

More sophisticated receivers have automatically adjusting filters which continuously assess the interference levels and adjust the filtering for optimum effect.

If too many filters are used, or if the filters are set too close to the 100 kHz centre frequency, the pulse may be distorted by the irregularities in the receiver's frequency response, resulting in errors in the time difference measurement.

8.7.2 *Cycle matching*

The accuracy with which a Loran C receiver makes a time difference measurement is governed by the accuracy and reliability with which the receiver selects a precise point of reference on the master and slave pulses. The usual point of reference is thirty microseconds from the start of the pulse, that is, after the third cycle of the 100 kHz carrier. Of particular importance is the reliability with which a Loran C receiver identifies this third cycle reference point, even in adverse conditions of signal reception. If the receiver incorrectly identifies the reference point as being at twenty or forty microseconds after the commencement of either the master or slave pulses, the resulting time difference will be ten microseconds in error, and the plotted position incorrect by one nautical mile or more. Three different methods of third cycle identification are outlined below.

8.7.3 *Third cycle identification*

1 Front edge location

The simplest of all schemes used for locating the third cycle tracking point is achieved by first hard limiting the received master and slave pulses (Figure 8.14). The pulses are then fed to logic circuitry which senses the commencement of the pulse (t_1), and consequently the tracking point thirty microseconds after this (t_2). The main disadvantage of this arrangement is that since the peak amplitude of the first half cycle of the radiated pulse is only one per cent of the

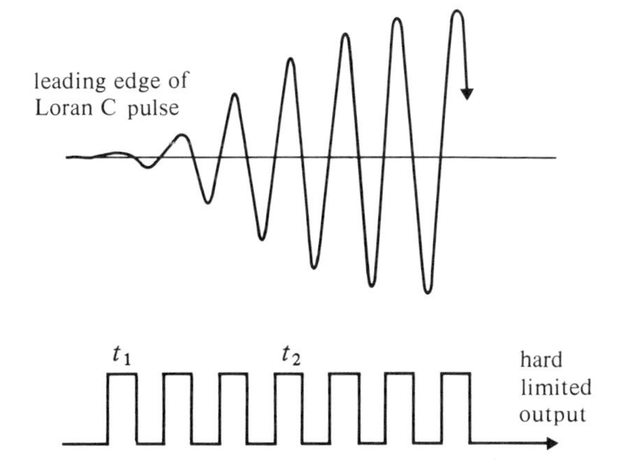

Figure 8.14 *Simple case of front edge location by hard limiting. This is not a practical solution since the first (and second) cycle may be lost in noise.*

maximum amplitude of the pulse, it is often below the atmospheric noise level, making the precise starting point of the pulse ill-defined.

2 Phase reversal
With this method of third cycle identification, the received pulse is delayed by one half cycle (5 μs); it is then summed with the undelayed pulse which has been reduced in amplitude by a factor of 0.84 (Figure 8.15). Prior to the 30 μs point, the undelayed pulse is larger in amplitude than the delayed pulse, but after the 30 μs point it is the delayed pulse which is the larger. Consequently a 180 degree phase change occurs in the summed waveform at 30 μs (Figure 8.16), and after hard limiting it is this point of phase change which is detected as the tracking point by the logic circuits.

This arrangement is less sensitive to the effects of noise on the first part of the pulse. It is dependent, however, on the accurate setting of the analogue circuits which produce the 5 μs delay and the 0.84 attenuation factor, and the stability of these adjustments with time and temperature variations.

3 Linear envelope processing
The third method of determining the third cycle tracking point differs from the previous two in that the pulse is not hard limited, but instead the unique shape of the pulse is retained and used. It is necessary to retain the pulse amplitude within close limits by automatic gain control circuits (AGC) (Figure 8.17). The pulse is

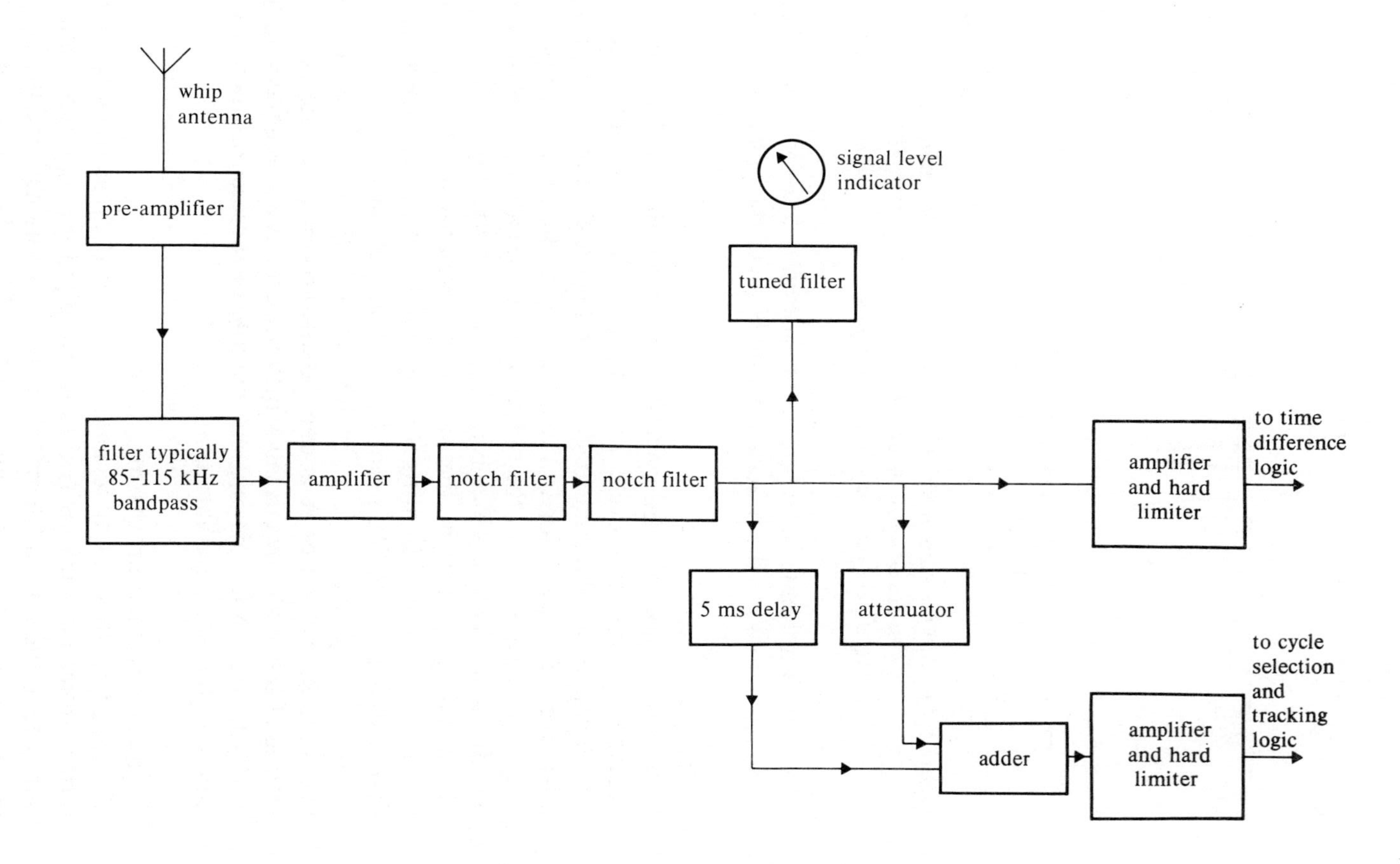

Figure 8.15 Block schematic of a Loran C receiver which detects the tracking point by the phase reversal method.

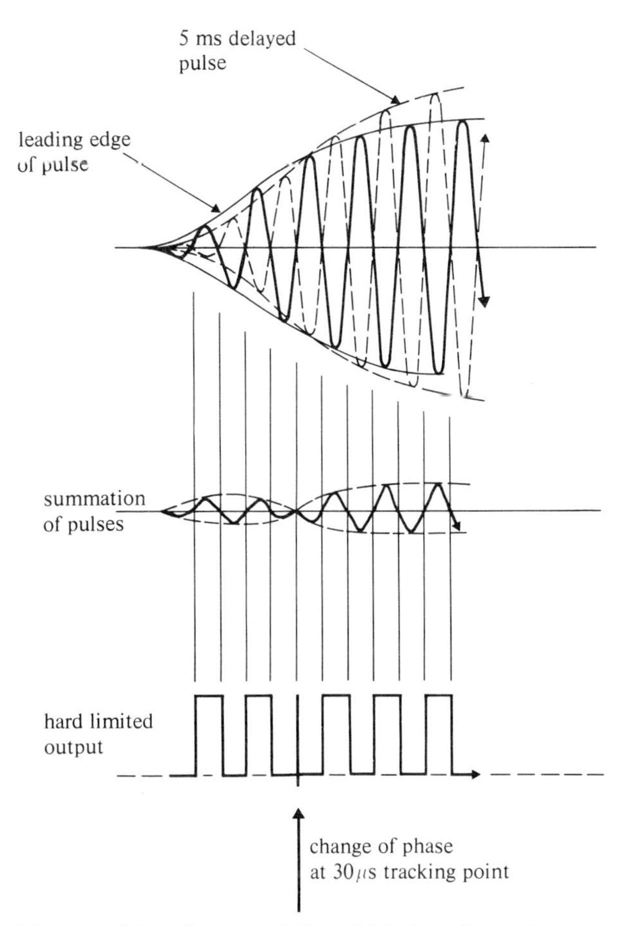

Figure 8.16 The principles of third cycle tracking by phase reversal.

sampled at 5 μs intervals (Figure 8.18), and the result fed via an analogue to digital converter to the receiver's central processor. The processor generates the ratio of amplitudes E_2/E_1 for each 5 μs step, and looks for the samples taken at 22.5 μs and 27.5 μs by detecting the unique ratio of 1.29. Once these points are identified it is then a simple matter to locate the 30 μs tracking point. This method of identifying the third cycle tracking point is probably the most reliable in adverse reception conditions.

8.7.4 Manual cycle selection

Under adverse signal conditions the cycle selection circuits may

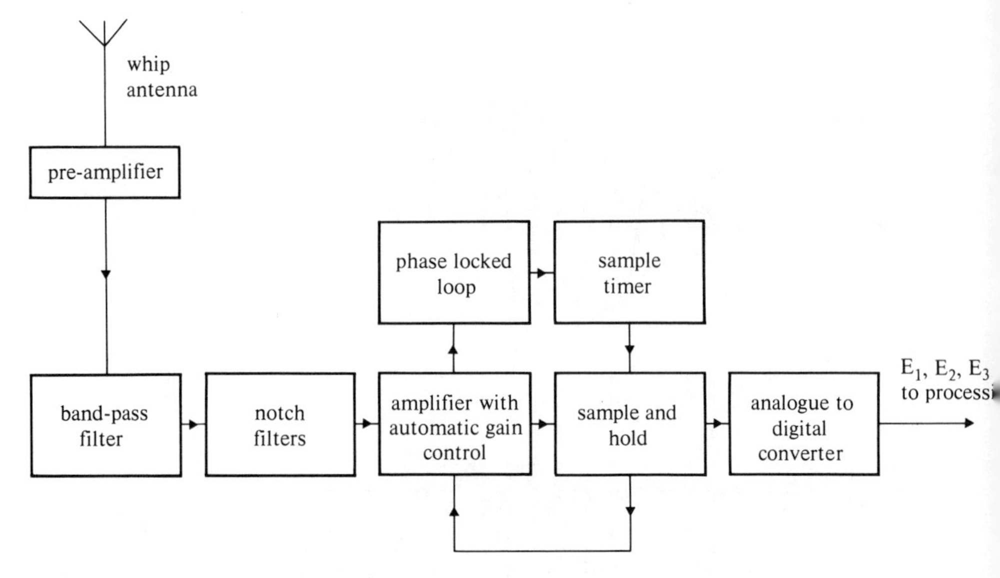

Figure 8.17 Block schematic of Loran C receiver using linear envelope processing for cycle matching.

incorrectly select the tracking point by one half cycle or more. At a known location such an error may be apparent and so some automatic Loran C receivers have the facility for the user to over-ride the automatic cycle selection, and correctly align the receiver to the tracking point.

8.7.5 *Extended range operation*

The receivers which have the facility to manually over-ride the automatic cycle selection can be used to advantage by the operator in poor signal conditions, by making the receiver track not at thirty microseconds but at a point closer to the peak amplitude of the pulse. Although there can be some advantage gained by this procedure, there is always the problem of skywave contamination, which is far more likely to occur at points beyond the thirty microsecond tracking point. A note should always be made of the amount by which the tracking point has been moved, and to which transmissions (master or slave) it applies. The time differences can then be corrected by the appropriate amounts.

8.7.6 *Latitude-longitude receivers*

With suitable digital processing, provided by either a computer or microprocessor, the Loran C time different position lines can be converted to a position fix displayed directly as a latitude and a longitude. The algorithms used for converting hyperbolic position lines into rectangular co-ordinates are basically the same for all hyperbolic systems. Amongst the constants stored in the processor are an appropriate propagation velocity, and the co-ordinates of the

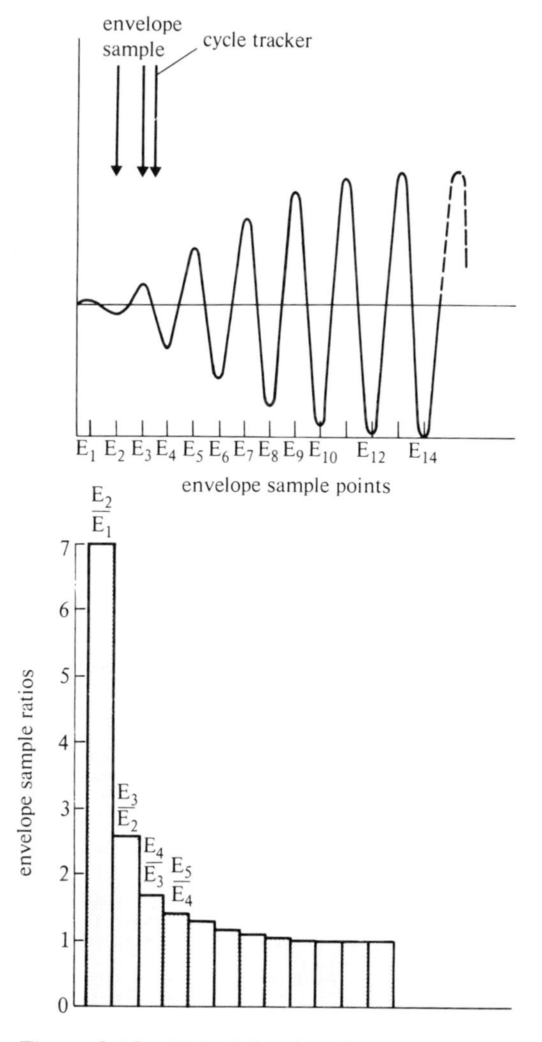

Figure 8.18 Principle of cycle matching by linear envelope processing.

transmitting stations. All but the most sophisticated of this type of receiver would not include propagation corrections for either land paths or for skywave delays, but they may have the facility for the user to enter the corrections applicable to a vessel's current location.

8.8 Using Loran C

The basic procedures for operating the time difference type of automatic Loran C receiver (Figure 8.19) are usually very similar for all the various models which exist. These procedures can be summarized as:

- suppression of interference;
- selection of appropriate chain;
- selection of appropriate slave stations;
- allow time for receiver to acquire signals and identify tracking points;
- note the displayed time differences;
- apply skywave corrections – if applicable;
- plot position lines onto Loran C charts;

In addition, the prudent navigator should be continuously aware of:

- the possible presence of interference;
- lost signal alarms;
- blink alarms;
- skywave interference;
- errors due to long overland signal paths.

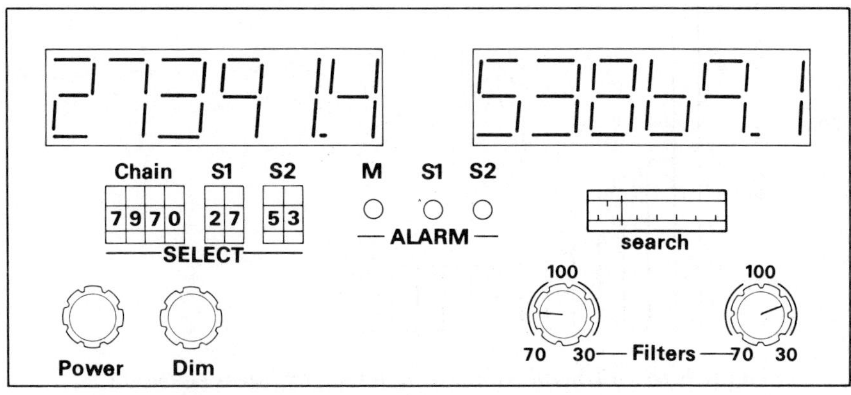

Figure 8.19 Typical automatic Loran C receiver.

9 Decca Navigator

Decca Navigator is a hyperbolic navigation system providing a position-fixing capability adequate for coastal navigation. The system is organized in chains, each comprising a master and usually three slave stations, providing a coverage of up to at least 240 nautical miles from the master station. The first Decca chain became operational in 1946 and since that time the system has grown to almost fifty operational chains throughout the world.

9.1 The system

As in all hyperbolic navigation systems, the hyperbolae of the Decca system are the loci of constant time differences between the arrival of signals from two transmission sources (master and slave). The Decca system is similar to the Omega system in that time differences are obtained by comparing the phases of the two received signals (Figure 9.1). This technique is basically ambiguous, since the same range of observed phase differences occur for each whole cycle difference through which the receiver is advanced along the baseline, between the two transmitting stations. A full cycle of phase difference occurs every half wavelength in distance, since in moving from point P to Q (Figure 9.2), the receiver has moved away

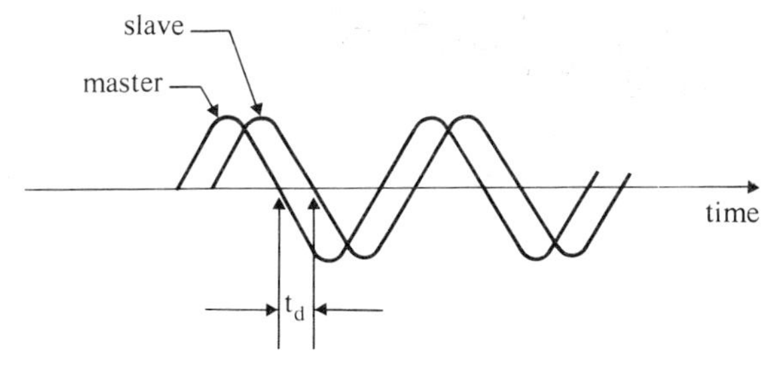

Figure 9.1 Determination of time difference from difference of signal phase.

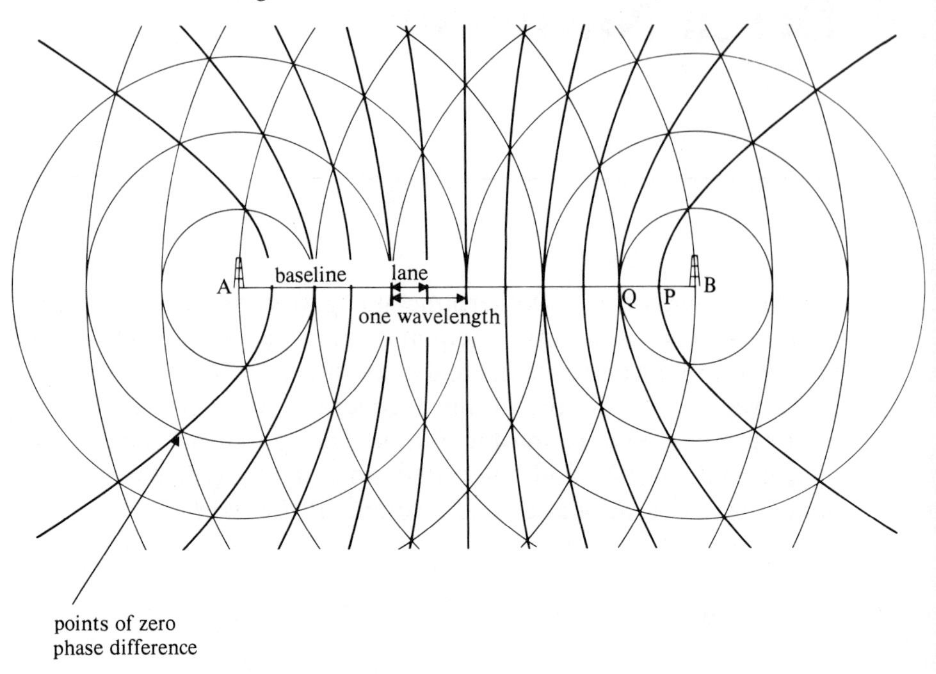

Figure 9.2 Hyperbolic pattern, lines corresponding to zero phase difference occur at half wavelength intervals.

from B by one-half wavelength, and towards A by the same amount, making one wavelength change.

The area between lines of zero phase difference are termed 'lanes', and the zero phase difference lines of position are numbered in sequence from the master station. A lane can be considered as being subdivided into one hundred lane fractions, each corresponding to 3.6 degrees of phase difference. In practice there are several hundred lanes in a typical pattern and these are grouped into 'zones' denoted by letters (Figure 9.3).

To differentiate between the hyperbolic lattices produced by the master and each slave station, each pattern is designated a colour – red, green, or purple – and the patterns are then printed on to charts in their appropriate colours. Different lane numbers are allocated to the zones of each master-slave pair, it is therefore possible to assign a position line to a lattice without reference to its colour:

	Designation	
Lattice	*Zones*	*Lane numbers*
Red	A-J	0–23
Green	A-J	30–47
Purple	A-J	50–79

A position line is fully described by colour, zone, lane number and fractional lane, e.g. Green D 36.70, but omitting the reference to Green still gives a uniquely defined position line.

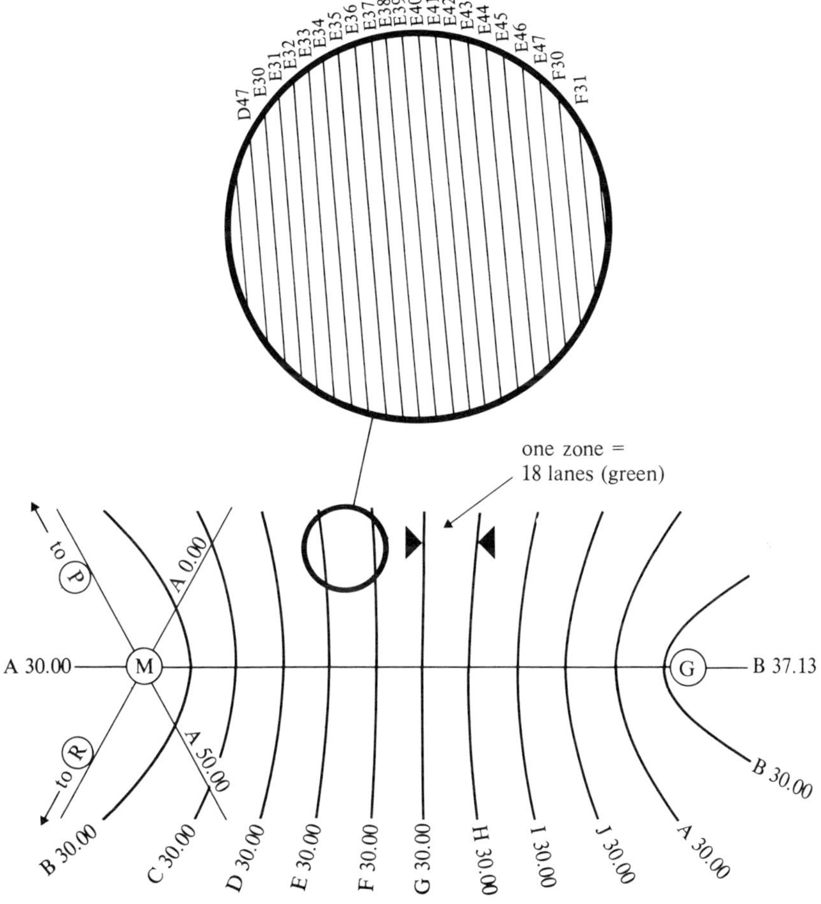

Figure 9.3 *Pattern phasing and zone labelling. This example represents a 'green' baseline 120 km long. Assuming a lane width on the baseline of 585 metres, the total lane number would be 205.13, i.e. 11 zones, 7.13 lanes (courtesy Decca Navigator Co.).*

It should always be remembered that it is only the fractional lane which is determined by measurement (phase difference); the zone and lane number are initially set at a known location, and then change as a result of the fractional lane increasing beyond 0.99 or decreasing below 0.

9.1.1 *Transmission frequencies*

It has been assumed so far that the master and slave transmissions all occur on the same frequency, but in reality it is not possible to receive simultaneous transmissions on the same frequency and still distinguish between them for the required phase comparisons. In the Decca Navigator system this particular difficulty is overcome by designating different frequencies for the master and slave transmissions, but choosing frequencies which are harmonically related to a common fundamental frequency (approximately 14 kHz). Within the receiver the master and slave signals are converted to a common frequency at which the phase comparison is then made. For each chain the fundamental frequency, transmitted frequencies, and comparison frequencies, differ from those of the other chains. To give an illustration of the frequency relationships, two of the frequencies of the English chain (code 5B) are:

master transmission frequency, 85.0 kHz ($= 6f$), where f is the fundamental frequency of 14.16667 kHz;

red slave transmission frequency, 113.333 kHz ($= 8f$).

Within the Decca Navigator receiver the frequency of the master signal is multiplied by a factor of four, and the frequency of the slave is multiplied by a factor of three, resulting in a frequency of 340 kHz ($24f$) in each case. The phases of the two signals can now be compared as if this common frequency had actually been radiated. The lane width on the baseline is equal to one half wavelength at the comparison frequency, which in this example is 440.1 metres.

Phase comparisons of the master signal with the other slave signals occur at different common frequencies:

	Radiated frequencies		*Comparison frequencies and lane widths*		
Station	Harmonic	Frequency (kHz)	Harmonic	Frequency (kHz)	Lane width
Master	$6f$	85.000			
Purple slave	$5f$	70.8333	$30f$	425	352.1 m
Red slave	$8f$	113.3333	$24f$	340	440.1 m
Green slave	$9f$	127.5000	$18f$	255	586.8 m

The master transmissions of other chains occur on different frequencies within the range 84.00 to 86.00 kHz and the slave transmissions maintain the same harmonic relationships.

9.2 Lane identification

A master-slave phase comparison produces a fractional lane position line which could exist within any lane, but by setting the lane number to the correct value at a known location and then maintaining the lane count, the position line is unambiguously defined. However, as a result of power failure, loss of signals, or skywave interference, the integrity of the lane count may come into question, so it is therefore of value to have an independent means of confirming the lane count. The Decca Navigator system therefore incorporates a means of checking the lane count by creating broad lanes, which are equivalent to a zone. This is achieved by, in effect, transmitting the fundamental frequency f, but without actually doing so, since the need to radiate VLF frequencies would add a major complication to the transmitting stations. Instead the master and slave stations in turn transmit simultaneously on all four frequencies – their own plus the frequencies of the other three stations. Each station therefore radiates on the frequencies $5f$, $6f$, $8f$ and $9f$, for about half a second in turn during specific two-and-a-half second lane identification (LI) transmission periods, within the twenty-second transmission cycle (Figure 9.4). Within the receiver the four frequencies are summed, and from the resultant the fundamental frequency (f) is derived. Phase comparison of the fundamental frequency derived from the master and slave transmissions produces phase differences which are displayed sequentially on the LI indicator, and thus provide unambiguous position lines within a zone.

An earlier technique of lane identification (V mode) produced the fundamental frequency by taking the difference of two transmitted frequencies (e.g. $6f$-$5f$), but the limitation was that skywave signals could adversely affect the phase of the resultant difference frequency by interfering with the groundwave from which the hyperbolic lattice is derived, resulting in possible errors in lane identification. The use of four frequencies in the 'multipulse' (MP) technique, results in a coarse pattern stability which is so much improved that at extreme range it can be more stable than the fine pattern.

The master and slave transmitters radiate an $8.2f$ signal which is used by airborne receivers to provide zone identification. It is also used to transmit commands and data for chain control and surveillance.

The 0.1 second breaks in the master transmission ($6f$) (Figure 9.4) are used to trigger the receiver to process the subsequent lane identification signals.

9.3 Areas of coverage

The Decca Navigator system is continually being extended, with the construction of new chains in many different parts of the world. Figure 9.5 shows the present coverage and also the chains presently under construction.

Charts for use in the areas covered by Decca Navigator are produced by various charting authorities; usually they are standard navigation charts, over-printed with the appropriate hyperbolic lattices.

9.4 System performance

9.4.1 Day-time

The factors which affect the accuracy of the Decca Navigator system are basically the same as those which affect other hyperbolic navigation systems. Inadequate signal strength in relation to the ambient noise level is the main cause of random error by day, although provided that use of the system is kept within the

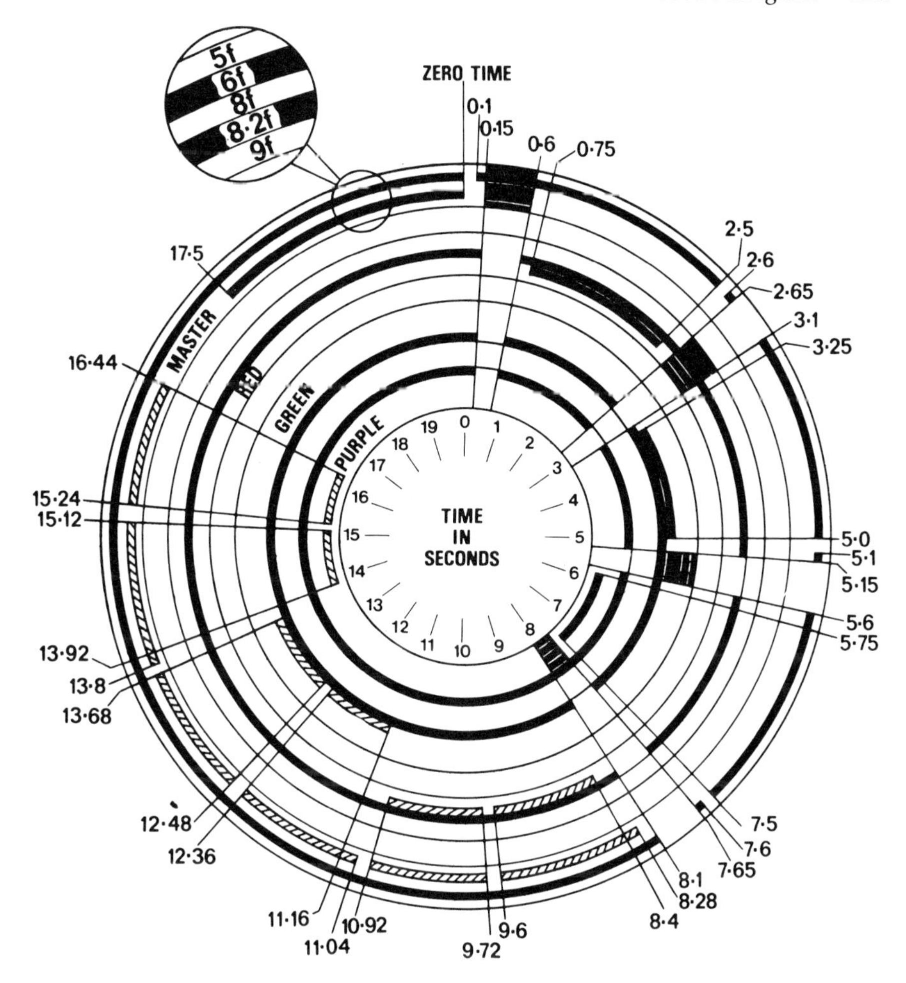

Figure 9.4 *Twenty-second transmission sequence for MP Decca Navigator chain. Shaded periods denote 8.5 f command and data transmissions for chain control and surveillance (Decca Navigator Co.).*

prescribed coverage area the random error can be described by accuracy contours (Figure 9.6).

A systematic error exists if the actual propagation velocity differs from the value used in computing the overprinted lattices. The Decca Navigator Company issue data sheets which show these fixed errors (Figure 9.7).

One fixed error source which is not included in the data sheets is charting error, which can occur if a particular chart is based on a

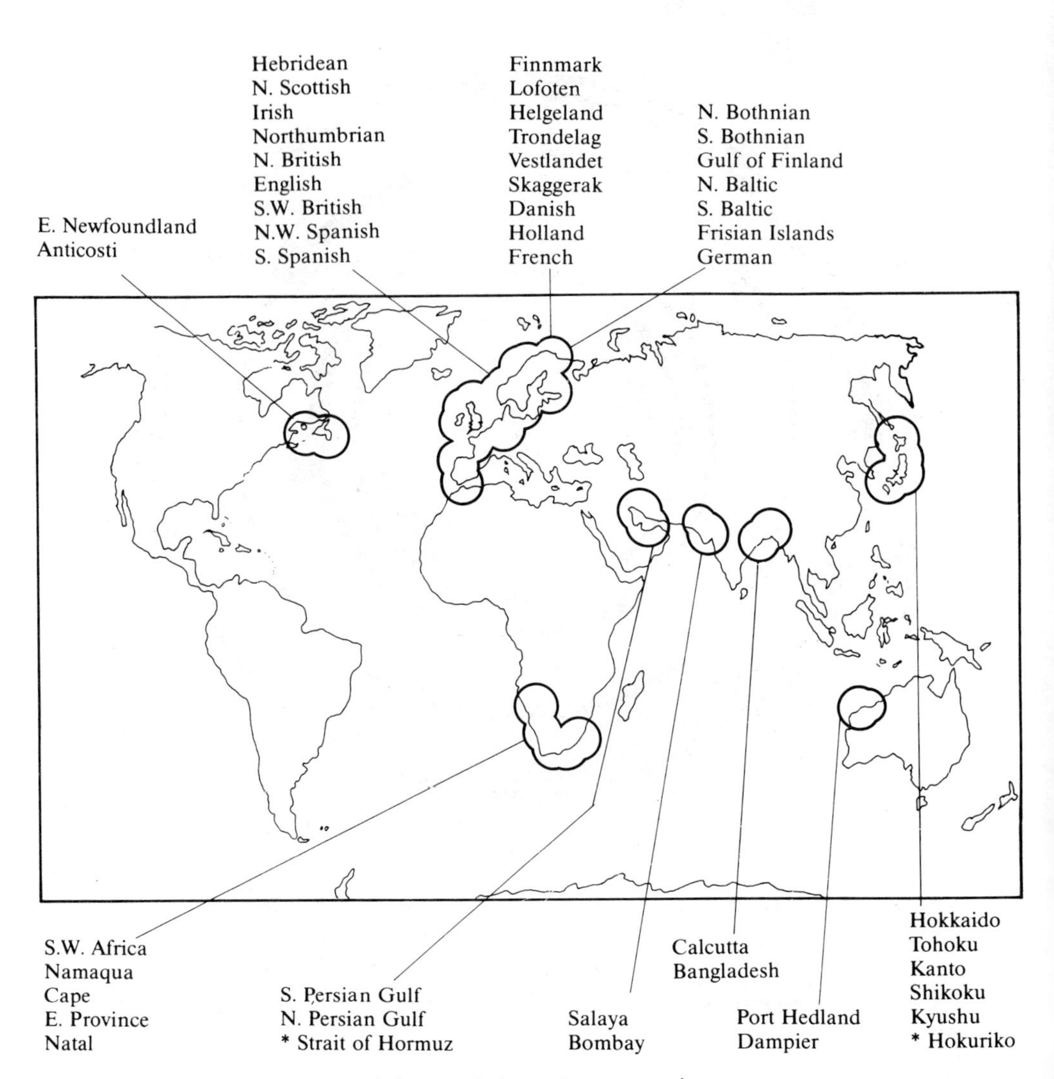
Hebridean
N. Scottish
Irish
Northumbrian
N. British
English
S.W. British
N.W. Spanish
S. Spanish
Finnmark
Lofoten
Helgeland
Trondelag
Vestlandet
Skaggerak
Danish
Holland
French
N. Bothnian
S. Bothnian
Gulf of Finland
N. Baltic
S. Baltic
Frisian Islands
German
E. Newfoundland
Anticosti
S.W. Africa
Namaqua
Cape
E. Province
Natal
S. Persian Gulf
N. Persian Gulf
* Strait of Hormuz
Salaya
Bombay
Calcutta
Bangladesh
Port Hedland
Dampier
Hokkaido
Tohoku
Kanto
Shikoku
Kyushu
* Hokuriko
* denotes chains under construction

Figure 9.5

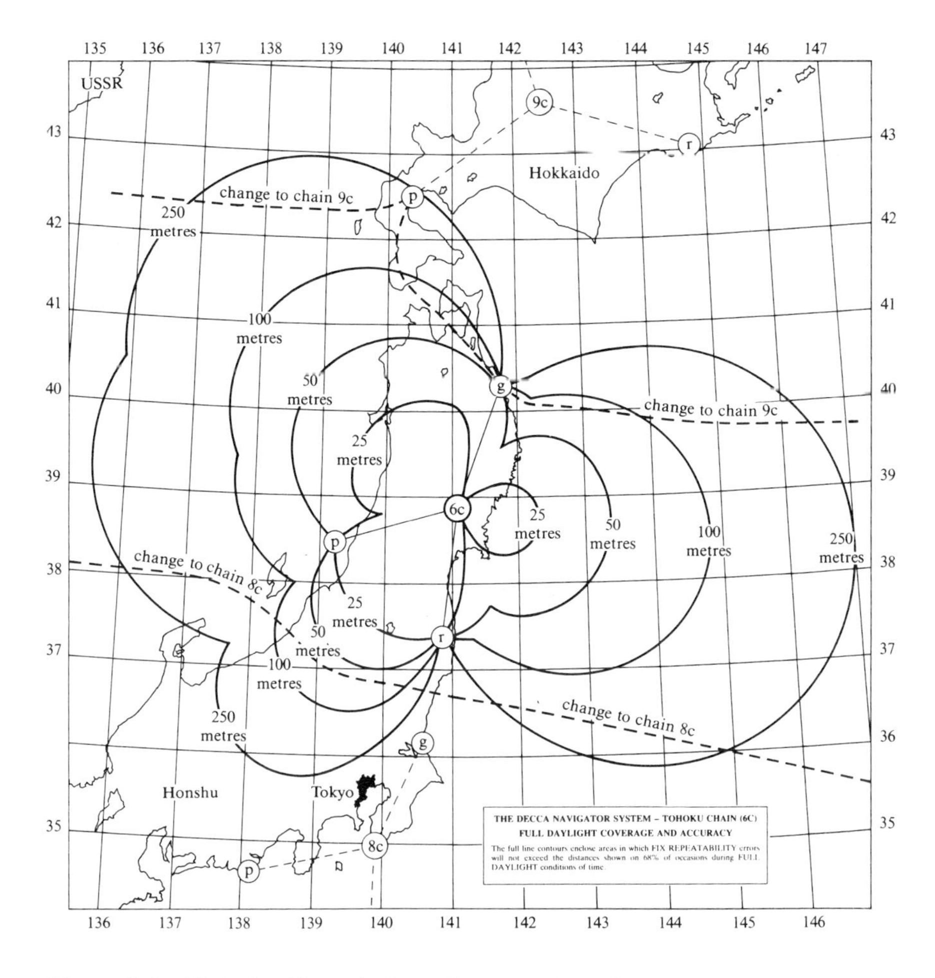

Figure 9.6 Tohoku (Japan) chain (courtesy Racal-Decca).

different survey datum from that used in determining the positions of the transmitting stations.

9.4.2 *Night-time*

The most significant error source is interference from the skywave propagated signal, which reaches maximum when the amplitude of the skywave signal equals or exceeds that of the groundwave signal. Interference between the two signals occurs during night-time and results in unwanted phase variations. In this case a correction

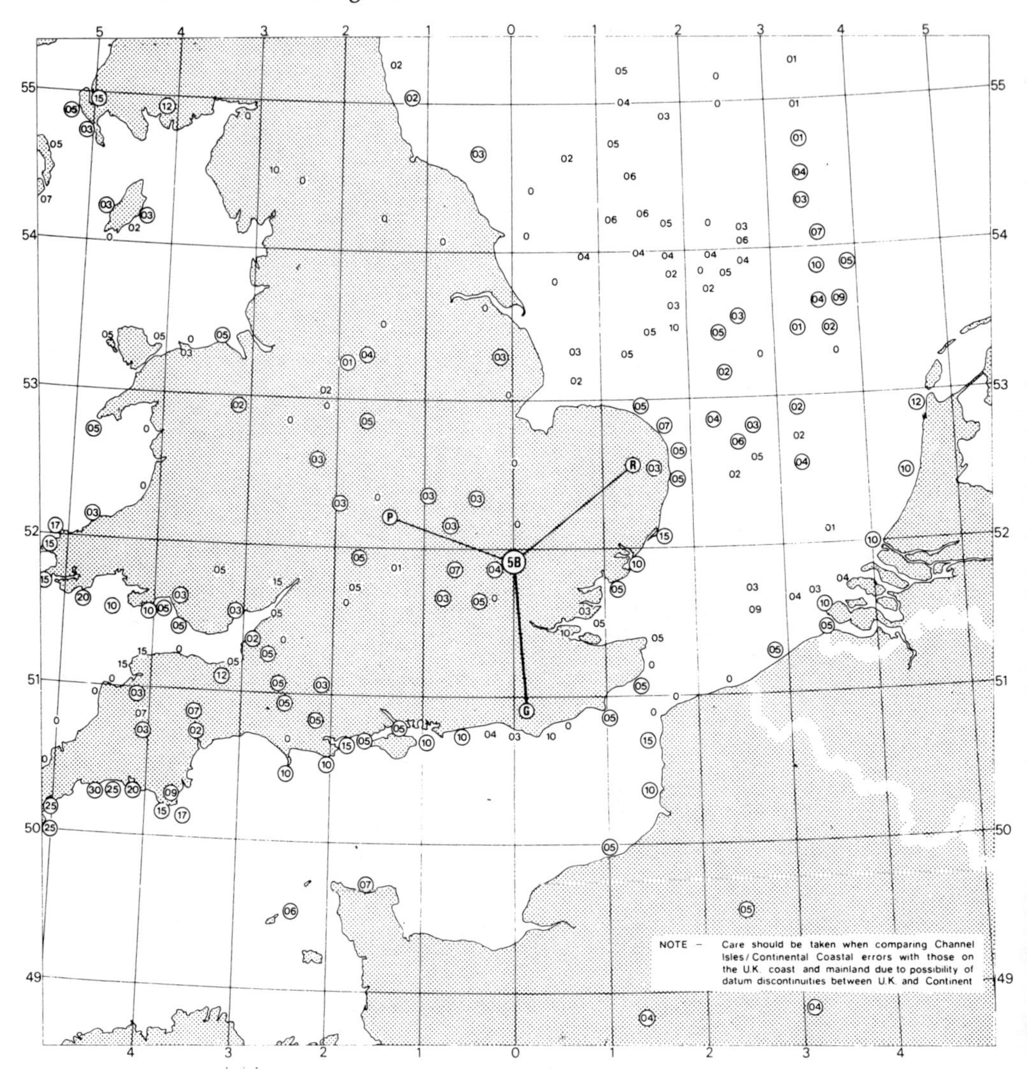

Figure 9.7 Example of systematic error map for green pattern – English chain (5B). Values shown are in hundredths of a lane unit, figures encircled should be subtracted, others are added (courtesy Racal-Decca).

cannot be applied, since the receiver is unable to differentiate between the ground and skywave components, and sees only the resultant phase R (Figure 9.8). Although it is not possible to predict discrete corrections for skywave interference, it is possible to predict statistically the magnitude of the resulting error (Figure 9.9 and Figure 9.10).

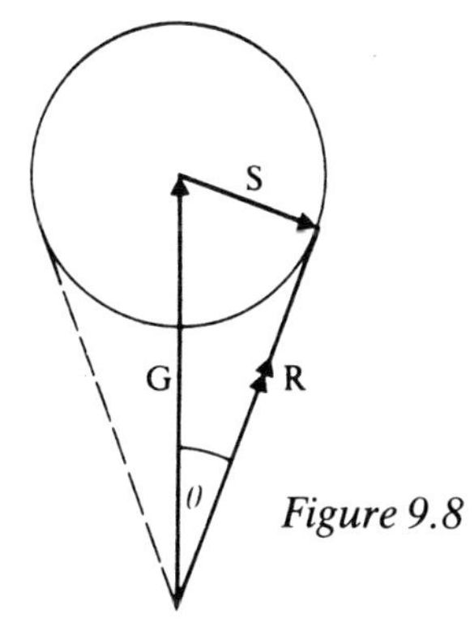

Figure 9.8 *(left) Vector diagram showing phase error θ of resultant received signal* R_1 *in presence of skywave component of random phase, interfering with groundwave G.*

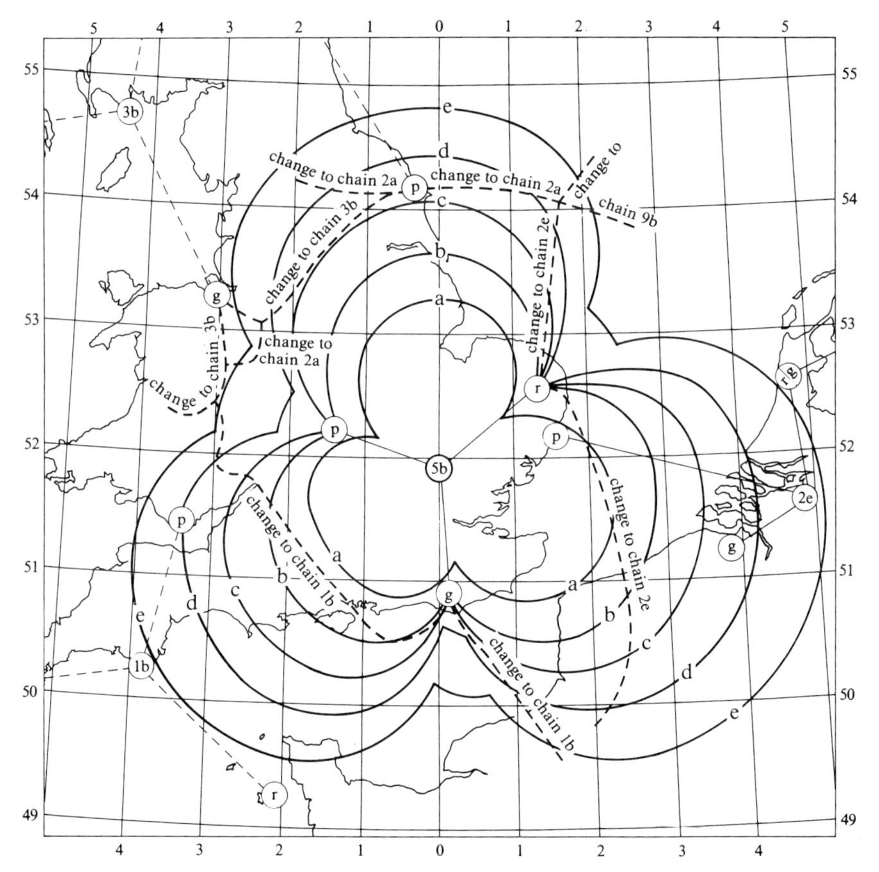

Figure 9.9 *Predicted coverage and accuracy diagram (68 per cent probability level) for times other than 'full daylight'. To be read in conjunction with Figure 9.10 (courtesy Racal-Decca).*

RANDOM FIXING ERRORS AT SEA LEVEL IN NAUTICAL MILES
68% PROBABILITY LEVEL

DECCA PERIOD	CONTOUR					
See Time and Season Factor Diagram below	a	b	c	d	e	
HALF LIGHT	<0·10	<0·10	<0·10	0·13	0·25	
DAWN/DUSK	<0·10	<0·10	0·13	0·25	0·50	
SUMMER NIGHT	<0·10	0·13	0·25	0·50	1·00	
WINTER NIGHT	0·10	0·18	0·37	0·75	1·50	

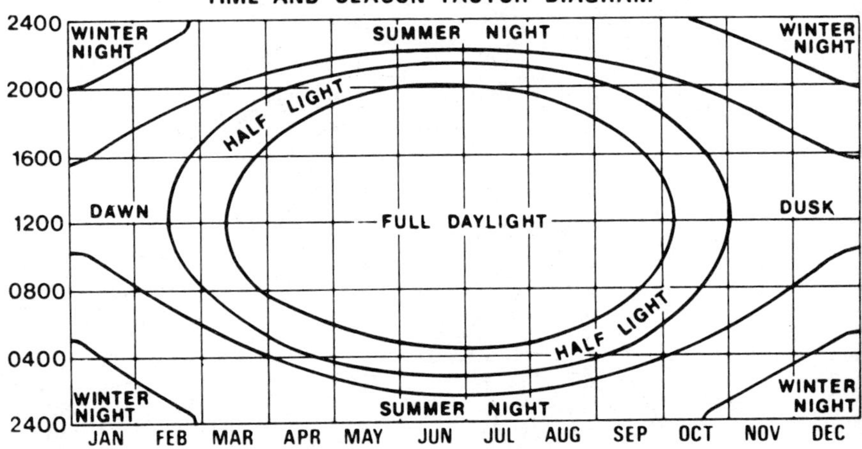

Figure 9.10 *Table and time/season diagram for use with Figure 9.9 (from* Decca Navigator Operating Instructions and Data Sheets).

In the extreme, skywave interference can cause an error in the lane count, but this can usually be detected and rectified by using the lane identification information. As already mentioned the multipulse lane identification display may be more accurate than the normal decometer readings (Figure 9.11). In general the transition from decometer to LI readings is made at distances greater than 100–150 nautical miles from the master and slave stations. Beyond this a fix from the LI reading tends to be more accurate than the basic pattern fix, and as range is further increased there comes a

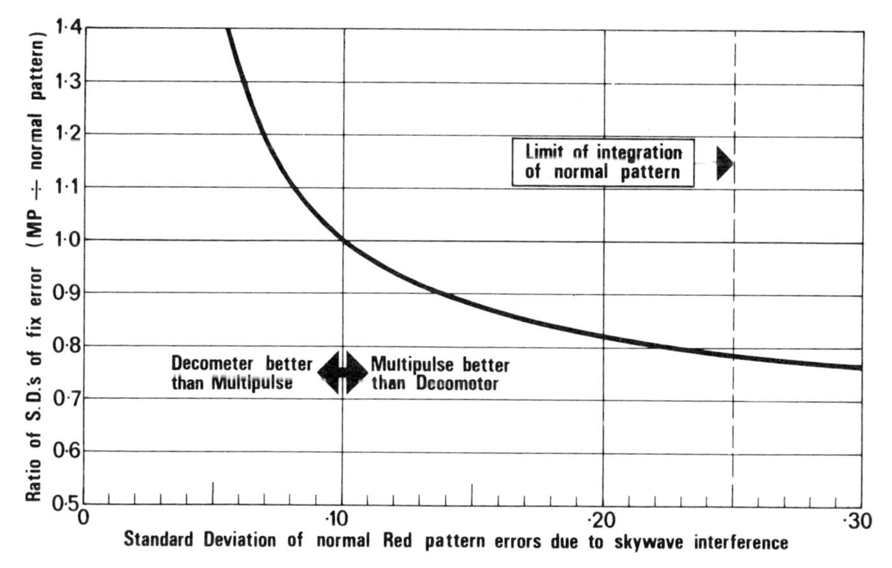

Figure 9.11 *Comparison between normal and multipulse position-fixing at night (courtesy Decca Navigator Co.).*

point at which the fine patterns become insufficiently reliable so that lane identification is then the sole means of fixing.

9.5 The receiver

The basic elements of a Decca Navigator receiver are shown in Figure 9.12, which deliberately simplifies the receiver functions so that the fundamental principles of operation can be clearly seen. It does not show for instance that the receiver is of the superheterodyne type, and that the phase comparisons occur at the intermediate frequency, since the end result is unaffected.

Within the receiver, master and slave signals are multiplied to the common harmonic frequencies, $30f$, $24f$ and $18f$, and the master-slave phases are compared in a discriminator circuit. The outputs from each discriminator are d.c. currents, proportional to the sine and cosine components of the phase difference angle. The outputs are fed to the orthogonal field coils of the decometer phase meter, causing the rotor to take up the angular position equal to the phase difference. This phase difference is displayed by the fractional lane pointer which is coupled to the decometer rotor, and which, through gearing, also drives the lane and zone indicators (Figure 9.13).

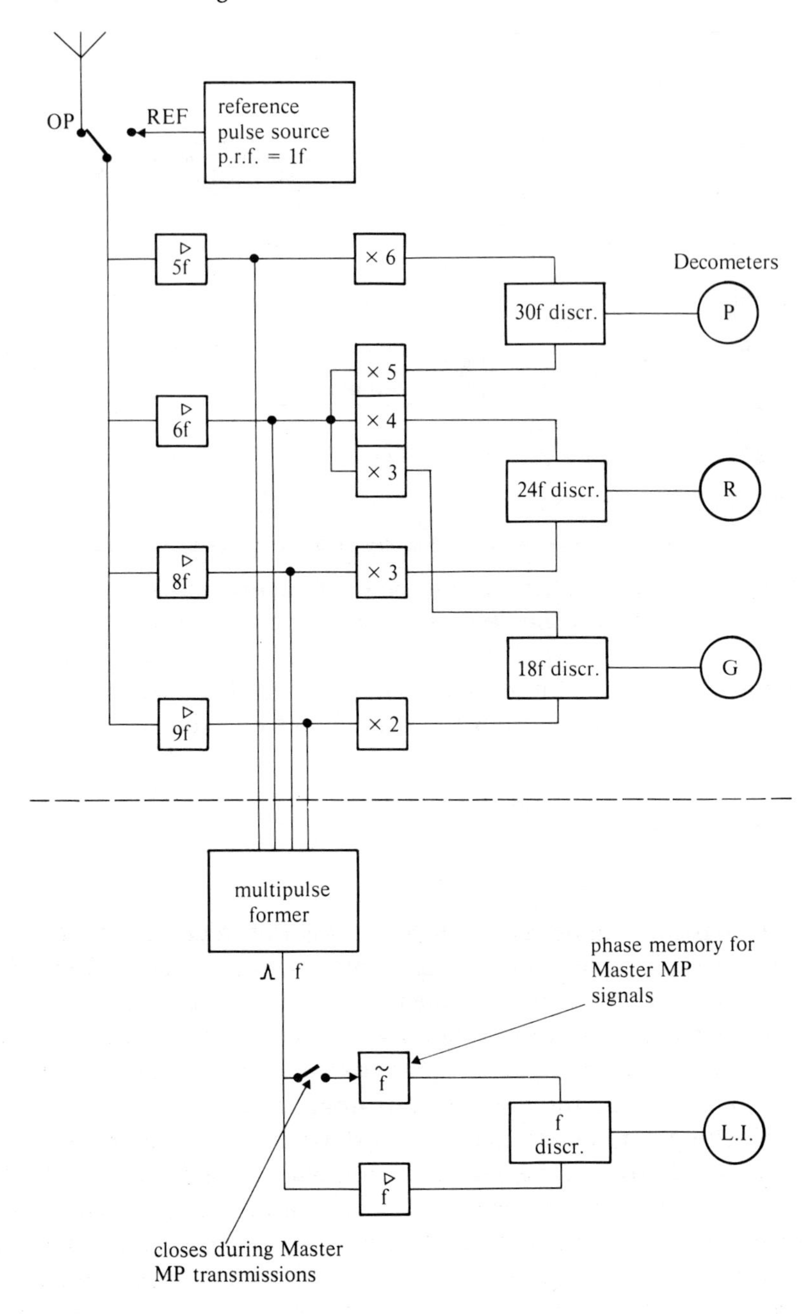

Figure 9.12 ***Basic elements of a Decca Navigator receiver with MP lane identification (courtesy Decca Navigator Co.).***

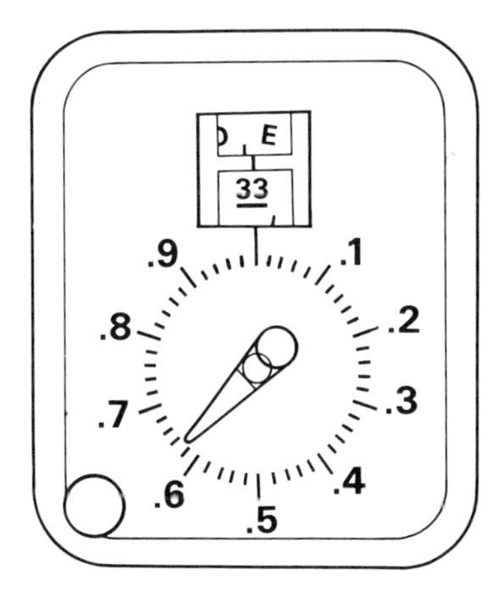

Figure 9.13 'Decometer' of Decca Mark 21 receiver.

Unequal phase delays in the receiver channels would cause a systematic error if left uncorrected, but use of the reference facility allows the appropriate adjustment to be made. The normal received signals are replaced by internally generated pulses, at the fundamental frequency f. The reference pulse has a strong harmonic content corresponding with the transmitted Decca frequencies, and when multiplied to the comparison frequencies, the resulting phase difference should be zero. A residual phase difference in the reference mode can be adjusted out by the user, hence removing a potential source of error.

To provide the lane identification capability already described, the normal master and slave transmissions are periodically interrupted for a multipulse transmission (Figure 9.4) from the master and each slave, at their allotted times. From the multipulse transmission, the receiver extracts the fundamental frequency f, and then by making the phase comparisons at this frequency a coarse hyperbolic pattern is generated, giving 'lanes' whose width is equal to 18 green, 24 red and 30 purple lanes. An additional phase meter (the LI indicator) gives a sequential indication of the correct lane (within a zone) of each pattern.

9.6 Using the Decca Navigator

Many Decca Navigator receivers which are presently in use on merchant vessels are of the Mark 21 type. The operation of both earlier and subsequent models is fundamentally the same as that of the Mark 21. As with any electronic navigation instrument, the instruction manual should be studied, and in the case of the Decca Navigator receiver the user should become fully conversant with the following operation phases: (1) initial setting up; (2) normal operation; (3) changing chain; and the additional useful procedure, (4) interchain fixing.

(1) The first step in initially setting up a Decca receiver is to select the appropriate chain, using the switches **1** and **2** (Figure 9.14). The equipment is then switched on by selecting *Lock 1* on the function switch **3**; this permits phase alignment of the receiver's phase locked oscillators with the received master and slave signals. Completion of phase alignment is indicated by a steady illumination of the Lock lamp **4** and steady decometer readings. In areas of poor signal/noise these two conditions may not be fully achieved.

The function switch is then set to *Ref.* which causes an internally generated reference signal to be injected into each of the receiver channels, and which should result in all lane fraction pointers

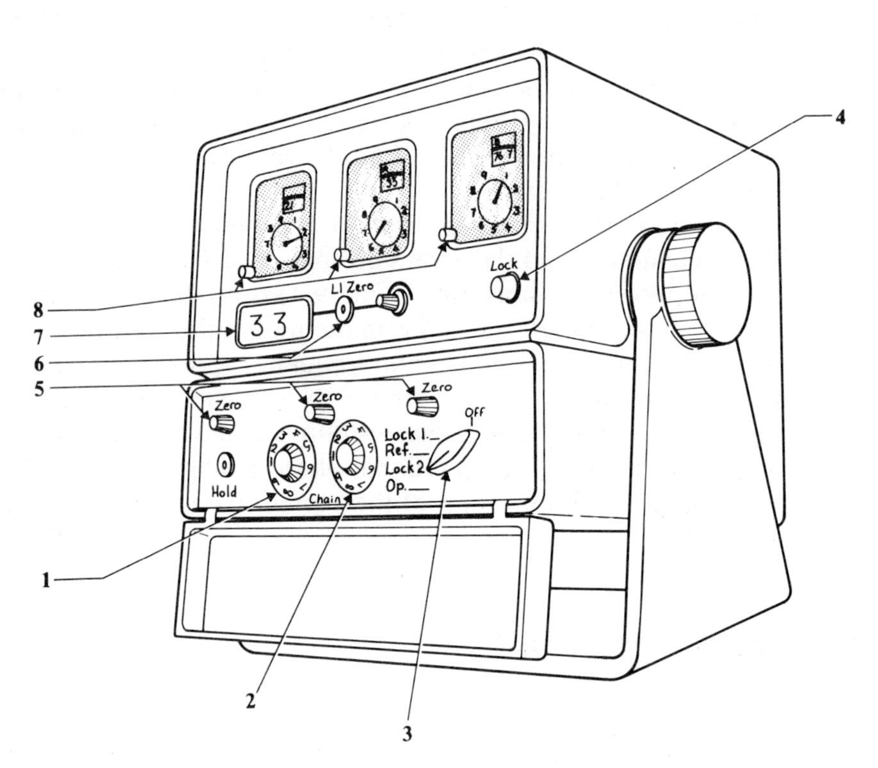

Figure 9.14 Operational controls of Mark 21 receiver.

reading zero. Any discrepancy from this can be removed by adjustment of the *Zero* controls **5**.

The function switch is then set to *Lock 2*, to allow the phase locked oscillators to resettle, before selecting the *Op*. position of the function switch.

The *LI Zero* control **6** is then used to set the LI display **7** to zero, during the master transmission.

The final adjustment is the setting of the decometers **8** to the correct zone and lane numbers. The lane numbers, and zone letters, change as a result of rotations of the fractional lane pointer; they are therefore ambiguous and need to be set to the correct values at switch-on. They also need to be checked at intervals against the possibility of lane slip.

The initial setting of the decometers is most easily achieved when the vessel is in port, or at any other precisely known location, since the appropriate Decca co-ordinates can be accurately determined from the latticed chart. Initial adjustment of the decometers can be more difficult when the setting up procedure is carried out when the vessel first enters an area of Decca coverage, and the vessel's position is not precisely known. In this situation the lane numbers are set from the best estimate of the vessel's position, and are then checked against the LI display and re-adjusted if necessary.

When using the lane identification facility in this way, or as a routine check for lane slip, particular care should be taken at night and at long range during daylight, since skywave interference can affect the accuracy of both the decometer and the LI reading. Under these circumstances, adjustments to lane numbers should only be made after several successive comparisons between decometer and LI.

(2) During normal operation, occasional adjustment of decometer and LI zero settings may be required, and should be carried out in the manner detailed in the operating instructions.

Plotting a position is simply a matter of reading the decometers, applying the corrections for system error where necessary, and plotting the position lines on the latticed charts, where necessary interpolating between whole lane numbers.

(3) Changing chain can be accomplished more quickly than initial setting up. The ship's position is transferred from the chart of the current chain to the chart for the new chain, after allowing for fixed error (if known) for both chains. The zone and lane numbers for the new chain can now be determined, and the decometers set to these values after the new chain has been selected, and the receiver has

phase locked. Finally the lane numbers should be checked against the LI indication.

(4) An inter-chain fix is made using one pattern from each of two adjacent chains, where two patterns provide a better angle of intersection than is possible using either chain independently. Special inter-chain charts are issued for certain areas, and a significant improvement over single chain fixing may thereby be achieved at longer ranges. The reading for the second chain is obtained from the LI readout, of which the reading accuracy of 0.1 lane is adequate for the purpose.

Full details for carrying out this, and other procedures briefly described here, are given in the equipment operating instructions.

Later, microprocessor-based receivers function on the same principle as earlier equipment but these receivers perform computations on the basic line-of-position outputs, and their rate of change, to provide a wide range of navigational information. This data, directly displayed in alpha-numeric form, may include (in addition to Decca Navigator hyperbolic coordinates);

Latitude and longitude
Course and speed
Range and bearing*
Distance and time to go, or distance and time gone*
Course to steer*

(**With respect to predetermined waypoint(s)*)

Automatic plotters are also available to provide a charted record of track-made-good. This facility is of particular value when the very high degree of repeatability inherent in the system is of greater interest than the absolute position fix (e.g. for fishing, cable laying, wreck location, etc.).

10 The Omega navigation system

Omega is a very low frequency (VLF) radio navigation operating in the band allocated for radio navigation between 10 kHz and 14 kHz. The first operational evaluation of the Omega system took place in 1961, when 10.2kHz signals were radiated from transmitters at New York, Hawaii, and Panama. Following this, experimental stations located in Norway, Trinidad, Hawaii, and New York commenced operation in 1966 to more fully evaluate and test the system. As a result of the success of these trials, worldwide system implementation commenced in 1968, aiming towards a global coverage with eight transmitting stations. Implementation of the system is being carried out by the United States Navy in conjunction with the respective nations hosting the transmitting stations. Since 1971 a US Coastguard unit, Omega Navigation Systems Operations Detail (ONSOD) has taken over management responsibility for operations, and provides engineering support for the transmitting stations as they become operational.

10.1 System description

In the Omega system, use is made of the long groundwave propagation distances of very low frequencies to create a hyperbolic navigation system with baseline distances of up to five thousand miles. Baseline distances of this magnitude give the Omega system a worldwide coverage using only eight transmitting stations. In consequence Omega cannot provide the position accuracy of the short-baseline hyperbolic navigation systems, but its global coverage makes it ideal for ocean navigation, and the aimed system accuracy of two to four nautical miles (2d*rms*) is adequate for this purpose.

In all hyperbolic navigation systems, the hyperbolae are loci of constant time difference between the arrival of signals from two transmitting stations. In the Omega system, the time difference is measured as a difference in phase of the two received signals (Figure

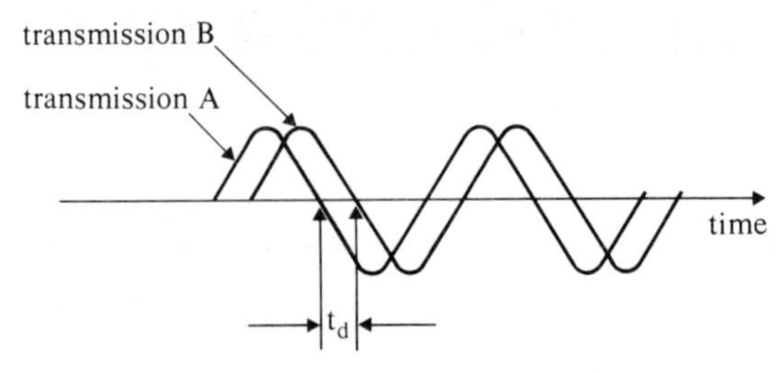

Figure 10.1 *Time difference* t_d *is determined by comparing the carrier phase of the two transmissions.*

10.1). This technique is basically ambiguous, since the same range of observed phase differences occur for each whole cycle difference through which the receiver is advanced along the baseline between the two transmitting stations. A full cycle difference occurs every half wavelength in distance, since in moving from point P to Q (Figure 10.2), the receiver has moved away from B by one half wavelength and towards A by the same amount, making one wavelength of change. Points of zero phase difference occur at every

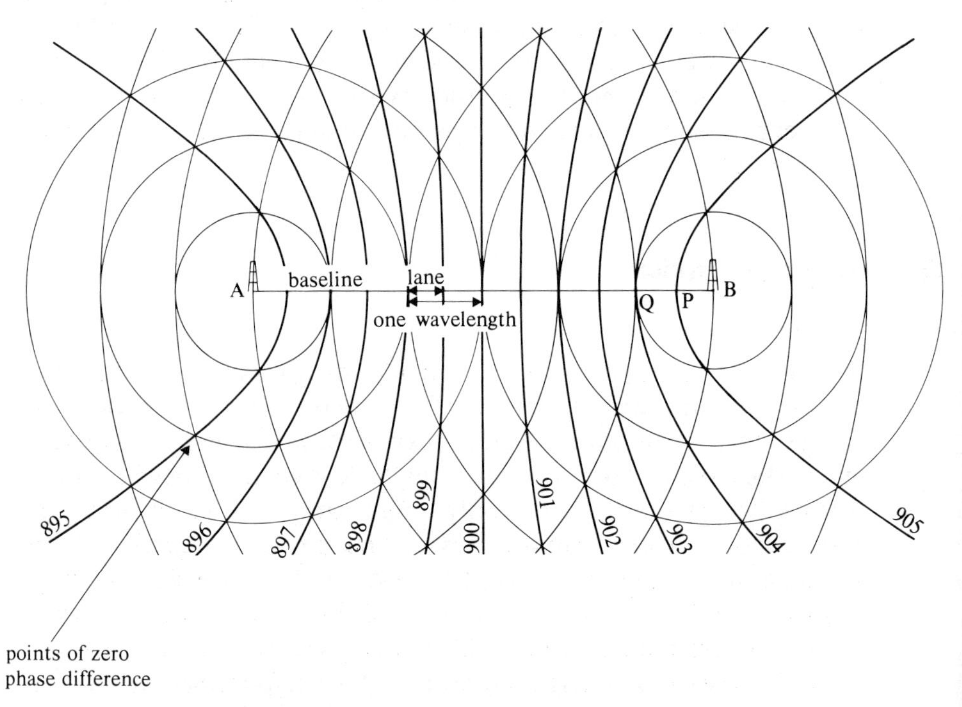

Figure 10.2 *The basic Omega hyperbolic pattern. The hyperbolic lines represent zero phase difference and occur at each half wavelength.*

half wavelength, which on the baseline is equal to a distance of approximately eight nautical miles at the basic Omega frequency of 10.2 kHz.

In the Omega system the zero phase difference lines of position (LOP) are numbered for identification, and the LOP equidistant from two transmitters is always designated as 900.

The space between two adjacent zero phase difference LOP is termed a 'lane', and is subdivided into one hundred centilanes each corresponding to 3.6 degrees of phase difference (Figure 10.3). A measured phase difference therefore defines a position line within a lane but does not specify the actual lane – this has to be determined by other means.

Assuming a vessel is at a known location, the appropriate lane numbers can be determined from an Omega chart and then set into the Omega receiver. As the vessel then progresses across the lanes a

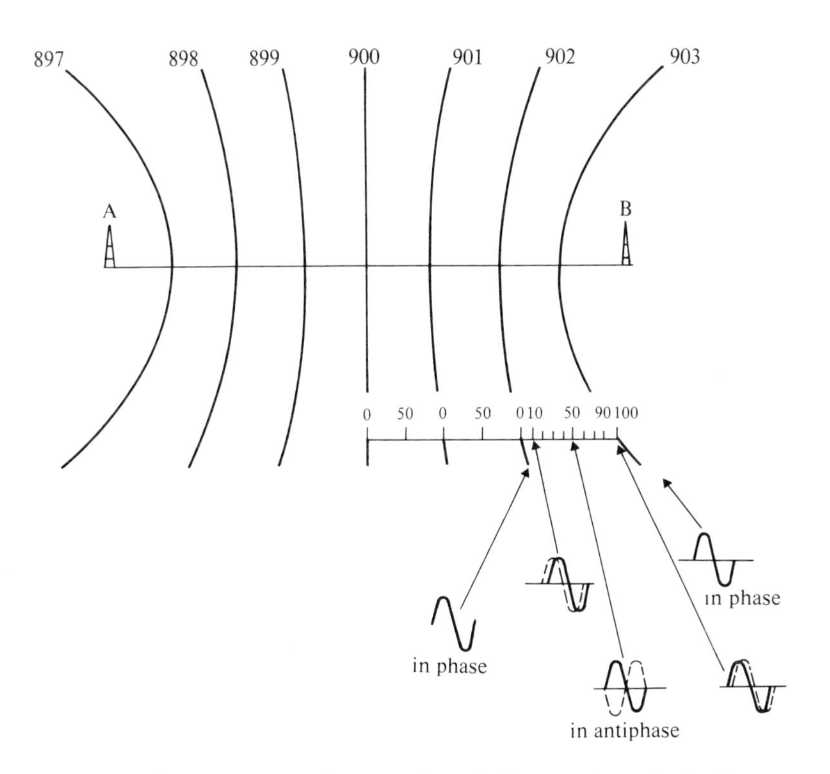

Figure 10.3 *Between each zero phase difference hyperbolic line the relative phase of the two transmissions changes by a full 360 degrees. Each 3.6 degrees of phase difference is termed a centilane.*

counter registers the corresponding increase or decrease in lane numbers. At any time the position lines are therefore given as lane numbers derived from counting lane crossings, plus the centilanes which are derived from the phase difference measurements. For example:

726 . 43
(lane number) (centilanes).

A position fix is obtained from a minimum of two LOP which can be derived from either three stations by making one station common, or from four stations. The propagation characteristics at VLF enable up to six stations to be usefully received at any location, and so the number of LOP actually used is most probably limited either by the number of LOP overprinted onto the Omega charts, or by the display limitations of the Omega receiver.

The eight Omega transmitters are independent, in the sense that they do not work in a master-slave configuration. Instead the phase of the transmission of each station is sufficiently precise with respect to that of all other stations that a position can be derived from the phase difference of any two transmissions.

The eight transmitting stations are designated alphabetically with the letters A to H (Figure 10.4), the convention being adopted that the phase difference will be taken alphabetically. For example: phase of A – phase of B; or phase of C – phase of H. The corresponding LOP would be designated AB and CH. It is also convention that lane numbers increase in an alphabetical direction.

10.1.1 Signal format

Since the basic Omega frequency of 10.2 kHz is common to all eight transmitting stations, problems of mutual interference prevent all stations transmitting continuously. Instead each station transmits for approximately one second, with an 0.2 second break between each transmission (Figure 10.5). With this transmission format, a problem arises at the receiver in identifying the signals from each transmitter. Various techniques exist for overcoming this problem and are described in section 10.7, Omega receivers.

The problem also arises that the phase comparisons have to be made between two signals which do not actually occur at the same time. This is usually achieved by comparing each with a reference oscillator within the Omega receiver:

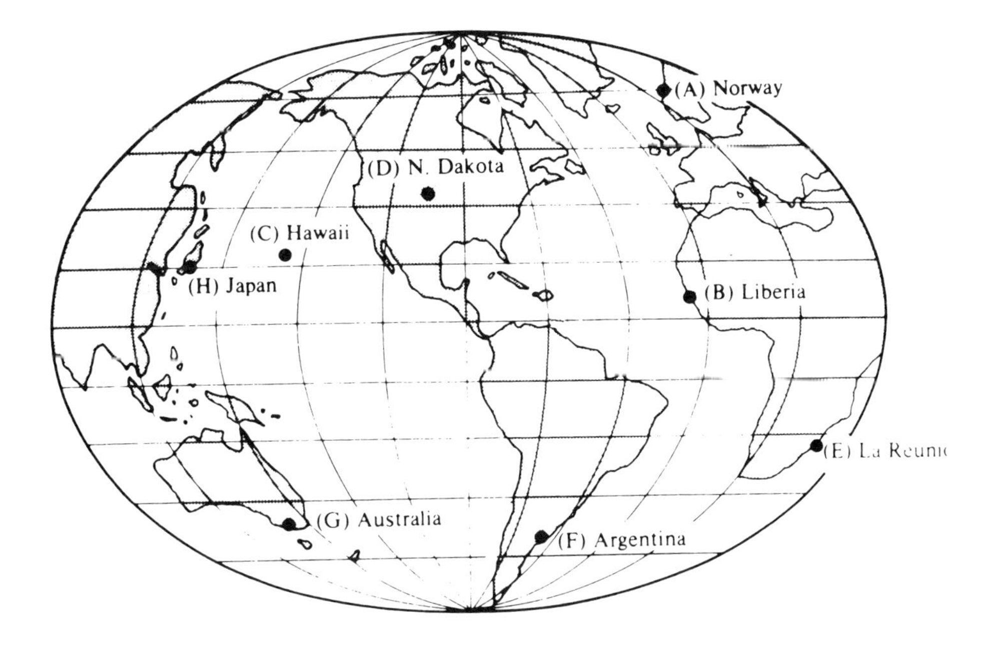

Station letter designation	*Location*	*Latitude*	*Longitude*
A	Aldra, Norway	66°25′N.	13°08′N.
B	Monrovia, Liberia	6°18′N.	10°40′W.
C	Haiku, Hawaii	21°24′N.	157°50′W.
D	La Moure, North Dakota	46°21′N.	98°20′W.
E	Le Reunion	20°58′S.	55°17′E.
F	Golfo Nuevo, Argentina	43°03′S.	65°11′W.
G	Australia	38°29′S.*	146°56′E.*
H	Tsushima, Japan	34°37′N.	129°27′E.

*Approximate.

Figure 10.4 The Omega transmitting stations.

$\phi A - \phi$ ref
$\phi B - \phi$ ref

Difference in phase, $\phi A - \phi B = (\phi A - \phi \text{ ref}) - (\phi B - \phi \text{ ref})$.

10.2 Lane ambiguity

The phase difference measurement made at the basic Omega

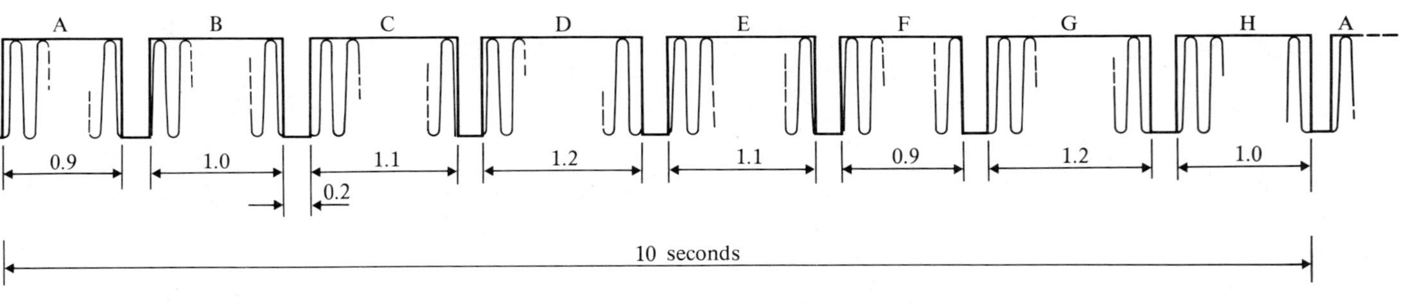

Figure 10.5 Representation of the basic Omega transmission sequence at 10.2 kHz.

frequency of 10.2 kHz produces a position line within a lane, but does not define the lane. This can usually be obtained by counting the lane crossings as previously described, but situations can arise where the lane count becomes incorrect. Alternatively it may not be possible to determine the lane numbers at the time of receiver initialization, if the vessel's position is not known within one lane (eight nautical miles on the baseline). In order to be able to deal with these two situations, each Omega stations transmits the additional frequencies of 11.05 kHz, 11.33 kHz and 13.6 kHz (Figure 10.6). Each station also transmits a unique frequency which at present plays no role in marine navigation.

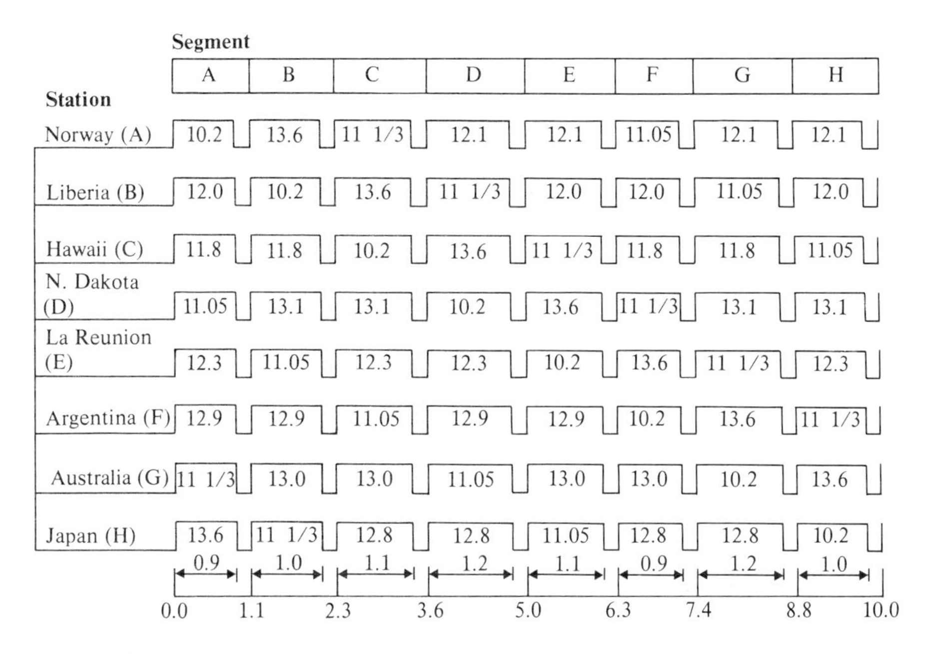

Figure 10.6 The full Omega transmission format.

Consider first the relationship between the 10.2 kHz and 13.6 kHz transmissions. Their frequency (and wavelength) are related in the ratio 3:4, and so the zero phase difference LOP of the two frequencies will produce a moiré pattern, in which every third 10.2 kHz lane is coincident with every fourth 13.6 kHz lane.

To see the value of this, take as an example a 10.2 kHz LOP determined as being twenty-five centilanes (Figure 10.7), but with doubt existing as to whether the lane number is 900, 901 or 902.

By determining the phase difference of the 13.6 kHz transmissions

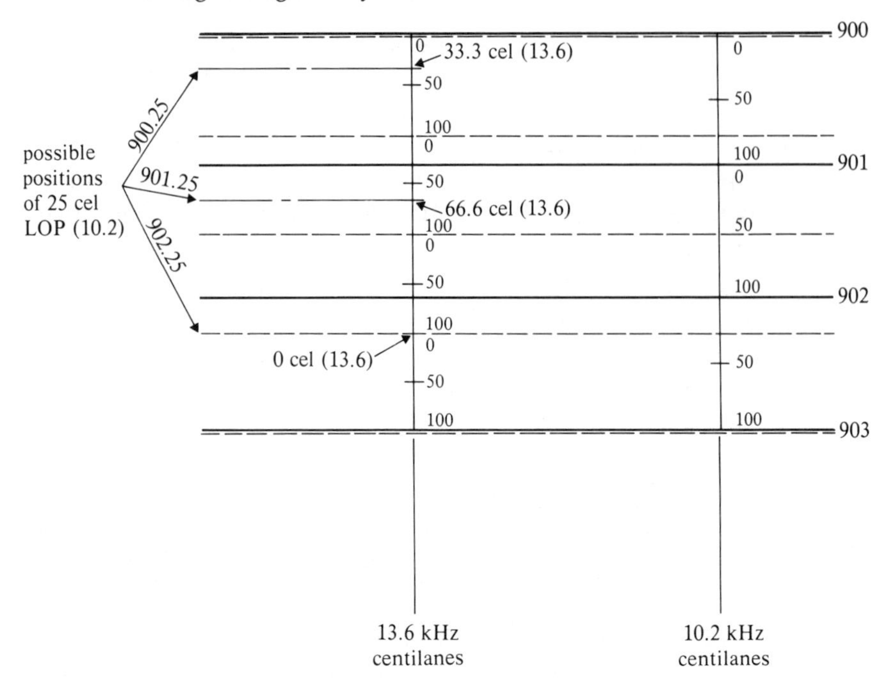

Figure 10.7 Resolving 10.2 kHz lane ambiguity using the 13.6 kHz transmissions and considering it as a moiré pattern.

of the same station pair, the ambiguity can be resolved. If the 13.6 kHz position line is 33.3 cel the correct 10.2 kHz lane is 900. If the 13.6 kHz LOP is 66.6 cel the correct 10.2 lane is 901, and finally a 13.6 kHz LOP of 0 cel indicates the correct 10.2 kHz lane to be 902.

An alternative way which is often used to consider this principle of lane determination is that of producing a 'difference frequency' phase difference, by subtracting the 10.2 kHz phase difference from the 13.6 kHz phase difference: 13.6 kHz − 10.2 kHz = 3.4 kHz. This can now be dealt with as a new frequency of 3.4 kHz which is precisely one-third of the basic Omega frequency of 10.2 kHz. It can be considered as if a frequency of 3.4 kHz has been radiated, producing lanes which are three times as wide as the basic 10.2 kHz lane (Figure 10.8).

Considering the original example, the uncertainty is now resolved from knowing the 3.4 kHz phase differences:

at 3.4 kHz 8.3 cel = 900.25 (10.2 kHz LOP)
41.6 cel = 901.25
75.0 cel = 902.25

The uncertainty in the vessel's position may be so gross that the lane ambiguity is greater than three lanes. In this situation the 11.33 kHz transmission could be used to produce a 1.13 kHz (11.33 − 10.2 kHz) difference frequency, which has an effective lane width nine times greater than the basic 10.2 kHz lane. Furthermore, the fourth frequency of 11.05 kHz can be used to produce a difference frequency of 0.283 kHz (11.333 − 11.050 kHz), which has a lane width equivalent to thirty-six 10.2 kHz lanes, but it is unlikely in

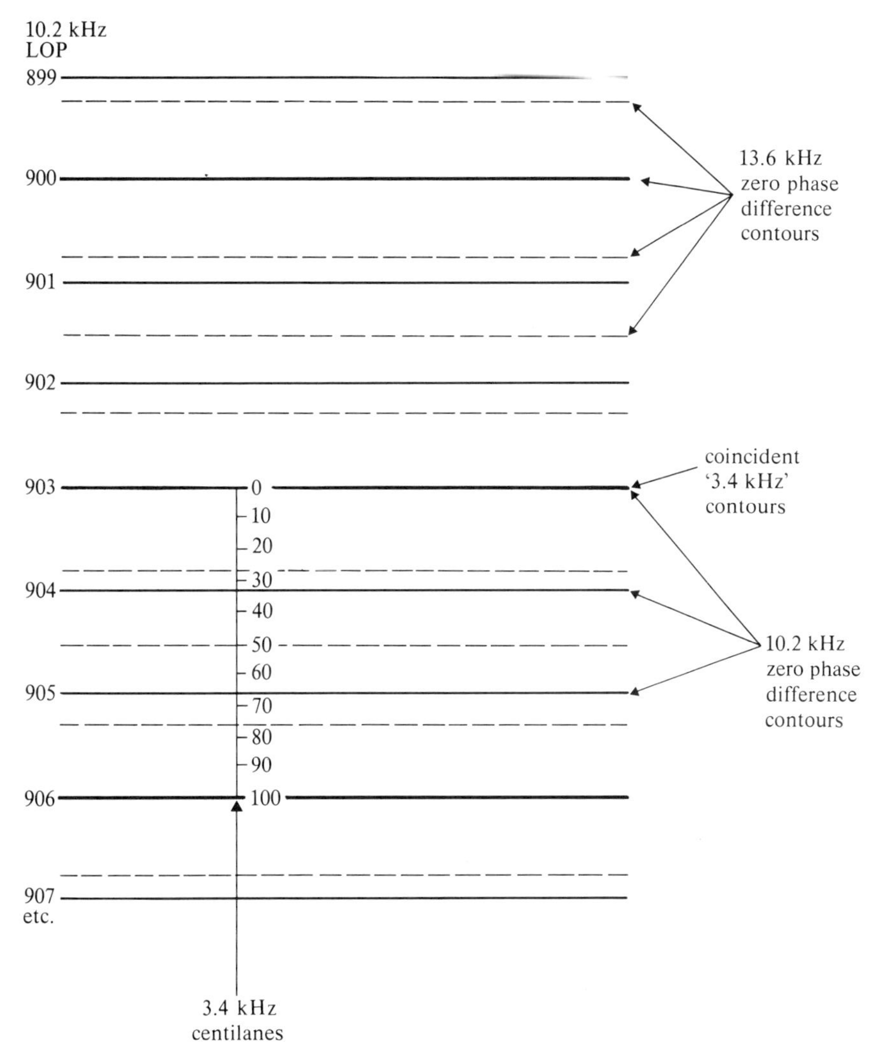

Figure 10.8 Resolving 10.2 kHz lane ambiguity using the 13.6 kHz transmissions and considering the difference as broad 3.4 kHz lanes.

marine navigation to require these coarser lane structures. The 11.05 kHz frequency was introduced late in the implementation of Omega to meet a requirement in air navigation.

10.3 The transmitting stations

The synchronization of all transmissions within the Omega system is precisely maintained so that the phases of each transmission are accurately related to each other, as if originating from a single coherent source. To achieve this precise synchronization all transmissions are controlled by cesium beam standards. Additionally the Japanese Maritime Safety Agency obtain weekly reciprocal path measurements from each transmitting station, and from these compute the adjustments to be made by each station.

The transmitters are designed to radiate 10 kW. The antenna, which is either a valley span or an umbrella array, is fed via the helix house which tunes the system to each of the transmitted frequencies.

The reliability of the transmitting equipment of any navigation system is of great importance, and this is true of the Omega system. Apart from the measures taken at the transmitting stations to ensure a high reliability, there is some inherent system redundancy since at most locations more than the required minimum of three stations can be received. To reduce the risk of two stations being off-air simultaneously an annual maintenance schedule is operated (Figure 10.9). During a station's allocated month for maintenance every endeavour is taken to keep the off-air time to an absolute minimum.

March	Argentina (F)
April	Liberia (B)
May	Hawaii (C)
June	La Reunion (E)
July	Norway (A)
February	Australia (G)
September	North Dakota (D)
October	Japan (H)

Figure 10.9 The designated months during which routine station maintenance can occur.

10.4 System performance

10.4.1 Proportional effects

The accuracy of positions which are determined by using the Omega system depends to some extent on the performance of the transmitting and receiving equipment, but the errors caused by various propagational effects are far more significant, and it is these which ultimately limit the performance of the system. In effect, the propagational anomalies cause the phase of the Omega signal to deviate from the theoretical phase value for a given location. The causes of these deviations are numerous, but their study is important if a navigator is to gain optimum performance from the Omega system.

10.4.2 Diurnal variations – propagation corrections

Fundamental to the accuracy of any hyperbolic navigation system is a precise value for the propagation velocity, which can then be used to relate a phase (or time) difference to a geographically described hyperbolic position line. The propagation of VLF signals is described in chapter 1 using the theory of waveguides. The actual propagation time is dependent on the dimensions of the waveguide, but one of the dimensions, the height of the ionosphere, changes from approximately seventy kilometres by day to approximately ninety kilometres by night (measured to lower D-region). The corresponding day to night variation in propagation velocity can vary the phase of an Omega signal at a given point by more than one hundred centicycles. This is equivalent to a movement of the whole imaginary pattern of hyperbolic lines over the earth's surface, and so to achieve a useful accuracy such a gross effect must be compensated for. It is fortunately a feature of propagation within the band of Omega frequencies (10.2 kHz to 13.6 kHz) that the phase variations are cyclic over a twenty-four hour period (Figure 10.10) and are also predictable. It is therefore possible to compute a correction for the diurnal variation which is accurate to within approximately five centicycles for an all-daylight path, and to within approximately ten centicycles for an all-night-time path.

To allow the users of Omega to make the appropriate corrections, the US Defense Mapping Agency Hydrographic/Topographic Center

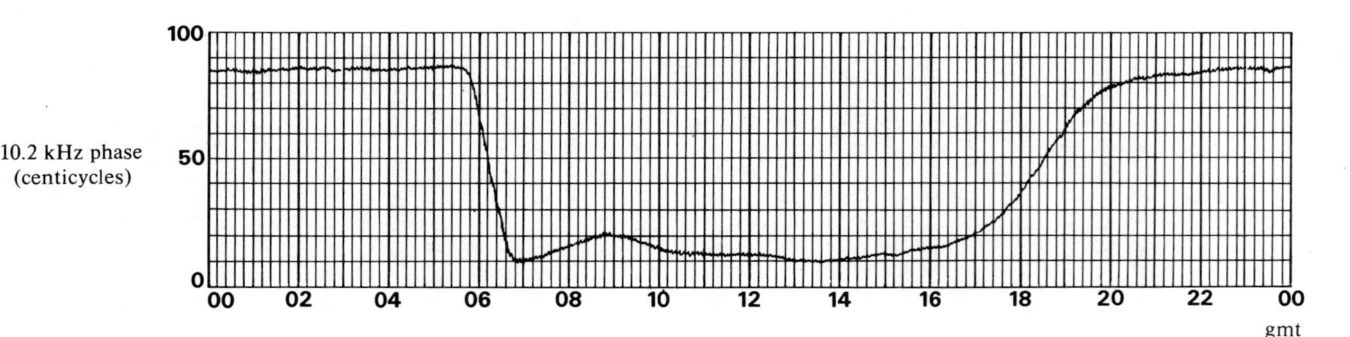

Figure 10.10 *The observed phase of a 10.2 kHz transmission over a twenty-four-hour period. The large changes occur during sunrise and sunset over the transmission path.*

publish volumes of *Omega Propagation Correction Tables* (Figure 10.11). These take into account variations in propagation velocity, due not only to variations in ionospheric height, but to other factors such as the orientation of the earth's magnetic field, and to surface

Pub. No. 224 (107-C) B

OMEGA PROPAGATION CORRECTION TABLES

FOR 10.2 kHz

AREA 07

MEDITERRANEAN

STATION B (LIBERIA)

Prepared and published by the
DEFENSE MAPPING AGENCY HYDROGRAPHIC CENTER
Washington, D.C. 20390
1976

Figure 10.11 One of the many volumes of Omega propagation correction tables (courtesy US Defense Mapping Agency Hydrographic/Topographic Center).

conductivity. The use of these correction tables is more fully described in section 10.10, Using the Omega navigation system.

In general the navigational accuracy of the Omega system is limited by the quality of these published propagation corrections (PPC), and so any improvement in the PPCs will therefore produce a corresponding improvement in navigational accuracy. To this end, a worldwide network of monitoring stations has been established. Data obtained from this monitoring is used to improve the published propagation corrections, which are updated periodically to incorporate changes resulting from the monitoring.

Figure 10.12 shows a typical diurnal variation in the phase difference of signals from stations A (Norway) and D (North Dakota). The second curve is obtained after applying the appropriate PPC for the two stations, as a result of which the total deviation in the LOP over the twenty-four hour period has been reduced; but in this particular example a mean error, or bias, remains even after applying the corrections. Corrections issued after a monitoring program in a given area should remove this type of error, as well as possibly further reducing the diurnal variations.

The published propagation corrections can only correct for predictable ionospheric behaviour. Unfortunately there are two phenomena which produce unpredictable disturbances to the propagation conditions.

10.4.3 Sudden ionospheric disturbances

The increased x-ray emission which accompanies a solar flare causes a sudden ionospheric disturbance (SID) on the sunlit side of the earth. This increase in ionization causes sudden phase anomalies (SPA), which affect the propagation of Omega and other VLF signals over sunlit paths. The effect is seen as a shift in phase at the Omega receiver, which lasts typically for about fifty minutes. The onset of the disturbance is rapid, reaching a maximum effect after five or six minutes, then a period of slow recovery through the remaining forty-five minutes. The frequency of occurrence of an SID is related to the eleven-year sunspot cycle, in a year of maximum sunspot activity the occurrence of SIDs may average one per day.

A typical SID causes a shift in phase of some ten to twenty centicycles, although occasional disturbances have been recorded which produce phase shifts of up to one hundred centicycles. If only

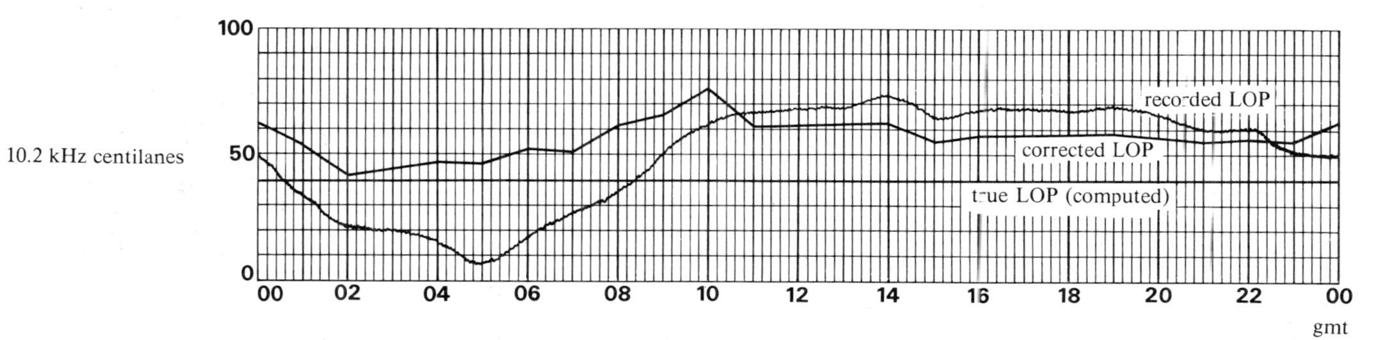

Figure 10.12 Typical reduction in diurnal variation by applying propagation corrections.

one station of a LOP pair is affected by an SID, the resulting maximum error would typically be one to two nautical miles. In reality it is likely that both propagation paths will be partially if not totally affected by the SID, and so the net result may be less.

10.4.4 *Polar cap absorption (polar cap disturbance)*

In addition to the x-rays associated with solar flares, some flares also emit protons which arrive in the vicinity of the earth several hours after the x-ray emissions. Because of the protons' charge, and the earth's magnetic field, they affect the ionosphere only in the polar regions. Here they cause an increase in ionization of the D-region, which results in an effective lowering of the D-layer, and the consequential changes in phase of Omega signals. The time scale of the polar cap absorption (PCA) effect is much longer than the SID, taking several hours to reach a maximum and several days to return to normality. The maximum magnitude of the effect varies from only ten centicycles, to more than fifty centicycles. The PCA effect occurs less often than the SID, typically ten PCA events occurring in a year of sunspot maximum. Since only high-latitude regions are affected, users of Omega should be particularly aware of signal paths over polar routes. Such signals cannot always be ignored though, since in many areas polar path signals must be used to make up two LOPs, particularly during transmitter off-air periods.

10.4.5 *Permafrost regions*

The ground conductivity of permafrost regions (such as parts of Greenland and Iceland) is typically one-fifth that of seawater. The high attenuation of VLF signals propagating over these regions reduces the useful range of some Omega signals, and consequently there are some areas in the shadow of permafrost regions where the number of available Omega signals is reduced to a level less than would be normally expected. The conductivity in these regions is also less well known, resulting in less accurate PPC corrections for polar paths.

10.4.6 *Long path interference*

A point exists at the antipode of each Omega transmitter where all great circle paths cross. The net result is usually a large increase in signal amplitude, but the resultant signal is unusable due to its indeterminate phase.

The rate at which the ionosphere attenuates VLF signals is related to the earth's magnetic field, and consequently to the direction of propagation. Propagation to the magnetic west is attenuated more severely than propagation to the east. Typical attenuation rates for an all-sea path during daytime are:

propagation east to west	4.2 dB/Mm
north to south	2.7 dB/Mm
west to east	1.9 dB/Mm.

Consequently, for each transmitter a point exists (typically 5500 nautical miles to the west of low-latitude transmitters) at which the amplitudes of the short and long path signals are equal. In these areas the signals tend to be unusable.

10.4.7 *Modal interference*

It has previously been explained (in chapter 1), that at the Omega frequencies two modes of propagation can exist, TM1 and TM2. Close to the transmitter the TM2 mode predominates, but the attenuation rate of this mode is greater than that of the TM1 mode. This latter mode therefore usually begins to predominate beyond a distance of several hundred miles from the transmitter, although at night-time this distance may be considerably further. The TM1 and TM2 modes propagate with different phase velocities, and so within the region in which their amplitudes are comparable the two modes will produce interference effects, since their phase relationship will change with distance from the transmitter. Signals which suffer severe modal interference are unusable. For transmitters near to the geomagnetic equator, modal interference exists out to considerably greater distances than for high latitude transmitters. Typically, modal interference can be experienced from the Liberian transmitter as far away as the east coast of the United States, during night-time.

Modal interference can cause large errors. The worst case is when one mode predominates during day-time and the second mode predominates at night. In this case whole lane errors are possible.

10.5 Charts

A series of Omega plotting charts is published by the US Defense Mapping Agency Hydrographic/Topographic Center (Figure 10.13). These charts are overprinted with up to five LOP, based on a standard propagation velocity (Figure 10.14). To prevent confusion through an excessive number of overprinted lines, it is usual to show every third LOP only. The LOP shown are in fact those which are coincident 10.2 kHz and 13.6 kHz LOP, namely the 3.4 kHz LOP.

Limited ranges of British and French charts are published with overprinted Omega LOP.

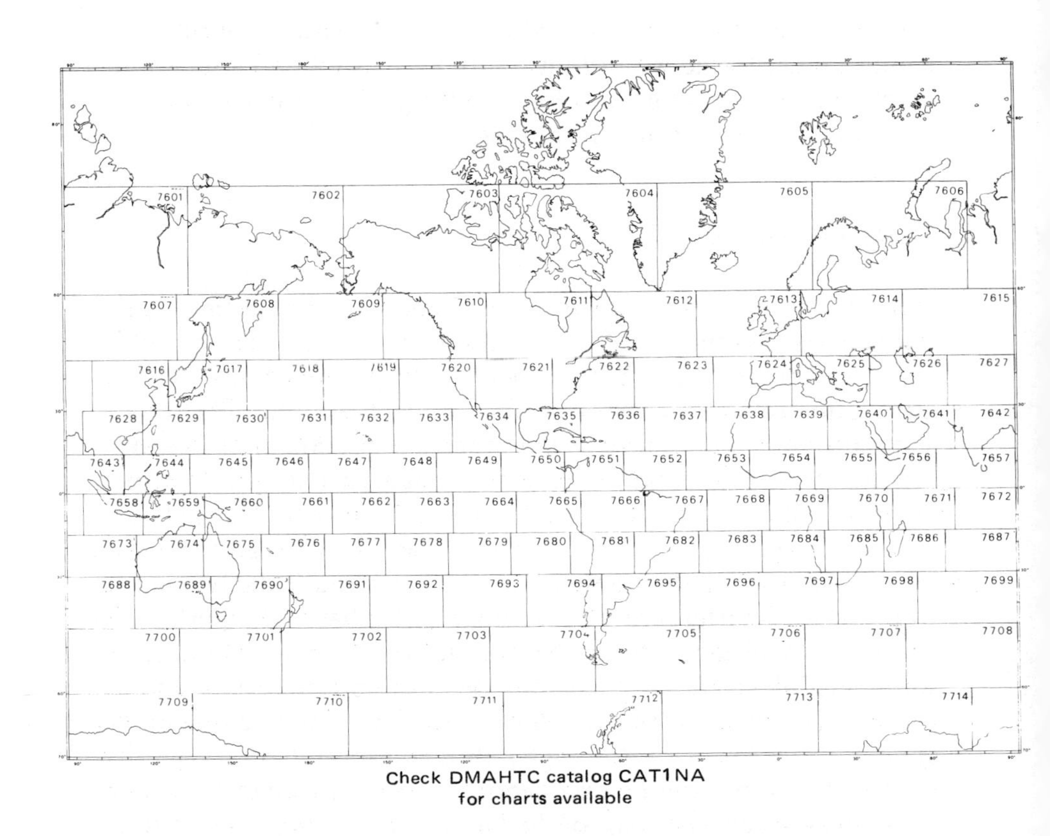

Figure 10.13 *The basic series of Omega plotting charts published by the US Defense Mapping Agency Hydrographic/Topographic Center.*

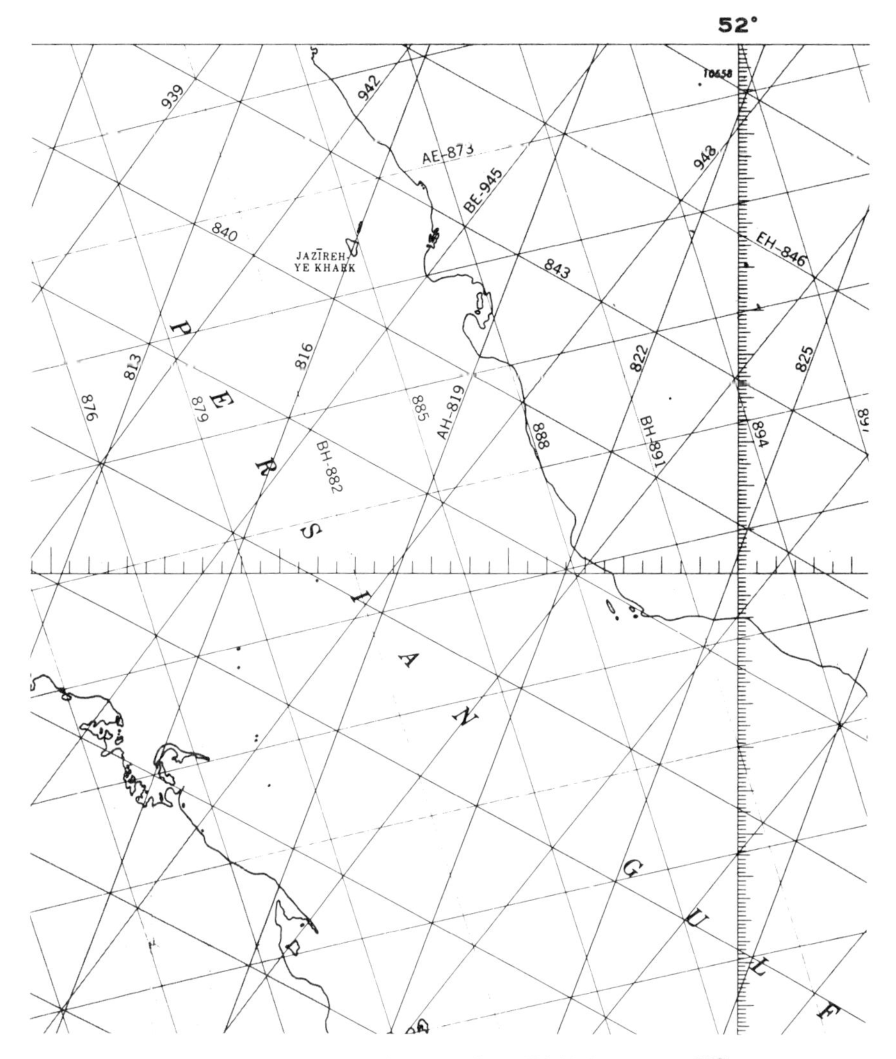

Figure 10.14 Part of Omega plotting chart 7641 (courtesy US Defense Mapping Agency Hydrographic/ Topographic Center).

10.6 Omega tables

Tables are published (Figure 10.15) which give the co-ordinates through which Omega LOP pass (Figure 10.16). From these tables LOP can be hand drawn onto any navigation chart.

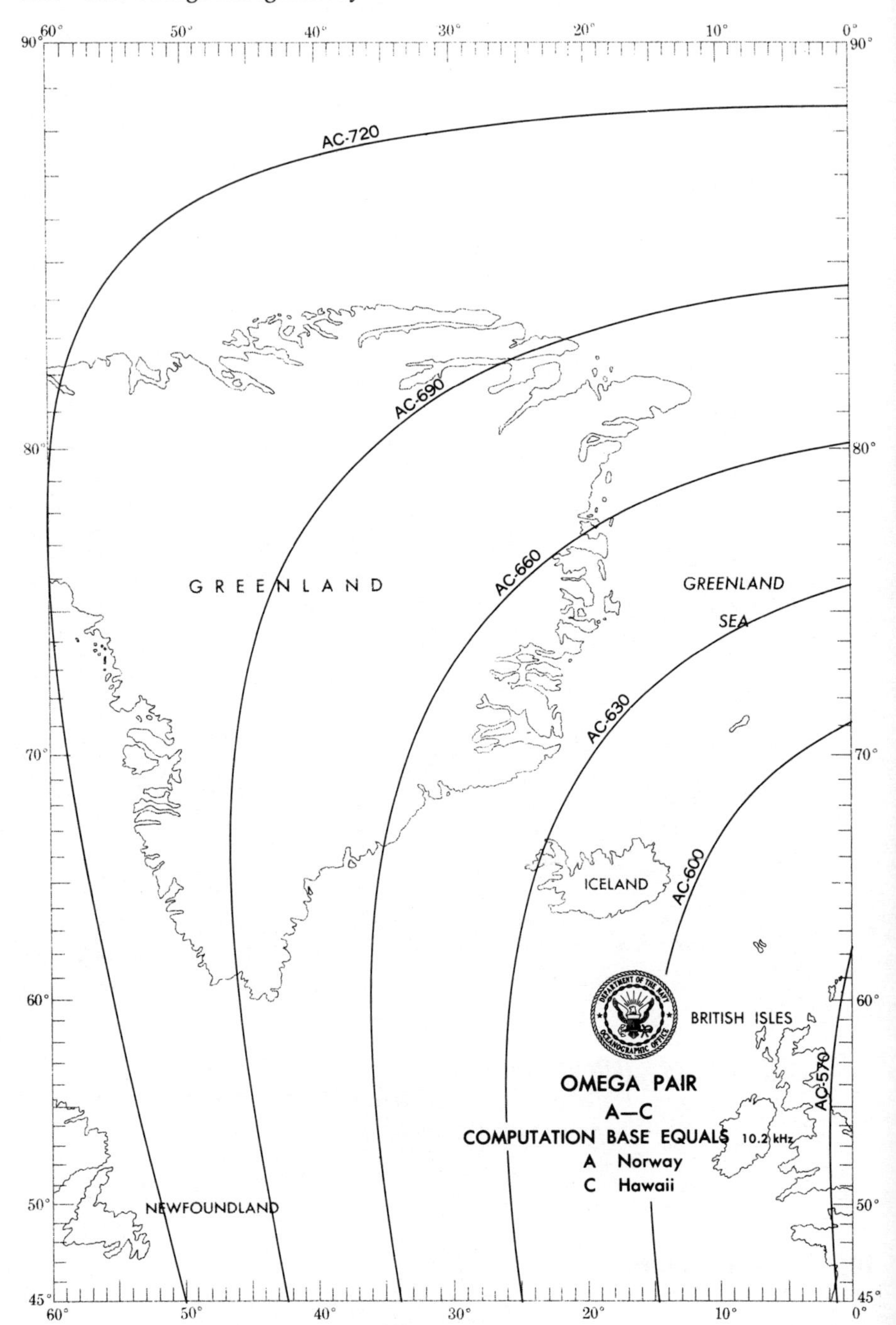

Figure 10.15 *Each volume of Omega plotting tables contains an illustration of the area covered by the tables (courtesy US Defense Mapping Agency Hydrographic/Topographic Center).*

19

T	A-C 650		A-C 651		A-C 652		A-C 653		A-C 654		T
Lat ° ′	° ′	Δ	° ′	Δ	° ′	Δ	° ′	Δ	° ′	Δ	Long ° ′
45 N	31 00.7W	175	31 18.2W	174	31 35.5W	173	31 52.9W	173	32 10.2W	172	
46 N	31 12.8W	175	31 30.3W	175	31 47.9W	175	32 05.3W	174	32 22.8W	174	
47 N	31 24.4W	177	31 42.2W	176	31 59.8W	175	32 17.3W	175	32 34.9W	175	
48 N	31 35.6W	178	31 53.4W	177	32 11.2W	177	32 28.9W	177	32 46.6W	176	
49 N	31 46.1W	179	32 04.1W	179	32 22.0W	178	32 39.8W	178	32 57.6W	178	
50 N	31 56.1W	181	32 14.1W	180	32 32.2W	180	32 50.2W	179	33 08.2W	179	
51 N	32 05.3W	182	32 23.5W	182	32 41.7W	181	32 59.9W	181	33 18.0W	180	
52 N	32 13.8W	184	32 32.1W	183	32 50.5W	183	33 08.8W	183	33 27.1W	182	
53 N	32 21.3W	186	32 39.9W	185	32 58.4W	185	33 16.9W	184	33 35.4W	184	
54 N	32 27.9W	188	32 46.7W	187	33 05.4W	187	33 24.1W	187	33 42.8W	186	
55 N	32 33.4W	190	32 52.4W	189	33 11.3W	188	33 30.2W	188	33 49.0W	188	
56 N	32 37.7W	192	32 56.9W	191	33 16.0W	191	33 35.1W	191	33 54.3W	190	
57 N	32 40.6W	194	33 00.0W	194	33 19.4W	193	33 38.8W	193	33 58.1W	193	
58 N	32 41.9W	197	33 01.6W	197	33 21.3W	196	33 40.9W	196	34 00.5W	195	
59 N	32 41.6W	199	33 01.5W	199	33 21.4W	199	33 41.3W	199	34 01.2W	199	
60 N	32 39.0W	203	32 59.3W	203	33 19.7W	202	33 39.8W	201	34 00.0W	201	
61 N	32 34.4W	206	32 55.0W	205	33 15.5W	205	33 36.1W	205	33 56.7W	205	
62 N	32 27.0W	210	32 48.0W	209	33 08.9W	209	33 29.9W	209	33 50.9W	209	
63 N	32 16.5W	215	32 38.1W	214	32 59.5W	214	33 20.9W	214	33 42.3W	213	
64 N	32 02.7W	219	32 24.7W	219	32 46.6W	219	33 08.6W	218	33 30.4W	218	
65 N	31 44.9W	225	32 07.4W	224	32 29.8W	223	32 52.2W	223	33 14.6W	223	
66 N	31 22.3W	232	31 45.5W	231	32 08.5W	230	32 31.6W	230	32 54.6W	230	
67 N	30 54.3W	239	31 18.2W	238	31 41.9W	237	32 05.7W	237	32 29.4W	237	
68 N	30 19.7W	247	30 44.4W	246	31 09.1W	246	31 33.7W	245	31 58.3W	245	
69 N	29 37.4W	257	30 03.1W	256	30 28.8W	256	30 54.3W	255	31 19.9W	255	
70 N	28 45.9W	268	29 12.7W	268	29 39.5W	267	30 06.2W	267	30 32.9W	267	
71 N	27 43.0W	283	28 11.3W	282	28 39.4W	281	29 07.6W	280	29 35.6W	279	
72 N	26 26.3W	300	26 56.1W	298	27 26.0W	298	27 55.9W	297	28 25.5W	296	
73 N	24 51.8W	322	25 23.9W	321	25 56.1W	319	26 27.8W	317	26 59.5W	317	
74 N	22 54.8W	350	23 29.7W	348	24 04.5W	346	24 39.0W	344	25 13.4W	342	
75 N	20 27.8W	388	21 06.4W	385	21 44.8W	382	22 22.9W	379	23 00.7W	376	
76 N	17 18.7W	443	18 02.8W	438	18 46.3W	433	19 29.4W	428	20 12.0W	423	
77 N	13 06.5W		13 59.1W		14 50.8W		15 41.4W		16 31.3W	494	
	76 48.8N	110	76 59.8N	110	77 10.8N	109	77 21.7N	108	77 32.6N	108	14 W
	77 13.0N	104	77 23.5N	104	77 33.9N	104	77 44.3N	103	77 54.6N	102	12 W
	77 34.1N	100	77 44.1N	99	77 54.0N	99	78 04.0N	99	78 13.8N	98	10 W
	77 52.4N	96	78 02.0N	95	78 11.6N	95	78 21.2N	95	78 30.7N	95	8 W
	78 08.5N	92	78 17.7N	92	78 27.0N	92	78 36.2N	92	78 45.4N	92	6 W
	78 22.4N	90	78 31.4N	89	78 40.4N	89	78 49.4N	89	78 58.3N	89	4 W
	78 34.6N	87	78 43.3N	87	78 52.1N	87	79 00.8N	87	79 09.5N	87	2 W
	78 45.0N	86	78 53.7N	85	79 02.2N	85	79 10.7N	85	79 19.3N	85	0
T	A-C 650		A-C 651		A-C 652		A-C 653		A-C 654		T

Figure 10.16 An extract from a volume of Omega plotting tables.

10.7 Omega receivers

Marine Omega receivers range in complexity from the simple manually synchronized two LOP receiver, to the sophisticated fully automatic equipment, displaying positions directly as latitude and longitude. This section discusses the more important aspects of these receivers.

10.7.1 *Phase difference measurement*

Since the transmissions in the Omega system are sequential, the usual method of comparing the phase of any two transmissions to produce a phase difference is to compare the phase of each transmission in turn with an internal reference oscillator. From each of these comparisons the difference in phase of any two transmissions can be derived.

For example, to produce the centilane number for the LOP A-B:

- During the transmission time of station A, the phase of the reference oscillator is subtracted from the phase of the A signal: $\phi A - \phi$ ref. The resultant is stored, usually in a digital form.
- During the transmission time of station B, $\phi B - \phi$ ref. is derived and stored.
- At any time the phase difference of A − B can now be derived, since $\phi A - \phi B = (\phi A - \phi \text{ ref}) - (\phi B - \phi \text{ ref})$.
- The resultant can be converted to centilanes and displayed.

10.7.2 *The reference oscillator*

By present-day standards of crystal oscillators and of atomic standards, the reference oscillator in an Omega receiver need be of only very modest performance. In the hyperbolic mode it is only required to hold a given stability (for the specified receiver performance) for the period between A transmission and H transmission, which is less than ten seconds. Long-term stability can be achieved by using automatic frequency control (AFC), derived from one or more of the Omega transmissions.

10.7.3 *Lane counting*

As described in the basic Omega system description (section 10.1), the measured phase difference between two received signals produces a position line within a lane. The lane width at the basic navigation frequency of 10.2 kHz is approximately eight nautical miles on the baseline. An absolute line of position (LOP) is therefore a composite of the large number and the phase difference in centilanes. The lane numbers are initially adjusted to be the correct values when the vessel is at a known location. After this the lane counters follow the centilane numbers, incrementing after 99 cel or decrementing after 00 cel, depending on the direction of travel across the lane.

At any time, a lane number may become incorrect through the occurrence of a lane slip (see section 10.2, Lane ambiguity). In this situation the same front panel control which is used to initially set the lane numbers would be used to adjust the number back to its correct value.

10.7.4 Display of LOP

As a minimum requirement two LOP are necessary to derive a position, and so a basic receiver must be able to produce two LOP from four stations. (Three stations can produce two LOP, but not necessarily with the best geometry.) The basic receiver has the facility for the user to select the LOP to be tracked, after consulting Omega charts and signal coverage diagrams.

An instrument which displays only two LOP does not make full use of the Omega system, since signal availability will often allow use of five or more LOP, and up to five may be shown on the Omega chart. A greater level of confidence will be obtained from an Omega receiver which provides three or more LOP, as it then becomes possible to identify signals affected by propagational anomalies.

While some Omega receivers can only simultaneously track and display two or three, others can track and display all the twenty-eight possible LOP, but of course of these only a few will be usable at any given time.

10.8 Synchronization

Before the phase of any Omega signal can be measured against the reference oscillator, the receiver must first identify the signal with respect to its origin. For example, at any point on the earth's surface usable signals may be received from typically five or six transmitters, albeit with widely varying amplitudes (Figure 10.17). The task of the receiver is to assign the appropriate designation letters, A to H, to

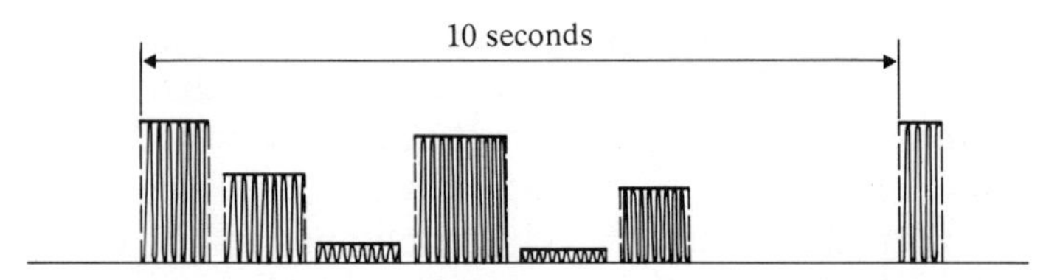

Figure 10.17 *A ten-second interval of received Omega signals illustrating the wide variation in amplitude; until synchronization is achieved the identity of each received signal is unknown.*

each one-second (approximate) burst of signal as it arrives at the receiver. This it must do without there being any unique coding assigned to each of the eight transmissions. This process is termed 'synchronization' and is clearly one which must always be performed correctly by the receiver, since incorrect synchronization would result in the displayed LOP not being those selected by the user.

There are a number of ways in which synchronization can be achieved; those methods which require intervention from the user are generally termed manual synchronization. More sophisticated receivers carry out the whole process unaided, and this is usually referred to as automatic synchronization. Three different methods of synchronization are discussed here, two of which are manual, the other being automatic. There are other methods, but in general they can be considered as variations on these basic three.

Irrespective of the method of achieving synchronization all receivers usually generate a continuous internal commutation pattern. This is a facsimile of the Omega transmission pattern (Figure 10.18), with the receiver knowing the identity of each station represented by the internal pattern as it occurs. The aim in any synchronization process is to identify a point (or points) in the real (received) Omega transmission sequence, and having done this to then bring the commutator sequence into step (synchronization) with it. From this point in time onwards, the Omega signal which is being received at any given time can be identified, since the identity of the corresponding segment of the commutator pattern is known.

10.8.1 *Time synchronization*

Use can be made of the very precise timing of the ten-second Omega sequence. Fundamentally the system is timed such that the

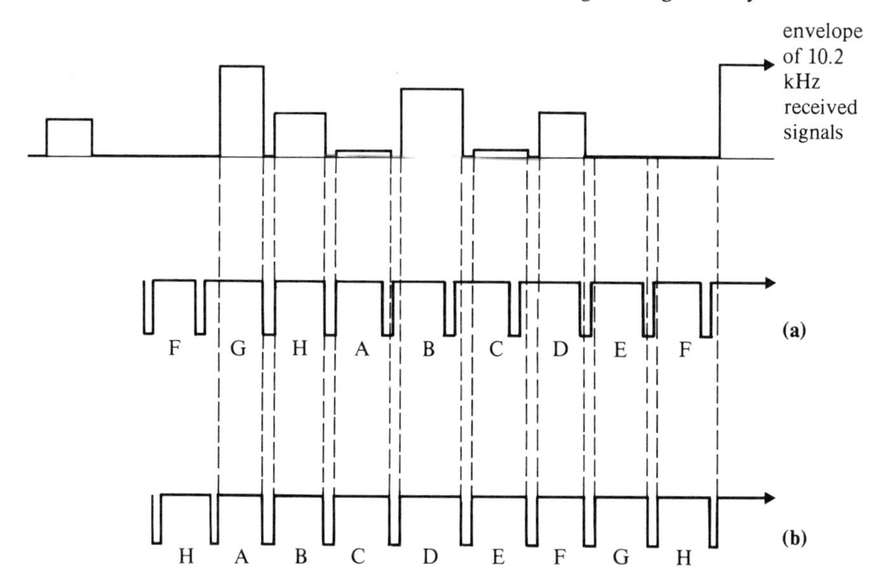

Figure 10.18 *The Omega receiver's internal commutation pattern:* (a) *incorrectly synchronized with the received signal;* (b) *correctly synchronized.*

start of the Omega cycle (start of station A transmission) is on the 00-, 10-, 20, 30-, 40-, and 50-second mark of every minute. However Omega time is not synchronized with Co-ordinated Universal Time (UTC), meaning that the leap second adjustments which are periodically made to UTC are not made to Omega time. At the time of writing (September 1978) this difference is seven seconds; the start of station A transmission thus occurs at 07, 14, and 21 seconds past the minute of UTC.

Synchronization of an Omega receiver can be achieved by referring to a chronometer or radio time signal, and at any of the precise moments in time when station A commences transmission, the operator activates the synchronizing switch of the Omega receiver. This will interrupt the internal commutator sequence and cause it to start at the beginning of the cycle. Henceforth the commutator will be synchronized with the Omega cycle, and the signal received at any time can be identified by reference to the commutator.

10.8.2 *Synchronization to a specified station*

The general approach of this type of synchronization is for the user to determine which station is currently his closest, and is thus likely to produce a strong received signal. Using a station selector control he sets the identity of this station. For example, in the case of a ship close to the west coast of Africa the control would be set to station B (Liberia). The synchronizing circuits would then examine the amplitude and length of each transmission as it is received, until it finds a large amplitude signal of precisely 1.0 second in length. There will always be one other transmission with the same length as that selected, but it should always be of significantly lower amplitude – in this case the other 1.0 second transmission is from station H, Japan. Since the synchronizing circuits have now identified a specific point in the Omega cycle, the commutator can be interrupted and brought into synchronization.

10.8.3 *Automatic synchronization*

In general automatic synchronization is taken to be the facility of an Omega receiver to bring its commutator into line with the Omega sequence without any operator initialization or intervention. To do this, use is usually made of the unique sequence of transmission lengths. The precise means of utilizing this unique quality are numerous.

One method might be to measure the transmission length of any two consecutive received signals. For example, a 1.2 second transmission followed by a 1.1 second transmission uniquely identifies stations D and E. The commutator can then be brought into synchronization.

More sophisticated methods of automatic synchronization continuously cross-correlate the commutator with the received signal while adjusting the timing of the commutator until synchronization is achieved.

In the various descriptions of synchronization methods it has been assumed that each would be carried out on signals received on the basic 10.2 kHz navigation frequency. Omega receivers which receive one or more of the secondary frequencies may well use these in addition to the basic frequency in their automatic synchronization circuits, since this extra information reduces the risk of a correlation error.

10.8.4 *Lane ambiguity*

A large error (greater than eight nautical miles) can occur in an Omega position if the lane count becomes incorrect, and the lane number of one or more of the displayed LOP is in error by one or more lanes. An error of this magnitude is usually shown up by regular plotting; nevertheless a means of determining a lane count error can be an advantage, and under some circumstances essential.

The use of the additional Omega frequencies (11.05 kHz, 11.33 kHz and 13.6 kHz) to resolve lane ambiguities by creating broader lanes, has previously been explained (section 10.2). It is unusual for a marine Omega receiver to utilize other than the first stage of lane identification, which is the use of the 13.6 kHz signal to produce the 3.4 kHz lanes. The precise form in which the lane identification information is displayed varies with different manufacturers' receivers. Suffice to say that information is presented which allows the user to determine whether the lane number is correct.

10.8.5 *The chart recorder*

Many marine Omega receivers incorporate a chart recorder which can be used for a number of purposes, but is primarily used to verify the displayed lane numbers. The chart recorder records the centilanes of each LOP selected, so each lane traversed by the vessel is then clearly shown on this recording (Figure 10.19). It is usual to display the previous four (or more) hours of recording, and for the navigating officers to regularly annotate the chart with a running total of the lanes traversed. At all times these running totals should agree with the displayed lane counts.

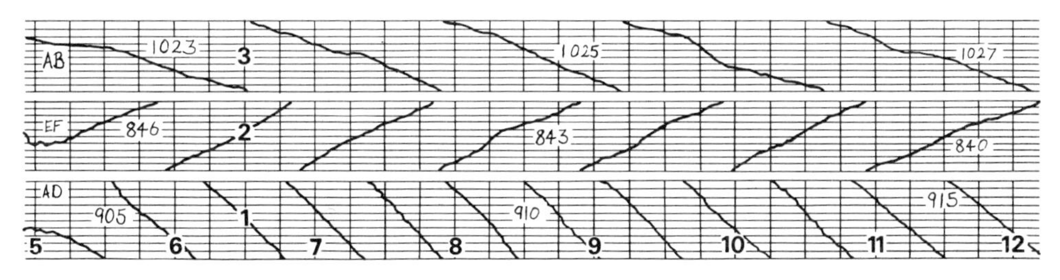

Figure 10.19 *Typical Omega receiver chart recording; the annotations assist in the identification of lane count errors.*

10.9 Latitude-longitude Omega receivers

To remove the requirement for the special Omega plotting charts, the LOP representing a vessel's position can be converted directly into a position displayed as a latitude and longitude, by the more sophisticated Omega receivers which have an inbuilt computing capability. The algorithms are similar to those used for the conversion of Loran C positions from hyperbolic to geographical co-ordinates, but in the Omega case it is only necessary to hold in the computer memory the co-ordinates of the eight Omega transmitting stations – compared with the greater number of Loran C stations. This is a simplification in favour of Omega but the need to compute the appropriate propagation corrections is an added complication over Loran C. As a basis for computing the propagation corrections it is necessary to store a conductivity model of the earth's surface within the computer.

The amount of data which is used within the receiver to produce the position fix varies from one instrument to another. At one extreme, a simple latitude-longitude receiver may convert two LOP into a position, whereas the most sophisticated receiver will use many LOP derived from all usable signals. In addition, such a receiver may use signals received on one or more of the other Omega frequencies to increase the accuracy of the fix, and to reduce the possibility of lane slip. Such a receiver may also automatically ignore signals which for the vessel's current position could be suffering from effects such as modal and long path interference.

10.10 Using the Omega navigation system

Completion of the eighth Omega transmitting station in Australia will give users of the Omega system a global navigation capability. The potential accuracy of the system will give position fixes of two nautical miles (95 per cent of the time) during day-time, and four nautical miles during night-time. This potential accuracy may not be fully achieved until all the planned monitoring stations are in operation, and have provided data for inclusion in the published propagation corrections. Notwithstanding this, large areas of the world have Omega coverage which is providing a position fix accuracy more than adequate for ocean navigation. The following sections describe the basic aspects of gaining optimum use from the Omega system.

10.10.1 *Initial procedures*

After switching on an Omega receiver, the first step is to synchronize the receiver with the Omega transmission cycle, or, if this is performed automatically, to establish that it has occurred correctly. The precise means of achieving this will be described in the particular equipment manual, which should always be read carefully before attempting to use an unfamiliar instrument.

10.10.2 *LOP selection*

The choice of LOP to use at the commencement of a voyage may be limited to the four or five printed on the chart, or limited to two or three by the capability of the receiver. In the case where the number of LOP is limited by the receiver it is important that the optimum choice is made. This choice is made by considering all the LOP reproduced on the chart and then eliminating those which contain

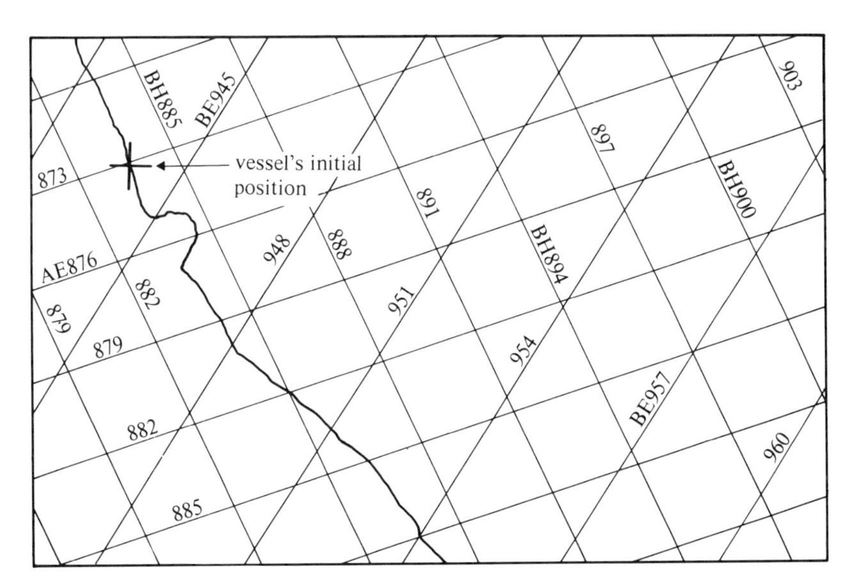

Figure 10.20 To set up an Omega receiver the lane numbers of the selected LOP are determined for the vessel's current position.

any stations whose signals may be suffering effects such as modal interference, long path interference, polar cap absorption, or an imminent scheduled maintenance period. Other criteria to use in arriving at the optimum choice of LOP are received signal strength and hyperbolic lane expansion. To assist in station selection, the US Defense Mapping Agency Hydrographic Center publishes charts of field strengths and areas of modal interference. In the case where the receiver can track all possible LOP, as many as possible should be used for the fix, but the above factors should be considered when determining the most probable position within the plotted position lines.

10.10.3 *The initial lane count*

After the receiver has synchronized and the selection of LOP is made, the displayed LOP numbers will have correct centilane values (as determined by the phase difference measurements), but at this time the lane numbers will be purely arbitrary. Therefore it is necessary to determine the appropriate lane numbers for the vessel's present position. The Omega co-ordinates for the plotted position in Figure 10.20 are:

LOP	AE	BE	BH
(1)	873.10	944.00	833.60

Before the receiver's lane numbers are adjusted, these charted LOP must be converted to the values which the receiver would be expected to be displaying at this moment in time. That is, by applying the published propagation corrections in reverse, by subtracting the appropriate correction from each station. For example:

Station	A	B	E	H
PPC	−90	−25	−32	−74
Correction to apply	− −58(A–E)	+7(B–E)	+49(B–H)	
LOP	AE	BE	BH	
Subtracting from charted LOP (1) gives expected Omega readings	873.68	943.93	883.11	

At this time actual Omega readings might typically be	000.81	000.02	000.09

(It is assumed that lane numbers are set at zero at switch-on)
The lane numbers are now adjusted to give the correct lane counts:

873.81 944.02 883.09

Note that the correct lane number for BE is 944 not 943, since the reading is obviously nine centilanes high and not ninety-one centilanes low.

It is good practice to now plot these LOP to obtain a 'feel' for the accuracy of each of the position lines.

.10.4 *Propagation corrections*

An important part of the procedure for plotting an Omega position is that of applying the appropriate corrections from the published propagation correction tables (Figure 10.11). Each volume contains the corrections for one station, for a specific area (Figure 10.21). Each page of the volume contains the corrections for two subdivisions of that area (Figure 10.22). In this illustration the upper table is for an area centred 52.0 E and the lower table is centred on 48.0 E.

When plotting a position a volume of correction tables is required for each station utilized. A vessel in area 3 (shaded) of Figure 10.21 may be using the LOP, BE, EF, and AF, and so correction tables will be required for this area for stations A, B, E and F. For example:

To plot the position line BE at 1600 gmt on 14 August.

The BE LOP reading is noted: 1056.27.

To this must be added: B correction − E correction.

By reference to time and date, B correction is found to be −53 (Figure 10.22); from this is subtracted the correction for station E, which is found to be −32 in the corresponding volume of corrections for E.

In this example the total correction to apply is:

$$\begin{aligned} B - E &= (-53) - (-32) \\ &= -21 \end{aligned}$$

The corrected LOP is therefore 1056.27 + (−21) = 1056.06.

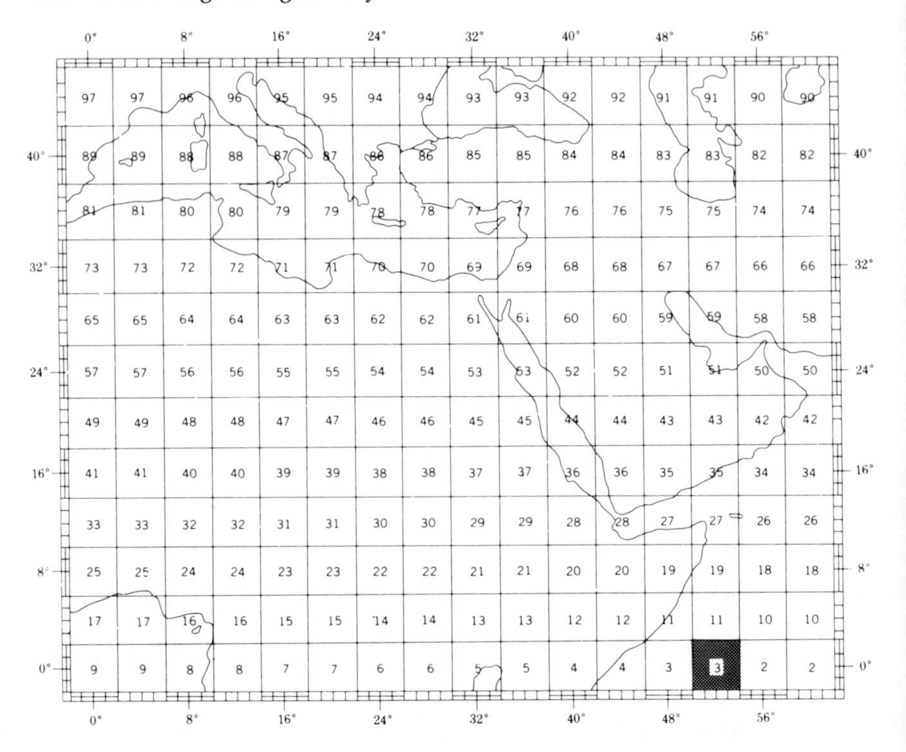

Figure 10.21 The area covered by one volume of Omega propagation corrections; the numbers refer to pages (courtesy US Defense Mapping Agency Hydrographic/Topographic Center).

In looking along the lines of corrections (Figure 10.22), the effect of the changing height of the ionosphere can clearly be seen in the rapidly changing values at sunrise and sunset. It is at these times when the corrections are likely to be at their least accurate, since any error in predicting the precise timing of the effects of sunrise and sunset can produce a large error. Considering the example of Figure 10.23, at 0200 gmt a correction of −19 cec should be applied according to the correction tables, whereas the appropriate amount of correction in this case is actually −70 cec. The resulting 51 cec error caused by applying the wrong amount of correction would produce an error in position of the LOP of at least four nautical miles.

The sunrise and sunset periods of each signal path should wherever possible be avoided, particularly when the effect is rapid, as in the case of the sunset transition of Figure 10.23.

10.10.5 Plotting a fix

The plotting of an Omega position is basically a three-stage process: (1) note the LOP readings; (2) apply the corrections; (3) plot the position. To facilitate the first two, the use of a work sheet is recommended (Figure 10.24).

To plot each LOP requires the use of an interpolator (Figure 10.25). This is divided into three lanes, since it is usual for only every third lane to be printed onto Omega plotting charts.

The prudent navigator using the Omega system should always be

10.2 KHZ OMEGA PROPAGATION CORRECTIONS IN UNITS OF CECS — LOCATION .0 52.0 E — STATION B LIBERIA

DATE		00	01	02	03	04	05	06	07	08	09	10	GMT 11	12	13	14	15	16	17	18	19	20	21	22	23	24
1-15	JAN	-106	-107	-107	-90	-70	-52	-34	-24	-26	-23	-21	-20	-21	-23	-27	-38	-55	-74	-91	-101	-105	-106	-107	-107	-106
16-31	JAN	-106	-107	-107	-92	-72	-54	-35	-24	-26	-23	-21	-20	-21	-23	-26	-36	-53	-72	-89	-101	-105	-106	-107	-107	-106
1-14	FEB	-106	-107	-107	-92	-72	-54	-36	-24	-25	-22	-20	-19	-20	-22	-25	-35	-52	-70	-88	-100	-104	-106	-107	-107	-106
15-29	FEB	-106	-107	-107	-92	-72	-54	-33	-24	-25	-22	-19	-19	-19	-21	-25	-35	-52	-70	-88	-100	-104	-106	-107	-107	-106
1-15	MAR	-106	-107	-107	-92	-69	-52	-33	-24	-24	-21	-19	-18	-19	-21	-25	-35	-52	-70	-88	-100	-104	-106	-107	-107	-106
16-31	MAR	-106	-107	-107	-90	-69	-50	-31	-24	-24	-21	-19	-18	-19	-21	-25	-36	-53	-71	-88	-100	-104	-106	-107	-107	-106
1-15	APR	-106	-107	-107	-87	-66	-48	-27	-25	-24	-21	-19	-18	-19	-21	-26	-37	-54	-72	-89	-101	-105	-106	-107	-107	-106
16-30	APR	-106	-107	-107	-85	-64	-47	-26	-25	-24	-21	-19	-18	-19	-22	-26	-38	-55	-72	-90	-101	-105	-106	-107	-107	-106
1-15	MAY	-106	-107	-107	-84	-64	-45	-26	-25	-24	-21	-19	-19	-20	-22	-27	-39	-55	-73	-90	-101	-105	-106	-107	-107	-106
16-31	MAY	-106	-107	-106	-84	-64	-44	-25	-26	-24	-21	-19	-19	-20	-22	-27	-39	-55	-73	-89	-100	-105	-106	-107	-107	-106
1-15	JUN	-106	-107	-107	-84	-64	-45	-26	-26	-25	-21	-20	-19	-20	-22	-27	-38	-55	-72	-89	-100	-104	-106	-107	-107	-106
16-30	JUN	-106	-107	-107	-86	-64	-46	-26	-26	-25	-22	-20	-19	-20	-22	-26	-38	-54	-71	-88	-100	-104	-106	-107	-107	-106
1-15	JUL	-106	-107	-107	-87	-66	-47	-27	-26	-25	-22	-20	-19	-20	-22	-26	-37	-53	-70	-87	-99	-104	-106	-107	-107	-106
16-31	JUL	-106	-107	-107	-88	-66	-47	-28	-25	-25	-21	-20	-19	-20	-22	-26	-36	-52	-70	-87	-99	-104	-106	-107	-107	-106
1-15	AUG	-106	-107	-107	-88	-67	-48	-28	-25	-25	-21	-19	-19	-19	-22	-25	-36	-53	-70	-88	-99	-104	-106	-107	-107	-106
16-31	AUG	-106	-107	-107	-87	-66	-47	-28	-25	-24	-21	-19	-18	-19	-22	-26	-37	-54	-71	-89	-100	-104	-106	-107	-107	-106
1-15	SEP	-106	-107	-107	-85	-66	-46	-27	-25	-24	-20	-18	-18	-19	-22	-26	-38	-55	-73	-90	-101	-105	-106	-107	-107	-106
16-30	SEP	-106	-107	-106	-85	-64	-46	-26	-25	-23	-20	-18	-18	-19	-22	-27	-40	-57	-75	-92	-102	-105	-107	-107	-107	-106
1-15	OCT	-106	-107	-104	-83	-62	-45	-26	-24	-23	-20	-19	-18	-20	-22	-28	-42	-59	-77	-94	-102	-105	-107	-107	-107	-106
16-31	OCT	-106	-107	-104	-81	-63	-45	-24	-25	-24	-20	-19	-19	-20	-23	-29	-43	-61	-79	-95	-102	-105	-107	-107	-107	-106
1-15	NOV	-106	-107	-103	-81	-63	-45	-26	-25	-24	-21	-20	-20	-21	-23	-29	-44	-61	-80	-95	-103	-105	-107	-107	-107	-106
16-30	NOV	-106	-107	-104	-83	-64	-46	-27	-26	-25	-22	-20	-20	-21	-24	-29	-43	-61	-80	-95	-103	-105	-107	-107	-107	-106
1-15	DEC	-106	-107	-104	-85	-65	-48	-29	-26	-25	-22	-21	-20	-21	-24	-29	-42	-60	-78	-94	-102	-105	-107	-107	-107	-106
16-31	DEC	-106	-107	-107	-87	-68	-50	-31	-25	-26	-23	-21	-20	-21	-24	-28	-40	-58	-76	-93	-102	-105	-107	-107	-107	-106

LOCATION .0 48.0 E — STATION B LIBERIA

DATE		00	01	02	03	04	05	06	07	08	09	10	GMT 11	12	13	14	15	16	17	18	19	20	21	22	23	24
1-15	JAN	-101	-101	-102	-91	-69	-52	-33	-24	-25	-22	-20	-20	-20	-22	-25	-34	-50	-68	-85	-96	-99	-101	-102	-102	-101
16-31	JAN	-101	-101	-102	-93	-71	-52	-34	-24	-25	-22	-20	-19	-20	-21	-24	-32	-48	-66	-84	-95	-99	-101	-102	-102	-101
1-14	FEB	-101	-101	-102	-93	-72	-53	-34	-23	-25	-22	-20	-19	-19	-21	-24	-31	-47	-65	-83	-94	-99	-101	-102	-102	-101
15-29	FEB	-101	-101	-102	-93	-71	-51	-33	-23	-24	-21	-19	-18	-19	-20	-23	-31	-47	-65	-82	-94	-99	-101	-102	-102	-101
1-15	MAR	-101	-101	-102	-93	-69	-51	-30	-23	-24	-21	-19	-18	-18	-20	-23	-31	-47	-65	-83	-95	-99	-101	-102	-102	-101
16-31	MAR	-101	-101	-102	-91	-67	-50	-29	-24	-24	-20	-18	-18	-18	-20	-23	-32	-48	-66	-83	-95	-99	-101	-102	-102	-101
1-15	APR	-101	-101	-102	-89	-65	-48	-27	-24	-24	-20	-18	-18	-18	-20	-24	-33	-49	-67	-84	-95	-99	-101	-102	-102	-101
16-30	APR	-101	-101	-102	-86	-64	-45	-26	-24	-23	-20	-18	-18	-19	-21	-24	-34	-50	-67	-84	-95	-99	-101	-102	-102	-101
1-15	MAY	-101	-101	-102	-85	-63	-44	-25	-25	-24	-20	-19	-18	-19	-21	-24	-34	-50	-68	-84	-95	-99	-101	-102	-102	-101
16-31	MAY	-101	-101	-102	-85	-62	-44	-25	-25	-24	-21	-19	-18	-19	-21	-25	-34	-50	-67	-84	-95	-99	-101	-102	-102	-101
1-15	JUN	-101	-101	-102	-86	-63	-44	-25	-26	-24	-21	-19	-19	-19	-21	-25	-34	-50	-67	-83	-95	-99	-101	-102	-102	-101
16-30	JUN	-101	-101	-102	-87	-64	-45	-26	-25	-24	-21	-19	-19	-19	-21	-24	-33	-49	-66	-82	-94	-99	-101	-102	-102	-101
1-15	JUL	-101	-101	-102	-88	-66	-46	-27	-26	-24	-21	-19	-19	-19	-21	-24	-33	-48	-65	-82	-94	-99	-101	-102	-102	-101
16-31	JUL	-101	-101	-102	-89	-65	-47	-27	-25	-24	-21	-19	-18	-19	-21	-24	-32	-48	-65	-82	-94	-99	-101	-102	-102	-101
1-15	AUG	-101	-101	-102	-89	-66	-47	-27	-24	-24	-21	-19	-18	-19	-21	-24	-32	-48	-65	-82	-94	-99	-101	-102	-102	-101
16-31	AUG	-101	-101	-102	-89	-65	-46	-27	-24	-24	-20	-18	-18	-18	-20	-24	-33	-49	-66	-83	-95	-99	-101	-102	-102	-101
1-15	SEP	-101	-101	-102	-86	-65	-46	-25	-24	-23	-20	-18	-18	-18	-21	-24	-34	-50	-68	-85	-95	-99	-101	-102	-102	-101
16-30	SEP	-101	-101	-102	-84	-64	-43	-26	-24	-23	-20	-18	-18	-19	-21	-24	-35	-52	-70	-87	-96	-100	-101	-102	-102	-101
1-15	OCT	-101	-101	-102	-84	-62	-44	-23	-24	-23	-20	-18	-18	-19	-21	-25	-37	-54	-72	-88	-97	-100	-101	-102	-102	-101
16-31	OCT	-101	-101	-101	-82	-62	-42	-25	-25	-23	-20	-18	-18	-19	-22	-26	-38	-55	-74	-89	-97	-100	-101	-102	-102	-101
1-15	NOV	-101	-101	-101	-82	-62	-44	-24	-25	-24	-21	-19	-19	-20	-22	-27	-39	-56	-74	-90	-97	-100	-101	-102	-102	-101
16-30	NOV	-101	-101	-101	-82	-62	-45	-27	-25	-24	-21	-20	-19	-20	-23	-27	-38	-56	-74	-90	-97	-100	-101	-102	-102	-101
1-15	DEC	-101	-101	-102	-84	-65	-47	-28	-25	-25	-22	-20	-20	-21	-23	-26	-37	-55	-73	-89	-97	-100	-101	-102	-102	-101
16-31	DEC	-101	-101	-102	-87	-67	-49	-30	-25	-25	-22	-20	-20	-20	-22	-26	-36	-53	-71	-87	-96	-100	-101	-102	-102	-101

Figure 10.22 A typical page of Omega propagation corrections; each table corresponds to one square of Figure 10.21 (courtesy US Defense Mapping Agency Hydrographic/Topographic Center).

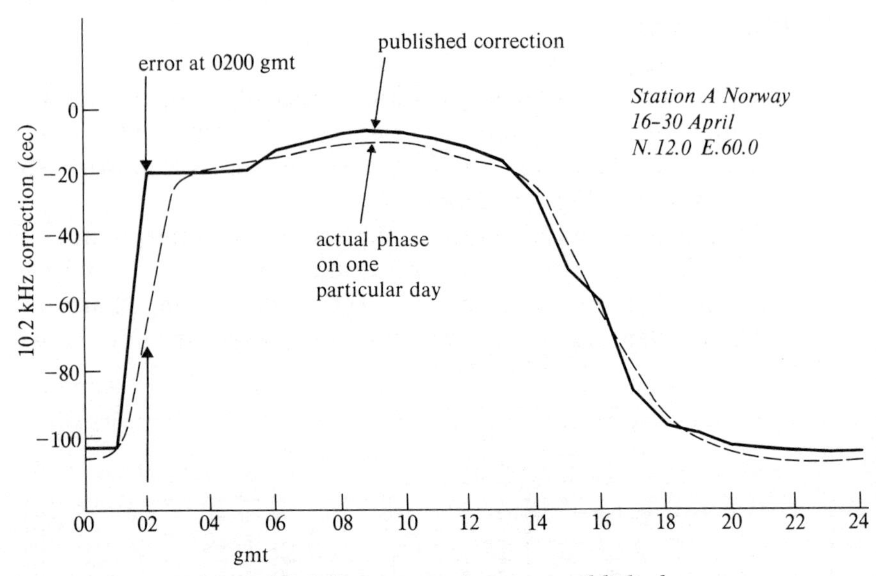

Figure 10.23 Example of discrepancy between published propagation corrections and observed phase. Greatest difference is often during sunrise and sunset periods.

Time	LOP	Correction			LOP	LOP	Corrections					Date 23 OCT. 77
(gmt)	1 2	1	2	1−2	(Reading)	(Corrected)	A	B	D	E		Comment
1300	DE	−12	−17	+5	636.41	636.46	−14	−11	−12	−17		
	BD	−11	−12	+1	834.73	834.74						
	AD	−14	−12	−2	757.96	757.94						
	AB	−14	−11	−3	823.22	823.19						
1400	DE	−15	−29	+14	668.89	669.03	−17	−12	−15	−29		
	BD	−12	−15	+3	835.97	836.00						
	AD	−17	−15	−2	756.88	756.86						
	AB	−17	−12	−5	820.90	820.85						
1500	DE	−13	−71	+58	702.17	702.75	−21	−14	−13	−71		
	BD	−14	−13	−1	835.28	835.27						
	AD	−21	−13	−8	756.77	756.69						
	AB	−21	−14	−7	821.41	821.34						
1600	DE	−13	−120	+107	733.32	734.39	−23	−17	−13	−120		
	BD	−17	−13	−4	836.52	836.48						
	AD	−23	−13	−10	755.76	755.66						
	AB	−23	−17	−6	819.24	819.18						

Figure 10.24 An extract from an Omega log book.

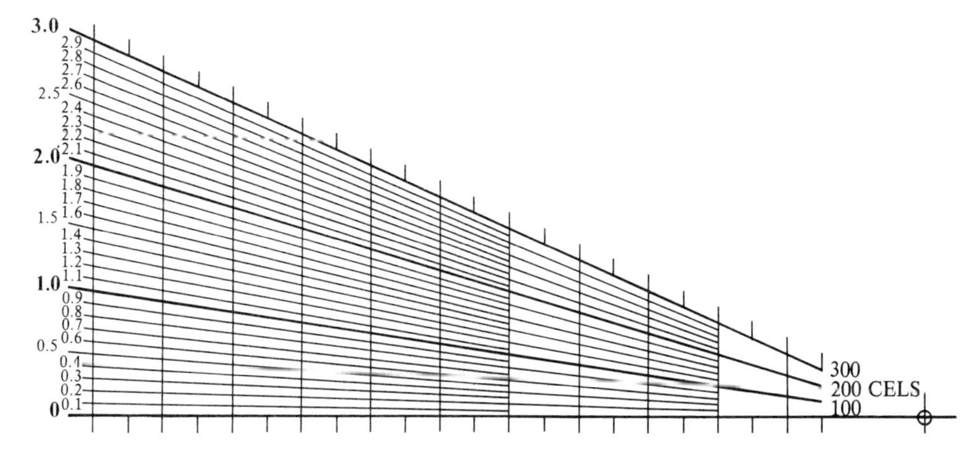

Figure 10.25 *An interpolator is required to plot position lines between the overprinted LOP (courtesy US Defense Mapping Agency Hydrographic/Topographic Center).*

aware of the following: (1) signals which are liable to suffer modal interference; (2) signals which are received via polar routes; (3) signals which are liable to suffer long path interference; (4) signals suffering rapid diurnal changes; (5) station off-air periods, scheduled and unscheduled.

10.10.6 *Lane slip*

The largest single cause of errors in Omega positions is an undetected lane slip, meaning that the lane count of one or more LOP has become incorrect by one or more lanes. As a result of a lane slip the lane count may be either higher or lower than the correct value. A lane width is eight nautical miles on the baseline, and so the position error caused by a single lane error will be at least this, and usually more.

The most common causes of lane slip are: (1) short unscheduled station off-air periods; (2) loss of signal due to high atmospheric noise; (3) modal interference; (4) long path interference. Once detected, a lane slip is easily remedied by using the appropriate front panel controls to increment or decrement the lane count as necessary.

It is of course important that lane slips do not go undetected, and many, if not most, will be observed through the regular plotting of

Omega positions since an eight-mile error accrued in one hour would clearly be seen.

10.10.7 *Use of coarse lane structure*

To assist in detecting and correcting for lane slip situations, some marine Omega receivers utilize the 13.6 kHz transmissions to produce the wider 3.4 kHz lanes from which a lane loss or gain can be detected. The precise form in which the 3.4 kHz information is presented varies between the different equipments, but in essence it must describe the position in the 3.4 kHz lane either in units of 3.4 kHz centilanes or 10.2 kHz centilanes, i.e. 0–99 cel or 0–2.99 cel. Taking the example illustrated by Figure 10.26, a vessel displays its AB LOP as 901.50 but the lane identification for AB is shown as 83 cel. This clearly shows a lane slip.

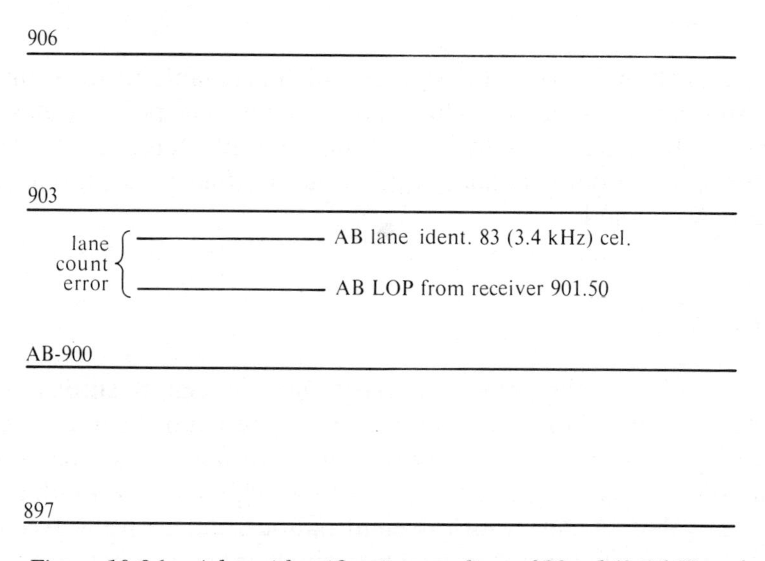

Figure 10.26 A lane identification reading of 83 cel (3.4 kHz cel units) clearly indicates a lane error in this example.

This comparison must always be with corrected readings; that is, the normal 10.2 kHz published propagation corrections are applied to the basic LOP, and a similar correction is applied to the lane identification reading. Separate volumes of 3.4 kHz corrections are published for this purpose.

Since published corrections are only predictions of phase deviations, it is unlikely that even after applying them the lane identification will give a position line precisely equivalent to the 10.2 kHz LOP (or be precisely one lane different). Nevertheless, there should be a general indication as to whether the lane number is correct, or in error by one lane.

The above scheme can only resolve errors of one lane. To resolve greater ambiguities, the transmissions of 11.33 kHz and 11.05 kHz are employed to produce lanes equivalent to the width of nine and thirty-six 10.2 kHz lanes, respectively. However it is unusual for these to be used in marine Omega receivers, since it is assumed that ambiguities of more than one lane can be resolved by conventional DR techniques.

10.10.8 *The chart recorder*

Many Omega receivers incorporate a chart recorder which continuously records the centilanes of each selected LOP, so that each lane crossed is clearly shown on the chart recorder. The chart should be annotated regularly with the identity of the LOP and running totals of the lane numbers. The behaviour of an Omega receiver during periods of lost signal may depend upon the particular characteristics of the receiver's tracking circuits, but in general loss of signal should clearly be seen on the chart recorder. In Figure 10.27 the centilane readings of one LOP become totally random during the off-air period of station B. The possibility of a lane slip can therefore be readily seen.

The chart recordings will indicate signals which become weak. It may also be possible to detect abnormal deviations such as those caused by a sudden ionospheric disturbance, although this is usually only possible when the Omega receiver is stationary.

10.11 Promulgation of information

In an endeavour to quickly reach all users of the Omega navigation system with information relating to the status of the Omega system, messages are broadcast by various time service stations. The US National Bureau of Standards Station WWV Boulder, Colorado, transmits a forty-seconds duration message, at sixteen minutes past the hour, and Kauai, Hawaii, broadcasts the same message at forty-seven minutes past the hour on 2.5, 5.0, 10, 15 and 20 MHz. Other

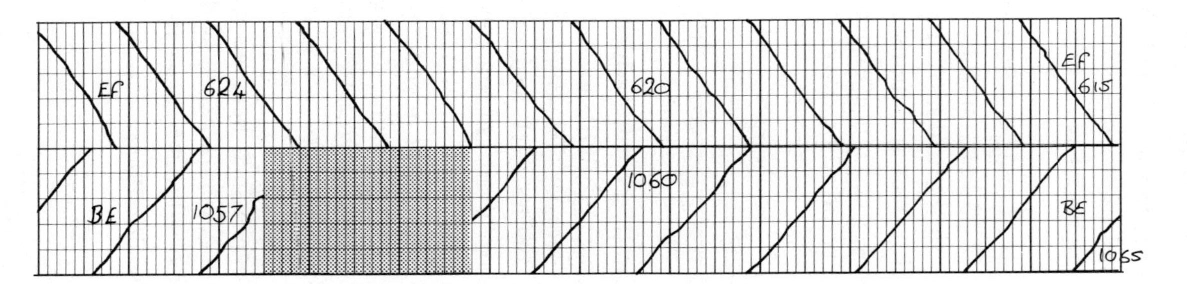

Figure 10.27 *The chart recording allows the correct lane count to be maintained after periods of signal loss.*

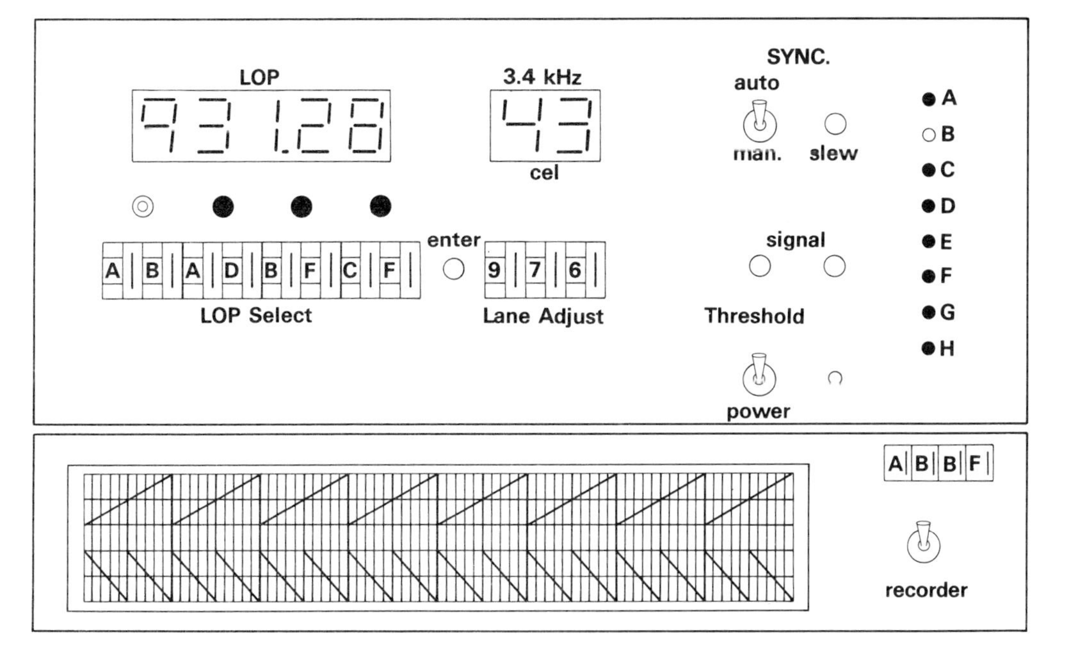

Figure 10.28 *A typical LOP Omega receiver.*

stations in the global time service and maritime information network are expected to commence similar transmissions.

10.12 Composite Omega

Composite Omega is a technique in which a composite signal is developed from measurements made at two or more of the Omega frequencies. The combination is made in such a way as to reduce the effect of diurnal ionospheric variations, and sudden ionospheric disturbances.

The underlying theory is that for certain waveguides, the product of group and phase velocity remains constant. VLF signals can be considered as being within the earth-ionosphere waveguide. To a first approximation therefore, the composite signal travels at a constant velocity midway between the phase and group velocities of the earth-ionosphere waveguide, and will be insensitive to moderate changes in height and conductivity of the reflecting layer.

There are many views regarding the optimum method of

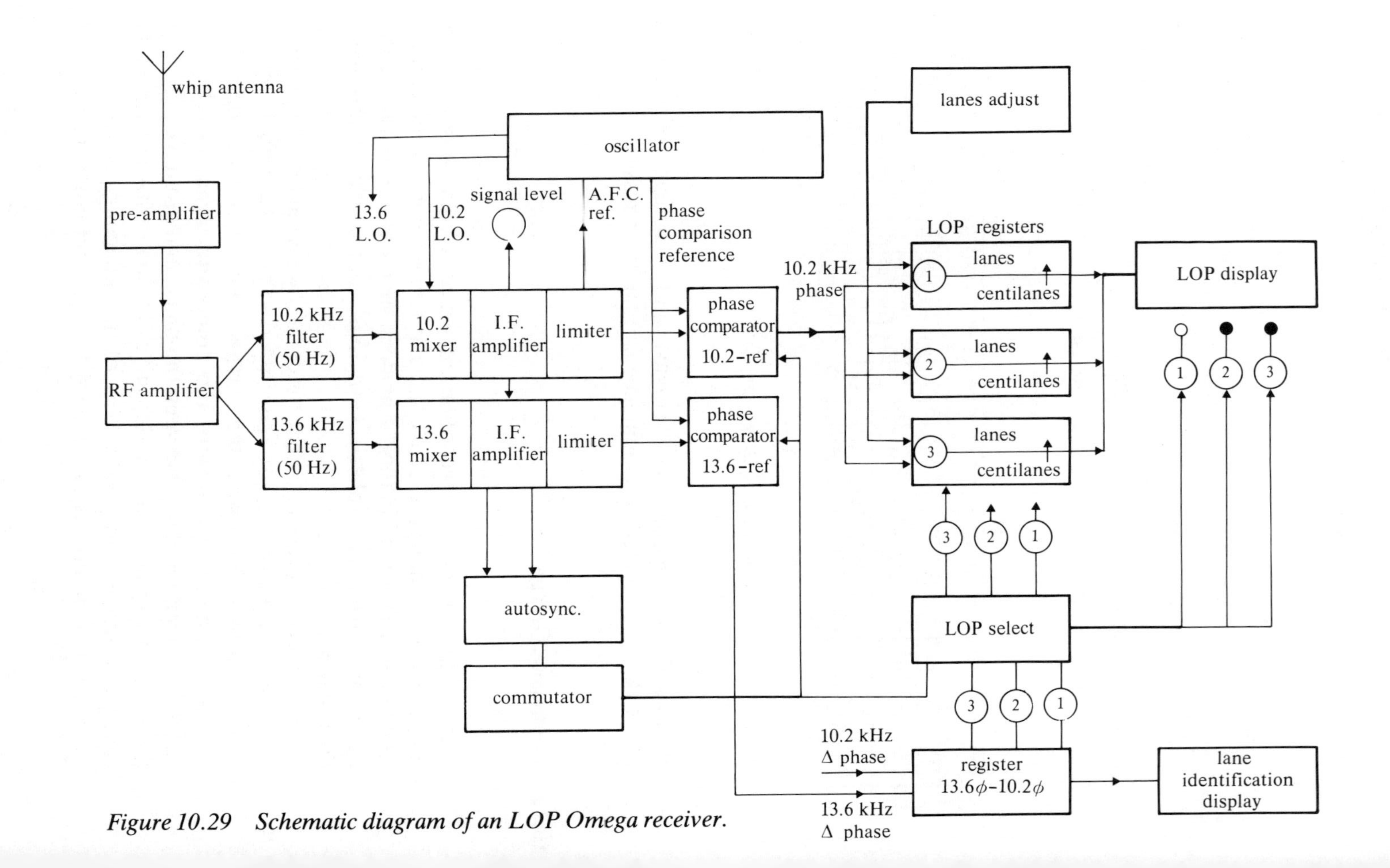

Figure 10.29 Schematic diagram of an LOP Omega receiver.

combining the signals and to the actual benefit gained from using composite Omega.

Considering the 10.2 kHz and 13.6 kHz Omega signals, the most general form of composite signal is defined by the linear equation:

$$\phi_c = A\phi_{13.3} + B\phi_{10.2}$$

where A and B are the weighting factors.

One specific form of this equation developed by Professor J. A. Pierce is:

$$\phi_c = \frac{27}{16}\phi_{13.6} - \frac{5}{4}\phi_{10.2} \quad (\phi_{13.6} \text{ has been normalized to } \phi_{10.2})$$

The composite technique has been incorporated into some automatic Omega receivers.

10.13 Differential Omega

Differential Omega is basically a means of improving the position-fixing accuracy of the Omega system through providing a measured correction to be used in place of the published propagation corrections.

The principles of Differential Omega operation are summarized in Figure 10.30. An Omega receiver located at a fixed monitor site produces uncorrected lines of position (LOP) for a number of selected station pairs. These are compared with corresponding LOP which have been computed for the site, based on the standard propagation velocity used in the production of Omega charts. The difference represents the error occurring at that moment in time, and so could effectively be used to correct the LOP of other Omega receivers in the immediate vicinity, since the Omega signals which they receive will have travelled by the same propagation paths and consequently will have been affected by the same disturbing influences.

Differential Omega as a navigation system has not yet been adopted on a wide scale, although its performance potential has been closely studied through numerous experimental projects.

computed LOP for monitor site	*LOP indicated by monitor*	*transmitted corrections*
EH 772.90	772.85	+0.15
AH 846.05	846.32	−0.27
BH 868.30	868.58	−0.28
AE 973.15	973.47	−0.32
BE 995.40	995.73	−0.33

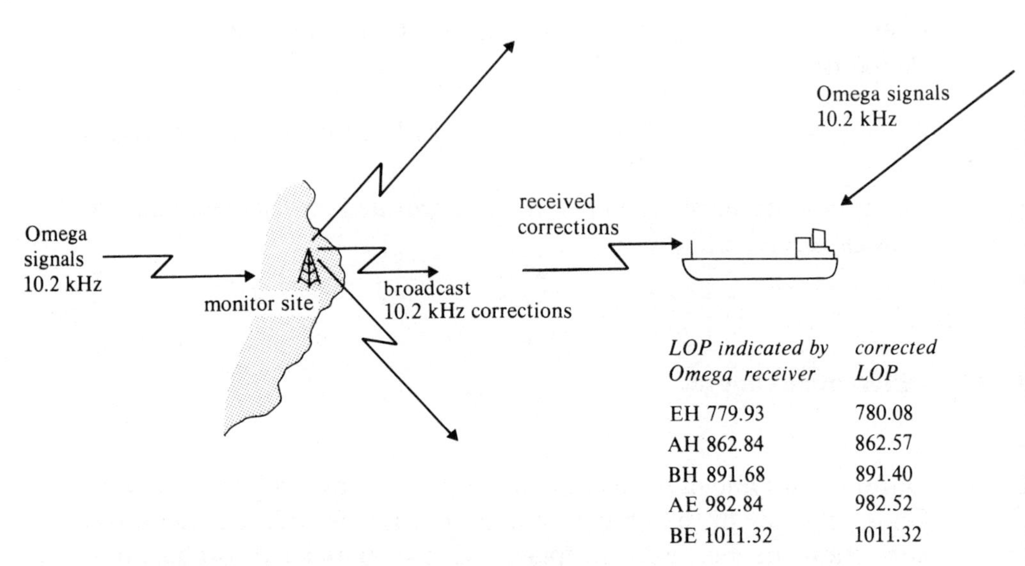

Figure 10.30 An example illustrating the principles of Differential Omega.

10.14 Performance

A Differential Omega user who is very close to the monitor site (within a few kilometres) will obtain the greatest benefit from the system, since the propagation paths will be almost identical. This is illustrated by the chart recordings in Figure 10.31, in which an LOP at the monitor site has been affected by a sudden ionospheric disturbance (Figure 10.31a), the effect of which is included in the transmitted corrections. An Omega receiver, located in this case thirty kilometres from the monitor site, is able to fully correct for the effect of the SID (Figure 10.31b). In this example the corrections were transmitted continuously (although this is not always possible) and so one area of investigation has been concerned with the allowable time lapse between successive

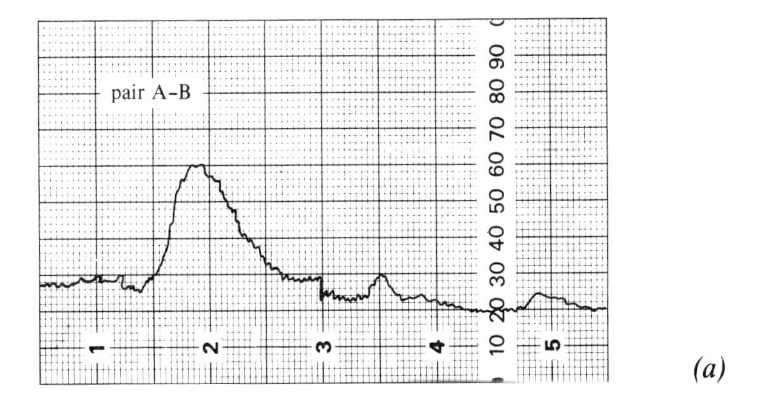

(a)

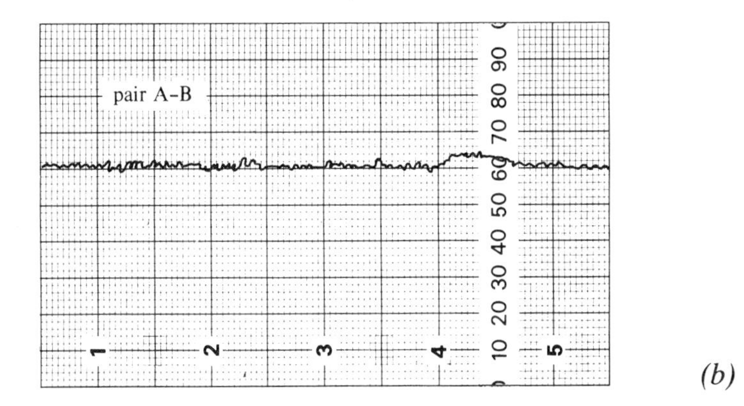

(b)

Figure 10.31 The effect of a sudden ionospheric disturbance (a) *can be smoothed by the application of Differential Omega corrections* (b).

correction transmissions. Considering the example of Figure 10.31, a transmission interval of ten minutes could have caused an error of twenty centilanes immediately prior to the transmission of the new correction.

The aspect of Differential Omega which is receiving the most attention is the relationship between the distance of the user from the monitor site and the corresponding performance improvement (over normal Omega operation). This is by no means a simple relationship, and the graph in Figure 10.32 is intended to give merely an indication of the order of accuracy which can be expected from Differential Omega. At a distance of 200–300 nautical miles the accuracy of the differential mode has deteriorated to that of basic Omega, using the published propagation corrections. Where

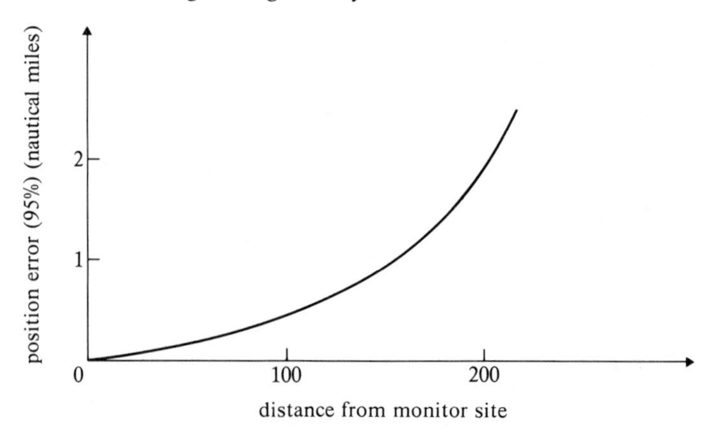

Figure 10.32 *An illustration of the order of accuracy improvement which can be expected from the application of Differential Omega techniques.*

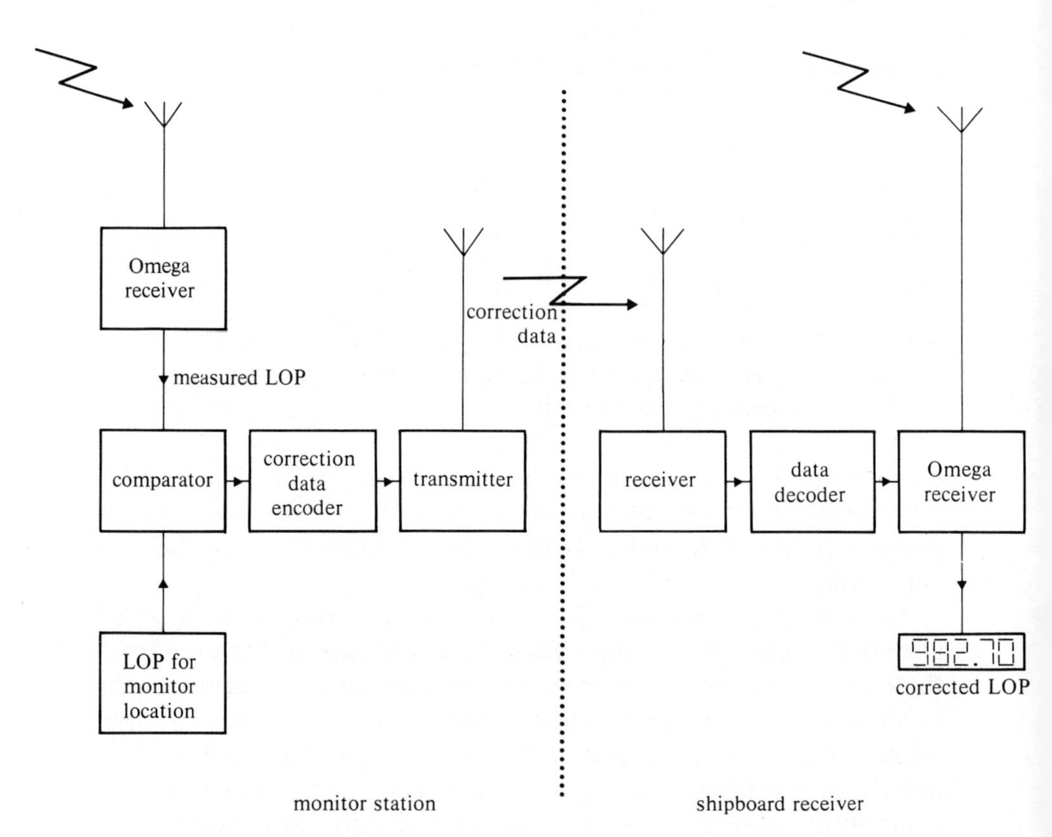

Figure 10.33 *Schematic illustration of a Differential Omega system.*

these longer distances are mainly in an east-west direction, a significant error can occur when the day/night transition passes between the monitor site and the user. During this time, the transmitted correction could reflect the fact that the Omega signal propagation path to the monitor site is almost totally in darkness, whereas the path to the user still remains sunlit.

One further consideration is the form of the transmitted corrections, in particular whether the broadcast data should be as individual station corrections, or corrections relating to selected station pairs (corresponding with the LOP overprinted on the local Omega chart). Before Differential Omega becomes an international system, the precise transmitted data sequence and modulation format must be agreed universally, and the transmissions must occur in a common frequency band.

The selected frquency band must be able to facilitate a continuous reception of corrections (irrespective of propagation conditions), up to a range of at least 100–200 nautical miles, if a useful landfall and coastal navigation capability is to be provided by Differential Omega. One suitable frequency band is that already occupied by the marine DF beacons (285–315 kHz), and the possibility exists that the DF beacons themselves could be used for the transmission of the corrections, provided that this does not interfere with their prime function.

11 The Navy Navigation Satellite System

The US Navy Navigation Satellite System (NNSS), alternatively known as Transit, was developed for the US Navy between 1958 and 1963 by the Applied Physics Laboratory of the Johns Hopkins University. It became available to non-military users in 1967. The system consists of several orbiting satellites and a network of tracking stations which monitor the satellites, and at intervals update their transmitted information. Use is made by the receiving apparatus of the observed Doppler shift of the satellite's transmission frequency to determine the relative velocity between the satellite and the receiver. From its transmitted data, a passing satellite's orbit is precisely determined. From this, and the rate of change of range to the satellite derived from the Doppler shift, the position of the receiver can be accurately computed.

Transit will ultimately be replaced by the newer GPS Navstar system (chapter 12). It will be maintained operational until after the full deployment of the GPS system and after the US Navy has been fully equipped with GPS receivers. It appears that thc end of Transit will occur in 1992.

11.1 The satellite system

The satellites are launched into orbits which are as nearly as possible circular and polar, and at a height of approximately seven hundred miles. Each satellite orbits the earth in approximately one hundred and seven minutes, continuously transmitting a two-minute message on each of two frequencies (150 MHz and 400 MHz). The message contains information which accurately defines the satellite's orbit. To precisely determine its present and future orbits each satellite is tracked as it passes within 'sight' of each of the four tracking stations. These tracking stations, which are located at Hawaii, California, Minnesota, and Maine, pass their monitoring data to a computing centre which determines the precise orbital shape. Then for the following twelve- to sixteen-hour period, the specific position of the satellite at two-minute intervals is computed. This data, along with time corrections, is passed to three injection stations located in California, Minnesota, and Maine. The injection stations store this information until the time of injection, when the antenna of the appropriate station locks-on to the satellite, and data is transmitted to it. The two minute transmission message of the satellite is then refreshed with the new orbital data.

1.1.1 *The satellites*

For the navigator and most other users of the Transit system the ground support organization is only of general background interest. Whereas an understanding of the satellite orbits is of more fundamental importance, since it is only when a satellite rises into view (radio line-of-sight) that a position fix can be made. At the time of writing (August 1985) there are six operational satellites:

Designation

30130
30140
30190
30200
30480
30500

Satellites 30130, 30140, 30190 and 30200 are of an older 1963 design referred to as OSCARS, they have been in operational service for periods ranging from 12–18 years. Satellites 30480 and 30500 are of a newer design designated NOVA. Among NOVA's improved features is drag compensation, where a suspended proof mass is used to detect the effect of atmospheric drag or solar radiation, which is then corrected for, in order to maintain a more predictable satellite trajectory.

If the orbits are perfectly polar they will be inert in space, and a symmetrical orbital configuration (Figure 11.1) could be maintained. The earth rotates within the orbits so that different satellites are seen throughout a twenty-four hour period. Satellite passes do not occur at uniform intervals, the average interval is a function of latitude, being at a maximum on the equator and a minimum at the poles.

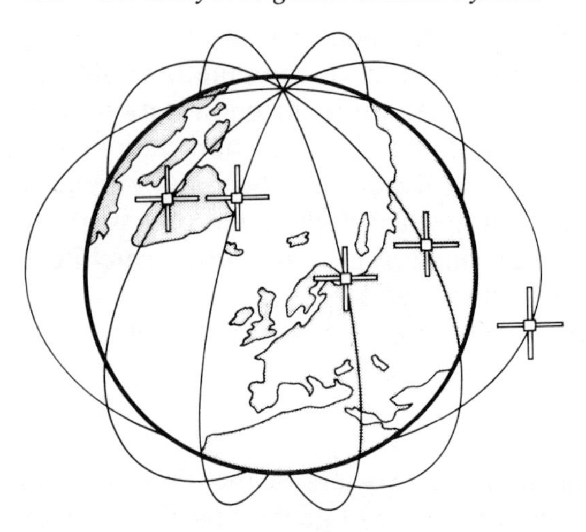

Figure 11.1 The ideal satellite orbit configuration.

In reality, satellites have not been launched into perfect polar orbits, and as a result each orbit is precessing by an amount dependent on the inaccuracy of the launch. The situation that can arise from an imperfect satellite configuration with precessing orbits should be understood, and is best illustrated by the situation that prevailed towards the end of the 1970s when there were five satellites – illustrated by Figure 11.2. From this it can be seen that

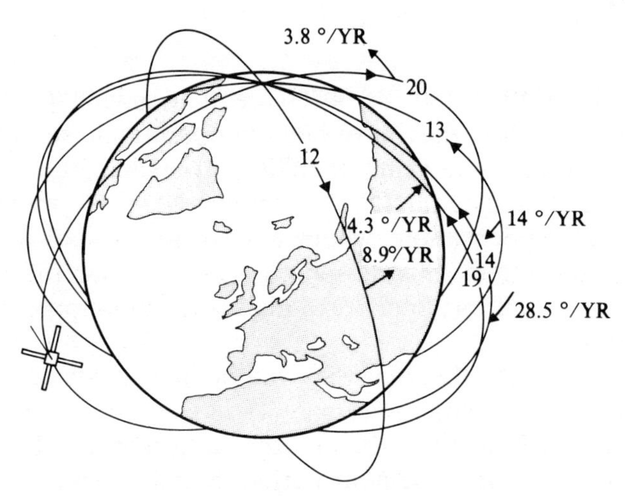

satellite numbers 12 = 30120
13 = 30130 etc.

Figure 11.2 The actual orbit configuration of April 1978.

the orbital planes of satellites 30140 and 30190 are almost the same, whereas there is a large gap between satellites 30120 and 30200. Due mainly to this gap, intervals between satellite fixes of up to twelve hours can occur near to the equator.

As an example of the typical daily distribution of satellite passes as seen by users of the Transit system, Figure 11.3 illustrates the result of one day's monitoring of satellite passes at a fixed location, on latitude 45°N. (during 1973). This diagram shows that each satellite makes three or four 'visible' passes with intervals of one

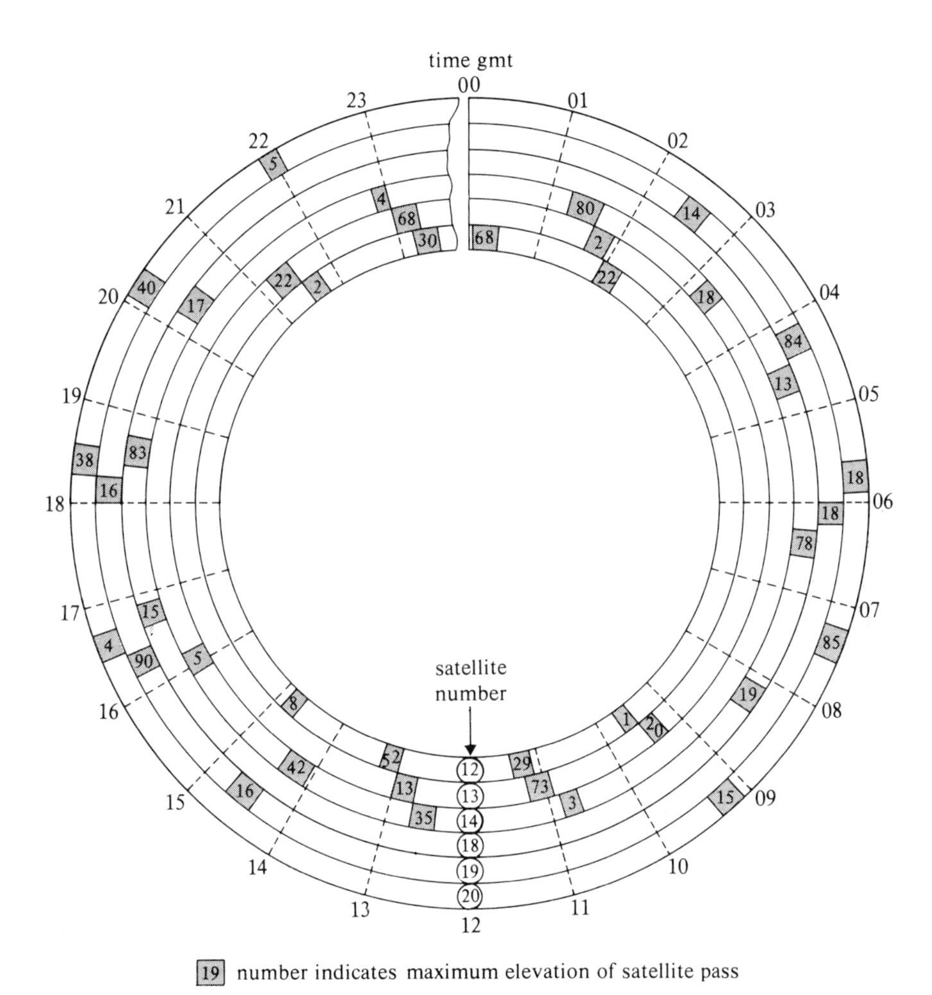

Figure 11.3 Typical satellite pass distribution throughout a 24-hour period at a mid-latitude location.

orbit time. There is then an interval of six to seven hours when the satellite is always out of sight, followed by another three or four visible passes at various maximum elevations. This overall pattern is a function of the earth rotating within the constellation of satellite orbits.

Figure 11.3 also illustrates the arbitrary relationship of the satellites with each other. Throughout a twenty-four hour period approximately forty satellite passes are seen at mid-latitudes, with an average time between passes of thirty-six minutes, although in reality the satellites do not appear with this regularity. Using the illustration of Figure 11.3, no satellites were seen between 1500 and 1600 gmt, four were seen between 1600 and 1700 gmt, and none between 1900 and 2000 gmt. At this particular time the situation is potentially even worse than it first appears, for the following reason. Satellite passes which have either a high or a low maximum elevation in general produce less accurate position fixes than mid-elevation passes. It is generally the case that passes with maximum elevations which are either below fifteen degrees or above seventy-five degrees are susceptible to large errors. For this reason many satellite navigation systems are programmed to produce position fixes only from passes of acceptable elevation.

Considering again the example of Figure 11.3, an acceptable fix would probably result from the sixteen degree pass at 1430 gmt, but then the next four passes are all of extreme elevation (8, 5, 90 and 4 degrees). Satellite 18 would probably produce a good fix from the pass commencing at 1641 gmt, except that the receiver will track satellite 20 until 1647 gmt (from which it will not produce a good fix). Having therefore missed the first six minutes of satellite 18, it is unlikely that this satellite will be tracked by the receiver, or if it is, it would probably have insufficient Doppler data to produce an acceptable fix. The next good pass occurs at 1800 gmt, resulting in a total period without an acceptable fix of almost three-and-a-half hours. It is this mechanism of satellite bunching, which can cause the long periods of up to twelve hours without a fix at equatorial latitudes.

11.2 The navigation message

The transmitted data which precisely defines the satellite orbit is modulated onto the two carrier frequencies, using a phase code

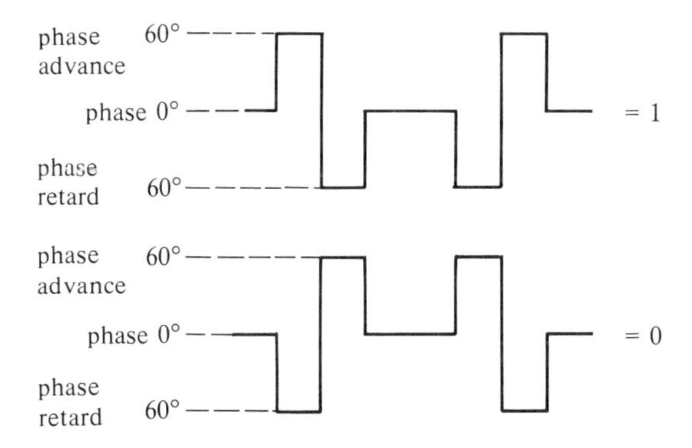

Figure 11.4 Modulation format of the satellite signal.

which does not affect the continuous use of the carrier for Doppler shift measurements (Figure 11.4).

The two-minute satellite message consists of 156 digital words of 39 bits each, and one word of 19 bits, making a total of 6103 binary bits. Words 1, 2 and 3 provide timing information, and of the rest only twenty-five are used for navigation (Figure 11.5). The other words contain undisclosed military information and are not decoded by the normal navigation receiver.

The message length is precisely two minutes, the beginning of which occurs at the instant of the even minute. This can be used to maintain an accurate display of time within the navigation receiver.

The decoded navigation message consists of twenty-five nine-digit words (Figure 11.6). The first eight words define small perturbations in the satellite orbit and are termed ephemeral parameters. The last seventeen words are fixed parameters, remaining unchanged between the twelve hourly injections of data. Eleven of these seventeen fixed parameters define the mean ellipse of the satellite orbit and are referred to as mean orbital parameters. The remaining fixed parameters provide information which is not directly used in the fix computation.

The ellipse defined by the mean orbital parameters is an approximation of the actual satellite track. The satellites deviate about their mean orbit due to variations in the earth's gravity, the density of the upper atmosphere, radiation pressure, and several other influences. The ephemeral parameters describe these deviations about the mean orbit. At the time of the injection sufficient ephemeral parameters are stored in the satellite's memory to define the deviation from the mean orbit at two minute intervals. Each transmitted satellite message contains the ephemeral parameters for

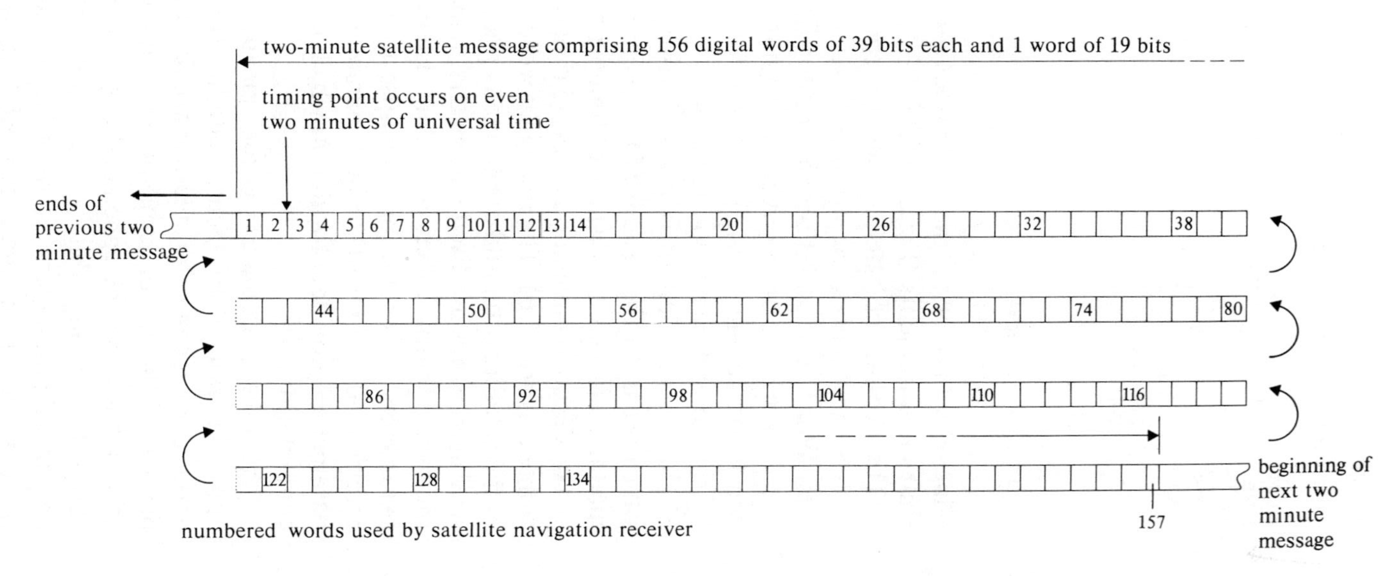

Figure 11.5 Illustration of the two-minute satellite message showing data words actually used.

Decoded navigation message	Word	Message, time t_o		
040571295	8	*Ephemeral parameters*	Data for time $t_o - 6$ mins.	
050611020	14		Data for time $t_o - 4$ mins.	
060640746	20		Data for time $t_o - 2$ mins.	
070650470	26		Data for time t_o	
080650217	32		Data for time $t_o + 2$ mins.	
090630000	38		Data for time $t_o + 4$ mins.	
500600168	44		Data for time $t_o + 6$ mins	
510550270	50		Data for time $t_o + 8$ mins.	
009343030	56	*Mean orbital parameters*	Time of satellite perigee	093.4303 min.
841033870	62		Mean motion of satellite	3.4103387 deg./min.
807448290	68		Angle of perigee	074.4829 deg.
800199980	74		Precession rate of perigee	0.0019998 deg./min.
800179560	80		Eccentricity	0.017956
807398840	86		Mean semi-major axis	7398.846
811627250	92		Angle of ascending node	116.2725 deg.
800001330	98		Precession rate of node	0.0000133 deg./min.
900029170	104		Cosine of orbit inclination	–0.002917
803581610	110		Inertial longitude of Greenwich	035.8161 deg.
800302000	116		Satellite number	3020 (20)
815500230	122		Time of last satellite injection	310 min. gmt day 65
809999960	128		Sine of orbit inclination	0.999996
799830000	134		Satellite frequency offset from 400 MHz	–79.983 PPM
000000000	140		Indicates when data injection occurs	
000000000	146		Indicates when data injection occurs	
000000000	152		Indicates when data injection occurs	

	Word	Message, time t_1 (= t_o + 2 mins.)	
050611020	8	*Ephemeral parameters*	Data for time $t_1 - 6$ mins.
060640746	14		Data for time $t_1 - 4$ mins.
070650470	20		Data for time $t_1 - 2$ mins.
080650217	26		Data for time t_1
090630000	32		Data for time $t_1 + 2$ mins.
500600168	38		Data for time $t_1 + 4$ mins.
510550270	44		Data for time $t_1 + 6$ mins.
520490319	50		Data for time $t_1 + 8$ mins.

[*continues*] [*continues*]

Figure 11.6 A typical satellite message, decoded and showing the meaning of each data word.

eight two-minute time intervals, from time (t_0 – 6) minutes to (t_0 + 8) minutes, where t_0 is present time. During the next two-minute period (time t_1), all ephemeral parameters move into new word locations, t_0 becoming (t_0 – 2) minutes. A new ephemeral word is introduced into the message, corresponding to (t_0 + 8) minutes and one ceases to be broadcast (t_0 – 6) minutes. Each ephemeral parameter is thus broadcast eight times, occupying in turn each of the ephemeral word locations. The ephemeral data defines the position of the satellite in space with a resolution of ±5 metres.

11.3 The position fix

The determination of a position fix from Transit satellites is based upon the Doppler effect, which in this case is the observed shift in frequency when the relative distance between a radio transmitter and receiver is changing. This change in frequency can be observed and measured by a satellite receiver located on a vessel when a satellite passes within range. The observed Doppler shift is caused by three sources of relative velocity: (1) the orbital velocity of the satellite; (2) the velocity of the vessel on which the receiver is located; and (3) the rotation of the earth about its axis. For the period of time that the satellite is observed above the horizon, Doppler shift measurements are made at regular intervals, and at the end of a satellite pass a Doppler shift curve can be constructed (Figure 11.7). This curve uniquely relates the position of a vessel to the satellite orbit, and from this the position of the vessel can ultimately be determined (Figure 11.8).

The actual Doppler shift measurement is made by comparing the received frequency (f_r) with a 400 MHz reference frequency (f_0) within the receiver, to produce a difference frequency ($f_0 - f_r$). The number of cycles of difference frequency are then counted over specific time intervals, the two minute message time being a convenient accurately timed interval. The total number of cycles in each interval are termed two minute Doppler counts. The observed frequency difference consists of two parts: one part is due to the Doppler shift, and the other is a fixed amount of frequency offset due to the fact that the satellites transmit not at 400 MHz, but at a frequency approximately 32 kHz (80 ppm) below this. This fixed offset ensures that the received frequency does not cross 400 MHz due to the Doppler shift, and so the receiver does not have to

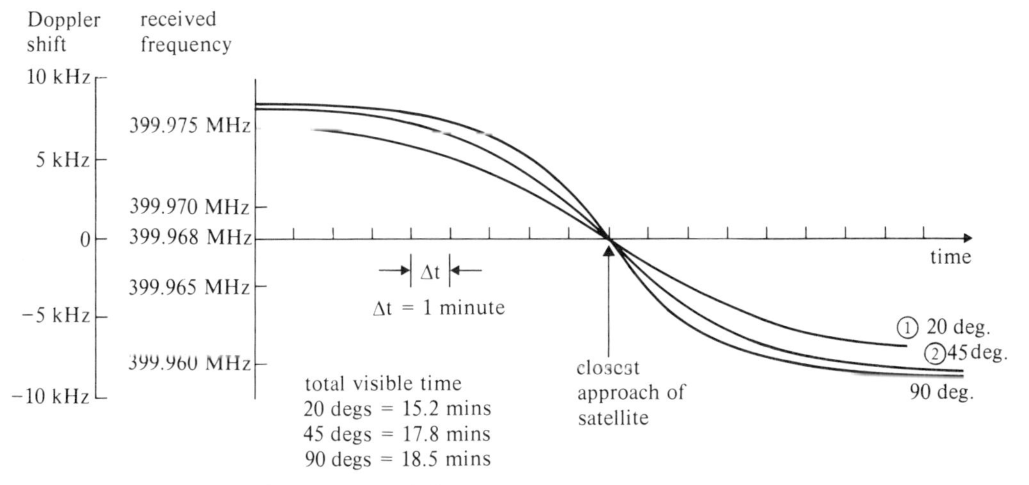

Figure 11.7 The Doppler shift curves resulting from observing satellite passes.

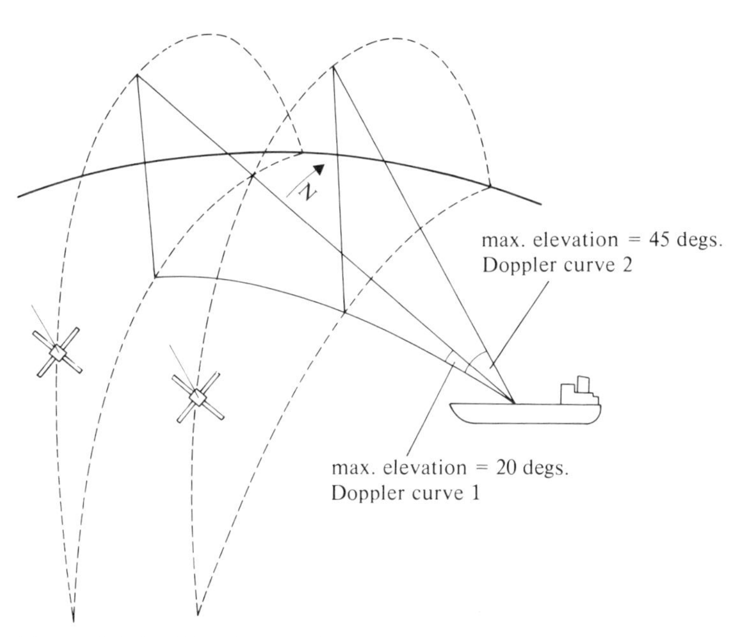

Figure 11.8 The shape of the Doppler curve relates to elevation and hence vessel's longitude. The point of zero Doppler shift relates to latitude.

concern itself with positive and negative Doppler counts. The precise amount of fixed frequency difference between the satellite transmission frequency and the receiver frequency, is determined during the fixed computation.

During a satellite pass a number of two-minute Doppler counts are obtained, N_{01}, N_{12}, N_{23} . . ., up to a maximum of nine (Figure 11.9). These counts are a direct measure of the amount by which the slant range from the satellite to the receiver has changed during the count interval $(R_1 - R_0)$, $(R_2 - R_1)$,. . . . This is a very precise measure of range change since each count represents one wavelength, which at 400 MHz is 0.75 metres.

The number of Doppler counts in a count interval (e.g., N_{12} of Figure 11.9) is given by the expression:

$$N_{12} = \int_{t_1 + R_1/C}^{t_2 + R_2/C} (f_0 - f_r).\mathrm{d}t. \qquad [\mathbf{11.1}]$$

(C = propagation velocity)
This can be resolved into the two components of the count:

$$N_{12} = \int_{t_1 + R_1/C}^{t_2 + R_2/C.} f_0.\mathrm{d}t. - \int_{t_1 + R_{1/c}}^{t_2 + R_{2/c}} f_r.\mathrm{d}t. \qquad [\mathbf{11.2}]$$

The first term represents the constant frequency difference between the transmitted frequency and the receiver reference frequency. The second term is the changing frequency f_r due to Doppler shift.

$$\int_{t_1 + R_1/C}^{t_2 + R_2/C} f_r \,.\, \mathrm{d}t. = \int_{t_1}^{t_2} f_t \,.\, \mathrm{d}t. \qquad [\mathbf{11.3}]$$

Substituting in equation [**11.2**] gives:

$$N_{12} = \int_{t_1 + R_1/C}^{t_2 + R_2/C} f_0.\mathrm{d}t. - \int_{t_1}^{t_2} f_t.\mathrm{d}t. \qquad [\mathbf{11.4}]$$

Since the transmission frequency f_t and the receiver reference frequency f_0 remain constant throughout a satellite pass, the integrals of equation [**11.4**] become:

$$N_{12} = f_0\,[(t_2 - t_1) + {}^1\!/\!_C\,(R_2 - R_1)] - f_t\,(t_2 - t_1) \qquad [\mathbf{11.5}]$$

Re-arranging gives:

$$N_{12} = (f_0 - f_t)(t_2 - t_1) + f_0/\mathrm{C}(R_2 - R_1) \qquad [\mathbf{11.6}]$$

$f_0/\mathrm{C} = 1 - \lambda_0$ (λ_0 = wavelength of the reference frequency)

$f_0 - f_t$ is the constant difference frequency f_c($\simeq$ 32 kHz).

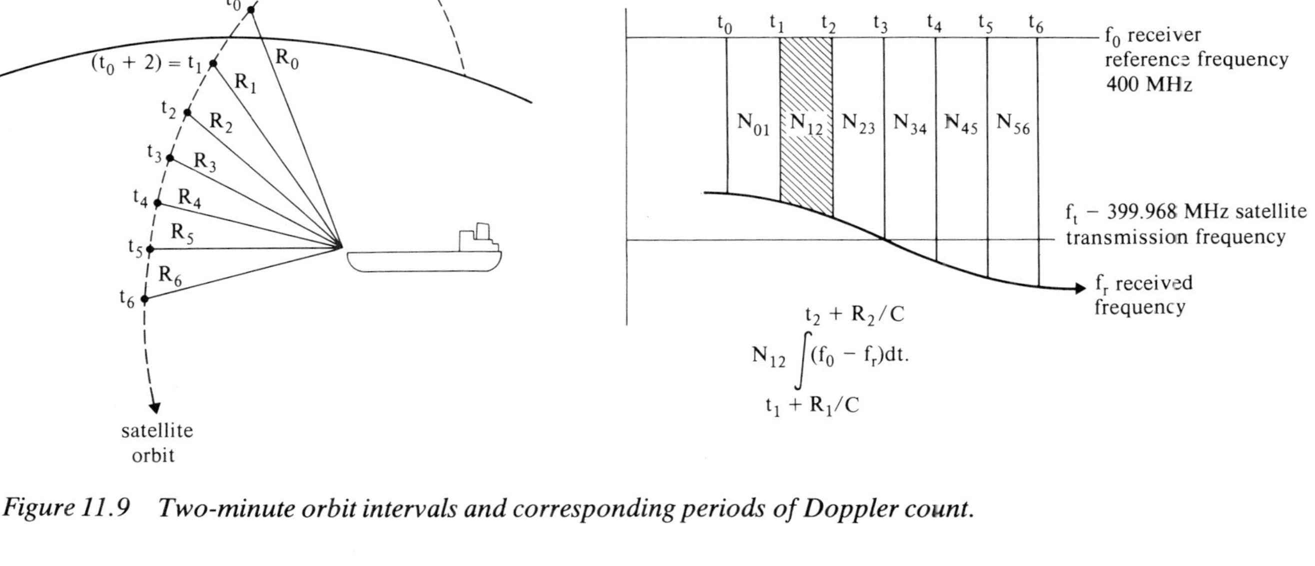

Figure 11.9 Two-minute orbit intervals and corresponding periods of Doppler count.

Therefore

$$N_{12} = f_c (t_2 - t_1) + 1/\phi_0 (R_2 - R_1) \quad [11.7]$$

Using this expression, the slant range changes $(R_3 - R_2)$, $(R_4 - R_3)$ . . . for Doppler count periods $N_{23}, N_{34}, \ldots$ can be derived.

The magnitude of the frequency offset f_c is assumed constant throughout a satellite pass, but its absolute value is not precisely known, due mainly to the unknown amount of drift of the receiver reference frequency f_r. The magnitude of f_c must therefore be determined along with the two co-ordinates defining the receiver location. The solution of three unknowns requires a minimum of three equations.

The basic mechanism for computing a position fix is for the satellite receiver to use its knowledge of the satellite orbit (derived from the two-minute message), along with an estimate of the vessel's current position, to produce corresponding slant ranges at two minute intervals. From these are obtained slant range changes which are then compared with the measured slant range changes derived from equation **[11.7]**.

Unless by coincidence the estimated position is precisely correct, there will be a 'residual' when each estimated slant range change is subtracted from the corresponding measured range change. The position fix process then consists of changing the co-ordinates of the estimated position by small increments. The difference frequency must also be changed since this represents a third unknown. The three unknowns are changed until the sum of the squares of the residuals is minimized. At this point the revised position estimate becomes the vessel's position as determined by the satellite receiver, and is displayed as such. This final position is usually arrived at after only a few iterations even when the original estimate of position is in error by several miles.

In the above explanation it was assumed that the satellite receiver was stationary so that all slant range measurements were made to a point, and it was to this point that the fix solution converged. In the practical case of a vessel underway, the computed range differences for the estimated position must take into account the changing position of the vessel. Therefore the ship's velocity must be fed to the computer, to provide revised estimates of position at the end of each count interval (Figure 11.10). These changes in position will be taken into account during the fix computation, although the displayed position will relate to one specific moment in time.

11.4 Errors

In considering the magnitude of errors in positions determined from the Transit system, the performance potential is discussed from two viewpoints: (1) the high accuracy of the survey system; and (2) the underway accuracy of a simple single frequency navigation receiver.

(1) The Applied Physics Laboratory of the Johns Hopkins University have published an error budget for individual Transit position fixes. This shows the optimum performance available to the user requiring the highest accuracy from the system:

	Error source	*Error (metres)*
1	Uncorrected propagation effects (ionospheric and tropospheric)	1–5
2	Instrumentation (receiver and satellite oscillator phase jitter)	1–6
3	Geodesy (uncertainty in the geopotential model)	5–10
4	Incorrectly modelled surface forces (effects of drag and radiation pressure on the satellites)	10–25
5	Unmodelled UT1-UTC effects and uncorrected co-ordinates of the pole	1
6	Ephemeris rounding error (last digit of transmitted ephemeral parameter is rounded)	5
	RSS	12–28

This error budget is probably slightly optimistic, since in a practical situation the refraction correction models would not be as sophisticated as those used by the Applied Physics Laboratory. A typical scatter plot of single pass fixes is shown in Figure 11.11.

(2) Having illustrated the high accuracy achievable from single

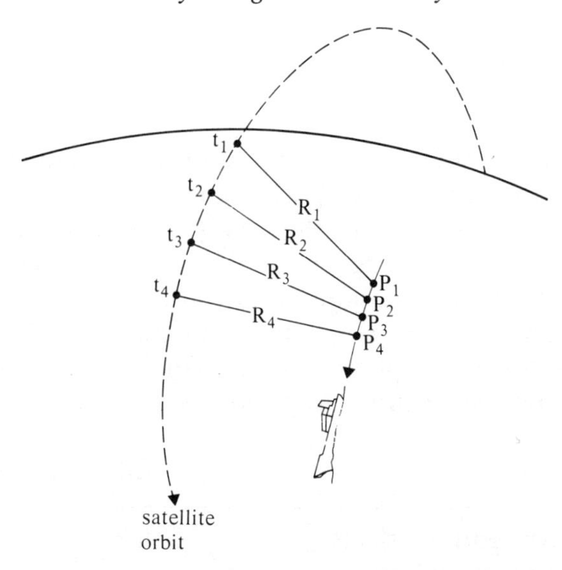

Figure 11.10 During the satellite pass slant ranges are related to up-dated positions P_1, P_2 etc.

passes in the static survey situation, the normal working accuracy of the typical navigation receiver is now considered.

11.4.1 *Refraction error*

The ionosphere has the effect of refracting radio waves which pass through it, and so all signals received from satellites will be affected in this way. As a result of this refraction the Doppler shift is altered from what it would be in a vacuum, thus introducing an error into the final position fix (mainly in longitude). The amount by which the wave is refracted is dependent upon its frequency, and for frequencies above 100 MHz the refraction effect on the Doppler shift is inversely proportional to frequency squared. This fact can be used to provide a means of compensating for the effects of ionospheric refraction. For this reason the Transit satellites transmit on two frequencies, 400 MHz and 150 MHz, and by mixing the measured Doppler shifts at these two frequencies the unrefracted Doppler shift can be determined. To take advantage of this method of compensating for ionospheric refraction, the receiver must be capable of simultaneously receiving and processing the Doppler shift on the two frequencies. This obviously adds to the complexity and hence cost of the satellite receiver. The magnitude of the position

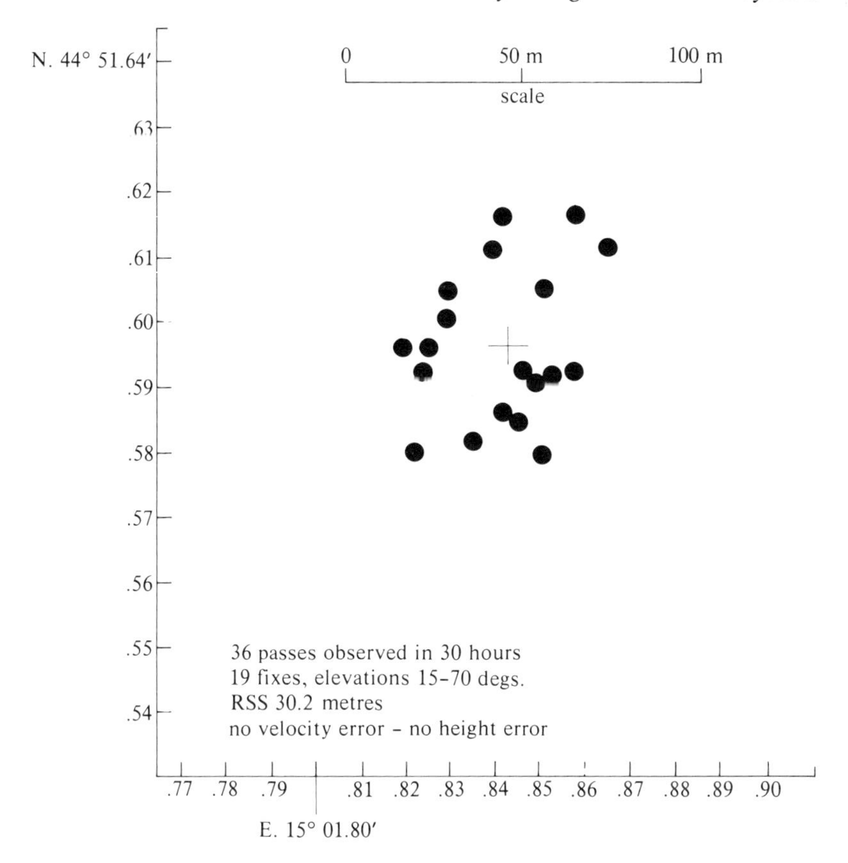

Figure 11.11 Typical scatter plot of dual-frequency receiver.

error resulting from ionospheric refraction varies, daily, seasonally, and over the eleven-year sunspot cycle. At night-time the error is small, but during day-time it can be as great as five hundred metres, although it is usually less than one hundred metres.

Since this error is typically less than a ship's length, and since it is significantly smaller than errors which can be caused by incorrect velocity inputs, it is usually accepted that for normal navigation purposes the equipment costs of a dual frequency receiver are not justifiable. Most navigation receivers are therefore 'single channel', receiving the 400 MHz transmissions only. Typical results from a single channel receiver are shown in Figure 11.12.

There exists a second source of refraction error, caused by the troposphere, which is the region of the earth's atmosphere extending upwards for approximately ten kilometres. The effect of tropospheric refraction is directly proportional to frequency, as is

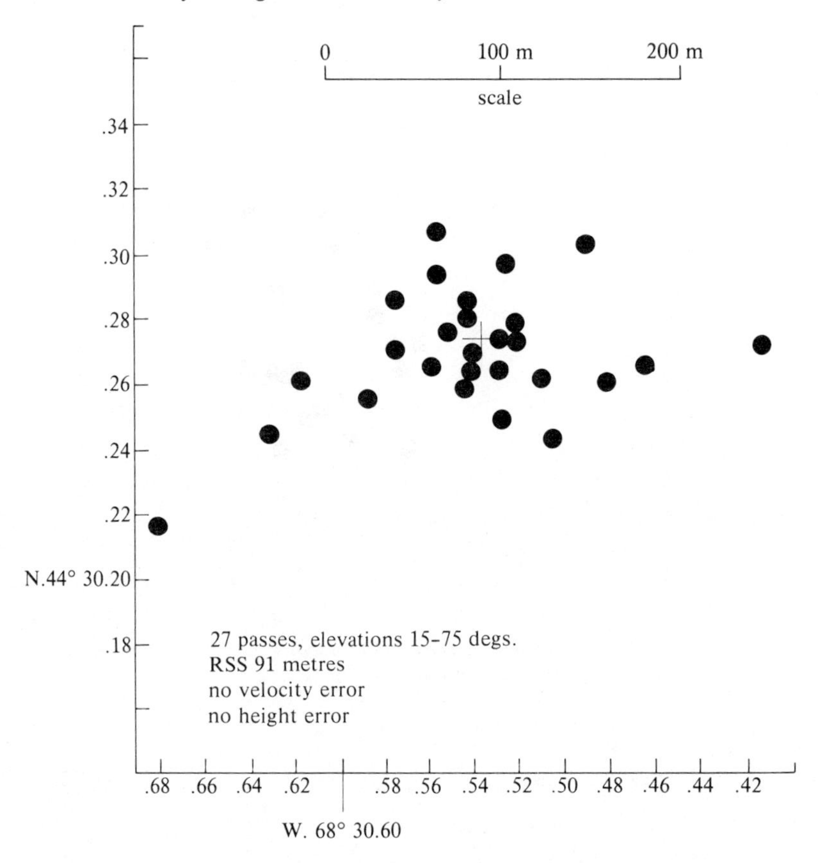

Figure 11.12 Typical scatter plot of single-frequency receiver.

the Doppler effect, and so the magnitude of the tropospheric refraction cannot be determined by using two frequencies. In accurate survey systems the effect is compensated by mathematically modelling the tropospheric refraction, but in navigation systems it is neglected.

11.4.2 Antenna height error

In computing the position fix, the relationship of the earth's surface with the satellite orbit must be precisely known. For this reason the computation includes a reference ellipsoid, which is a geometrical shape best fitting the shape of the earth. The ellipsoid is uniquely defined by specifying two dimensions, the semi-major axis (the radius at the equator) and the flattening (f), which is given by:

$$f = \frac{a - b}{a}$$

where a = semi-major axis
b = semi-minor axis.

In the satellite system the reference ellipsoid is taken as WGS 72 (World Geodetic System ellipsoid of 1972); this has a semi-major axis of 6,378,135 metres and a flattening of 1/298.26.

In the position-fix process slant ranges to points on the satellite orbit are determined. If these are subsequently related to the reference ellipsoid the resulting position will be incorrect, since the actual earth's surface deviates from the ellipsoid. A major precise method of describing the earth's surface is by reference to the 'geoid'. (If mean sea level was established everywhere it would define the shape of the geoid.) For the slant ranges to produce the correct position on the earth's surface, the difference between the geoid and the ellipsoid must be included in the computation. This is usually obtained from a geoidal height map (Figure 11.13), which shows the differences as height contours. In addition to this, height above the geoid must also be included in the computation, which for the land surveyor means height above sea level, and for the navigator is height of the antenna above the water-line. The effect of neglecting the deviation from the ellipsoid, or applying an inappropriate amount of correction, is illustrated by Figure 11.14. The resulting position error is a function of the satellite's maximum elevation, being greatest for high elevation satellites. The error is almost totally one of longitude and its magnitude is given approximately by: position error (longitude displacement) = height error × tangent of maximum satellite elevation angle. For example, a position fix derived from a satellite pass with a maximum elevation of seventy degrees and a height error of forty metres, would produce a longitude error of approximately: $40 \times \tan 70 = 110$ metres. The geoidal height map was derived from the observation of the earth's gravity influencing satellite orbits, the map is known to be in error in some areas by up to twenty metres.

In a survey situation where high accuracy is the prime requirement, it is possible to determine true geoidal height by making height a variable in the position fix solution. A height estimate is initially entered which is iterated to produce the best solution for antenna height. After several satellite passes are observed at a fixed location, the antenna height can be resolved to a high degree of accuracy. Figure 11.15 shows how use of this type of three-dimensional

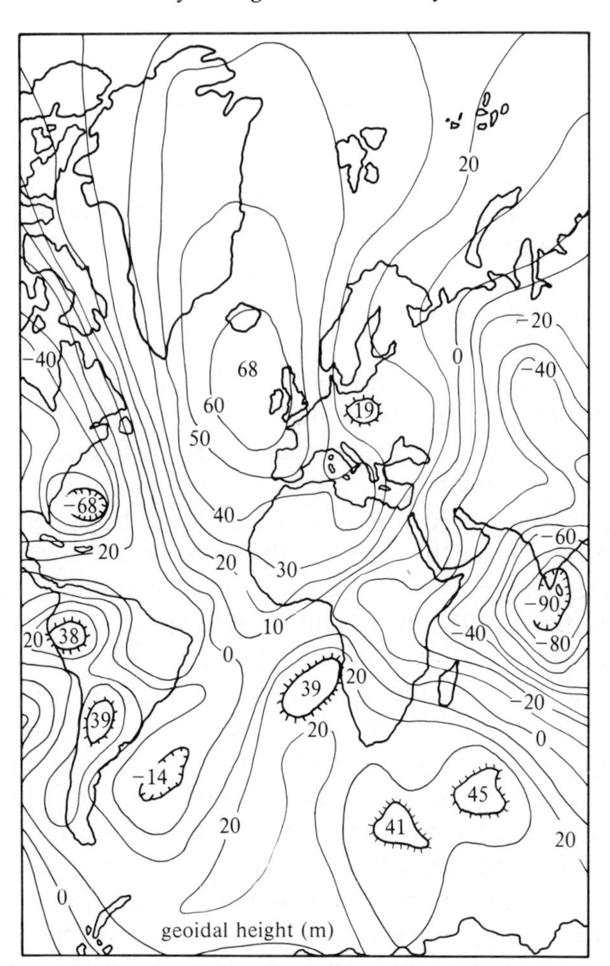

Figure 11.13 Geoidal height map.

solution provides an increase in the accuracy of the final fix. Each time a satellite is tracked and a position determined, it is then combined with the previous positions to produce an improved overall solution.

When related to the various error sources inherent in the single channel navigation system, the use of the geoidal height map provides a more than adequate accuracy. In fact during ocean passages, many navigators deliberately neglect to insert changes in height. Accepting instead that provided satellite passes with maximum elevations of greater than seventy degrees are not used, the resulting position error contribution should not exceed three hundred metres when the height input is maintained at zero.

Some satellite navigation receivers have stored in their computer a simplified geoidal height map, which automatically applies an approximate amount of height correction.

11.4.3 *Chart errors*

In the preparation of charts, ellipsoids have been chosen which are best suited to the local shape of the geoid. As a result of using differing ellipsoids and each referenced to a different datum, the charts resulting from these local surveys will vary from satellite-derived co-ordinates which are based on the WGS 72 ellipsoid and a global datum. This difference can be of the order of several hundred metres, and can be seen as a bias when plotting a large number of position fixes obtained while a vessel is in port. Details of various ellipsoids and datums are included in the computer programs of most dual channel survey systems, so that a satellite position can be computed in co-ordinates directly applicable to the local chart.

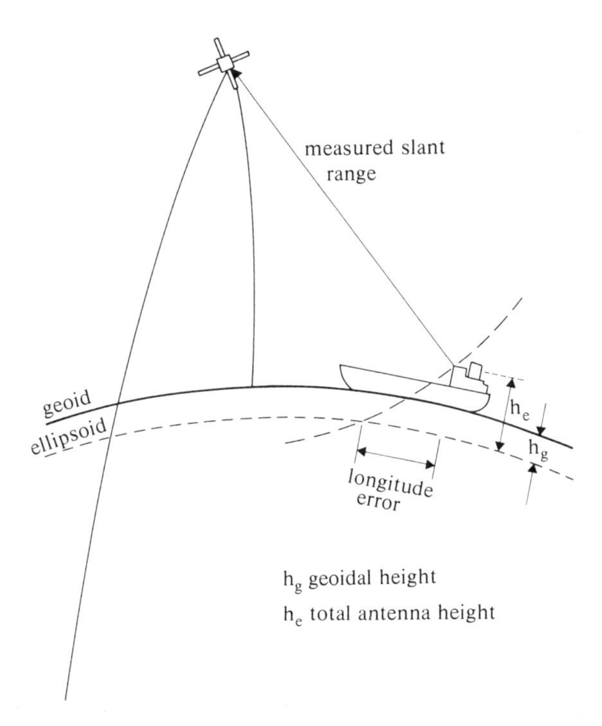

Figure 11.14 The effect of height error on the position-fix solution.

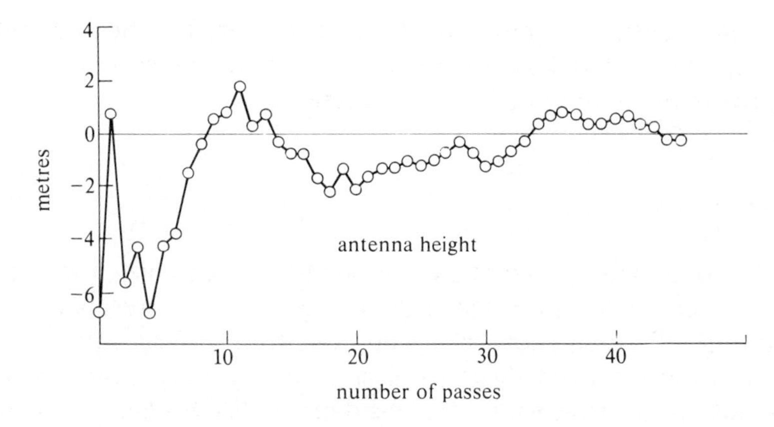

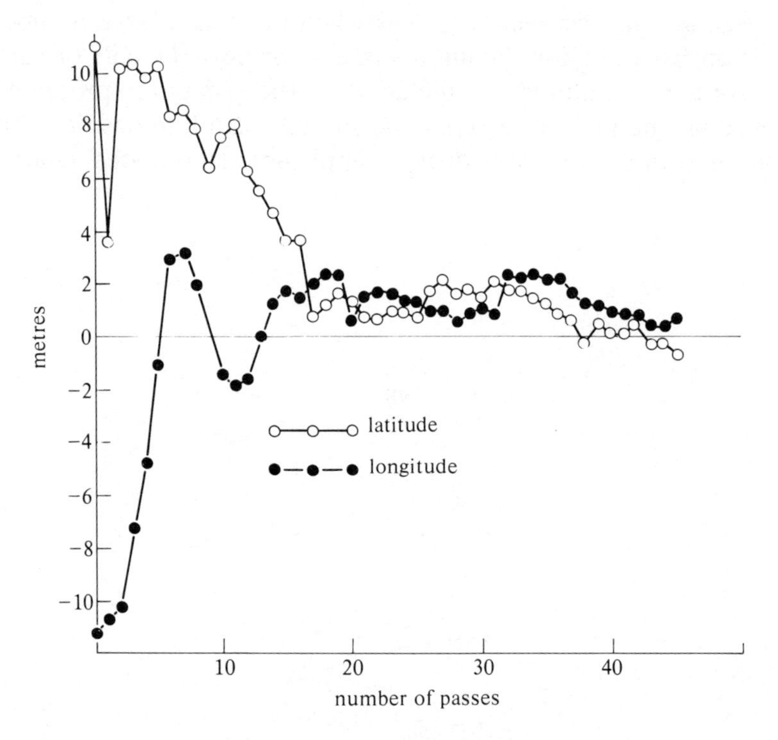

Figure 11.15 Typical accuracy improvement achieved by using many satellite passes at a fixed location.

11.4.4 Velocity error

Since the position fix is derived from the Doppler curve resulting from a satellite pass, any contribution to this curve due to the vessel's own velocity must be removed before the fix is computed. Additionally the velocity must be known in order to calculate the position of the vessel at each Doppler interval, from which to relate the slant range changes. It is therefore usual to interface the satellite receiver directly to the ship's speed log and gyro-compass, so that the processor is fed continuously with up to date speed and heading information. Any error in this data will ultimately produce an error in the position fix. Velocity errors are quite often the largest single source of error in satellite navigation receivers. The velocity error usually occurs because the speed log is measuring ship's speed relative to the water, and not relative to the earth. As a very general guide, one knot of velocity error can produce up to one quarter mile error in position; in fact the position fix is most sensitive to velocity north errors which cause an error in position largely of longitude. The general effects of velocity north and velocity east errors are illustrated in Figure 11.16 and Figure 11.17.

The velocity error causes an increasing position error after the satellite fix has been computed, since the velocity input is used to compute dead reckon positions until the time of the next fix.

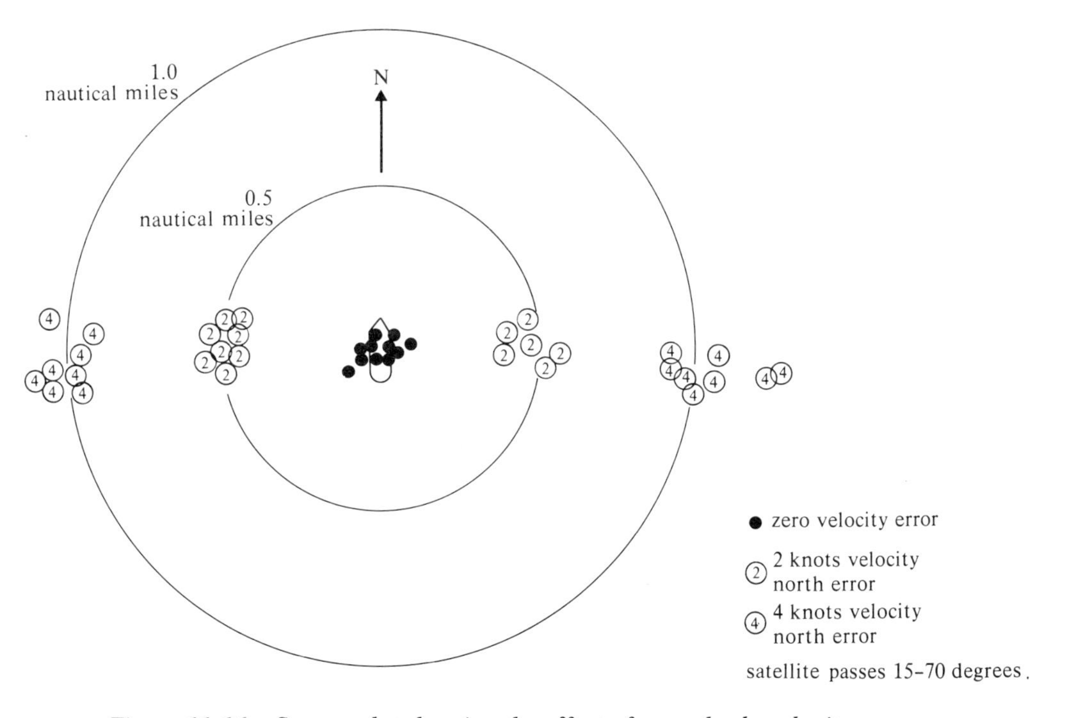

Figure 11.16 Scatter plot showing the effect of a northerly velocity error.

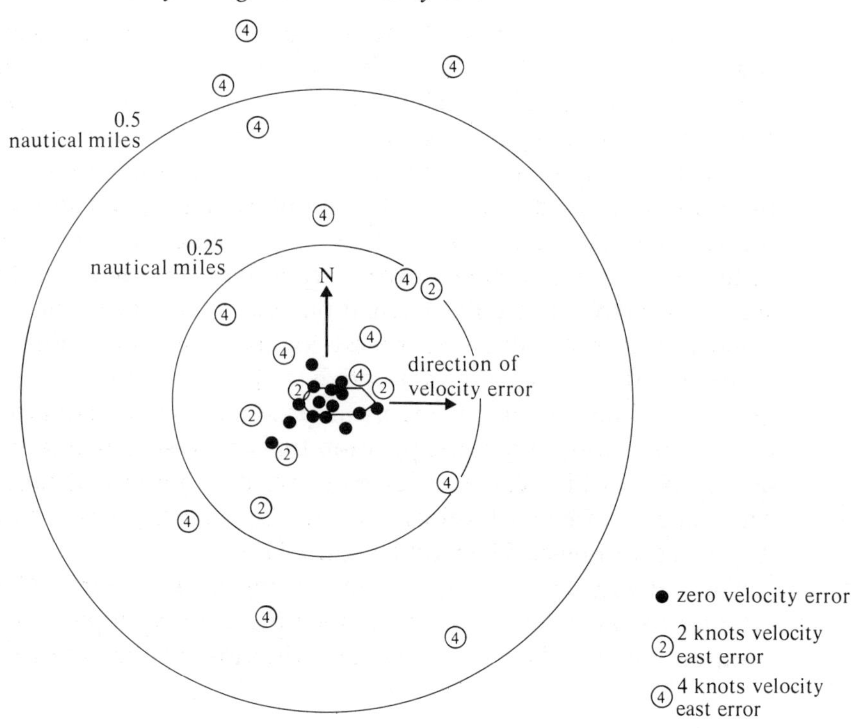

Figure 11.17 *Scatter plot showing the effect of an easterly velocity error.*

11.5 User equipment

There are basically three distinct groups of Transit satellite receiving equipment: the land survey system, the oceanographic survey system, and the single channel navigation system. Only the latter is described in detail here, since the first two are not used in the navigation of merchant vessels.

11.5.1 The lland survey system

This consists basically of an antenna with mounting tripod, a dual channel receiver, and a data recorder – usually either magnetic or paper tape (Figure 11.18). The whole system is designed to be readily transportable. The major difference between this and the other two system types is that the satellite orbital data and Doppler

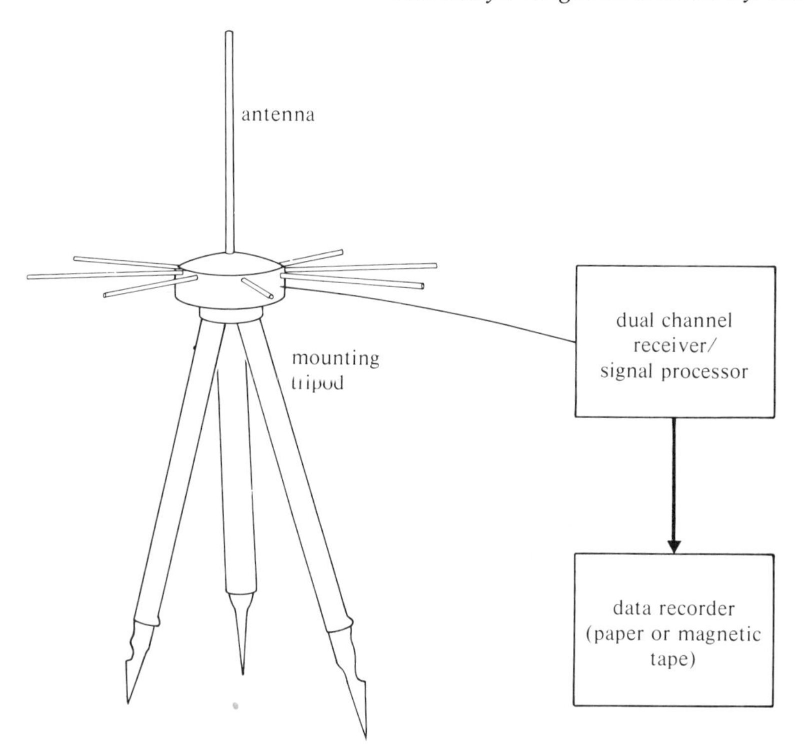

Figure 11.18 The main units of a land survey system.

shift counts are not used immediately to produce a position fix. Instead the total data collected at each site (usually over a two- to three-day period) is sent to a computing centre, where the precise co-ordinates of each site are determined by using a sophisticated three-dimensional program.

1.5.2 *The oceanographic survey system*

In using the Transit system for oceanographic surveys, or any other application requiring high accuracy navigation, great efforts are made to keep velocity error to an absolute minimum. In addition to, or instead of, deriving velocity from the conventional speed-log and the gyro-compass, it is determined by more accurate means. These include a two-axis Doppler sonar operating in the bottom track mode, and land-based systems such as Decca, Omega, or Loran C, operating in either the hyperbolic or the range-range mode. Figure 11.19 illustrates the main constituent parts of such a system.

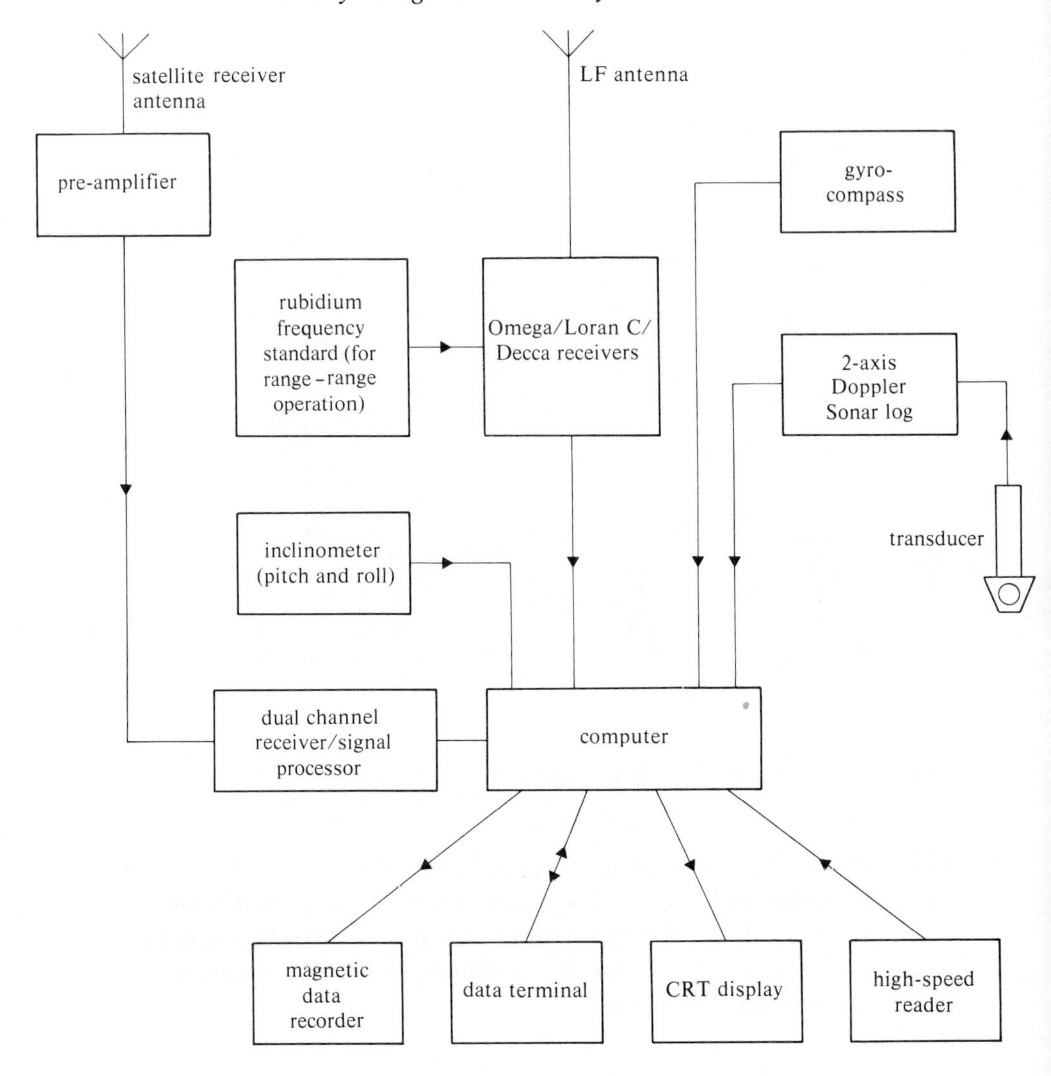

Figure 11.19 An oceanographic survey system.

11.6 The single channel navigation system

The single channel navigation system is usually designed to be low cost and simple in operation. The low-cost goal is achieved at the expense of the loss in accuracy associated with using only one frequency (400 MHz), and only simple velocity inputs (log and gyrocompass). Figure 11.20 illustrates the basic format of commercial

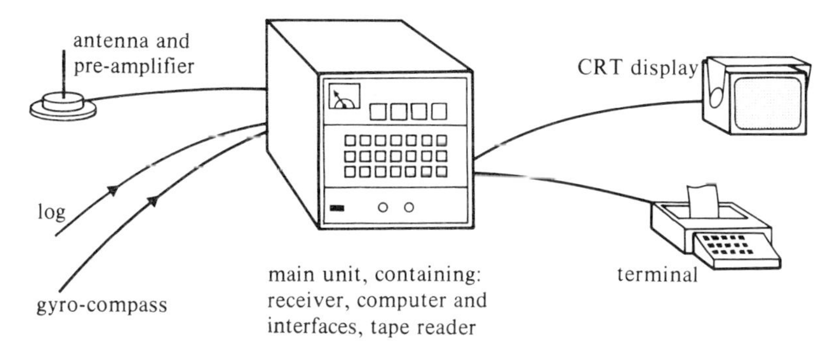

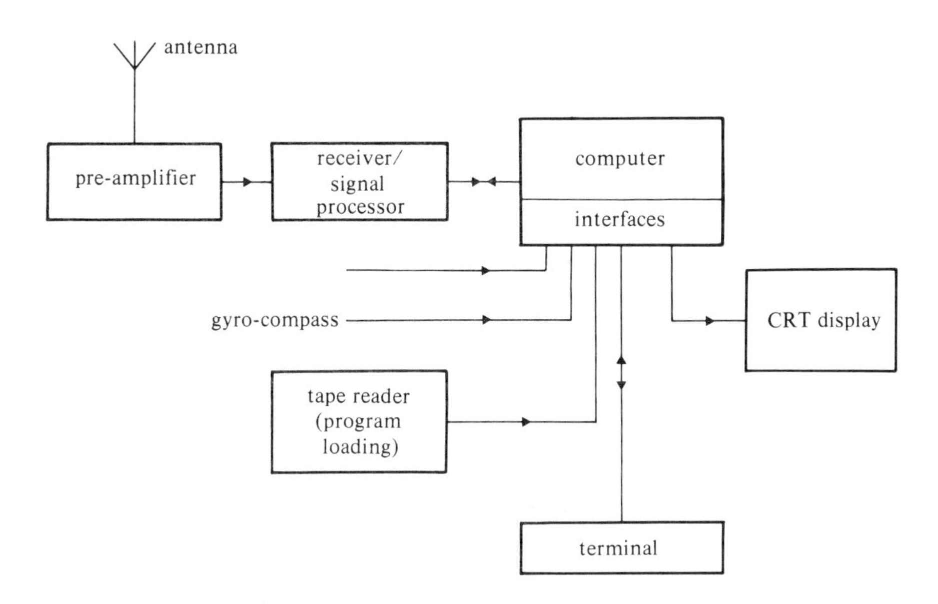

Figure 11.20 *An early type of single-channel satellite navigation system.*

single channel navigation systems since their inception in 1967–8 until recent years (typically 1976–7). The receiver/signal processor and the antenna/pre-amplifier units were specifically designed and built for marine application, whereas other units such as the computer terminal, program loader and CRT display tended to be standard products of the computer industry. As such these were probably not ideally suited to the severities of the shipboard environment, and in consequence reliability problems occurred too frequently. With so many constituent parts the system cost remained prohibitively high. By 1976–7 microprocessor technology had developed to the point where they could now be used in place of the mini-computer to solve the position fix computation, and to exercise overall system control. This then opened the way to the totally purpose-built, single unit (excluding antenna) satellite navigation system, typified in Figure 11.21 and Figure 11.22. The comparative simplicity of this receiver is reflected in its low cost, so that satellite navigation has now come within the reach of many more users.

The performance differences between the various manufactured navigation receivers are mainly due to variations in the computer/microprocessor programs. The position fix solution is basically the same for all equipments, but there are numerous variations in program detail. Some may have a number of simplifications made to bring the computation within the capability of the selected processor, whereas others may contain many additional sub-routines, devised by the particular manufacturer to enhance his system's performance. Some of the relevant aspects of the navigation receiver's operation are described below.

11.6.1 Doppler counts

In the previous description of the determination of the Doppler curve for a satellite pass, it was stated that the Doppler count is integrated over a full two-minute message period, resulting in eight or nine counts for a full satellite pass. However, it can be shown that there are advantages in integrating the Doppler count over a shorter interval, such as a period of five message words (twenty-three seconds), which results in a total of forty to forty-five counts for a full pass. Clearly a greater number of points on the Doppler curve means that the curve is more accurately defined, and by use of these 'short' Doppler counts a given satellite pass usually results in a more accurate fix.

Not the full story. Introduced because of wave wash in submarines.

11.6.2 Majority voted data

The content of the two-minute satellite message is basically a repetition of the previous two-minute message, except for a change in one ephemeral word. After a full satellite pass the same data has therefore been repeated eight or nine times. It is obvious that the orbital data must be accurately decoded, and so use can be made of

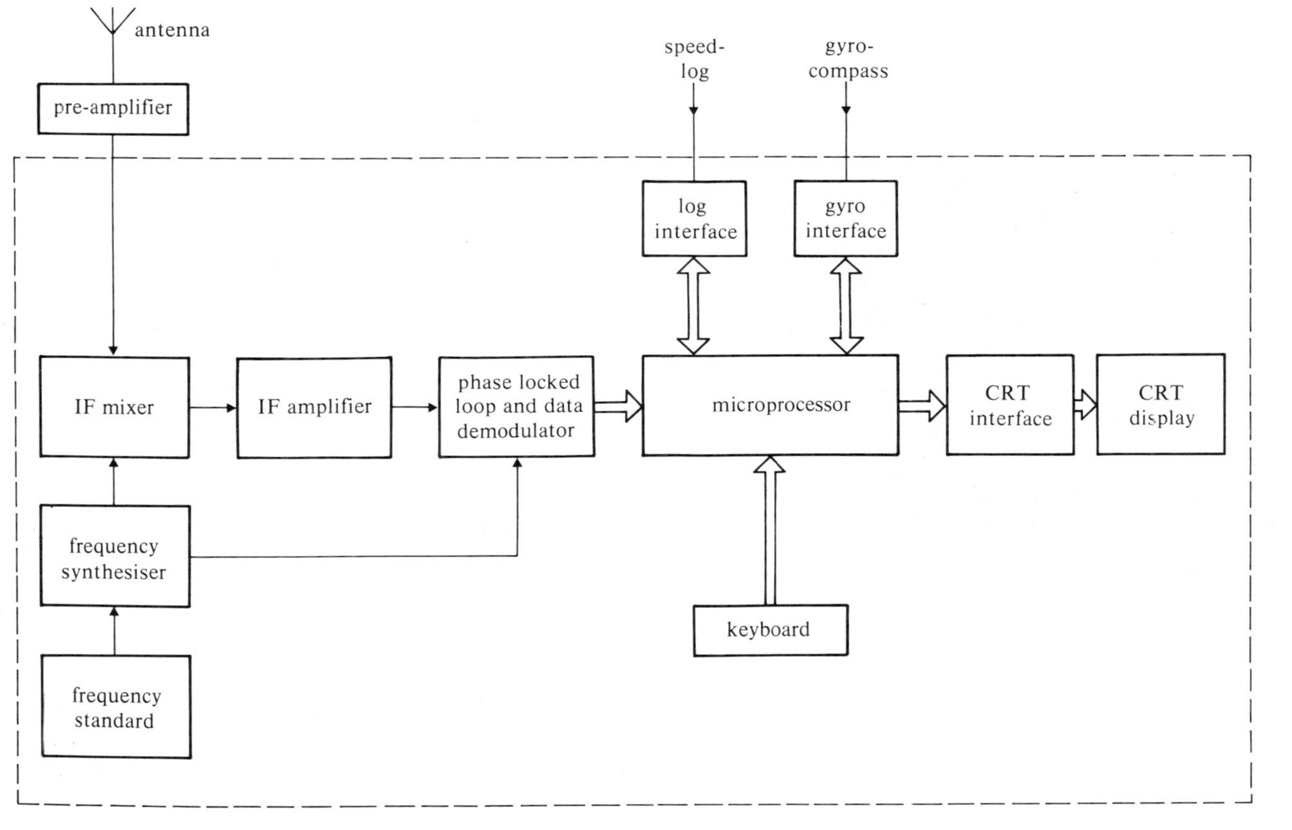

Figure 11.21 Schematic diagram of a modern single-channel system.

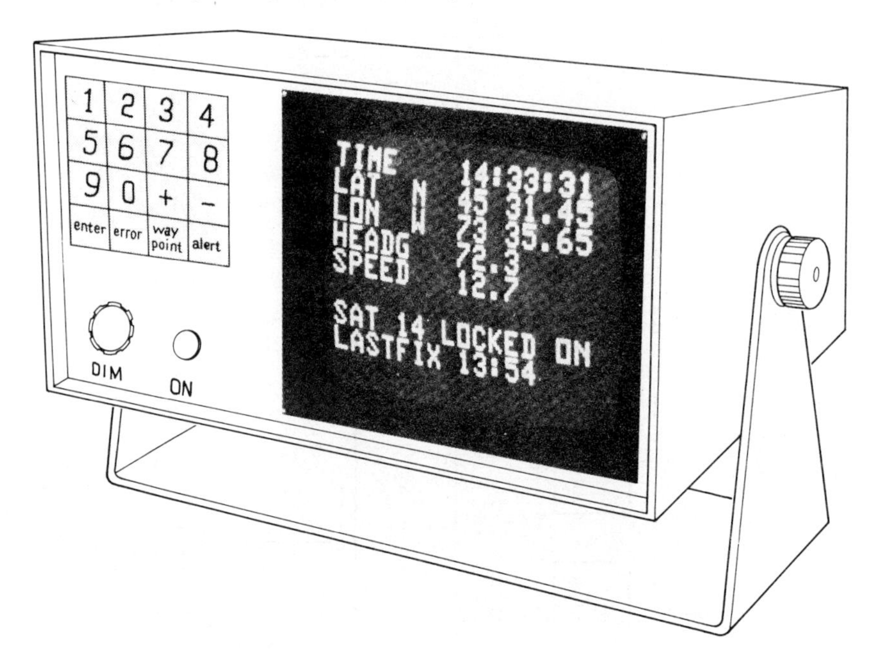

Figure 11.22 Typical single-channel satellite navigation receiver.

the repetitions to compare the content of subsequent messages. This majority voting technique allows message decoding errors to be identified and corrected, which consequently reduces the probability of an erroneous fix.

11.6.3 *Editing criteria*

Most satellite navigation receivers will only display the position fix resulting from a satellite pass if certain criteria are satisfied. The precise choice and numerical values of these editing criteria may vary from one manufacturer's equipment to another, but in general they include:

• Satellite elevation – both low- and high-elevation satellite passes may result in inaccurate position fixes. Limits are usually set, typically at fifteen and seventy-five degrees elevation, beyond which the fix is not displayed.

• Doppler counts – a minimum of three Doppler counts is sufficient from which to compute a fix, but with this minimum number the fix is liable to be inaccurate, particularly if it represents only the first part of the pass. Editing criteria may be set which require a greater number than three counts.

- Doppler count symmetry – the accuracy of the computed position fix is enhanced if the received Doppler data is symmetrical, that is, if the same number of Doppler counts are received after the satellite's closest approach (zero Doppler shift) as before it. Some equipments therefore do not display positions which have been derived from grossly asymmetrical Doppler data.
- Iterations – the position fix is obtained by changing (iterating) the original position estimate, until the estimated slant range changes correspond (within limits) to the slant range changes derived from the Doppler measurements. A limit is usually put on the number of iterations within the program, at which point it is assumed that the estimated and measured slant ranges will not converge sufficiently, due perhaps to noisy Doppler measurements. Alternatively the computation may undergo a fixed number of iterations, and then the acceptability of the fix is judged on the magnitude of the residual.

1.6.4 *Satellite alerts*

After a satellite has passed, and from it a position fix computed, the orbital data received from that satellite can be retained in the processor's memory and used to compute the future rise times for that satellite. Once the orbital data from all satellites has been accumulated, the navigator can be advised in advance of the times and maximum elevations of all forthcoming passes. This may prove of value, particularly as a warning of a lengthy period without a usable pass.

1.6.5 *Crossing satellite passes*

If, shortly after the receiver locks-on to a satellite at the commencement of its pass, a second lower-elevation satellite rises, the received frequency of the second satellite will at some point be the same as the frequency of the first (Figure 11.23). At the crossing point of the two Doppler curves there is a risk that the receiver will lock-on to the second satellite in preference to the first. Safeguards should be incorporated to either minimize the risk of this event occurring, or if it does occur, to ensure that a spurious position fix does not result.

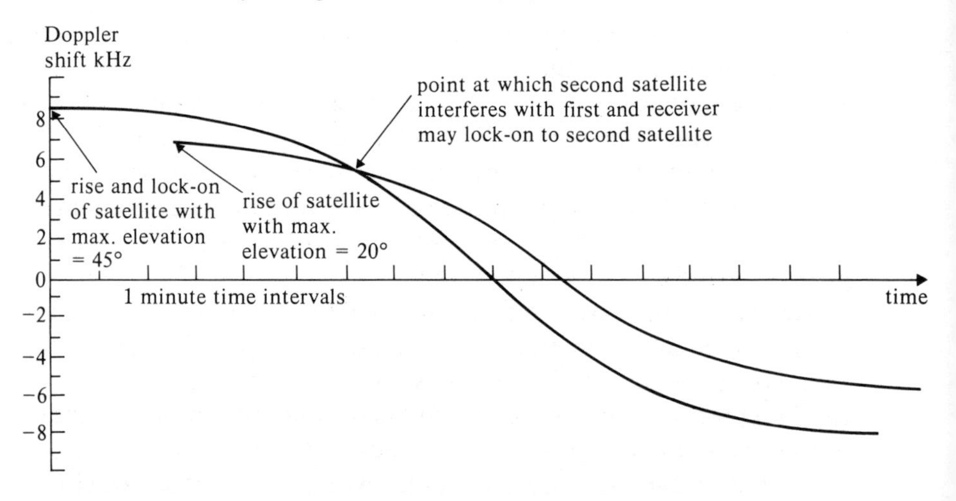

Figure 11.23 Doppler curve of a low-elevation satellite crossing that of a higher-elevation satellite.

11.6.6 Navigational computations

Since the satellite navigation receiver incorporates a relatively powerful computational capability, it can be programmed to perform related navigational calculations. These include distance, ETA and course (great circle or rhumb line), to specified way-points.

Since a satellite-derived fix occurs at irregular intervals, it is usual to compute and display the vessel's current position based on simple dead-reckoning, by using the heading and speed derived from the gyro-compass and speed-log. The displayed position is usually updated with the new DR position at two-minute intervals. By comparing the next satellite-derived position with the DR position at that time, it is possible to calculate set and drift. Then, if required, the DR computations up to the time of the next fix can be set and drift compensated.

11.7 Using the satellite navigation receiver

Of all radio navigation instruments, the satellite navigation receiver is probably the simplest to use, requiring minimal initialization and then virtual hands-off operation. However it is just as important with satellite navigation, as with other navigation systems, to be fully aware of the limitations of one's own equipment, and of the system in general.

11.7.1 *Initialization*

With the earlier satellite navigation systems the first step of the initialization procedure was usually to load the program into the computer memory, from either punched paper or magnetic tape. Navigation receivers with integral microprocessors now have the program permanently held in non-volatile memory, removing the program loading requirement. Initialization is therefore simply a matter of inserting (using the keyboard) an estimate of present position, the antenna height, and the present time (gmt). The time need be set only to an accuracy of within fifteen minutes, since the internal clock will be accurately adjusted during the first satellite pass.

11.7.2 *Operation*

The only adjustments to be made during operation are changes in geoidal height, although with some equipments even this is not required since a geoidal height model is held within the computer.

The first good satellite pass after initialization will change the displayed position from the original estimate, to a position computed from the pass. At regular intervals after this (usually two minutes), the position is updated to a dead-reckoned position, derived from the speed and heading input. The position will therefore degrade in accuracy up to the time of the next satellite pass, and the subsequent position fix.

An indication that a satellite has risen above the horizon and its transmission is being received is usually indicated on the receiver's display.

Basic navigational information is usually displayed in a form similar to that depicted in Figure 11.22 but other pages of display can be selected which show secondary information, such as set and drift, way-points and satellite alerts.

In using Transit satellite navigation the prudent navigator should be particularly aware of the fact that (1) the accuracy of the satellite-derived position fix is dependent on the accuracy of the velocity data which is fed to the receiver during the satellite pass; that (2) displayed positions between satellite fixes are based on simple dead-reckoning; and that (3) on occasions there may be an interval of several hours between satellite fixes (depending on latitude).

12 Navstar-Global Positioning System

There are now many electronic position-fixing systems in current operation, but each is deficient in either its accuracy or area of coverage, and so none can be regarded as a universal coastal and ocean navigation aid. Even an integrated system utilizing two or more existing navigation systems will not provide a totally global capability. Research and development continues towards perfecting the ideal system, with most attention being directed towards more advanced satellite systems. One in particular is the Navstar-Global Positioning System (GPS) which is currently under development by the United States Department of Defense. As with the existing Navy Navigation Satellite System (Transit), Navstar's development is primarily to meet military requirements, but also like Transit it is expected that once in operation it will be made available to civil users.

12.1 The Navstar system

Since GPS will be a very important navigation aid of the future, it is described here in as much detail as practicable, bearing in mind that there may well be changes in both the substance and the time-scale of the system's implementation.

The Navstar system will have a total of eighteen satellites, uniformly spaced in six orbital planes inclined at 55 degrees. The satellites are in 12 hour circular orbits at a height of 20,200 kilometres above the earth (Figure 12.1). The complete configuration will provide world-wide continuous three-dimensional navigation – except for small regions, which will have short periods of degraded performance each day.

Until 1986, between four and six prototype satellites are being used as the basis of extensive testing; their orbits are designed to emulate full system coverage for two hours each day, at the US Army Proving Grounds in Yuma, Arizona. Limited coverage,

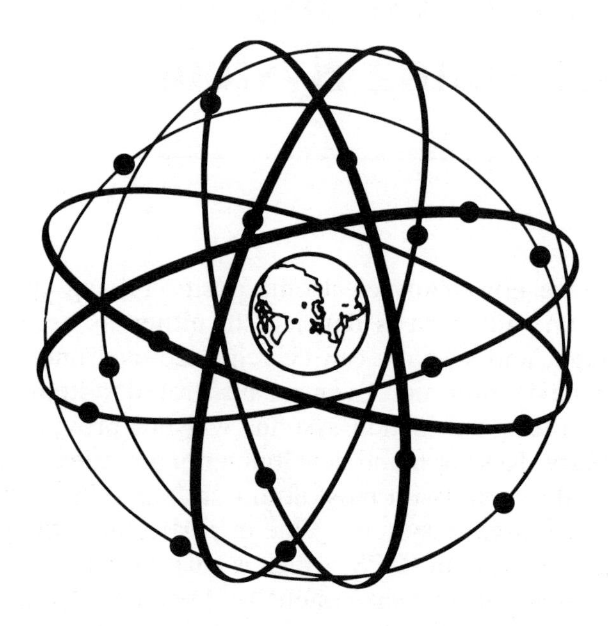

Figure 12.1 The eighteen Navstar satellites will be in six orbital planes.

sufficient to test GPS receivers, will also occur in many other parts of the world up to 1986.

The launching of production satellites is scheduled to commence in 1986 and by the end of 1987 there could be twelve operational satellites, sufficient to provide continuous two-dimensional, world-wide navigation. The full complement of eighteen satellites (plus three orbiting spares) should be in existence by 1989 – yielding the final objective of continuous, world-wide, three-dimensional navigation.

As with Transit, the GPS satellites are supported by the Ground Control Segment, consisting of five monitor stations (three of which also upload), and a Master Control Station at Colorado Springs. The monitor stations track all satellites in view, accumulating ranging data, which is processed to determine the satellite ephemerides, clock drifts and propagation delay. The navigation data contained within the satellites' transmission is updated at least three times a day, from the information computed at the Master Control Station and relayed via the upload stations.

The main constituents of the satellites' hardware are: the atomic frequency standard for accurate timing, the processor to store navigation data, the pseudo-random noise signal assembly for generating the ranging signal, and the transmitter. The satellites transmit on two L-band frequencies, L_1 is the primary navigation frequency of 1575.42 MHz and L_2 at 1227.6 MHz is used primarily for the measurement and correction of ionospheric refraction error.

12.2 The position fix

The position fix is achieved basically by determining the distance from the user to each of three selected satellites, and then solving for these three ranges will define the position in three dimensions (Figure 12.2). For a user who is restricted to the earth's surface two range measurements are sufficient to define a position. The range measurement is achieved by measuring the propagation time from satellite to the user. This implies that the receiver clock is precisely synchronized with the satellite clock. However in reality this is not so; instead it is assumed that the receiver clock is in error, and the range measurements are consequently termed 'pseudo-ranges'. The

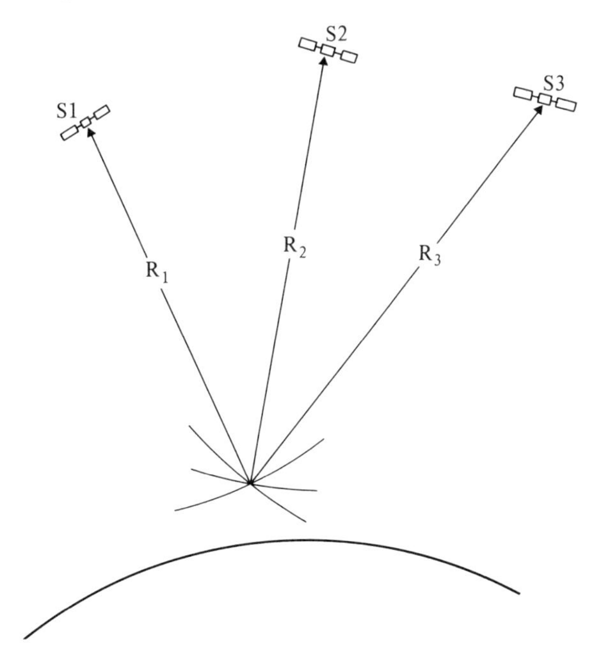

Figure 12.2 Measured distances to three satellites define the position of the observer in three dimensions.

pseudo-range from satellite to user will contain errors due both to the receiver clock, and to propagational delays caused by the ionosphere and the troposphere.

The pseudo-range (PR) for a given satellite is defined as:

$$PR = R + c.070t_A + c(\Delta t_U - \Delta t_S)$$

where: PR = pseudo-range to the satellite
R = true range
c = the speed of light
Δt_S = satellite clock offset from GPS system time
Δt_U = receiver clock offset from GPS system time
Δt_A = propagation delays and other errors.

The receiver clock error becomes a fourth unknown (in a three-dimensional position fix), and so can be resolved from the four equations resulting from pseudo-ranges to four satellites. Similarly there will be three unknowns in a two-dimensional position fix, requiring three pseudo-ranges for a solution. This is illustrated conceptually in Figure 12.3, in which the three pseudo-ranges do not meet at a point but enclose the shaded area. A fixed range value

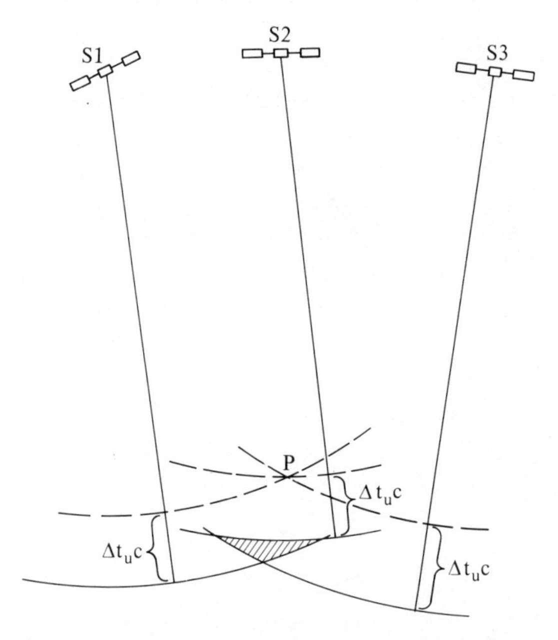

Figure 12.3 The measured ranges (pseudo-ranges) will not meet at a point until corrected for the effect of receiver clock error (Δt_u.C).

$\Delta t_U.c$ can always be found, which, when removed from the pseudo-ranges, will cause the radii to meet at a point P corresponding to the user's position.

Since the receiver computes one value of receiver clock offset Δt_U which is common to all satellites, it is important that the satellite clocks are maintained accurately to a common GPS time. Each satellite therefore has a highly stable atomic clock, with a known or predictable offset from GPS system time. The master control station monitors the satellites' time standards daily, and generates clock correction parameters which are subsequently included in each satellite's own transmitted message.

12.3 Satellite transmissions

As previously mentioned each satellite transmits on two frequencies, L_1 and L_2, (1575 MHz and 1227 MHz respectively). The L_1 signal is modulated with two pseudo-random noise codes, the P code (precision), and the C/A code (known either as clear/acquisition or coarse acquisition). The L_2 signal is modulated with the P code only.

The function of these codes is (i) for satellite identification (since each has a unique code), and (ii) for measurement of the propagation time (and hence the pseudo-range from satellite to user).

The P code operates at 10.23 M.bits/s and has a complete cycle of 267 days. Each satellite generates an exclusive seven-day-long segment of the code. Unless a Navstar receiver has a time standard synchronized with GPS time, and the user knows his position within two to four nautical miles, the P code is difficult to acquire. Therefore use is made first of the C/A code which is only one millisecond long and operates at 1.023 M.bits/s, and to which match and lock-on can easily be made. After this a transfer is made to the P code by using a hand over word (HOW) contained in the navigation message.

2.3.1 *The navigation message*

A navigation message is transmitted from each satellite in the form of a fifty bit/second data stream. The data message is contained in a thirty-second-long, 1500 bit data frame, which in turn has five subframes. Each sub-frame contains system time, and also information which permits the hand over from the C/A to the P code. In addition, the first sub-frame contains data relating to satellite clock correction and propagation delay correction. The second and third sub-frames contain the satellite's ephemeris, defining the position of the satellite in space. The fourth sub-frame contains a message of alpha-numeric characters, the use of which has not yet been specified. The fifth sub-frame contains data relating to the ephemerides, the clock corrections, and the propagation delay parameters for all satellites, thus allowing the optimum choice of satellites for the position-fix solution. After determining which satellites are optimum for the user's location, from the signals of just one satellite, the C/A codes of the selected satellites are generated by the receiver for matching with the codes of all incoming signals.

12.4 The Navstar receiver

All Navstar receivers are comprised of four basic elements, antenna, receiver, computer and the data display segment. Within this basic description there exists a number of possibilities for variation, depending upon the performance requirement.

One possible receiver variation is in the number of signals which can be processed simultaneously. Each satellite transmits three signals, the C/A and P on the L_1 frequency, and the P on the L_2 frequency. The most simple receiver would have only a single channel which utilized the C/A signal only of each satellite in turn. A greater level of performance is obtained from a five-channel receiver which simultaneously processes the L_1 frequency P signal from the four selected satellites, and also sequentially monitors the P signals of the L_2 frequency of each satellite. In each case the effect of the ionosphere is determined by receiving the L_2 frequency.

12.4.1 Receiver operation

The first stage in a Navstar receiver's operation is for it to determine the optimum satellites for use in the position-fix solution. To solve the position in two dimensions requires the solutions of three unknowns (the third being time), and hence three pseudo-range measurements are required. In order to be able to select the optimum satellites, and then to easily lock-on to their C/A codes, the receiver must have a knowledge of the location of each of the satellites, although for both these requirements the information need only be accurate to a few kilometres. Each satellite transmits the ephemeris of all satellites (in the fifth sub-frame), so after one satellite has been acquired the positions of all other satellites are known. It is the acquisition of the first satellite which takes the longest time, since this must be done without any prior knowledge. The acquisition of this first satellite can be accomplished by endeavouring to acquire the C/A signal of each satellite in turn, until synchronization with one of the satellites is successful.

Having once obtained an almanac of satellite ephemerides, the selection of the optimum satellites is made with the aim of achieving a large angular separation at the user, but avoiding low-elevation satellites which would be subject to excessive propagation errors, or which would shortly disappear below the horizon. The choice of satellites is re-assessed at regular intervals and new satellites are selected to maintain an optimum constellation. To further reduce the acquisition time of each satellite, it is necessary for the receiver to determine the magnitude of the Doppler shift component of the received signal. When there is no significant contribution to this from the user's vehicle motion, the amount of Doppler shift due to satellite motion will be within ± 5 kHz.

The receiver's synchronization with the C/A code of a particular satellite is achieved by generating within the receiver the C/A code for the satellite, and then slipping this past the received signal until correlation is achieved. After synchronization, both the carrier phase and the code sequence timing are continuously tracked. In the case of more complex receivers which utilize the longer P code to achieve high-accuracy time measurements, acquisition of the C/A code is simply a stage in the acquisition of the P code. In simple low-accuracy receivers it is the C/A code which is used directly to determine the pseudo-range. The arrival times of the satellite signals are determined within the receiver by comparing the phase of the synchronized internally generated code with the receiver clock. This produces the pseudo-ranges which must be adjusted for propagation effects and satellite clock errors before being used in the position-fix solution.

The propagation effects are mainly the delays in the ionosphere and troposphere. The ionospheric delay is determined from observing the difference in arrival time between the P signals of the L_1 and L_2 frequencies, and then utilizing the relationship of ionospheric delay varying inversely with the square of the carrier frequency. Tropospheric delay cannot be compensated for in this way; instead a simple model which relates delay with the satellite elevation angle is usually employed.

The satellite clock offset (from GPS time) is compensated for by using the correction coefficient which is included in the first sub-frame of the navigation message.

The corrected pseudo-ranges can now be used in the computation which yields position and receiver clock error.

12.5 Position accuracy

As mentioned earlier it is 'expected' that GPS Navstar will be made available for civilian use, but this does not necessarily mean that all the benefits of the system will be made available to everyone.

It is already a declared policy of the controlling authorities that the P codes will be encrypted, thereby denying civilian users access to the P codes (and hence to the L_2 frequency, for ionospheric refraction correction). The precision with which range can be determined is further reduced by using only the C/A code.

Accuracy figures based on limited trials with early receivers indicate that two-dimensional positions with errors better than 25–50 metres, 2d*rms* could be consistently and continuously achievable.

Further improvement on these figures could be achieved in local situations by using differential techniques, similar to those described for the Omega system in section 10.13.

13 Sonar navigation

The word sonar is derived from *SO*und *N*avigation *A*nd *R*anging, but in recent times has become applied to most underwater acoustic systems, even though they may not be strictly related to navigation and ranging. The two sonar systems described here are the echo-sounder, a device for measuring the depth of water, and the Doppler log, a device for measuring speed.

13.1 Propagation of acoustic energy

Sound propagation in water is used in marine navigation because it is the only form of propagation which is at all efficient, all forms of electromagnetic radiation being too heavily attenuated to be of any significant use. But even the propagation of acoustic energy in water is considerably less efficient than the propagation of radio waves in the atmosphere. The attenuation of acoustic energy in salt water due to absorption is shown by Figure 13.1, and in addition to this there is the attenuation with distance due to spreading of the energy. From the viewpoint of achieving a given transmission range with a minimum of radiated energy, the lower frequencies (typically 12–100 kHz) are usually employed, unless the nature of the system requires the use of a higher frequency for a specific reason.

The performance of sonar systems is usually related to the accuracy with which the velocity of propagation is known, but this is a function of water temperature, pressure, and salinity. The velocity is usually taken as being 1500 metres/second for seawater, with a temperature of 13°C, and at atmospheric pressure. The effects of changes in salinity and temperature on the accuracy of an echo-sounder are illustrated by Figure 13.2 and Figure 13.3. Since the echosounder determines depth by measuring the time for a pulse of energy to travel to the sea-bed and back, by taking the velocity as being 1500 m/s (at 13°C and salinity of 35 ppt), any deviation from this is most likely to result in a safe error, giving a measured depth

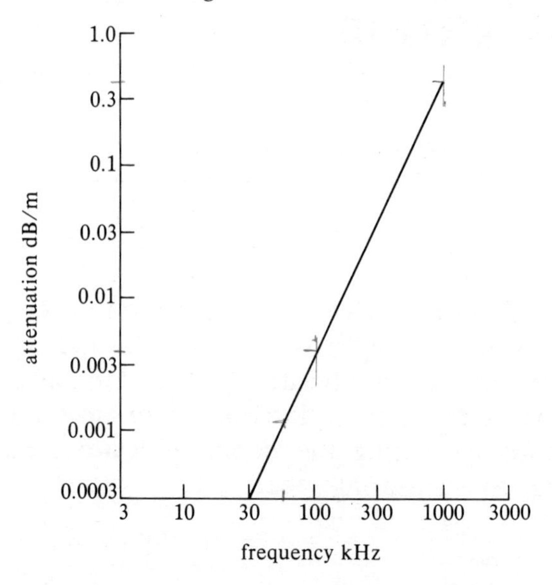

Figure 13.1 *The relationship of frequency and attenuation due to absorption of acoustic energy in salt water.*

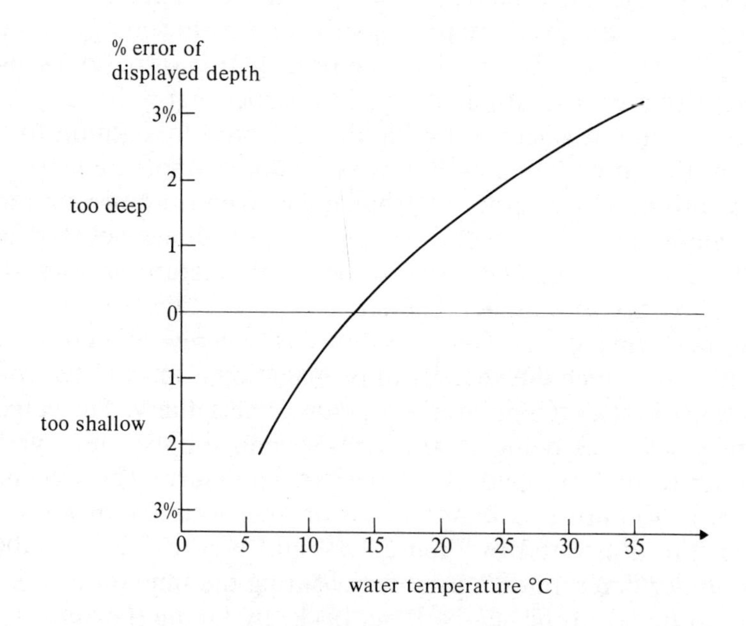

Figure 13.2 *The effect of water temperature on echosounder accuracy.*

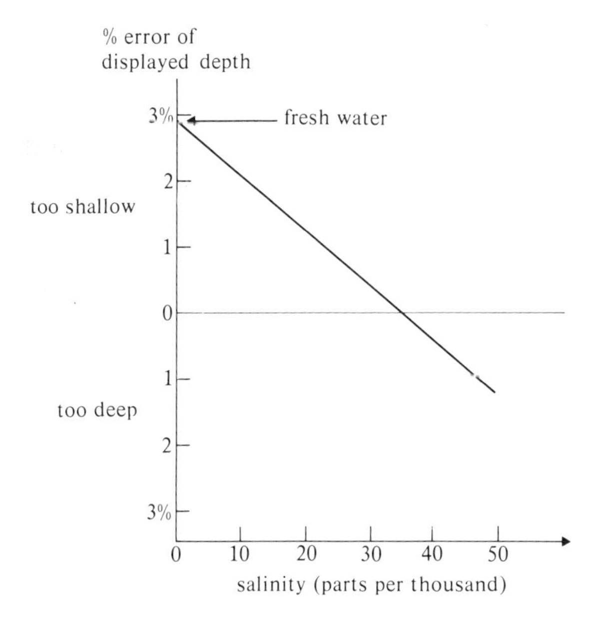

Figure 13.3 The effect of water salinity on echosounder accuracy.

which is less than the true depth. Typical variation of propagation velocity with depth is illustrated in Figure 13.4.

13.2 Reflection of acoustic energy

In some areas of the world the water temperature can vary suddenly with depth; associated with these thermal layers, and occurring at their boundaries, are changes in propagation properties which can cause refraction or reflection of the acoustic wave. A clearly defined division between layers of differing acoustic impedance is required between the water and the sea-bed to produce a strong bottom reflection. For this reason a rocky bottom produces strong reflections, but there are situations when a clearly defined boundary does not occur at the sea-bed, in which case there will be weak or even no bottom echoes. Situations of this sort can occasionally arise over beds of weeds, or over a bottom of freshly stirred-up mud. In both cases, the transition from the sound transmitting properties of the water to that of the bottom is only gradually altered.

When passing over a very dense shoal of fish the true bottom echo may be temporarily obscured, and it is then a matter of deciding from the depth indication whether it is indeed a shoal of fish, or a marked protuberance from the bottom.

When passing over a good reflecting bottom, sufficient energy may be returned to allow a reflection to take place, from either the hull of the vessel or the surface, and then again from the bottom. These second and even third reflections can be seen as false bottom lines on chart recording echosounders.

The final aspect of acoustic propagation to consider is noise, of which there are two prime sources. One is due to external sound sources such as fish, ship noise, and wave motion. The other kind of noise is 'reverberation noise' caused by reflections from air-bubbles, marine life, and mud and sand particles. Each of these causes a

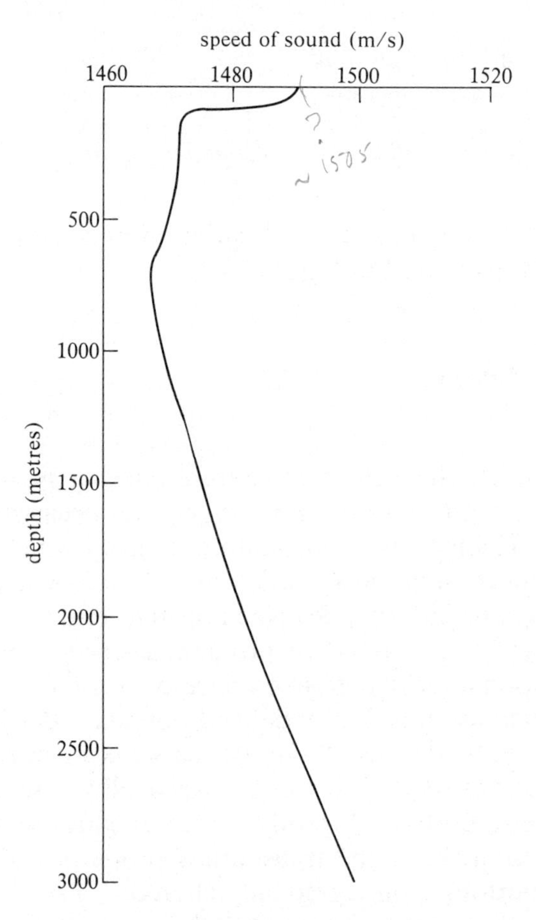

Figure 13.4 Typical variation of propagation velocity of acoustic energy with water depth.

small reflection which produces a continuous return signal, but, due to the attenuation of the signal through absorption and spreading, the reverberation signal will decrease with time after the end of the transmission pulse. The amount of reverberation noise is dependent on the particle density in the water, which will be greatest in fast-flowing muddy rivers, or where two currents mix.

13.3 The transducer

Fundamental to all sonar systems is the means by which electrical energy is converted to acoustic energy and then launched into the water. Equally fundamental is the means by which returned acoustic energy is converted to electrical energy for subsequent processing by the electronic circuits of the sonar system. The device which carries out these functions, the transducer, is the same or similar for both situations, since it is reversible in its operation. There are two basic types of transducer in general use in sonar systems, magnetostrictive and piezoelectric.

13.3.1 *The magnetostrictive transducer*

Magnetrostriction is an effect which occurs in all ferromagnetic materials, being particularly pronounced in iron, nickel and cobalt. When these materials are influenced by external magnetic fields the randomly-orientated magnetic domains within the material become aligned, and, depending upon the material, it will in consequence either expand or contract. A given ferromagnetic material will always either contract or expand in the presence of a magnetic field, irrespective of the direction of that field. The effect is, therefore, that the transducer will vibrate at a frequency of twice that of the electrical current which produced the magnetic field. The frequency doubling effect can be overcome by applying a permanent magnetic bias, either electromagnetically, or by suitably located permanent magnets.

The principal features of a magnetostrictive transducer are illustrated in Figure 13.5. The ferromagnetic material is usually nickel, since this exhibits a significant change (reduction) in length in the presence of a magnetic field. It is necessary to build the core of nickel laminations, each electrically insulated from its neighbour,

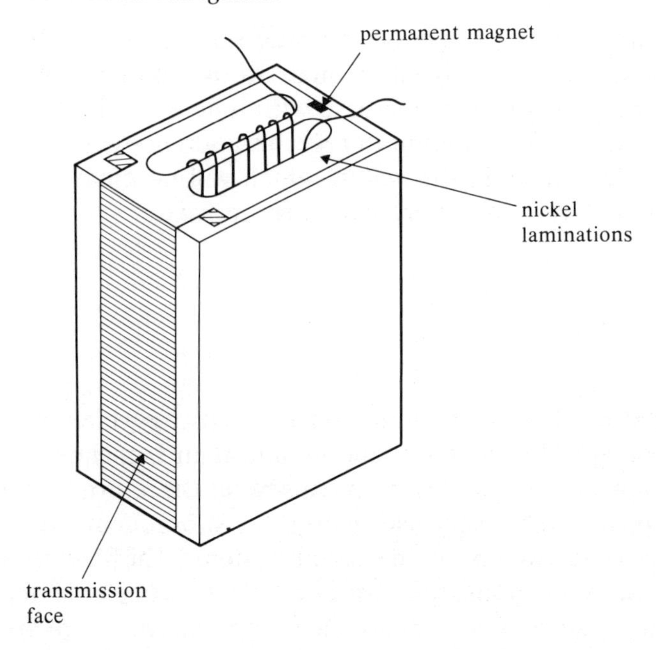

Figure 13.5 Principal features of the magnetostrictive transducer.

in order to reduce losses through eddy currents circulating within the nickel core. The nickel laminations are usually annealed, which reduces hysteresis loss and conveniently produces an oxide layer which provides the electrical insulation. The permanent magnet in Figure 13.5 provides the magnetic bias to prevent the effect of frequency doubling.

In operation, the electrical winding is connected to the transmitter, so that the transmission pulses cause the transducer core to vibrate. To achieve maximum efficiency, the frequency of mechanical resonance of the transducer is made equal to the transmission frequency. By placing the free face of the transducer into contact with the water, the vibrations will be transmitted to the water, thus setting up acoustic waves. The magnetostrictive transducer exhibits a reciprocal behaviour, hence acoustic waves which impinge on the transducer face will cause sympathetic movement within the transducer core. This generates an alternating magnetic field which will induce an alternating voltage in the core winding.

13.3.2 *The piezoelectric transducer*

Certain crystal materials such as quartz, barium titanate and lead zirconate exhibit the property that when mechanically stressed a separation of the positive and negative charges occurs. The reciprocal effect also exists: if an electric field is applied to the crystal its physical dimensions change, the amount of strain being proportional to the electric field intensity. A transducer is formed from piezoelectric material by cutting a slice of the material in such a way that expansion and contracton occurs in the required plane. The slice is then coated on opposite faces, producing, in effect, an electrical capacitor. A voltage applied across the capacitor produces an electrical field which causes either expansion or contraction of the crystal, depending upon the polarity of the field. In a practical piezoelectric transducer assembly the metallic contact surfaces are usually solid stainless steel or aluminium cylinders. The physical dimensions of these are such as to make the whole assembly resonant at the transmission frequency. The assembly is held together by a compression bolt which prevents the risk of failure of the crystal/metal interfaces, and also serves as a tuning adjustment (Figure 13.6).

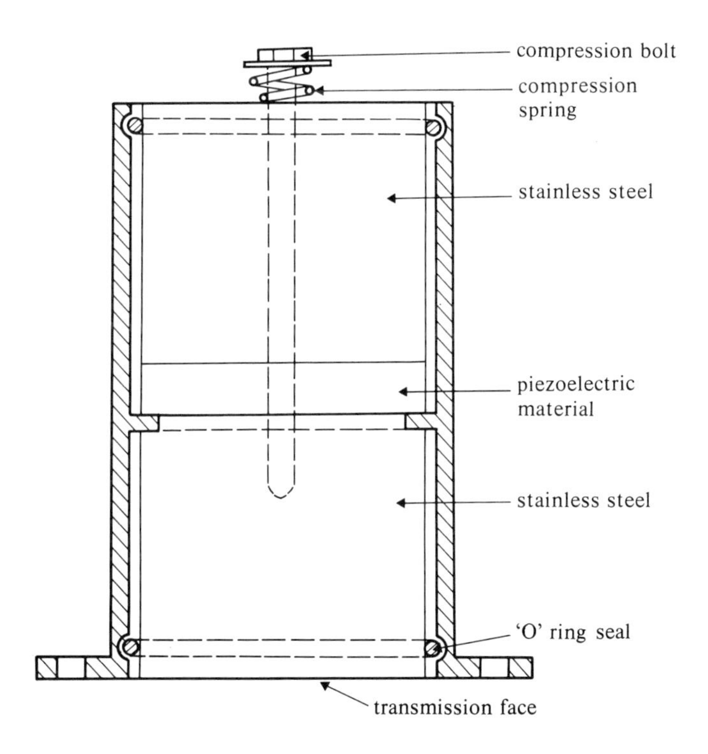

Figure 13.6 Principal features of the piezoelectric transducer.

The piezoelectric type of transducer will operate over a wide frequency range, typically from a few kilohertz to five Megahertz, whereas the magnetostrictive transducer has a limited frequency range of typically 10–100 kHz, although this is adequate for most echosounders.

13.3.3 *Transducer siting*

The correct design of transducer is of course essential, but equally important for optimum performance of any sonar system is the choice of transducer site. For instance, all areas close to noise sources should be avoided. The propeller is one such noise source, and so too are areas of turbulence caused by the discharge of water.

The main problem encountered is one of aeration – that is, air bubbles which pass close to the face of the transducer and act as a large reflector of the transmitted energy. There are usually two causes of aeration. The first is the bow wave, which becomes aerated while it is above the normal water line and is then forced under the hull of the vessel at a point about one-third of the ship's length from the bow. A transducer sited astern of this point would probably be affected by aeration. The behaviour of the bow wave associated with the bulbous bow differs from this; here the aerated water is forced under the vessel at a point very much closer to the bow, and the only point where the transducer can then be sited with any degree of confidence is actually in the bulbous bow.

The second source of aeration is cavitation along the surface of the hull, due to irregularities and protrusions on the hull surface. Sites which are aft of discharge orifices and log tubes should therefore be avoided.

For the reasons mentioned, there are few locations on a ship's hull suitable for the mounting of a transducer. Great care should always be exercised in the choice of transducer site, otherwise the performance of the echosounder or other sonar system may be seriously degraded, if not totally inhibited, by aeration.

13.4 The echosounder

The eschosounder is a device which measures the depth of water by measuring the time taken for a pulse of acoustic energy to travel to the sea-bed and back.

If the velocity of sound in water is taken to be 1500 m/s the depth is given by:

depth (metres) = ½ (1500 × propagation time (t_p)).

The basic functional elements of the echosounder are illustrated in Figure 13.7. The pulse is initially generated as electrical energy by the transmitter and then converted to acoustic energy by the transducer. After reflection by the sea-bed, the returned pulse is converted back to electrical power by the transducer and then amplified by the receiver. The time difference is measured between transmission and reception of the returned pulse and the depth is determined directly. The different types of echosounder vary mainly in the manner in which the time delay is measured, and the depth displayed.

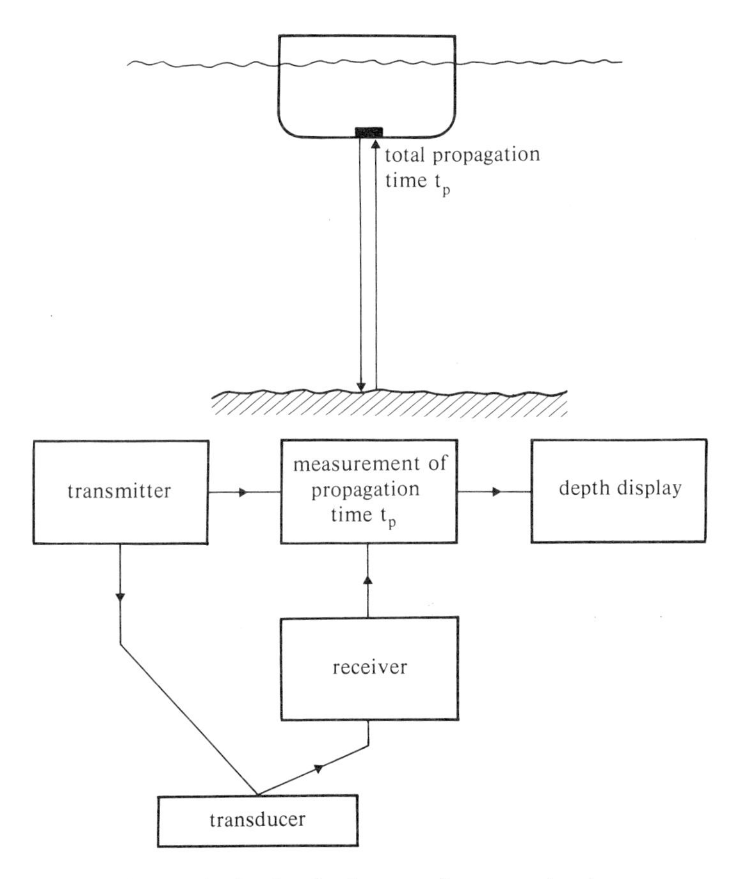

Figure 13.7 *The basis of echosounder operation is measurement of the total propagation time of the acoustic energy.*

13.4.1 The chart recording echosounder

This type of echosounder has a number of inherent advantages over other types. These can be summarized as follows:
- provides a permanent record of soundings;
- rate of change of depth can be clearly seen;
- the bottom line can be seen even if a proportion of the echoes are either false or are lost;
- the measurement of depth and its display is performed by one simple mechanism;
- all echoes can be displayed.

The simple mechanism referred to above is illustrated in Figure 13.8. The endless belt is rotated at a precisely set and regulated

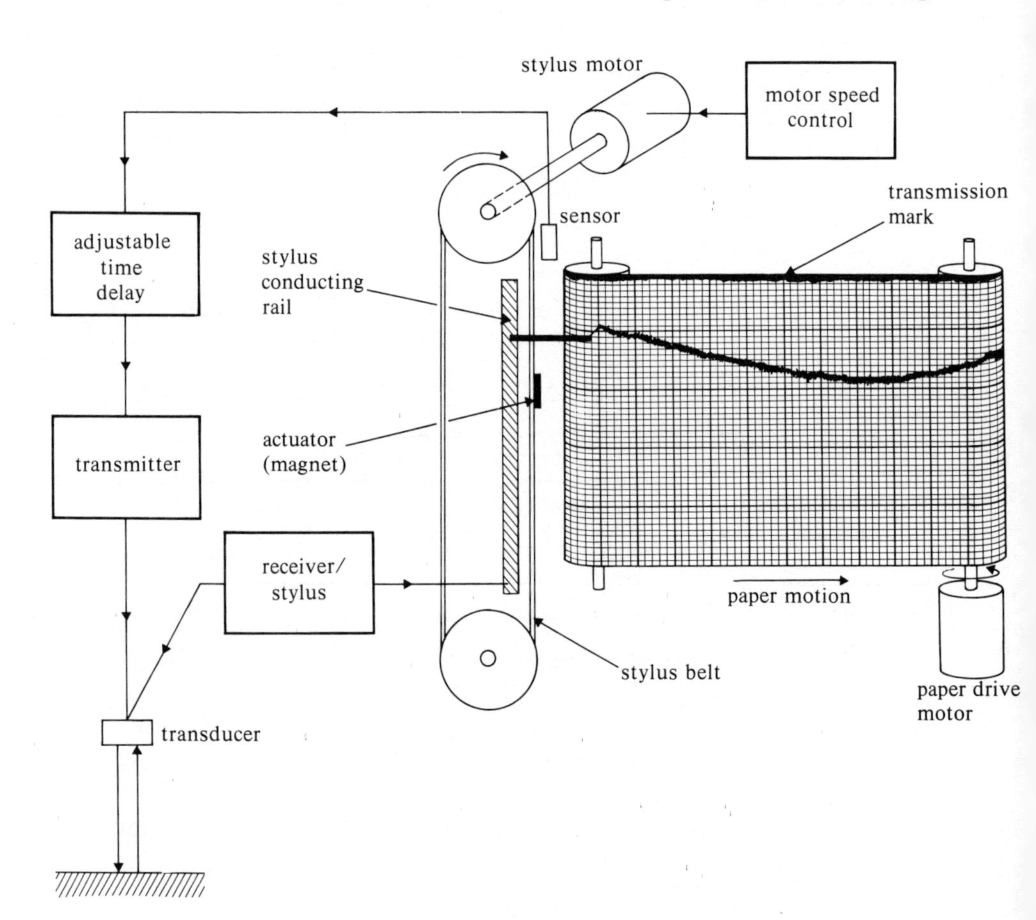

Figure 13.8 Illustration of the principles of operation of the chart recording echosounder.

speed, which relates the time for the marking stylus to travel across the recording paper with the time delay associated with the maximum depth to be displayed. Mounted on the belt, and slightly ahead of the stylus, is a means of activating the transmitter. This is either an electrical contact, or a proximity device such as a magnet or reed relay. The transmission pulse occurs as a result of this mechanism, but it is after a delay period which is precisely set so that at the time of transmission the stylus is aligned with the zero depth mark at the top of the recording paper. Provided that the correct depth range has been selected, at some point before the stylus reaches the bottom of the recording paper the sea-bed echo will return and cause a voltage to be applied to the stylus. This then marks the electro-sensitive paper at a point corresponding with the depth of the water.

The maximum recordable bottom depth is determined by the speed at which the stylus travels across the recording paper. It is usual to display at least two depth ranges, a shallow range typically 0–100 metres, and a deep range typically 0–1000 metres. In this case, on selecting the deep range, the stylus would travel at one-tenth of the speed of the shallow range.

An alternative method of providing different ranges of displayed depth, is to maintain a constant stylus speed but change the point at which the transmission pulse occurs. This method, known as phased ranges, is illustrated in Figure 13.9. In this case there are four transmission sensors positioned symmetrically around the stylus belt. Selecting sensor 1 is equivalent to the basic configuration of Figure 13.8. The delay circuit adjustment is set so that transmission occurs precisely as the stylus reaches the zero graduation on the paper. The bottom graduation of the paper will correspond to a depth which is dependent on the stylus speed (for example 100 metres). On selecting range 2 the transmission pulse occurs before the stylus reaches the paper; the interval is such that the top graduation corresponds to the same depth as the bottom graduation of range 1, that is, 100 metres; the bottom graduation of this range will then correspond to 200 metres. Similarly range 3 will be 200–300 metres and range 4 will be 300–400 metres.

The operation of a typical chart recording echosounder of the first type (change of stylus speed) is shown in more detail by Figure 13.10. This diagram shows that in changing range three separate functions are affected:

(1) The stylus speed is changed by the appropriate factor. Within the equipment there will be pre-set adjustments which are used to

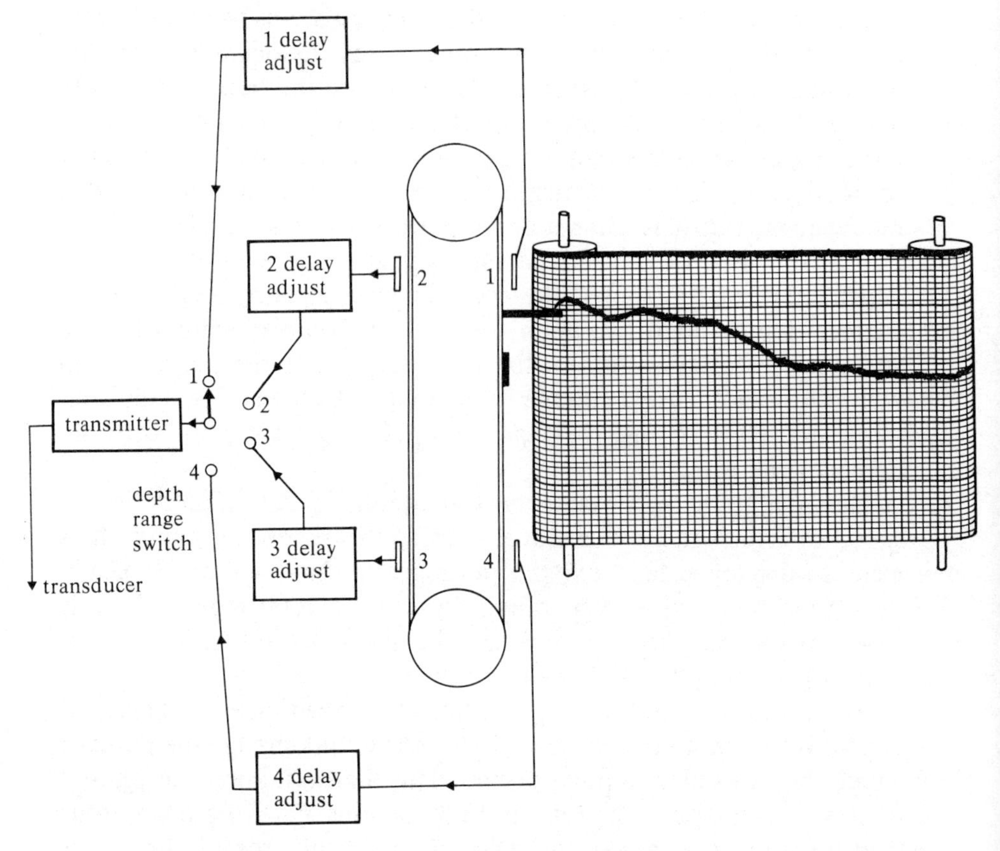

Figure 13.9 The principle of phased ranges.

initially set the motor speed on each range, so as to accurately calibrate the depth graduations on the recording paper.

(2) The pre-set time delay, which ensures that transmission occurs precisely as the stylus passes the zero graduation, is changed to allow for the change in stylus speed.

(3) The transmission pulse length is usually changed with change of depth range. The generated transmission pulse is basically a number of cycles of electrical energy, usually at the frequency of resonance of the transducer. For a conventional echosounder this would typically be in the range of 20–100 kHz. For the shallow range a short pulse is desirable since the length of pulse limits the minimum depth which can be measured as the echo cannot be received until the transmission has ceased. A pulse length of say 1 ms means that the minimum measurable depth is theoretically 0.75 m. However, because the transducer is a resonant system, it

will continue to oscillate even after the transmitter ceases to deliver energy to it. This will effectively limit the minimum depth for a 1 ms pulse to typically 1.5 metres, so a shorter pulse is required if depths less than this are to be indicated. In the case of the deep ranges, the objective is to transmit sufficient energy for a good return signal. This is both a function of pulse length and pulse amplitude, and so on selecting the deeper ranges the pulse length is increased up to typically 5 ms.

Very often one transducer is used for both transmission and reception, and so to prevent the high voltages which exist across the

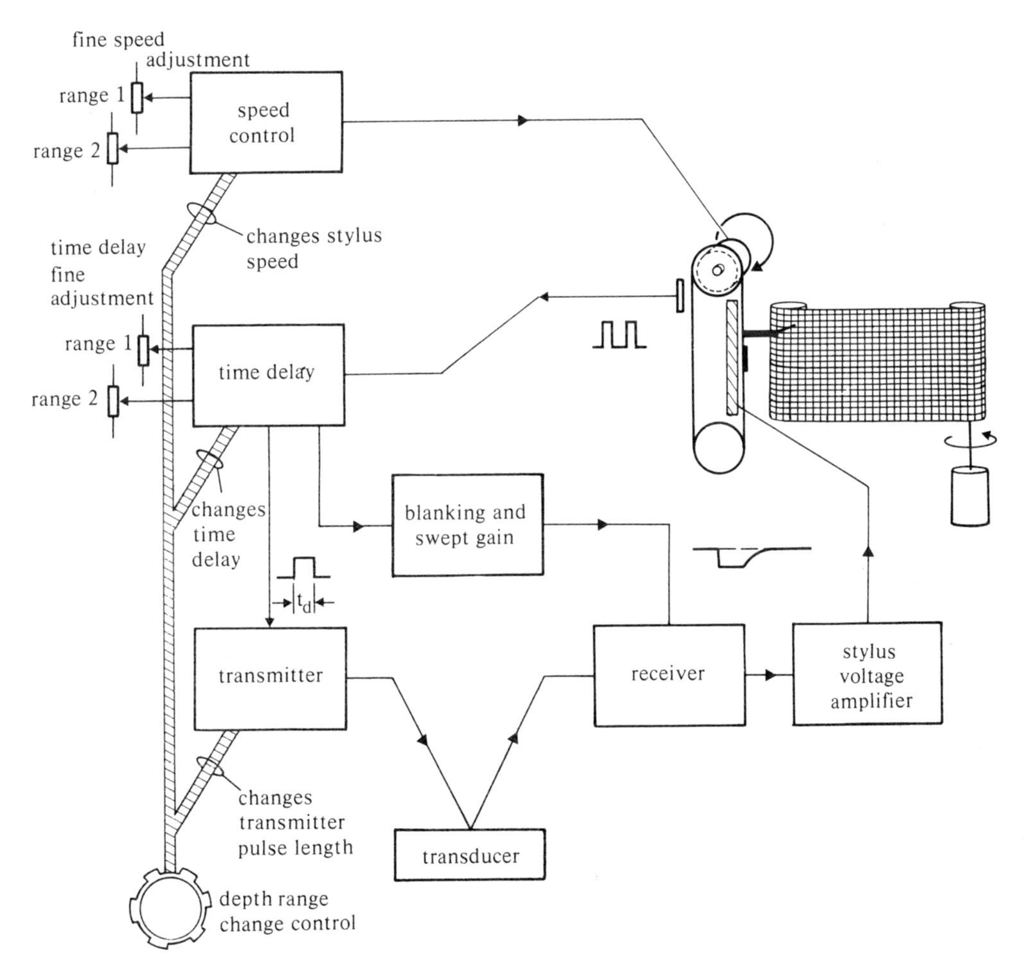

Figure 13.10 Functional diagram of the chart recording echosounder.

transducer terminals during transmission passing through the receiver, a blanking voltage is applied to inhibit receiver operation during this time. After cessation of transmission, the receiver gain is not immediately restored to normal since there will still be relatively large amounts of energy due both to ringing of the transducer and to reverberation noise. Instead automatic swept gain is applied, which returns the receiver to normal operation in a manner proportional to the decay of these effects.

13.4.2 The electronic echosounder

The chart recording type of echosounder has two disadvantages: (1) the rotating stylus mechanism can be considered an advantage in providing a simple means of measuring depth but it can also be considered a disadvantage, since moving parts require periodic servicing and adjustment; (2) the system does not provide a display of depth at locations remote from that of the echosounder itself. To overcome these two disadvantages a number of all-electronic echosounders have been developed. The more recent models of this type measure the time delay of the returned pulse using digital techniques, and then display the depth on one or more digital or analogue display indicators (Figure 13.11).

The main difficulty in designing an echosounder of this type arises from the fact that only one echo can be displayed, and so the electronic processing must unerringly select the bottom echo from other reflections. To this end, correlation techniques are used to discriminate consistent bottom echoes from random noise echoes. In the event of no detectable bottom echo occurring during a sounding cycle, the depth determined from the previous sounding cycle can remain displayed. However, after several sounding cycles without a detectable bottom echo, a warning of this fact should be clearly indicated.

13.4.3 Using an echosounder

The echosounder is basically a very simple navigation instrument to use, usually requiring the periodic adjustment of only two controls, gain and range selection, and occasionally requiring a change of paper if it is this type of instrument.

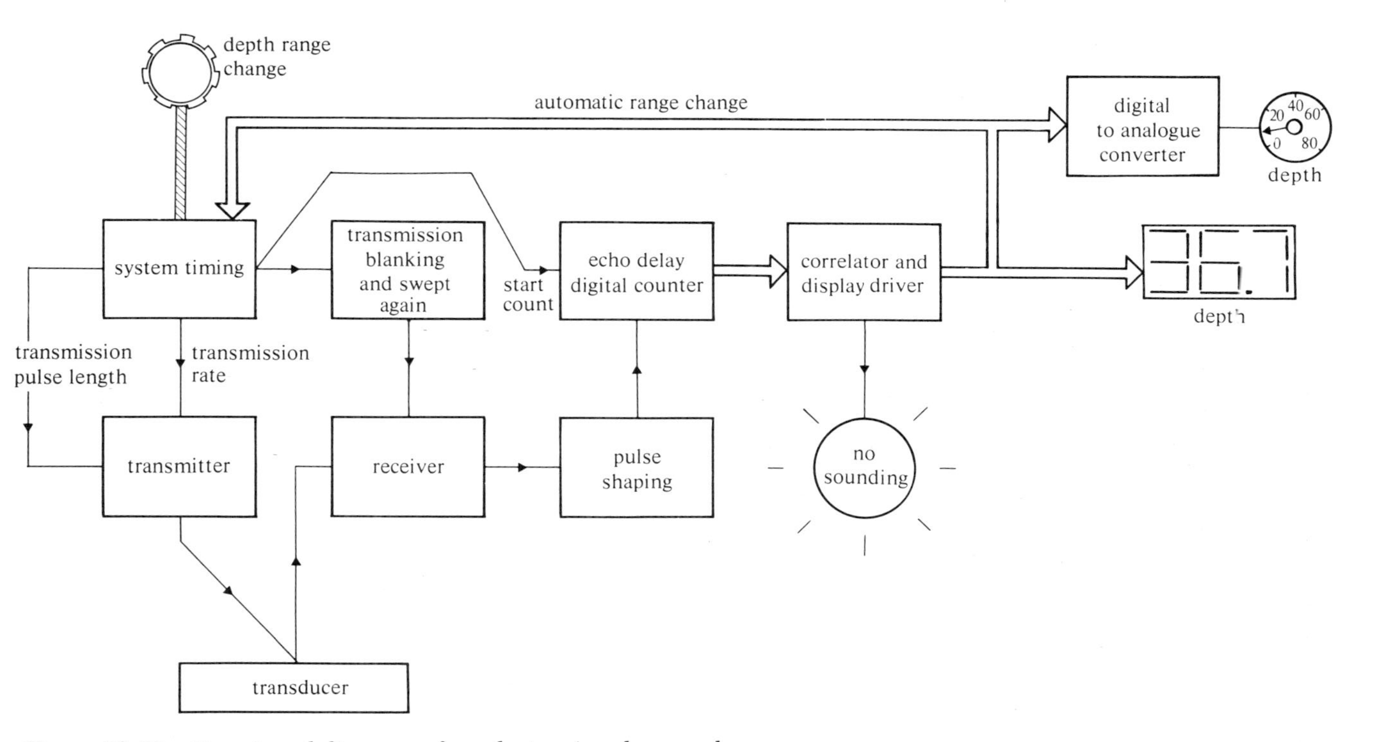

Figure 13.11 Functional diagram of an electronic echosounder.

13.5 The Doppler sonar log

A sonar beam can be used to determine a ship's speed by measuring the frequency shift (due to the Doppler effect) of an acoustic wave returned from a fixed reflector. Assuming the reflector to be directly ahead, and on the same plane as the transmitting/receiving transducer, the frequency shift is given by:

$$f_d = \frac{2Vf_0}{C} \qquad \textbf{[13.1]}^*$$

where $f_d = f_0 \neq f_r$

and f_0 = transmitted frequency
f_r = received frequency
V = velocity of vessel
C = propagation velocity of acoustic wave.

In reality, the only fixed reflector is the sea-bed and so the Doppler shift represents only a component of the horizontal motion (Figure 13.12) given by:

$$f_d = \frac{2Vf_0 \cos\theta}{C} \qquad \textbf{[13.2]}$$

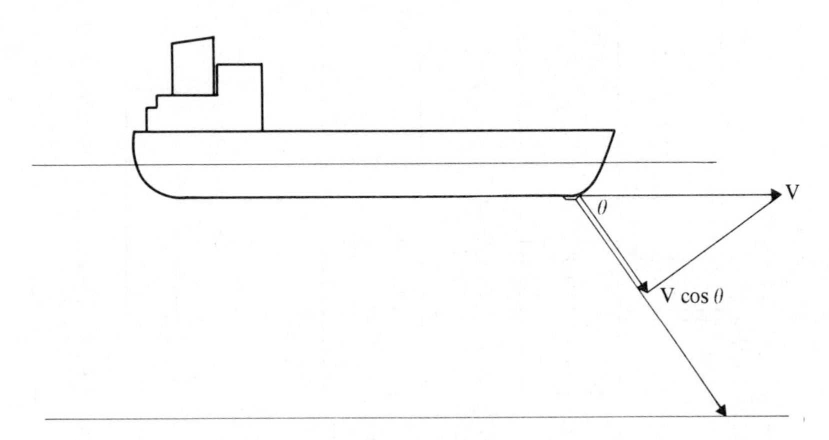

Figure 13.12 *The observed Doppler shift is proportional to the component of velocity along the sonar beam.*

*see Appendix: derivation of Doppler shift, on p. 239.

Any vertical motion (V_v) will have a component of velocity in the direction of the sonar beam resulting in a Doppler frequency shift error component of:

$$\frac{2V_v f_0 \sin\theta}{C} \qquad [13.3]$$

A means of overcoming this problem is to utilize the Janus configuration (Figure 13.13), in which a second transducer is inclined at the same angle θ but such that the beam is directed in the aft direction. This second sonar beam will still produce the vertical motion error component. However, since the horizontal motion relative to this beam is now $-V$, the Doppler frequency shift due to this motion is:

$$\frac{-2Vf_0\cos\theta}{C}$$

and so if the difference of the two Doppler shifted signals is developed, the components of vertical motion will cancel and the components of horizontal motion will be additive:

$$\left[\frac{(2Vf_0\cos\theta)}{C} + \frac{(2V_v f_0\sin\theta)}{C}\right] - \left[\frac{(2V_v f_0\sin\theta)}{C} - \frac{(2vf_0\cos\theta)}{C}\right]$$

resulting in:

$$f_d = \frac{4Vf_0\cos\theta}{C} \qquad [13.4]$$

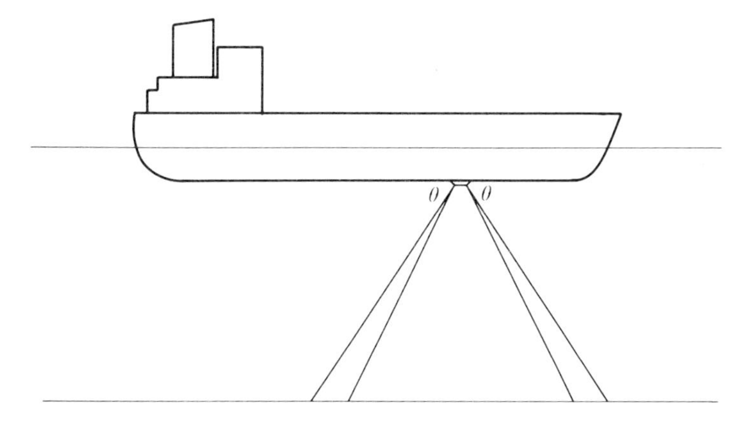

Figure 13.13 The dual sonar beams of the Janus configuration.

velocity is thus given by:

$$V = \frac{f_d\, C}{4 f_0 \cos\theta} \qquad \textbf{[13.5]}$$

for the forward looking transducer

$$f_d - f_f = f_f - f_0$$

where f_f = received frequency of forward transducer

and for aft transducer

$$f_d - f_f = f_f - f_0$$

where f_a = received frequency of aft transducer.

The difference of the two Doppler frequencies gives:

$$(f_f - f_0) - (f_a - f_0) = f_f - f_a$$

$$= f_d \text{ in equation } \textbf{[13.5]}.$$

That is f_d is the difference of the two received frequencies.

13.5.1 The effect of pitching

The Janus configuration does not compensate for the effect of pitching and rolling. If a ship pitches by an amount β (Figure 13.14), the horizontal components become:

$$V\cos(\theta + \beta) + V\cos(\theta - \beta)$$
$$= V(\cos\theta\cos\beta - \sin\theta\sin\beta) + V(\cos\theta\cos\beta + \sin\theta\sin\beta)$$
$$= 2V\cos\cos\beta \qquad \textbf{[13.6]}$$

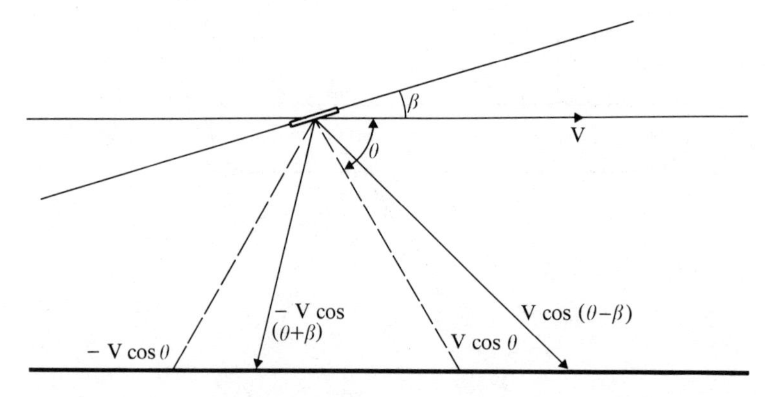

Figure 13.14 The effect of pitching (and rolling) on a Janus configuration.

The cos β term of equation [**13.6**] reduces the velocity vector for any direction of rotation by an amount proportional to the average pitch (or roll) angle.

13.5.2 Two-axis speed

The fore-aft transducer configuration so far described measures only the forward speed. Athwartships motion can be measured by a second pair of transducers in a Janus configuration producing port and starboard beams (Figure 13.15).

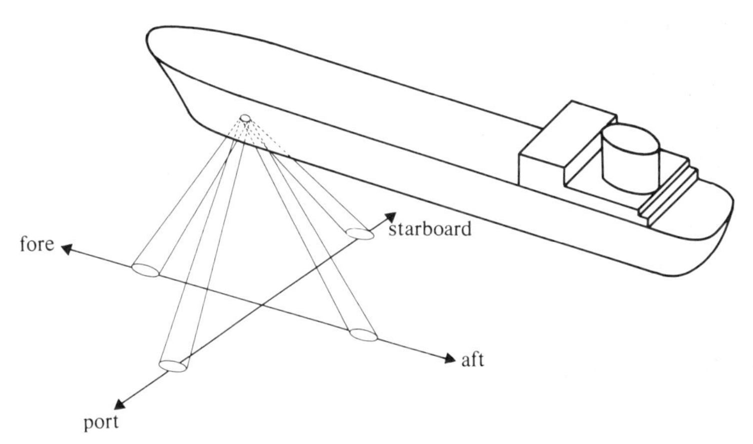

Figure 13.15 The sonar beams of a dual-axis Doppler system.

13.6 Bottom reflections

The transducers of a Doppler log system are usually inclined such that the beams are incident on the seabed at an angle of sixty degrees, this being a compromise between the two requirements of a usable return signal and a usable component of horizontal velocity. In the case of a perfectly smooth bottom the beam would be reflected away from the vessel at an angle equal to the angle of incidence. Fortunately there is usually some back scatter, although ultimately the working depth is limited by the nature of the seabed. At depths approaching the limit of operation (typically 200 metres) intermittent loss of signal can result from a vessel's roll and pitch, which causes longer transmission paths, and shallower reflection angles. In these situations it is normally arranged that the Doppler log remembers the velocity determined from the last usable signal. The full potential of a Doppler sonar log in measuring speed over the ground is only achieved while it is operating in this bottom track mode. When the water depth reaches the point at which the returned signal has attenuated to an unusuable level, the Doppler log can continue to function by making use of reverberation echoes. Even in the water track mode the Doppler log continues to have advantages over other types of log, since the Doppler measurement can be referenced to the layer of water ten to twenty metres below the vessel. This eliminates the effect of disturbances caused by the vessel itself, and the effect of surface currents caused by wind and wave motion.

The Doppler frequency shifts are dependent upon the velocity of sound in the water in the immediate vicinity of the transducer, but this velocity varies with the temperature and salinity of the water. A correction for temperature variation can be included by using a temperature probe built into the transducer housing. A more accurate correction can be derived from a velocimeter, which can also be built into the transducer housing.

13.7 Doppler sonar equipment

The choice of transmission frequency for a Doppler sonar system is a compromise between the transducer size, which increases as the frequency reduces, and the absorption loss, which at the higher frequencies is proportional to the square of the frequency. Transmission frequencies are usually within the range 150–600 kHz. If pulse transmission is employed the transducers for transmitting and receiving can be common. This is usually the case, since pulse transmission permits the use of a higher transmission power with resulting increase in the amplitude of the returned signals. Transmitting a continuous wave has the advantage of providing a more precise velocity resolution, and permits operation with little clearance between the hull and seabed. It is most suitable when the Doppler sonar system is required to measure the very low lateral velocity of a VLCC during its berthing. In using a Doppler sonar system as a berthing aid, athwartships transducers are often sited at both fore and aft positions on the hull, providing independent measurements of lateral motion.

Appendix: derivation of Doppler shift

In the case when the source and the observer are both moving towards the reflection the frequency (f_r) of the received wave is given by:

$$f_r = f_0 \frac{(C+V)}{(C-V)}$$

$$f_r = f_0 \frac{1+V/C}{1-V/C}$$

where: f_0 = transmitted frequency
f_r = received frequency
C = velocity of wave
V = relative velocity of observer and reflector.

Using infinite series $1/(1-x) = 1 + x + x^2 + x^3 + x^4 + \ldots$ gives:

$$f_r = f_0 (1 - V/C)(1 + V/C + V^2/C^2 + \ldots)$$

and since $C \gg V$* all terms above $2V/C$ can be neglected.

Therefore:

$$f_r = f_0 (1 - 2V/C)$$

and $f_d = f_0 - f_r = 2Vf_0/C$ where f_d = Doppler frequency shift.

* For water C is approximately 5000 ft/s, and V would not usually be greater than say 5 ft/s (= 30 kts).

14 Berthing systems

Various [illegible] during a vessel's channel approach and final berthing manoeuvres. The information which is usually of most value, particularly for very large vessels, is speed over the ground, since to maintain the kinetic energy of such vessels below that which can be absorbed by the berth requires a skilful judgment of velocity.

During the approach to the berth, distance-off can be adequately determined from radar, but during the final stages of berthing it can be of assistance to have independent measurements of distance from the bow and stern, to the berth. At least one type of berthing system provides this information.

Three systems are described here; the first, the Doppler sonar log, is mentioned only briefly, since it is described more fully in chapter 13. The other two systems are normally located on the berth, thereby reducing the risk of damage to the berth from all ships which use it.

14.1 Doppler sonar log

The Doppler sonar system is capable of giving a high accuracy, of the order of one per cent, and can detect movement over the ground down to approximately 0.5 cm/s, which makes it ideally suited as a berthing aid. It also has the significant advantage that the velocity information is presented visually on board the vessel. The Doppler sonar provides all the necessary velocity information during the whole of the ship's approach to the berth. This equipment is not limited by range or weather conditions, but it does have the limitation that there is no way of providing distance-off information. It also has the disadvantage that if the ship's own engines are being used for manoeuvring, and are run astern for longer than a few minutes, readings can be temporarily interrupted due to aeration interference under the transducers. Such interference can also occasionally be caused by tugs.

14.2 Jetty-mounted sonar systems

The basic elements of jetty-mounted sonar systems are sonar transducers mounted on the piles of the jetty three to five metres below the water line, and pointing outwards from the jetty in the direction from which the ship will approach (Figure 14.1). Typically the transmission occurs at a frequency of 100 kHz with a peak power of thirty watts, and a beam width of ten to twenty degrees. The measurement made can be either a direct speed measurement by determination of the Doppler shift of the signal reflected from the vessel, or it can be a pulsed range measurement, with change of range with time giving the speed. This type of system can provide a more than adequate level of accuracy, typically better than ten centimetres on range measurements if water velocity correction is made, otherwise better than two per cent of distance measured. The ability to determine both speed of approach and distance-off is a distinct advantage of this type of system. It suffers from the disadvantages of a limited range of 100–200 metres, and is liable to suffer interference from aeration produced by the propellers of tugs, and other small vessels in the vicinity during the berthing.

14.3 Microwave Doppler systems

Microwave Doppler systems utilize the Doppler frequency shift of the microwave signal reflected from a moving target, to determine the relative velocity between the source and the target. This is in the same way that a sonar Doppler system uses the Doppler shift in the frequency of the reflected acoustic signal. The source of the radiated microwaves can be located either on the vessel, or on the berth, since in each case the relative motion is the same, and each is an adequate reflector of microwave energy. Both arrangements are used in practice, but the description here relates specifically to the berth-mounted system, since many more vessels benefit from a single shore installation.

In principle, a microwave Doppler system could operate in any of

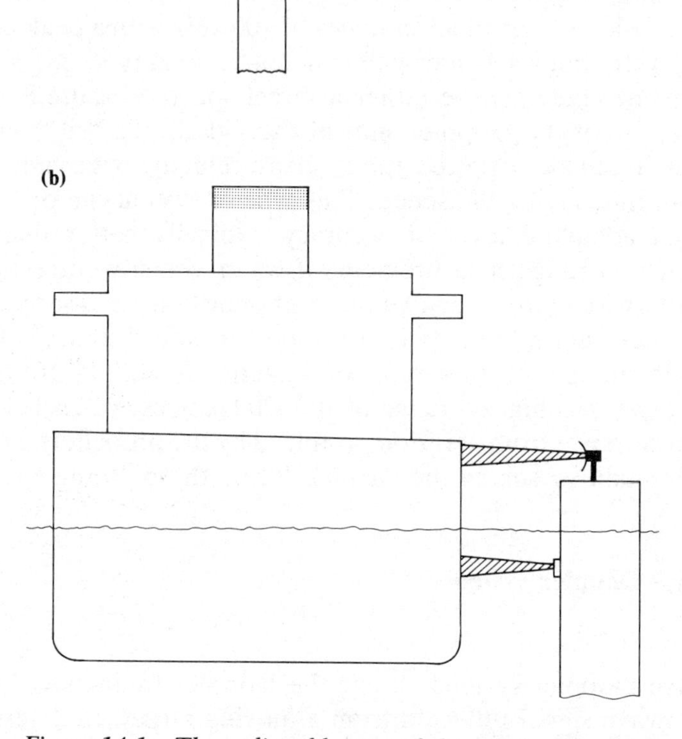

Figure 14.1 *The radiated beams of shore-based berthing systems:* (a) *sonar and microwave Doppler;* (b) *above-water microwave, below-water sonar.*

the frequency bands used by marine radar, and some do in fact operate in the radar X-band of 8.2–12.4 GHz, but in recent years frequencies have been specifically allocated within the 14 GHz band.

The maximum operational range is achieved by forming the transmitted energy into a narrow beam, using an antenna which

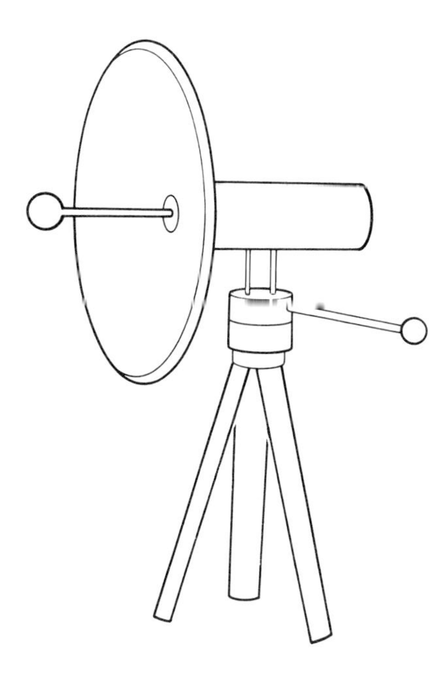

Figure 14.2 *The microwave unit of a Doppler system; the velocity display may be an integral part of the microwave unit, or part of a separate display unit.*

incorporates a parabolic reflector (Figure 14.2). The actual range at which usable returned signals are first received depends on many factors, including the area of the reflecting surface, the aspect of the approaching vessel, the sea state, precipitation, and the phase stability of the microwave source. With the maximum allowable effective radiated power, and all other conditions being favourable, the maximum range of a very large vessel is typically two to three kilometres. Whether or not use is made of this maximum range depends upon the nature of the approach to the berth, but undoubtedly there are many cases in which the full range is used to good effect. To use the system at long range requires the collaboration of shore personnel, both to maintain the beam on the approaching vessel, and to relay the velocity readings. When the system is used during this berth approach phase, care must be exercised to ensure that the measured velocity is not used when it only represents a component of the vessel's true forward speed (Figure 14.3).

During the final stages of the berthing operation it is usual for two systems to be used, in order to obtain independent measurements of

Figure 14.3 The measured velocity (V_L) must only be taken as representing the vessel's forward speed when the angle θ is small.

the vessel's lateral motion at both the bow and stern. It has also been found necessary to incorporate a display of relative direction of motion to the berth, since this is not always obvious when a large vessel is moving very slowly, and perhaps has components of both lateral and rotational motion.

In many cases the involvement of shore personnel is required throughout the berthing to relay velocity information, but in other cases the need for this has been eliminated by the use of large analogue or digital displays, which are visible from the bridge of the approaching vessel.

The microwave Doppler system is obviously unaffected by the disturbing influences which affect sonar systems, but erroneous readings can be caused by small boats and lines crossing the microwave beam. Also, with fully laden tankers the beam may not strike the side of the hull, but instead it may pass above the main deck. In such situations there is usually sufficient returned signal, but the readings could be affected by movement of personnel on deck.

Loss of signal can occur if the vessel subtends a large angle with the berth, since the reflected energy is no longer towards the source.

14.3.1 *Theory of operation*

The Doppler shift frequency is given by:

$$f_d = \frac{2Vf_0}{C}$$

(from Appendix: derivation of Doppler shift, on p. 239)

where C = velocity of electromagnetic radiation
V = relative velocity, source – reflector
f_0 = radiated frequency

Taking as an example a radiated frequency f_0 of 14.0 GHz, and a relative velocity V of ten knots, this yields a Doppler shift f_d of 476 Hz. For a relative velocity of only 20 cm/s, which is typically the maximum impact velocity of a very large fully-laden vessel, the Doppler shift is 18.6 Hz. In order to be able to display the very low velocities at which large ships are berthed, Doppler shifts of less than 1 Hz must be measured. The source of the microwave energy must therefore have sufficient short-term stability for it to be possible to measure the small frequency differences between the transmitted and received signals. In practice this proves to be possible only at close range when the propagation time is short, but this is acceptable since it is only when a vessel is close to the berth that low velocities become significant.

The source of microwave energy is usually a Gunn diode oscillator, which is basically a thin wafer of *n*-type gallium arsenide, mounted in a resonant cavity. The application of a voltage across the diode sets up domains which pass through the gallium arsenide, the transit time determining the frequency of the generated microwave energy. The microwave assemblies of Doppler systems of this type are usually similar to that depicted by Figure 14.4. The continuously generated microwave energy (10–50 mW typically) passes through an isolator and circulator to the antenna, where it is formed into a beam and radiated towards the approaching vessel. Prior to this a small amount of the energy is removed, to be used as the reference signal. The reflected energy P_R which is received by the single antenna, is now at a frequency f_r having suffered a Doppler shift. This received signal is routed via the circulator to the diode mixer assembly D1, which is also fed with the reference frequency f_0, and thus produces as an output the difference frequency f_d ($= f_r - f_0$).

The output from the mixer is at a very low level, and so requires amplification before it is converted to a measure of velocity, and displayed in a suitable form.

Figure 14.4 shows two mixer assemblies, where the purpose of the

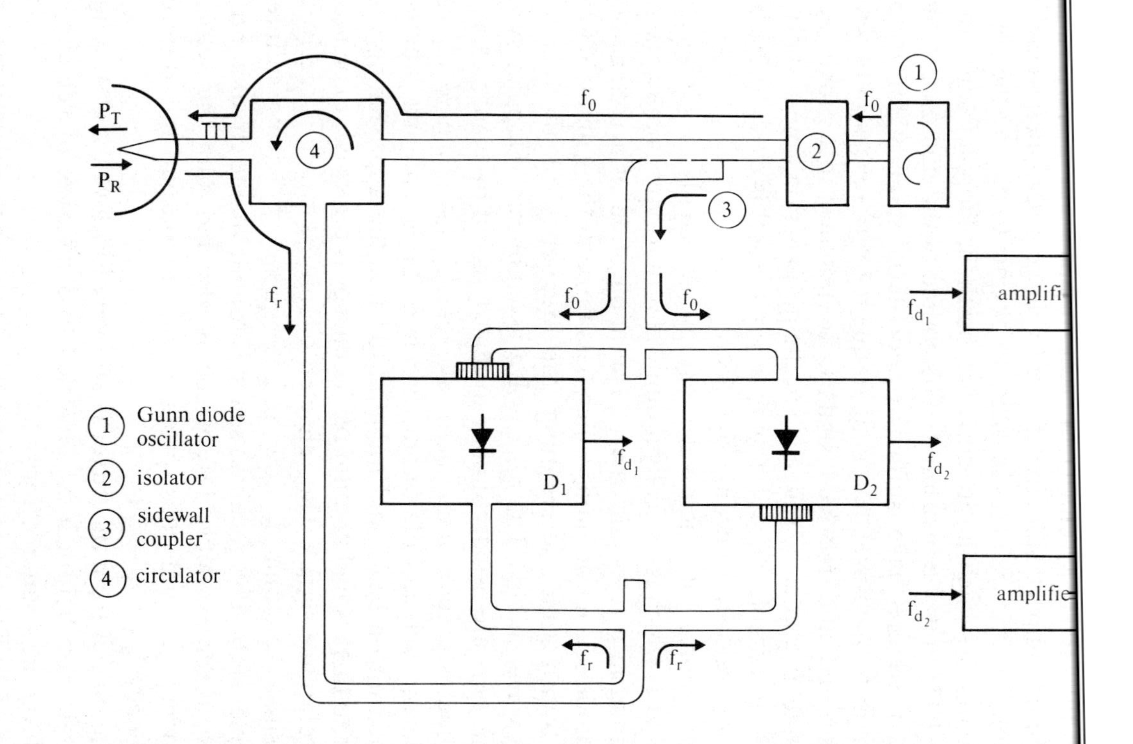

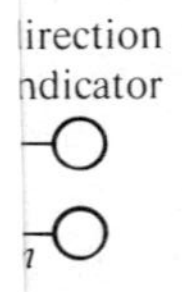

Figure 14.4 Schematic representation of a typical microwave Doppler berthing system

second is to determine whether the direction of the moving target is towards or away from the source. The path length of the reference signal is one-eighth of a wavelength longer to one mixer than to the other. Similarly, the received signal path lengths are unequal by the same amount. The two mixers produce the same difference frequency f_d, but they differ in phase by ninety degrees.

When the target is approaching the source, the received signal has a higher frequency than the reference, and the difference frequency from the first mixer lags that from the second by ninety degrees. When the target is receding, the phase relationship is reversed. From these two possible phase relationships the direction processing determines the relative direction of the vessel's motion, which is then displayed as either *to* or *from* the berth.

14.4 Microwave range measurement

The microwave Doppler system is limited by its ability to provide only speed information, and is not able to provide a measurement of distance from the berth. In practice it is usually the former which is the most difficult to judge, and so a precise measurement of speed is usually considered to be the prime requirement, but various methods of measuring distance using microwave techniques do exist.

In general the measurement of distance is more susceptible to extraneous effects than velocity measurement. Consider for example the situation where all or part of the microwave beam passes across the deck of the approaching vessel. The Doppler measurement will usually be unaffected, since all parts of the vessel within the beam are moving at the same speed, and so the many reflections occurring from deck structures will all have the same Doppler shift. When measuring range under these circumstances the returned signal is a composite of reflections occurring at many different ranges, and, depending upon the particular measurement technique, it may prove difficult to determine the precise distance from the berth to the nearside of the vessel.

The effects of multiple signal paths and antenna sidelobes present a greater problem in the case of distance measurement, since reflections from any target within the beam will be confused with the wanted signal, whereas with the measurement of velocity, all but moving targets can be immediately excluded.

Under the ideal conditions of sufficient freeboard to represent an unambiguous target, and no multiple path

transmitted pulse, and, from a knowledge of the pulse's velocity, the distance from the source to the target can easily be derived. The limitation of this technique is that time delays cannot be measured which are shorter than the pulse length, thereby limiting the minimum range to typically twenty metres.

An approach which is more suitable for this application of distance measurement is that employed in FM radar. The transmission is continuous but is linearly frequency modulated (Figure 14.5). The

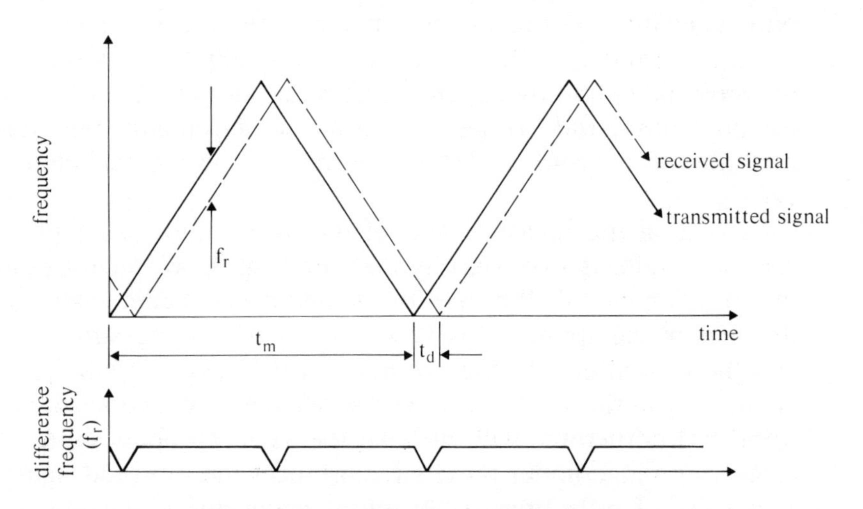

Figure 14.5 The FM radar technique produces a difference frequency (f_r) proportional to target range.

frequency of the received signal will vary in the same manner, but due to the round-trip time delay, the received signal will always be at a different point in the modulation cycle. So for a specific target range, there will be a specific and constant frequency difference f_r, given by:

$$f_r = \frac{4 . f_m . R \triangle f}{C}$$

where f_m = frequency of the modulation ($1/tm$)
$\triangle f$ = peak to peak frequency deviation
R = source-target distance.

For example:

typically f_m = 100 Hz and $\triangle f$ = 100 MHz
for a distance (R) = 100 metres
the difference frequency f_r is 13.33 kHz.

One microwave source which is suitable for this type of system is the varactor-controlled Gunn diode oscillator, since variation in the voltage applied to the varactor diode, produces a varation in the generated frequency. The rest of the microwave assembly can be exactly the same as that required for Doppler velocity measurement, since in each case the output is the frequency difference between the transmitted and received signals. Speed and distance can be easily obtained from one microwave assembly, by alternately transmitting steady carrier and frequency modulated carrier.

15 Integrated navigation systems

navigation data. The purpose of combining navigation aids in this way is to achieve a performance which is better than that of the individual constituent systems, and to provide this improved performance over the widest possible area. The scope of the larger integrated systems is often widened by including in the tasks of the central computer both radar anti-collision and general navigation computations. At the other extreme are hybrid systems, which have only two constituents, each chosen to complement the other.

15.1 Hybrid systems

The navigation aid which is most frequently used as a component of both large- and small-scale integration is the Transit satellite system. This is because Transit has a valuable contribution to make through its global coverage and high position-fix accuracy, but requires a complementary system to off-set its major limitation of only producing fixes at irregular intervals. In addition to the requirement of the complementary system to provide a navigation input during the interval between satellite passes, it is also necessary to provide the Transit system with velocity data during the pass, so that compensation can be made for the effect of the vessels' own motion on the observed Doppler shift of the satellite's transmission. The second system is therefore required basically to measure the vessel's velocity, since this can be integrated with time to provide position information during the intervals between the satellite fixes. Velocity is most usually derived from a speed-log and a gyro-compass (Figure 15.1), although velocity over the ground can be obtained more accurately from a bottom tracking dual-axis Doppler sonar system. Since the time between satellite fixes is accurately known, a precise average velocity can be computed for the interval between the fixes.

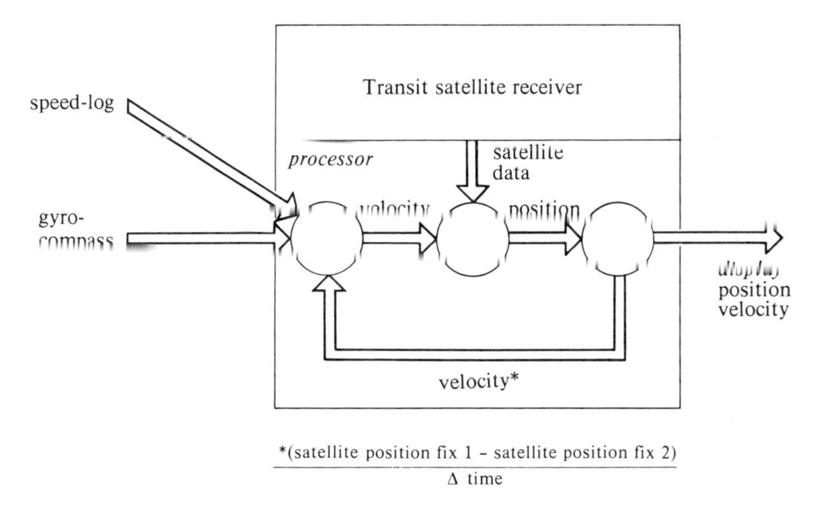

Figure 15.1 *A simple integrated system, Transit satellite, speed-log and gyro-compass.*

This can then be used to reduce the systematic errors of the velocity sensor, thereby improving the overall system performance. Therefore it could be considered that the velocity sensor is the prime navigation aid, and that Transit provides the system reference, and a means of reducing systematic error. From whichever viewpoint this particular hybrid arrangement is described, it is clear that the two constituent systems are ideally complementary.

Another navigation aid which is complementary to Transit, although perhaps less obviously so, is the Omega navigation system (Figure 15.2). Again the relationship can be seen from two

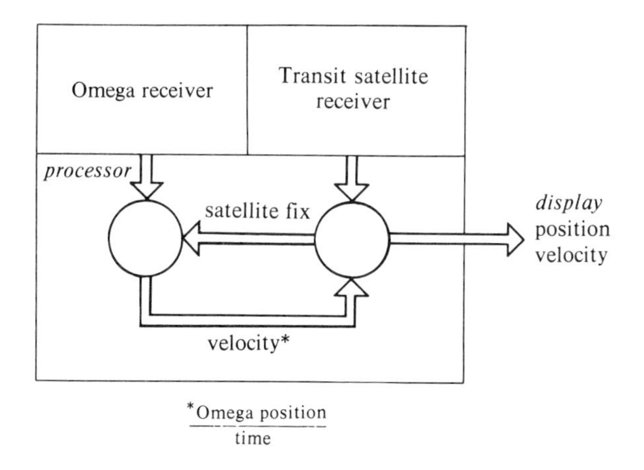

Figure 15.2 *A hybrid system, Transit and Omega.*

viewpoints, with each system having the prime role. Omega is a global system from which positions can be ...

context the system becomes a global differential Omega system, correcting both the computed propagation corrections, and the long-term errors which result from solar proton events (polar cap absorption). Problems associated with lane slip are also minimized, since these can be more easily detected and rectified.

In order to produce a usable velocity from Omega positions determined at ten-second intervals, it is necessary to smooth the result to reduce the effect of random errors. Much of the time, it is advantageous to have a long smoothing time constant (typically thirty minutes); this provides the additional benefit of reducing errors during short-duration sudden ionospheric disturbances. However, long time constants reduce the response to the system, to the extent that the displayed position will severely lag behind the actual position of a quickly manoeuvring vessel. Additional speed-log and gyro-compass inputs can be used to improve the short-term response of the system by rate aiding, and the smoothed Omega-derived velocity then basically provides a drift and set correction.

15.2 Integrated systems

A large integrated system in which many navigation aids feed a central computer is illustrated by Figure 15.3. As a separate function, the central computer is sometimes used to provide an anti-collision radar facility, by taking information from a conventional radar, reproducing it on a slave display, and then superimposing vectors and other data relating to selected targets.

Not all of the navigation aids will contribute all the time to the final position-fix solution, since some have only limited areas of coverage. The actual combining of data from the various sensors is performed by the computer or microprocessor. The mathematical

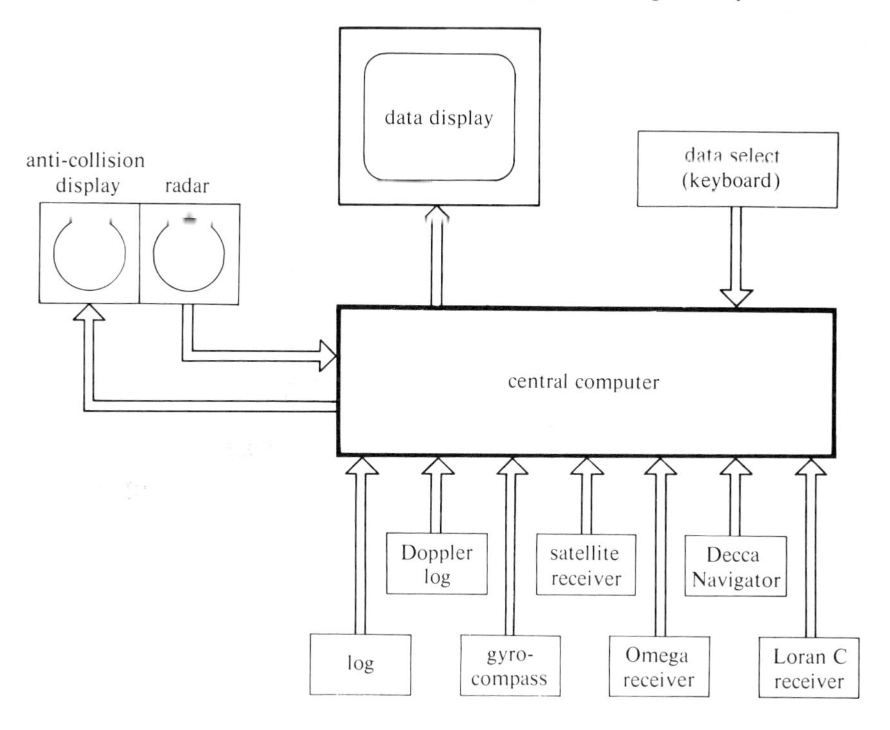

Figure 15.3 A large integrated navigation system.

method of combining the data is chosen by the system designer to achieve the optimum result from the data available; a weighted least-squares solution may be chosen, or possibly the use of Kalman filtering techniques.

It is usual with hybrid and integrated systems to include in the output display additional navigational data, since this represents a simple task for the computer. Such information would include course to steer (great circle or rhumb line), time of arrival and distance to selected way-points.

16 Speed measurement

fundamental measurements made in the process of navigation, and speed-logs were one of the first electrical and later electronic aids to navigation.

Electronic systems to determine speed (and distance travelled) continue to be prime navigation aids, producing an output that is used either as a basis for dead reckoning, or as an input to other systems such Transit Satnav and Radar.

The Doppler sonar log has already been described in chapter 13. This log is capable of measuring the speed of a vessel over the ground. Other logs measure the speed of a vessel relative to the water through which it is sailing. Two such systems are described here.

16.1 The electromagnetic speed-log

16.1.1 Theory

The fundamental principle that forms the basis of the electromagnetic log is Faraday's law of electromagnetic induction, namely – 'Whenever there is a change in the magnetic flux linked with a circuit an electromotive force is induced, the strength of which is proportional to the rate of change of the flux linked with the circuit.'

The most commonly understood application of this principle is that of an EMF induced in a wire when it is moved at right angles to a magnetic field. The directional relationship between motion and magnetic field can be determined from Fleming's right hand rule.

In the case of the electromagnetic speed-log, the conductor is the water (moving past the ship's hull) and the magnetic field is produced by a solenoid, installed in such a way as to allow the field to extend into the water. The EMF induced in the water is measured

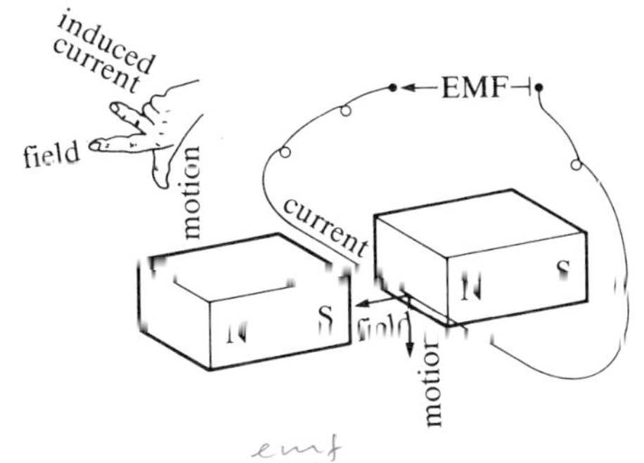

Figure 16.1 Current induced in a conductor moving through a magnetic field.

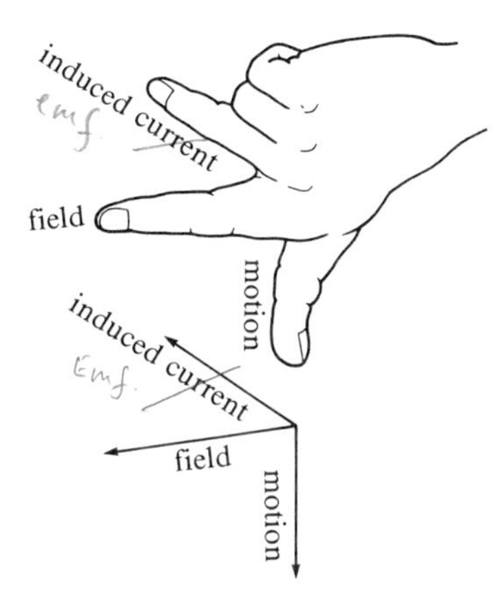

Figure 16.2 Fleming's right hand rule.

by determining the potential difference between electrodes, which project into the water that is moving through the magnetic field.

The induced EMF e is given by the following equation:

$$e = \beta lv$$

where β = the magnetic field
l = the length of the conductor
v = the velocity of the conductor through the magnetic field.

In the electromagnetic log β and l are maintained constant, therefore the induced EMF is directly proportional to the velocity v, which is the velocity of the vessel through the water.

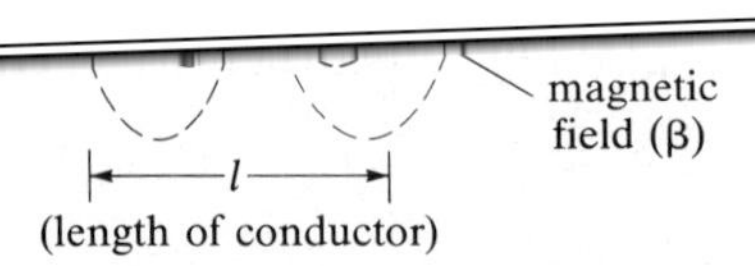

magnetic
field (β)

l
(length of conductor)

Figure 16.3 Principle of operation of the electromagnetic speed-log.

16.1.2 *Application*

In a practical electromagnetic log the solenoid and sensors are contained in a housing that mounts flush with, or projects a little way below, the hull. Even with a multi-turn solenoid and a relatively high magnetizing current, the potential difference across the sensors is only of the order of 100 μv per knot.

The solenoid energizing voltage is usually an a.c. voltage to reduce the effect of electrolytic action, and it is also simpler to amplify an induced a.c. voltage. However this inevitably causes a directly induced pick-up between the solenoid circuit and the sensor circuit, that has to be balanced out at the time of installation. Figure 16.4 shows in a simplified form the main constituents of an electromagnetic speed-log.

The first stage of the amplifier has a very high input impedance, so that variation in resistance of the water does not affect the potential difference across the sensors. As previously explained, a facility is provided to balance out any directly induced EMF.

So that the solenoid energizing voltage does not have to be precisely stabilized, it is usual to determine the velocity related voltage as a ratio of the original energizing voltage. This ratio will remain constant for a given velocity, even if the magnitude of the energizing voltage fluctuates.

As well as using the analogue output voltage to indicate speed directly, it can also be electronically integrated to display distance

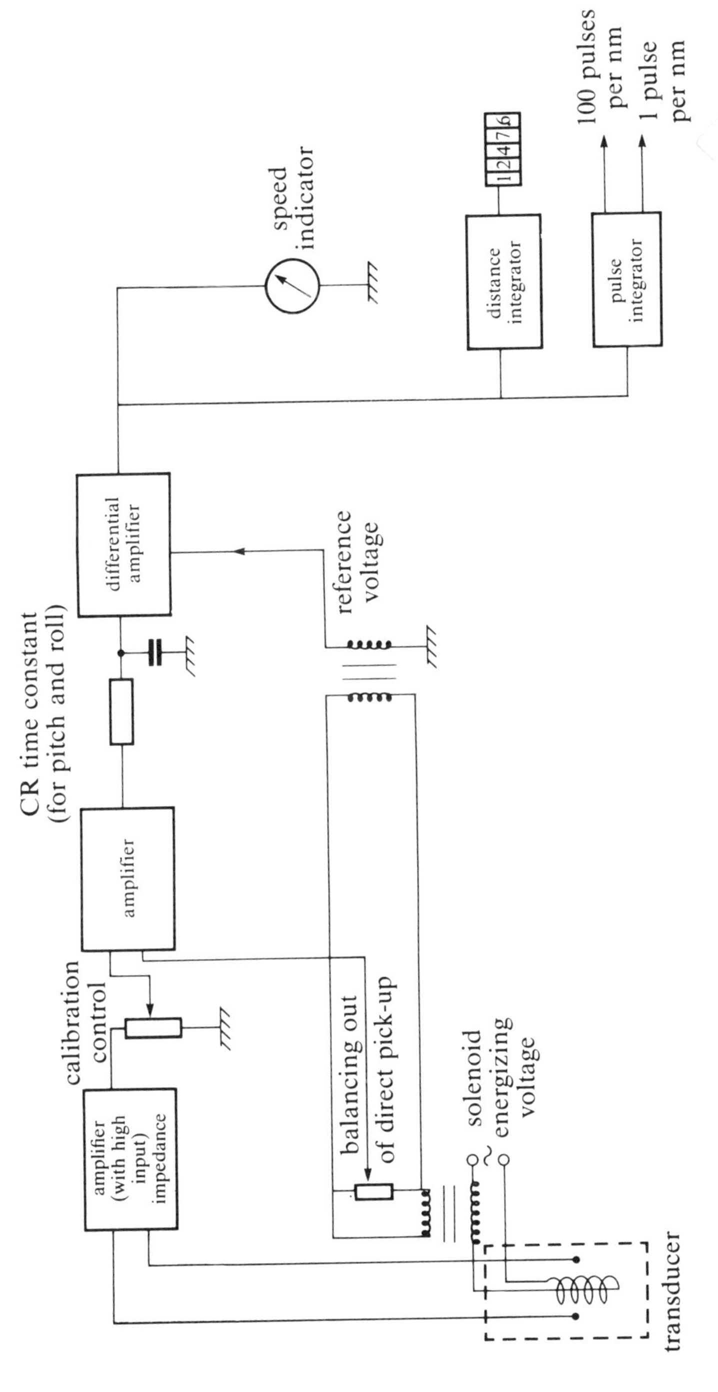

Figure 16.4 Simplified schematic diagram of an electromagnetic speed-log.

run. Additional electronic circuitry develops a given number of pulses per mile for feeding to satellite navigation, radar and other systems requiring a velocity input.

16.1.3 *Accuracy*

The EM log measures speed through the water and so its accuracy potential must be considered in this respect. The measured velocity should be relative to the general mass of water around the vessel, rather than the flow close to the hull which can vary due to the non-linearity of a hull design. This effect can be minimized by optimum siting of the transducer assembly and then by calibration of the log.

Pitch and roll will also induce an error, the effect of which can be lessened by introducing an electrical time constant that is longer than a period of vessel motion.

A well-adjusted and calibrated electromagnetic log can have an accuracy better than 0.1 per cent of the speed range in use (measured in the fore and aft direction).

16.2 The pressure tube (Pitot) log

16.2.1 *Theory*

Referring to Figure 16.5, the pressure tube *A* is protruding through the bottom of a vessel's hull, has an opening facing the direction of motion and is filled with water. The water in the tube acts on a diaphragm. The pressure on the diaphragm increases with the speed of the vessel (and tube) through the water, the relationship between speed and pressure being a square law:

$$p \;\alpha\; v^2$$

or $p = k\, v^2$

where k is a constant determined by the specific hull and log characteristics.

As well as the dynamic pressure that is used directly to determine speed, there will be a pressure which acts even when the log tube is at rest. This static pressure is proportional to the total depth of submersion of the tube (vessel's draught plus the length of the tube).

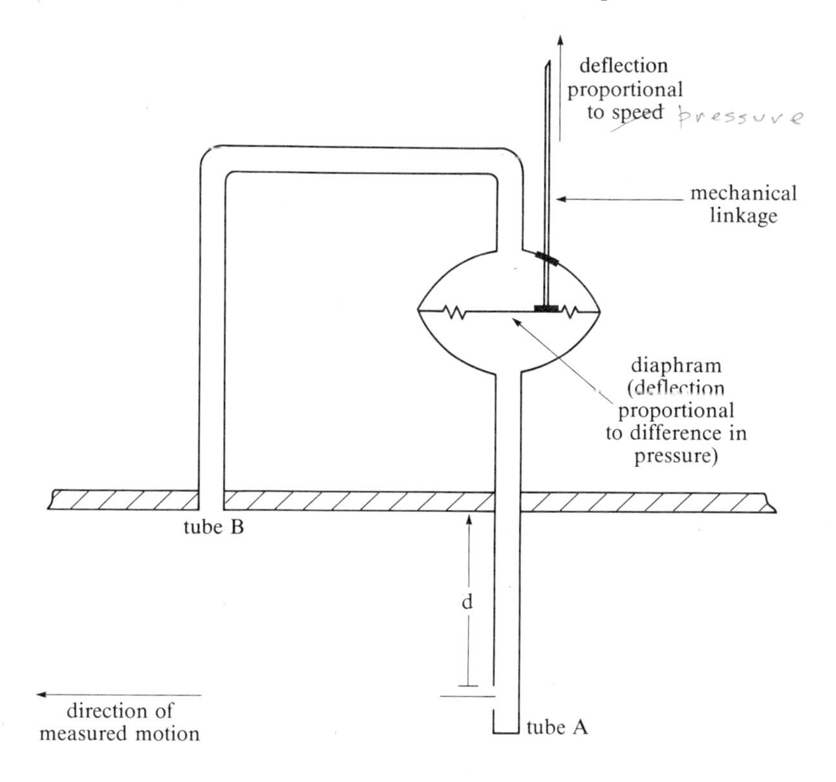

Figure 16.5 Principle of operation of the pressure tube log.

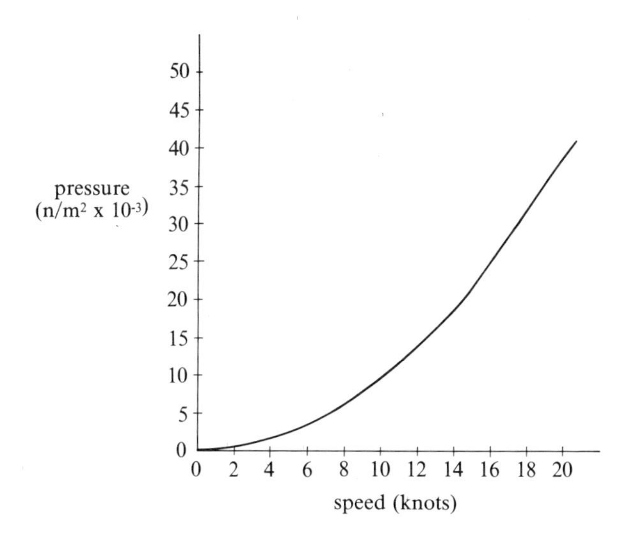

Figure 16.6 The pressure tube log; relationship between speed and differential pressure.

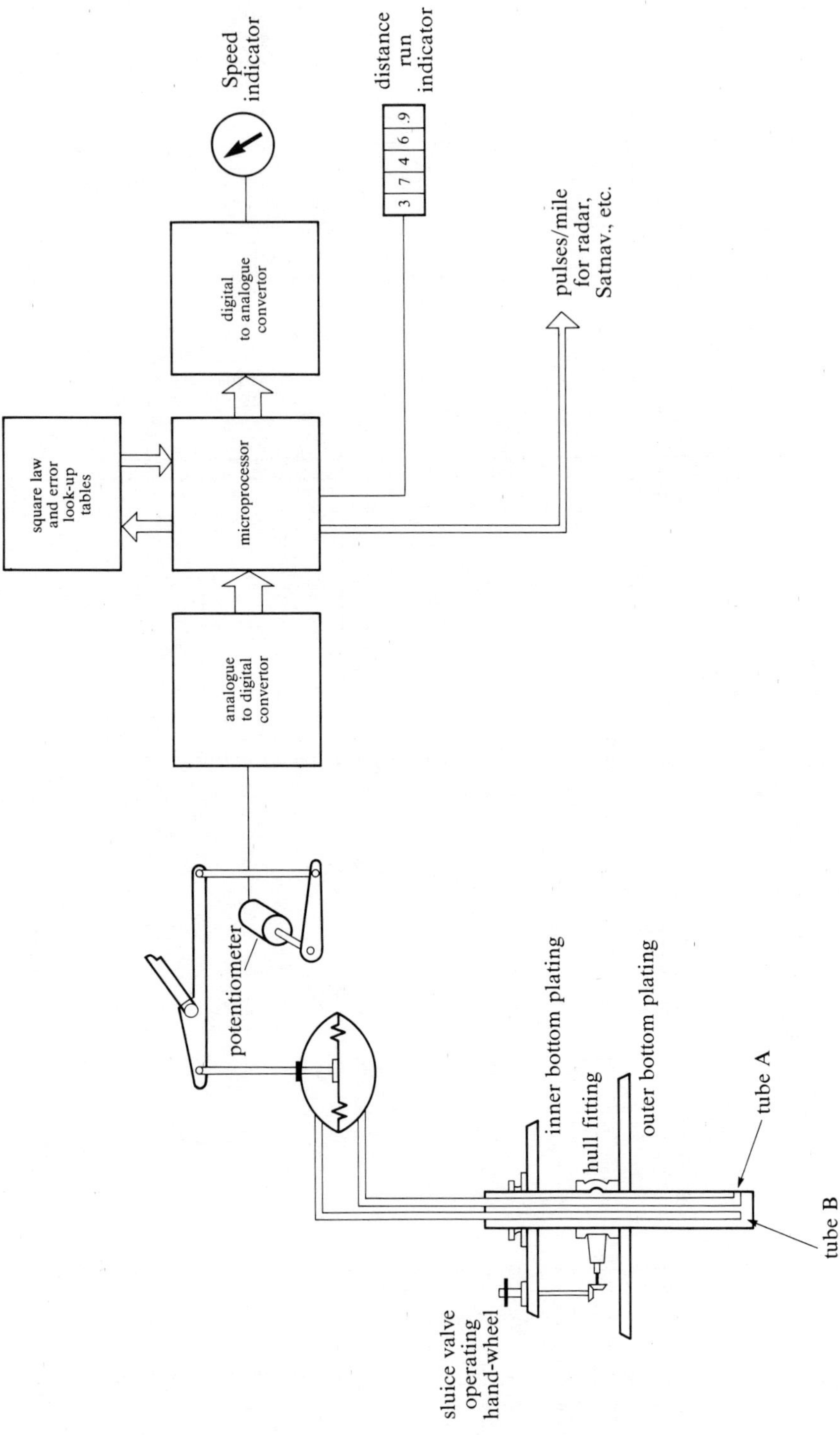

Figure 16.7 Simplified schematic diagram of an electronic pressure tube log.

The static pressure is balanced out by the use of a second tube with a horizontal aperture. The pressure in tube *B* does not change with speed, and so when applied to the opposite face of the diaphragm the resulting force on the diaphragm is dependent only on the dynamic pressure (and hence speed). The difference in submersion depths (*d*) of the two tubes represents a fixed pressure difference that can be compensated for within the log itself.

16.2.2 *Application*

The practical Pitot log must translate the diaphragm deflection caused by the vessel's motion into an indication of speed. The first stage is a mechanical linkage from the diaphragm; this will create a linear motion proportional to the square of the speed. Traditionally, pressure tube logs have been complex electromechanical devices, incorporating rods, pivots, levers, springs and cams, to create a linear indication of speed from the diaphragm movement. A spiral cam is used to translate the square law motion to the linear speed indication. Since such devices are not electronic navigation systems they are not described in detail here, although electronics have already been introduced into some Pitot logs and will continue to replace an increasing amount of the electromechanical operations. For example, the square law cam assembly can be replaced by a look-up table or a mathematical squaring function in a microprocessor. A look-up table can also hold the corrections for the system errors as determined during calibration trials.

The accuracy of a correctly installed and calibrated pressure tube log is better than 1 per cent of the range in use. There is, of course, no indication of sideways movement of the vessel.

17 The gyro-compass

G. A. A. Grant
J. Klinkert

17.1 Theory

17.1.1 The free gyroscope

The basis of marine gyro-compasses lies in the free gyroscope. It is a spinning wheel or rotor so mounted in a frame that the axis upon which the wheel spins may be pointed initially in any preferred direction. A study of Figure 17.1 shows that apart from the fairly obvious spinning axis the mounting framework contains two further axes mutually perpendicular. In practice one of these is invariably *vertical* because the gyro will eventually be used as a compass which must afford direction-indication about a vertical axis, or that axis

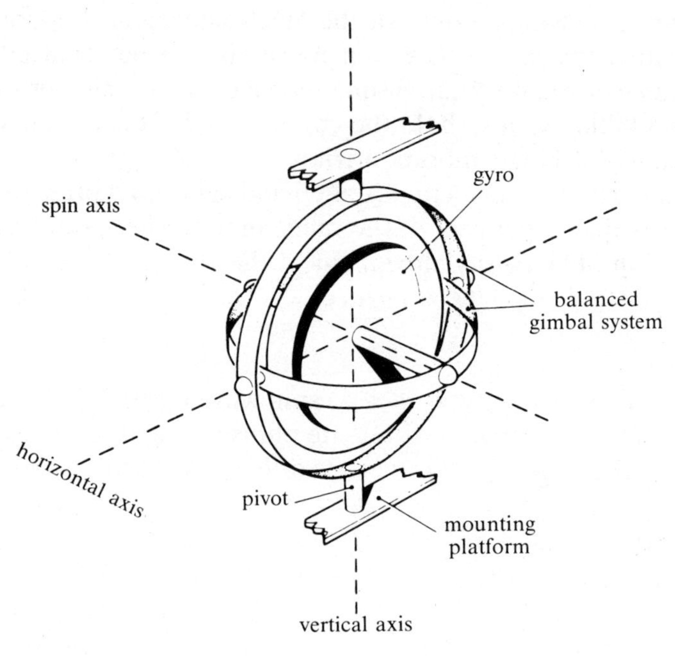

Figure 17.1

around which a ship turns from one course to another. It therefore follows that if one axis is to be substantially vertical the other will be horizontal, and this permits the gyro assembly to turn in azimuth about the former and tilt about the latter. If the gyroscope were needed for purposes other than as a gyro compass the problem of mounting would be satisfied with *any* two axes mutually perpendicular as might be required in space equipment where the terms vertical and horizontal have, of course, no meaning. Freedom for the rotor to spin and for the spin axis to turn and tilt are referred to as the 'three degrees of freedom'.

17.1.2 *Gyroscopic inertia*

When the rotor is stopped no effort is required to topple and turn it within the frame described and illustrated above. The situation is quite different if the rotor is set spinning. It exhibits a property which is popularly termed 'rigidity in space', or more correctly gyroscopic inertia. This property is assessed quantitatively from the angular momentum of the rotor which in turn is the product of its angular velocity and moment of inertia. Whenever possible gyro wheels are made to rotate very quickly, and the size, shape and distribution of weight is a matter of careful design to ensure adequate angular momentum without excessive frictional losses or wear at the supporting bearings.

Newton's First Law of Motion states: 'Every body remains in its initial state of rest, or uniform motion in a straight line, unless a force is exerted on it.' If this is applied to an elementary particle of the gyro wheel it confirms that, when the wheel is spinning, there must be a continuous force (centripetal) directed towards the centre of rotation which is balanced by an equal force (centrifugal) directed outwards to account for its circular track. Under these conditions, and provided that there is sufficient cohesion between constituent particles, the wheel will continue to spin within its initial plane, and still remain a wheel! The penultimate phrase is important because it means that the spin axis maintains its orientation in *space* and points to an imaginary star, often referred to as a 'gyro star'. This property of gyroscopic inertia is appropriate because if the gyro is left alone it provides a datum reference from which it does not deviate.

17.1.3 *Precession*

If a torque or couple is applied about the spin axis the gyroscopic inertia will be increased if the couple acts in the same direction as the spin of the wheel; it will be decreased if it acts in the opposite direction. The function of the rotor motor, apart from raising the speed of rotation initially, is purely to balance the frictional losses at the spin bearings.

A torque or couple which is applied about any other axis may be resolved into components about the three axes of freedom already mentioned. In this sense the applied couple, if it acts *perpendicularly* to the plane of the wheel also acts perpendicularly to the spin axis and may, for convenience, be resolved into components which act about the horizontal and vertical axes. Figures 17.2 and 17.3 show an applied couple (*T*) acting about the vertical and horizontal axis respectively. Since the wheel cannot rotate about two different axes simultaneously it endeavours to accommodate itself to the applied couple by *precessing*. The plane of the wheel turns about the vertical axis when the applied couple is made to act about the horizontal axis (Figure 17.3), and vice versa (Figure 17.2). The overwhelming angular momentum of the wheel makes this possible and from both the figures it will be seen that the *direction of precession is always*

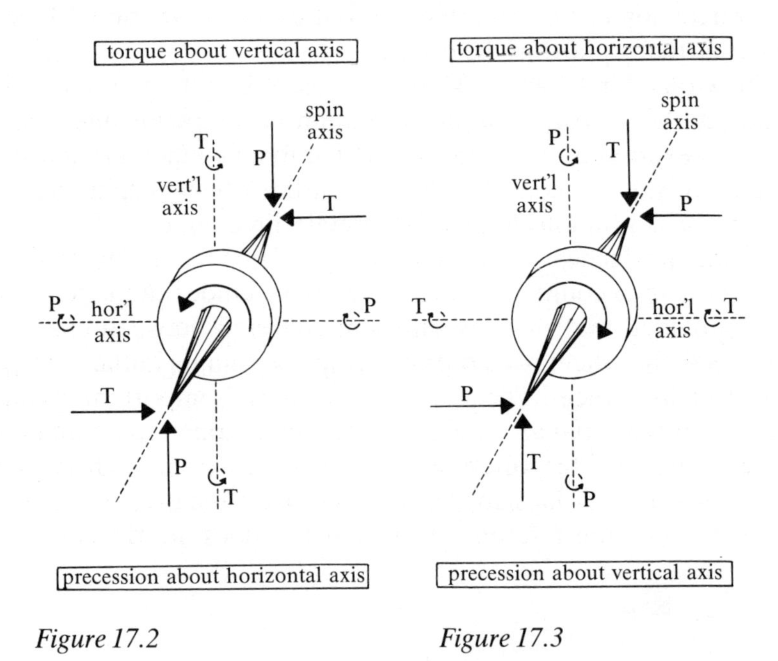

Figure 17.2

Figure 17.3

perpendicular to the applied torque measured in the direction of the gyro spin. If the arrows marked T are rotated in the sense of the wheel spin, S, they indicate the direction of precession, P.

The relationship between rate of precession (P), applied couple (T), moment of inertia (I), and rate of spin (S).

These quantities, when expressed in consistent units, are governed by the following relationship:

$$P = \frac{T}{I.S}$$

From this it follows that for constant angular momentum ($I.S$) the rate of precession is directly proportional to the magnitude of the applied torque (T). The rate of precession is also, separately (or together), inversely proportional to the rate at which the wheel spins (S) and inversely proportional to a function of the weight, shape, size and radius of gyration of the wheel (I), provided constant torque is considered. If the torque is applied continuously the gyro precesses continuously, and the relationship shows that in general heavier rotors which are spinning at high speeds are affected less by frictional couples which act about the horizontal and vertical axes.

17.1.4 *The earth's rotation*

Since the free gyroscope points its spin axis to a gyro star, and since the intention is to use the gyro ultimately as a compass on earth, it follows that the gyro will display to an observer on the earth an *apparent* motion within its frame identical with that of the star to which it points initially. This apparent motion can be described either graphically or quantitatively in terms of the usual nautical astronomy practised by navigators. However, it is sometimes preferred to treat with a horizontal plane tangential to the earth's surface first and to reverse this motion subsequently to account for the gyro's motion.

Figure 17.4 shows a vector which represents the rate of spin of the earth (ω) about its polar axis. In an intermediate latitude (λ) this is resolved into a component of rotation about the true vertical ($\omega \sin \lambda$) and a component about the north/south horizontal ($\omega \cos \lambda$). In Figure 17.5 the component about the northern point of the horizon has been further resolved into two components mutually at right

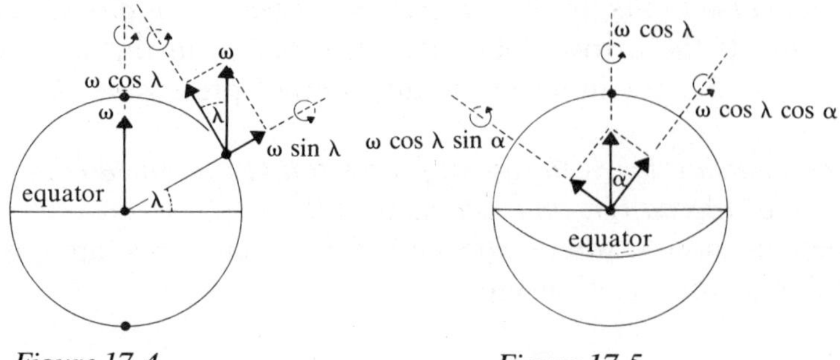

Figure 17.4 *Figure 17.5*

angles, namely in a direction α° east of north and (90 − α)° west of north. These are shown as ω cos λ α and ω cos λ sin α respectively.

17.1.5 *The apparent motion of the gyro*

The free gyroscope will exhibit an apparent motion in the opposite sense to that of the earth's horizontal surface. Figure 17.6 shows this motion reversed from that shown in Figure 17.4. The spin axis of the gyro, when substantially horizontal as shown, will 'drift' with its north end eastward at the rate of ω sin λ which, in conventional units is 15 sin λ°/hour. This is the rate of change of azimuth of a star near the horizon. The component of motion about the northern point of the horizon (ω cos λ) only subtracts an insignificant amount from the rate of gyro spin if this is of itself earthwise, or adds to it if it is of itself contra-earthwise.

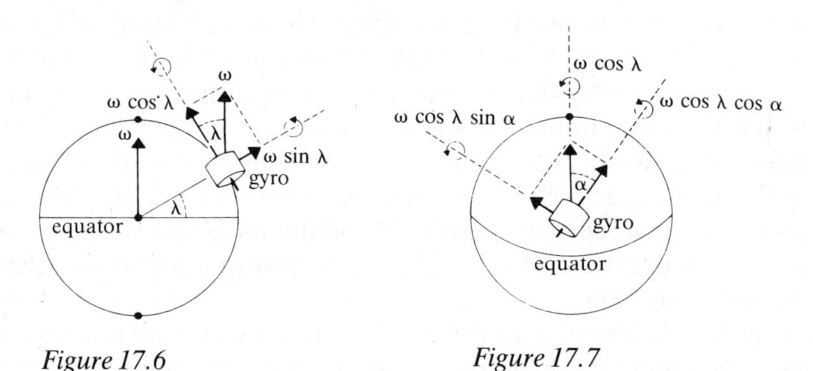

Figure 17.6 *Figure 17.7*

Figure 17.7 for the gyro corresponds to Figure 17.5 for the earth. The gyro spin axis is orientated α° east of north and this indicates that there is a component of motion about the horizontal axis of the gyro which creates an apparent 'tilt' at the rate of $\omega \cos \lambda \sin \alpha$. In conventional units this is expressed as $15 \cos \lambda \sin \alpha$°/hour and is identified with the rate of change of altitude of a star. The sense of this motion is in sympathy with the star to which the spin axis points, i.e. when the spin axis points eastward it tilts upward because stars rise in the east; when the spin axis points westward it tilts downward because stars set in the west. Of course, if one end of the spin axis points eastward the other end points westward and the conclusions are obvious. It should be noted that $\omega \cos \lambda \sin \alpha$ is maximum when α is 090° or 270°; this means that the spin axis of a gyro displays its maximum rate of tilting when it is orientated east/west and does not tilt at all when it points north/south. This is again identified with the rate of change of altitude of a star.

7.1.6 *The rates of tilting and drifting of the spin axis*

These may be summarized as follows:

(i) *Rate of tilting*
This occurs at $15 \cos \lambda \sin \alpha$°/hour; upward east of north; downward west of north and consistent in both hemispheres.

(ii) *Rate of drifting*
This occurs at $15 \sin \lambda$°/hour; north end eastward in the northern hemisphere and north end westward in the southern hemisphere.

It must be pointed out that the rate of $15 \sin \lambda$°/hour is valid only when the spin axis is substantially horizontal which it normally is. Excessive tilt (t°) never occurs in gyro-compasses; if it did then $15 \sin \lambda\ (1 - \cot \lambda \tan t \cos \alpha)$°/hour would have to be substituted for $15 \sin \lambda$°/hour.

The rates of tilting and drifting of the gyro spin axis are for an instant only under the specified conditions of latitude, tilt and azimuth and cannot be used for a long term forecast. This means that any estimate of the apparent motion of the gyro which extends over several hours must be made by the routine methods of nautical astronomy familiar to all navigators.

17.1.7 *Estimate of the apparent motion of a free gyroscope*

Since the gyro spin axis is identified with a fixed star to which it points it is only necessary to specify its initial conditions to appreciate its subsequent movement. Those conditions are listed as follows:

Gyro:	Latitude	tilt	azimuth
Gyro Star:	latitude	altitude	azimuth

The only difference is in the term altitude which, in the case of the gyro is its obvious tilt about the horizontal axis. The following examples make clear some of the kind of forecasts which can be made and in each use is made of the conventional diagram of the celestial sphere drawn in the plane of the rational horizon.

Example 1

A free gyroscope, in latitude 45° N has its spin axis horizontal and pointing east/west. Where does it point in 6 hours time?

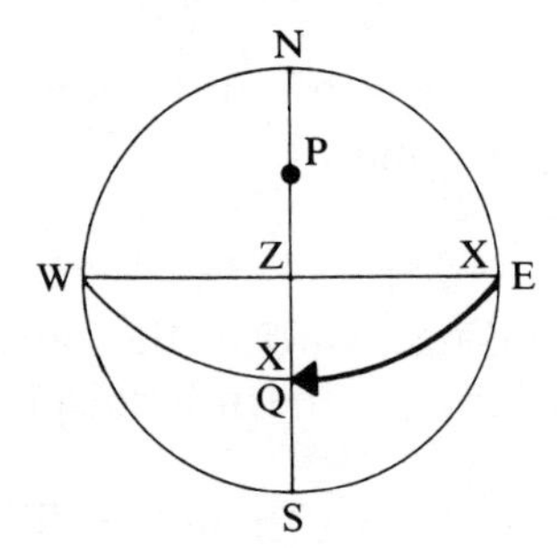

Figure 17.8

In this case the altitude of the gyro star is zero because the spin axis is horizontal. The latitude (45° N), the altitude (0°) and the azimuth (090°) define the initial position of the gyro star to coincide with the eastern point of the horizon (*X, E*). The star is located on the six o'clock hour circle and its declination is zero. In six hours' time it will arrive on the meridian at X' so that it will then bear 180° and its altitude will be 45° (SX'). Interpreted for the gyro spin axis:

The axis will point 000°/180° and the end pointing south will be tilted 45° above the horizontal.

Example 2

A free gyroscope in latitude 60° S has its spin axis pointing north/south with the south end tilted 60° above the horizontal. Where does it point in 6 hours' time?

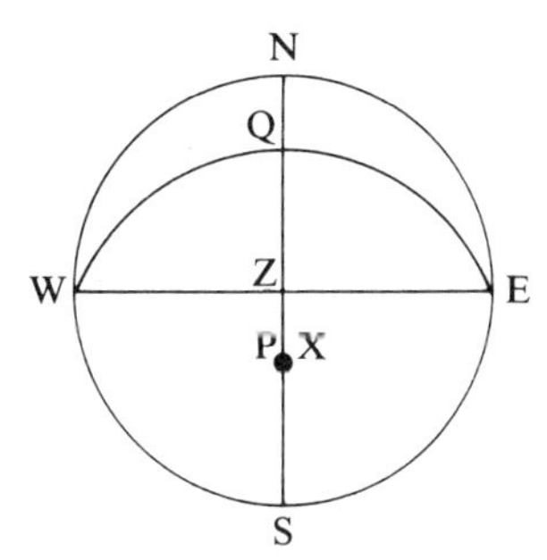

Figure 17.9

In this instance the initial conditions indicate the gyro star to occupy the same position as the south celestial pole. The gyro spin axis will therefore exhibit *no apparent motion at all*; the case is unique because the spin axis is parallel to the earth's rotational axis.

From the given information the gyro star occupies the northern (or southern) point of the horizon (*X*, *N*). It must therefore be considered as the limiting case of a circumpolar star whose parallel of declination encircles the celestial pole as shown in Figure 17.10. The trace of the gyro star indicates that in 12 hours (strictly, in half a sidereal day) it will rejoin the meridian south of the zenith at X' with an altitude of 40° above the horizon. Interpreted for the gyro spin axis:

This axis will again point 000°/180° having *reversed* its direction with the end which originally pointed north (*X*) now pointing south (X') and elevated 40° (SX') above the horizontal.

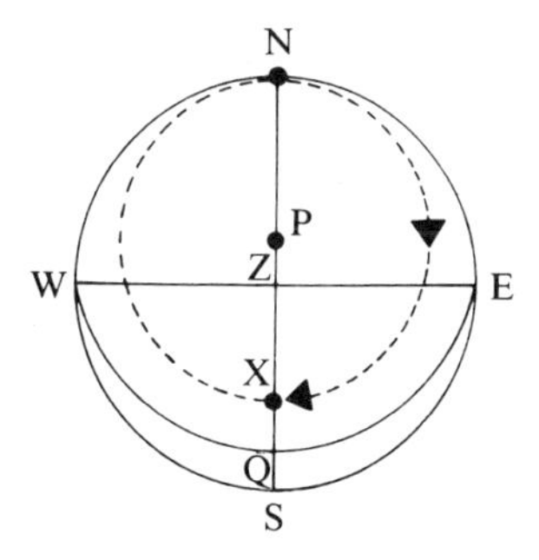

Figure 17.10

17.1.8 The requirements of a directional gyro

For a gyro to provide azimuth indication it is clearly necessary to eliminate or compensate for the tilt and drift of the spin axis caused by the earth's rotation.

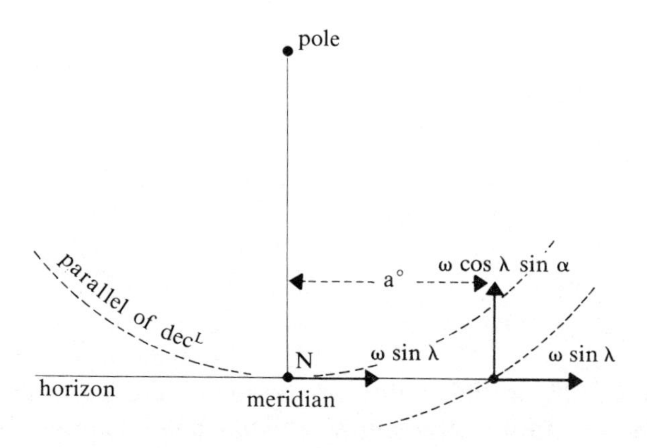

Figure 17.11

Figure 17.11 shows part of the parallel of declination associated with a gyro star which is just circumpolar. Apart from spurious wanders it can be seen that if the spin axis is set pointing at the northern point of the horizon (N) a correction for drift at the rate $\omega \sin \lambda$ (directed westward in the northern hemisphere) is all that is necessary to provide a north pointing directional device. Likewise, but not so simply, the spin axis could be secured to an azimuth $\alpha°$ from north provided simultaneous corrections for drift and tilt are applied at the rates $\omega \sin \lambda$ and $\omega \cos \lambda \sin \alpha$ respectively in the opposite sense to that illustrated. At this point it is important to realize that a free gyro which can be made to precess at the rates stated will point in a preferred direction only so long as frictional torques at the horizontal and vertical axes are either zero or sufficiently small to satisfy the needs of marine navigators. Such a *directional gyro* (DG) is very useful in aircraft, particularly in high latitudes. It is not a compass and has no north seeking property. Aboard ship, unlike aircraft, such an instrument needs to be used for many consecutive days and consequently wander due to friction would be considerable. Of course, a directional gyro having a typical wander rate of about 0.5°/hour can be checked by astronomical azimuths and periodically reset but this procedure, although familiar to air navigators, is not traditional at sea so that on balance the directional gyro is not favoured except for naval vessels which require to execute severe manoeuvres in high latitudes.

17.1.9 *The requirements of a gyro-compass*

The emphasis here lies in the instrument's ability to seek the north and maintain this (or any other preferred direction) in spite of frictional torques which tend to cause uncontrolled wander. The gyro must be *controlled* and not merely *corrected* for tilt and drift. It must be brought to an equilibrium direction and be restored to it immediately it tries to depart from it. This propensity must be part of its inherent properties. In order to satisfy this requirement it is necessary to provide the means of creating appropriate torques which will cause the gyro to precess whenever it deviates from some earthly datum direction. From this it follows that some sensing device must be incorporated to detect such deviation, and an earthly quality like the *earth's magnetic field* or the *acceleration due to gravity* subsequently used to initiate its elimination. All gyro-compasses are therefore gyros which are constantly monitored and in aircraft the earth's magnetic field is used for this purpose. Such compasses are in fact magnetic compasses basically, and are subject to the deviations caused by the vehicle's magnetism. Gyrocompasses aboard ship make use of the *gravitational influence* which, within the present day limitations of speed and latitude, is satisfactory enough but would become unsuitable if marine craft of the future travelled more quickly and/or operated in latitudes near the north pole, as for instance under the ice. Any compensation for ship's motion which might be incorporated would depend upon the accuracy of assessing that motion in the first instance; a problem not always easy to solve!

17.1.10 *Modes of gravity control*

There are basically only two methods of applying a gravity control to the gyro. The first is to cause the system to have the *effect of bottom heaviness* and the second to have the *effect of top heaviness*.

Each of these is illustrated below. For this purpose each gyro has a rigid structure or frame carrying a weight which, when displaced from either the *top dead centre*, or *bottom dead centre* positions, causes a couple which acts about the horizontal axis. Reference to Figure 17.12 shows that (in the northern hemisphere) the spin axis,

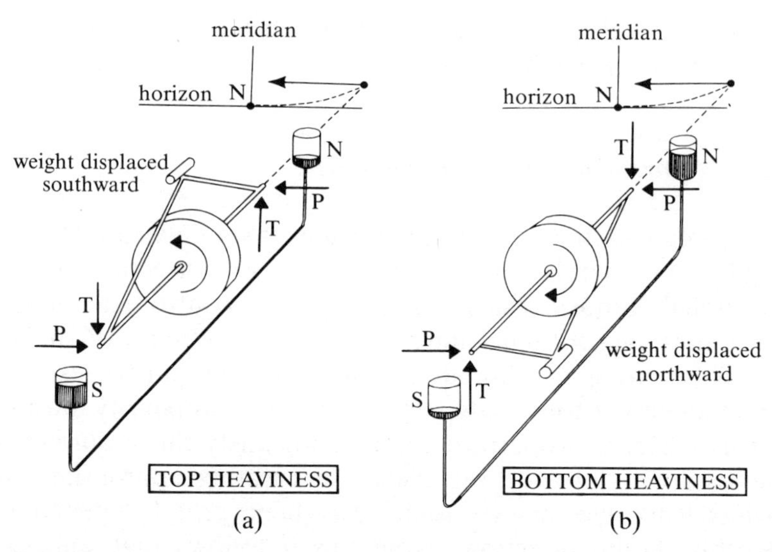

Figure 17.12

having been previously set pointing at the northern point of the horizon will, after an interval drift eastward and tilt upward in its endeavour to follow the circumpolar star. The torques (T) produced by the weight must clearly have the effect of restoring the spin axis towards north. This required precession is shown by the arrows marked P. A further study of Figure 17.12 shows that when the rule for the direction of precession is applied the direction of wheel spin is opposite for the top and bottom heavy weight effect types of control in the sense that:

(i) A top heavy weight effect control demands an anticlockwise (contra-earth) rotor spin viewed northward.
(ii) A bottom heavy weight effect control demands a clockwise (earthwise) rotor spin viewed northward.

In fact, about half of the world's marine gyro-compasses are actually bottom heavy and the remaining half provided with control systems which act in a manner to produce the *effect* of either top or bottom heaviness.

17.1.11 Method of applying gravity control in gyro-compasses

For a large number of years British gyro-compasses have been confined to those manufactured by the Sperry and Brown companies. Each of these compasses simulates gravity control by means of a liquid ballistic: Figure 17.12a shows that Sperry features the top heavy weight effect by fixing liquid containers (mercury) to the spin axis frame and allowing a natural gravity flow downhill from the high to the low container whenever the system is tilted. The conventional Brown compass features the bottom heavy weight effect with a similar liquid ballistic (containing oil) in which a transfer is effected uphill from the low to the high container. The oil is pumped against gravity by air pressure generated within the rotor case and directed onto the oil surface to force it through the inter-connecting pipes. Although many of these Brown compasses are still being used at sea their manufacture has now ceased in favour of the Arma Brown gyro-compass. This compass, similar to its predecessor, simulates a bottom heavy weight effect in quite a different way. Torsion wires are attached to the sensitive element of the system and when these are twisted they create reactionary couples having the same effect as a bottom heavy weight.

7.1.12 *Rate of precession from a liquid ballistic*

Figure 17.13 shows the displacement of liquid from one container to the other when the control system is tilted through a small angle β^c. If the containers are separated by a distance $2l$ and contain liquid of density p then the weight of liquid transported is given by:

$$\pi r^2 h p$$

where, r is the radius of the cylindrical container and h the mean depth of liquid which leaves one container and arrives at the other. Writing A for πr^2 the moment of weight transferred is $2Al^2 p\beta^c$.

Let $B = 2Al^2 p$, being a constant of the ballistic arrangement, so that the rate of precession (Ω) is given by,

$$\Omega = \frac{B\alpha}{H}$$

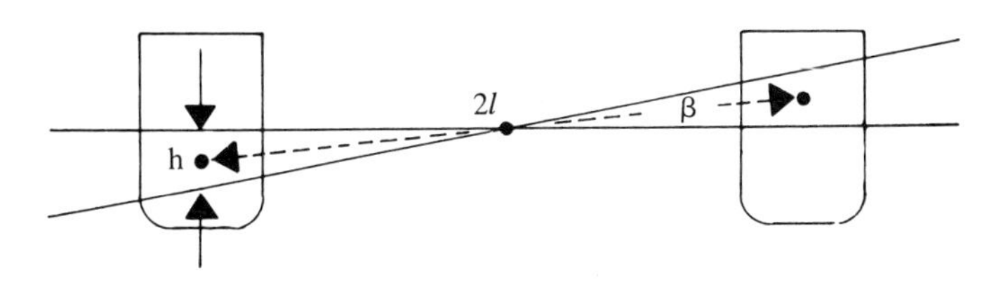

Figure 17.13

Since the Arma Brown compass does not employ a liquid ballistic B would contain parameters appropriate to the torsion wires and the amount by which they are twisted. In either case, if this is measured in consistent units its specific form may be appreciated when it is compared with the general relationship (see page 000) thus:

$$\Omega = P = \frac{T}{IS} = \frac{B\beta}{IS} = \frac{B\beta}{H}$$

where H is the angular momentum of the gyro.

17.1.13 *The trace of the north end of an undamped gravity controlled gyro-compass to north latitude*

Figure 17.14 delineates the trace which the north end of the spin axis makes on a vertical backcloth under the influence of a gravity control. The horizon is shown with its northern point marked N. The observer's meridian projected both above and below the horizon cuts it at right angles. The line RS provides a tentative guide for drawing vectors which represent rates of east/west precession ($B\beta/H$). When the tilt (β) is directed north end up precession is directed westward, and vice versa. The line PQ likewise provides a similar guide for drawing vectors which represent rates of tilting due

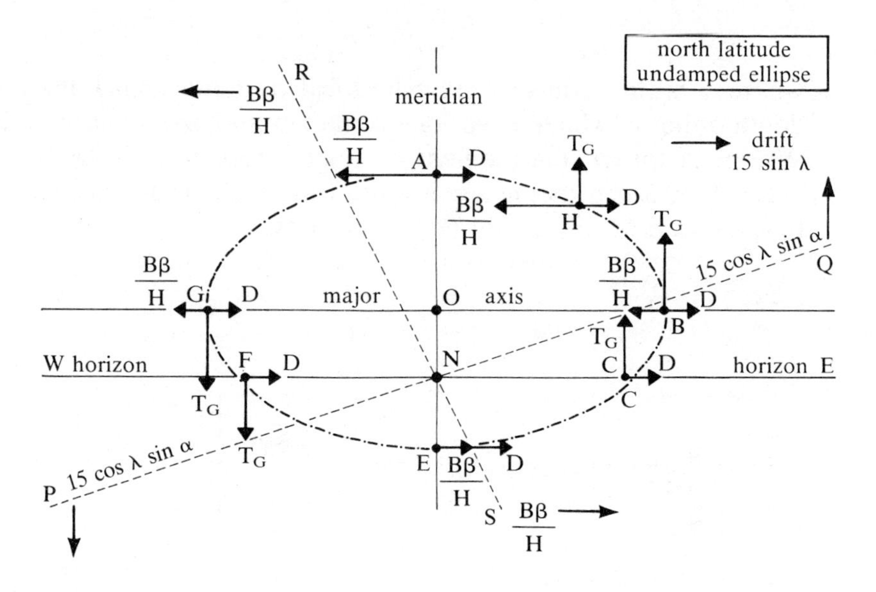

Figure 17.14

to the earth's rotation (15 cos λ sin α), i.e. in the sense upward when the azimuth is east, and downward when the azimuth is west. With these points in mind the following discussions are basic and do not refer to systems which include an automatic correction for drift.

The trace is an ellipse whose major axis is parallel to and a little above the horizon in the northern hemisphere. The minor axis coincides with an arc of the meridian. At the points *B* and *G* the movement is solely vertical; thus, for the east/west tendency to be zero the drift of the spin axis (*D*) which acts in the sense eastward (at 15 sin λ) in the northern hemisphere, must be eliminated by a rate of precession north end westward of equal amount. Since this can only occur when the north end is elevated it defines the location of the major axis as being above the horizon.

At the points *A* and *E* the movement is horizontal because no vertical motion can exist when the azimuth is zero (15 cos λ sin 0 = 0). At *A* the westward motion is determined by the excess of Ω westward over the eastward drift (*D*), while at *E* the eastward motion is the cumulative effect of a much reduced rate of precession and the drift which now both act in the same sense.

The unique positions *F* and *C* show the absence of any control precession because the spin axis is horizontal when crossing the horizon. The motion is due solely to the effect of the earth's rotation, i.e. tilt and drift. Position *H* on the trace shows an intermediate situation with the north end of the spin axis traversing westward in response to the excess of Ω over *D*, and upward in sympathy with rising stars to the eastward.

Left to run in this way the controlled gyroscope circuits the point *O* in Figure 17.14 and moves from one side of the meridian to the other to an extent which is determined by the initial conditions. The trace of the north end is therefore different to that of the apparent movement of a free gyroscope. The free gyroscope circles the celestial pole in one sidereal day; the controlled gyro performs an elliptical trace in the same direction the centre of which lies near the horizon and the period (*T*) of which may be shown to be such that $T = 2\pi\sqrt{[H/B\ \omega \cos \lambda)]}$.

.1.14 *The trace of the north end of an undamped gravity controlled gyro-compass in south latitude*

The trace which corresponds to the southern hemisphere is shown in Figure 17.15. Reference is again made to the north end of the spin axis. The rotation is once more anticlockwise but the major axis of the ellipse is parallel to and a little *below* the horizon. The positions *C* and *G* in Figure 17.15 should be compared with *B* and *G* in Figure 17.14. In the southern hemisphere the *westward* drift of the north end can only be eliminated by *eastward* precession, and since this can only occur when the north end is depressed it defines the major axis as being below the horizon. At positions *A* and *E* the movement is again solely in azimuth. At *A* both drift and precession act in the same sense; at *B* the eastward movement is shown by the excess of the precession eastward over the westward drift. Positions *H* and *B* show again the absence of precession due to the control because the tilt is zero. Position *F* shows the three vectors appropriate to this section of the trace.

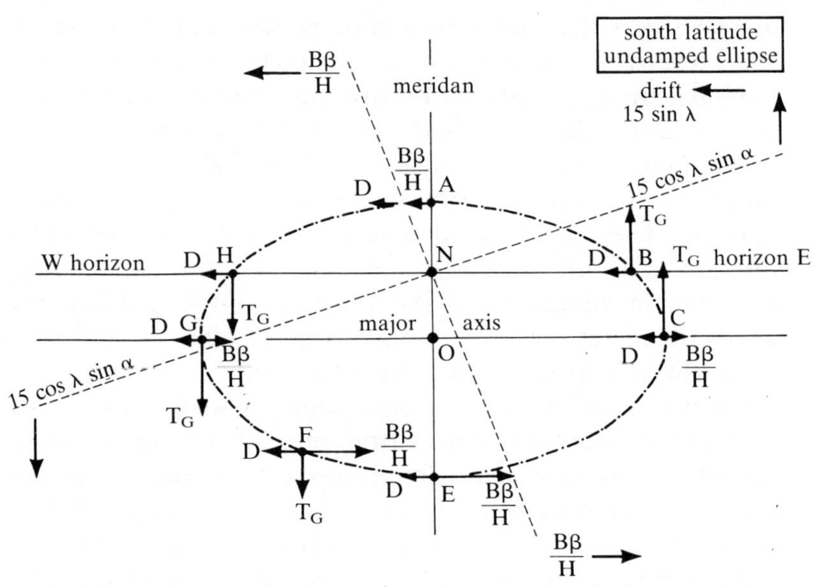

Figure 17.15

17.1.15 *The trace of the north end of an undamped gravity controlled gyro-compass at the equator*

At the equator the ellipse trace has its centre at O and the major axis coincides with the horizon. At near zero latitude the drift is zero also ($15 \sin \lambda = 0$) so that the only vectors causing movement are those due to tilt and ballistic precession. Clearly, the size of the ellipse will be governed by the azimuth at which the spin axis is set initially.

17.1.16 *Damping*

From this discussion of the effects of the gravity control it is clear that some means must be provided for damping the oscillations of the gyro from one side of the meridian to the other. The spin axis must be made to point steadily in some preferred direction. There are two basic methods of damping the unwanted oscillations and each relies upon imposed torques to create appropriate rates of precession. The one method makes use of a couple acting about the vertical axis to damp the oscillations initially about the horizontal axis, i.e. in *tilt*; while the other makes use of a couple acting about the horizontal axis to damp the oscillations initially about the vertical axis, i.e. in *azimuth*. It is unnecessary to effect damping couples about both perpendicular axes simultaneously because by choosing one of them it is inevitable that oscillations about the other are reduced in consequence. By these means the elliptical trace is transformed to a spiral, the oscillations vanish, and the system assumes the character of a direction seeking compass.

17.1.17 *Damping in tilt*

Figure 17.16 is a graphical representation of the damped spiral of a gyro-compass where the oscillations are initially damped about the horizontal axis. The figure is necessarily distorted for explanatory purposes and the vertical amplitudes are much exaggerated. The framework of the diagram is similar to Figure 17.14 with the addition of the approximate guide line AB, which provides a means of estimating the size of vectors which represent the rate of precession for damping purposes caused by a couple applied about the vertical axis.

In a gyro-compass which is damped in tilt the intention is to superimpose a rate of precession over the natural rate of tilting due to the earth's rotation ($15 \cos \lambda \sin \alpha$) in such a way that movement 'towards' the horizon is accelerated.

This is done by creating a simultaneous torque about the vertical axis. If the compass is controlled by a top heavy weight effect the imposed couple must act anticlockwise (viewed from above) whenever the spin axis north end is elevated above the horizon, and clockwise when the spin axis north end is depressed below the horizon. If the compass is controlled by a bottom heavy weight effect the imposed couple must act in the opposite sense; so that in either case the damping precession about the horizontal axis has a rate which shows a fixed proportion (r) to the azimuthal rate of precession. This is shown in magnitude by $\Omega_d = rB\beta/H$ and in direction by the arrows at the points A and B in Figure 17.16. The

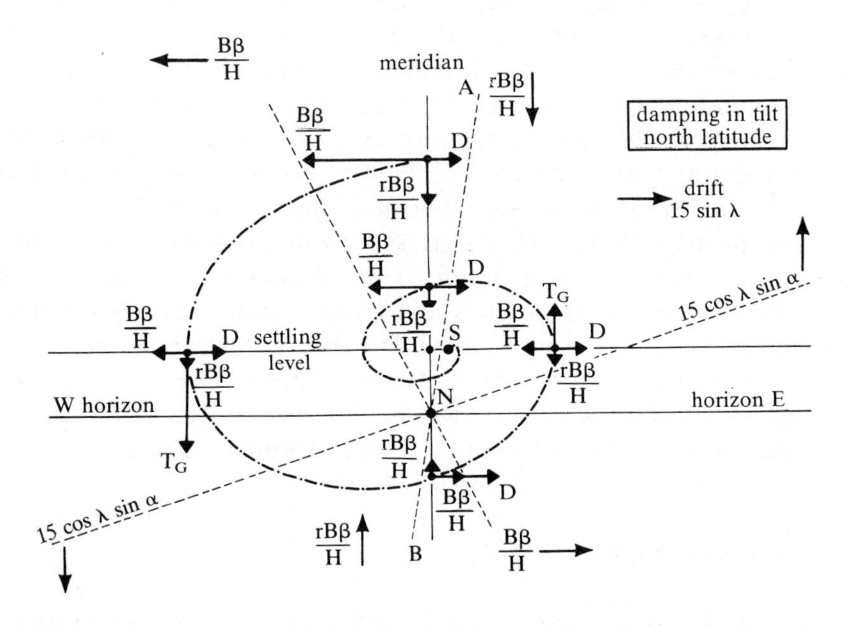

Figure 17.16

effect of this damping precession is to *encourage* movement of the north end of the spin axis *towards* the horizon and to *discourage* movement *away* from the horizon. All subsequent values of the tilt (β) are therefore smaller than they would otherwise be under the conditions of an undamped elliptical oscillation (Figure 17.14). It follows therefore that every succeeding rate of azimuthal precession ($B\beta/H$) must also decrease with respect to the undamped oscillations. The process, if continuous, which it is, must be cumulative and the trace takes the form of an equiangular spiral schematically illustrated. Left to run in this way the now controlled and damped compass finally settles at some position S.

17.1.18 *The settling position of a compass damped in tilt*

Figure 17.17 illustrates the position of the spin axis with reference to the horizon and the meridian in both the northern and southern hemisphere. The conditions necessary for equilibrium may be most easily established by considering separately the conditions which obtain for zero motion in azimuth and zero motion in tilt. The former can only occur when the rate of ballistic precession ($B\beta/H$) is equal and opposite to the rate of drifting (15 sin λ). This requires that the control system exhibits north end upward tilt in the northern hemisphere; north end downward tilt in the southern hemisphere. In each case, an inevitable rate of damping precession about the horizontal axis ($rB\beta/H$) accompanies precession about the vertical axis. Since motion in tilt is ultimately zero it follows that in the northern hemisphere the equilibrium position must lie east of true north so that an equivalent rate of tilting upward due to the earth's rotation (15 cos λ sin α) serves to eliminate the downward rate of tilting precession required for damping purposes. Figure 17.17 illustrates the comparable conditions which obtain in the southern hemisphere where the directions of motions are reversed. The position of equilibrium will lie below the horizon and to the west of true north.

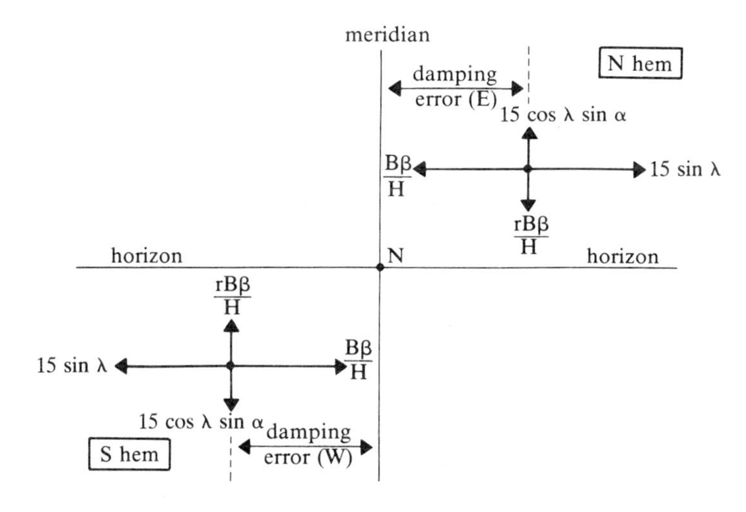

Figure 17.17

17.1.19 *Damping or latitude error*

The displacement of the spin axis east or west of the true meridian in the settled position is called the *damping* or *latitude error* and is a feature of gyro-compasses which are damped by a torque about the vertical axis. This error may be estimated and tabulated and subsequently applied; it may be eliminated by moving the lubber line (for steering courses only) or the compass card moved appropriately by the amount of the error; or the gyro system may be given artificial additional north or south heaviness sufficient to eliminate the drift so that the equilibrium position becomes, as in the case of the equator, the northern point of the horizon.

The magnitude of the damping error if it is allowed to persist may be calculated as follows:

Refer to Figure 17.17.

Since for equilibrium settling in azimuth (α):

$$\frac{B\beta}{H} = 15 \sin \lambda$$

and for equilibrium in tilt

$$\frac{rB\beta}{H} = 15 \cos \lambda \sin \alpha$$

then,

$$r\,15 \sin \lambda = 15 \cos \lambda \sin \alpha$$

i.e. $\sin \alpha = r \tan \lambda$

and, $\alpha^c = r \tan \lambda$

so that, $\alpha^\circ = r \tan \lambda\ 57.3$

If, for instance, the eccentricity of the ballistic connection on a Sperry type compass Mk. E.14 is about $1/40$, then

$$\alpha^\circ = \frac{\tan \lambda\ 57.3}{40} \simeq 1\tfrac{1}{2} \tan \lambda$$

which indicates that the error would be about 1½° in Latitude 45°, 3° in Latitude 63½° and 4½° in Latitude 71½°.

17.1.20 *Damping in azimuth (Brown)*

Figure 17.18 shows the damped spiral of a gyro-compass where the oscillations are initially damped about the vertical axis. The spiral trace is again necessarily distorted and exaggerated for explanatory purposes.

In a gyro-compass which is damped in azimuth the intention is to overlay a rate of precession about the vertical axis out of phase with the normal precession caused by the control system about that same axis. It is clear that, if damping precession in azimuth acts directly opposite and concurrently with normal control precession, the effect will not be to damp the oscillations but merely to reduce the effectiveness of the main control. The ellipse trace will be retained and the oscillations will increase in amplitude. To effect damping in azimuth the additional couples about the horizontal axis must be made to operate out of phase with the main control, i.e. they must exhibit a delayed action so that, in general, movement of the north end of the spin axis *towards* the meridian is further *encouraged* and movement *away* from the meridian is *discouraged*. All subsequent azimuthal displacements from the meridian are reduced and become smaller than they would be under the conditions of an undamped elliptical oscillation. It follows that all subsequent tilting vectors (15 cos λ sin α) also become smaller because α is decreasing. The process is continuous and cumulative so that the trace is once more a spiral but with its centre this time on the true meridian.

.1.21 *The settling position of a compass damped in azimuth*

Figure 17.18 shows the settled position of equilibrium with the rate of main control precession westward (northern latitudes) equated to the sum of the easterly drifting due to the earth's rotation and the rate of damping precession much less than and of opposite sign to that of the main control system. Clearly, the north end of the spin axis retains an upward tilt in the settled position in north latitudes and for similar reasons a downward tilt in the southern hemisphere. In each case the spin axis points true north and compasses which are damped in azimuth are unique to the extent that damping error in azimuth is absent.

.1.22 *Performance test calculation*

Figure 17.19 shows the damped periodic oscillation of a gyro-compass. The ratio between the second and first observed amplitudes of the oscillation is the damping factor (f).

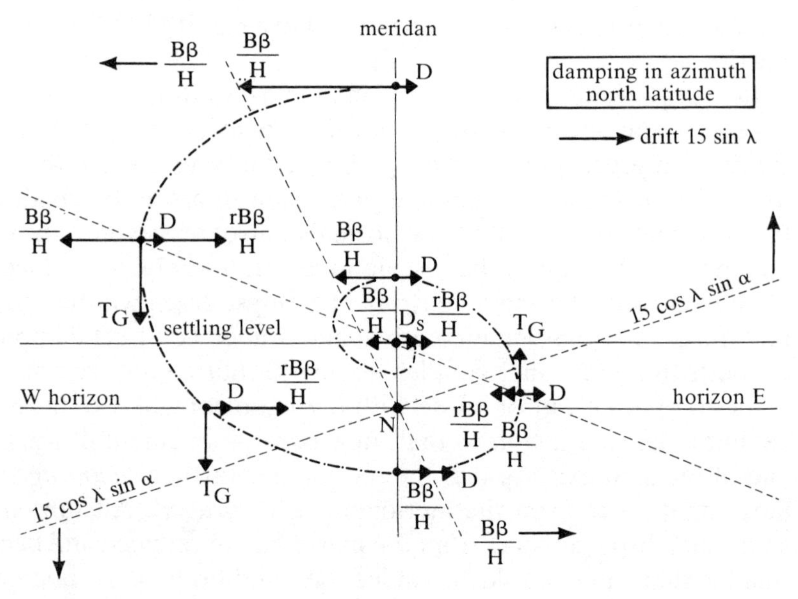

Figure 17.18

$$f = \frac{(B \sim C)}{(A \sim B)}$$

The damping percentage is given by $(1 - f) \times 100$.

Having determined the damping factor, and assuming the final settling heading to be x it follows that,

$$f = \frac{(x - B)}{(A - x)}$$

As example the following maximum readings of the compass have been taken during a settling performance test: $A = 326°$, $B = 026°$ and $C = 359°$. In such a case the observed headings oscillate about the 000° heading mark; it is therefore necessary to add 360° to B, i.e. $B = 026° + 360° = 386°$. From these observations,

$$f = \frac{(B \sim C)}{(A \sim B)}$$

$$f = \frac{(386 \sim 359)}{(326 \sim 386)} = \frac{27}{60} = 0.45$$

The damping percentage is $(1 - 0.45) \times 100 = 55$ per cent.

Also, $$f = \frac{(x - B)}{(A - x)}$$

$$\frac{27}{60} = \frac{(x - 386)}{(326 - x)}$$

$$8{,}802 - 27x = 60x - 23{,}160$$

From which, $x = 367.4° - 360° = 007.4°$

Similarly, $$f = \frac{(C - x)}{(x - B)}$$

$$\frac{27}{60} = \frac{(359 - x)}{(x - 386)}$$

From which again, $x = 007.4°$

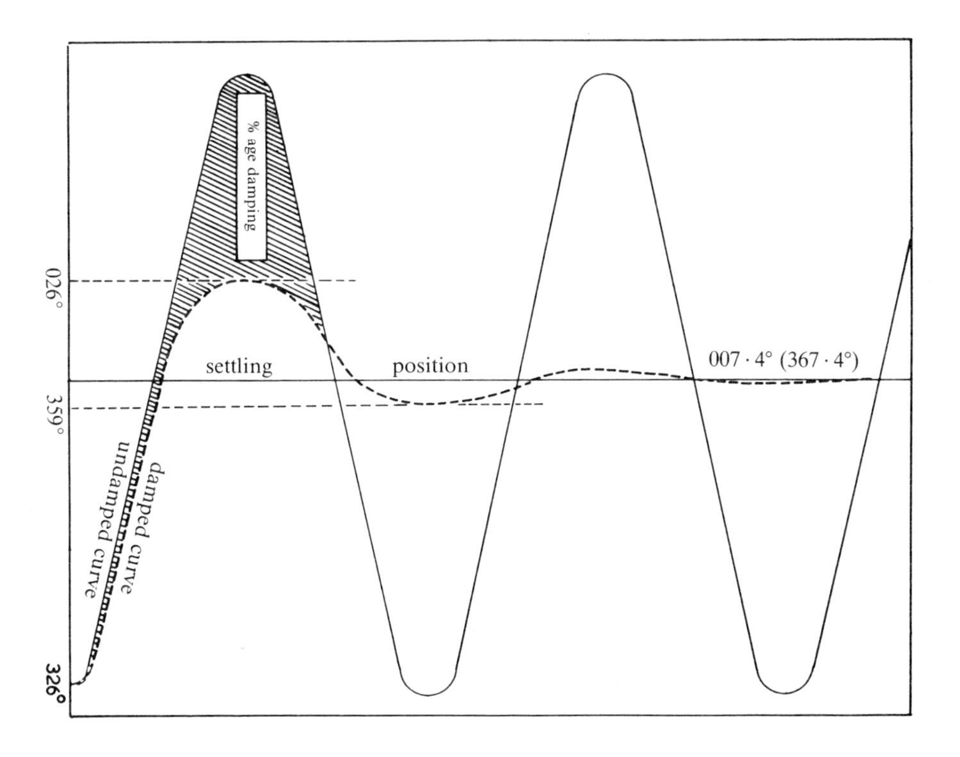

Figure 17.19

17.2 Gyro-compass errors

The following classification and description of the errors of gravity controlled gyro-compasses is given mainly in general terms but where necessary a specific arrangement is indicated so that a particular case may be examined.

17.2.1 Wandering errors

These are compass errors which cannot be predicted and which show little or no consistency either in sign or magnitude.

(*a*) In the event of the speed of the rotor not being perfectly steady it follows that the angular momentum of the system will not be constant either. In this event the required rates of precession ($B\beta/H$) exercised by the ballistic control would suffer to the extent of variations in H caused by the voltage supply to the rotor case being irregular. An inconsistent error or wander would occur and its remedy entails ensuring that the voltage regulator performs its proper function or is replaced.

(*b*) Since the most critical axis of a gyro-compass is its vertical axis about which the ship turns in azimuth it follows that any extraneous torques such as friction may cause unwanted couples which in turn would precess the compass and cause wander. The lower guide bearing might be suspect in such cases.

17.2.2 Permanent errors

These are clearly errors where the sign and amount remain constant. The likely cause is where there is some imbalance which causes an unwanted couple in consequence of which the compass seeks a new equilibrium position.

(*a*) *Constant torque about the vertical axis*
Figure 17.20 shows how this could happen in, for instance, the suspension wire of a Sperry type compass. If the wire contains an unwanted twist then a constant couple acts about the vertical axis which causes a constant rate of precession about the horizontal axis (P_{twist}). The compass settles with an azimuth (α) from the meridian

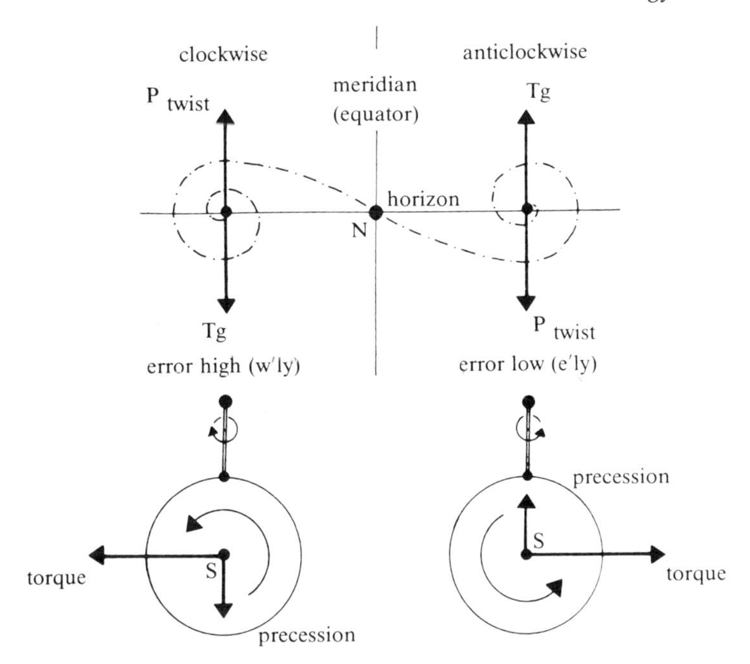

Figure 17.20

such that $\omega \cos \lambda \sin \alpha$ is equal to but of opposite sense to P_{twist}. If Figure 17.20 is assumed for the equator, where the compass would normally settle with its spin axis on the northern point of the horizon (N), it can be seen that the settling position lies *west* of the meridian if the twist is *clockwise*; *east* of the meridian if the twist is *anticlockwise* but in neither case does the compass assume more or less tilt than that which is appropriate for the latitude in which it is operating.

(*b*) *Constant torque about the horizontal axis*

Figure 17.21 shows the effect of north/south imbalance of the sensitive element on a top heavy weight effect compass. The rotor and/or rotor case being either north or south heavy will cause a constant torque about the horizontal east/west axis. This in turn causes an unwanted precession (P_{weight}) about the vertical axis as shown.

The compass, in reaching its equilibrium settling position, will assume a tilt which is sufficient to cause precession in azimuth ($B\beta/H$) of equal amount but opposite sign to P_{weight}. If the compass is damped in azimuth a change in the tilt alone will occur, but if the compass is damped in tilt then the compass will settle with an

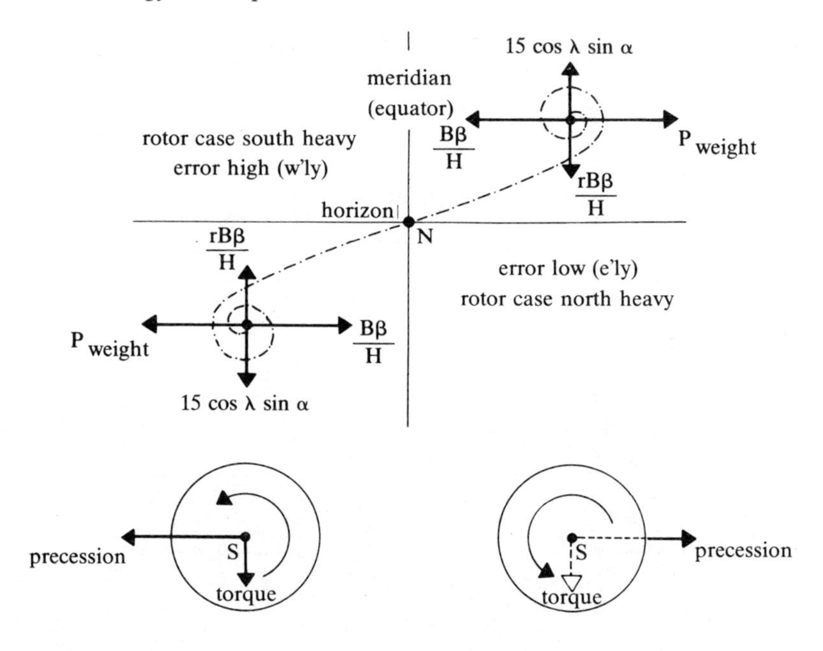

Figure 17.21

azimuth (α) such that ω cos λ sin α is equal to but of opposite sense to the inevitable rate of damping precession ($rB\beta/H$). Such a compass with *north heaviness* will settle *east* of its normal settling position and with slight upward tilt; with *south heaviness* it will settle to the *west* with the north end slightly depressed. Figure 17.21 which assumes the equator makes this clear.

(c) Constant torque about the east/west axis of the phantom

If this occurs on a top heavy weight effect compass which is damped in tilt it is more easily appreciated by considering the liquid ballistic to be imbalanced north/south without any transfer of liquid, e.g. at the equator. In the event of the ballistic frame being north heavy a damping torque about the vertical axis will cause damping precession north end upward. The compass will assume its equilibrium position when sufficient liquid has been displaced from north to south, i.e. the north end tilts upward, to create a damping precession of equal amount and opposite sign. The sole effect is to change the tilt slightly because the rates of azimuthal precession which accompany the rates of damping precession are of equal amount and opposite sign. An error in azimuth does not therefore occur due to a torque about the east/west axis of the phantom. Figure 17.22 shows the

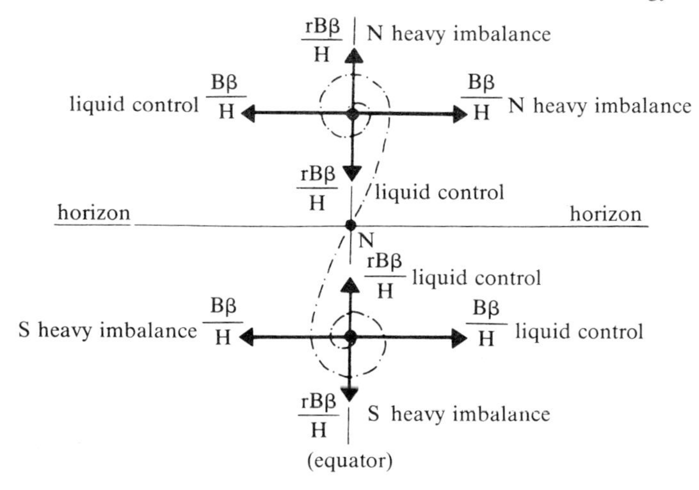

Figure 17.22

equilibrium position displaced above the horizon when imbalance is north heavy, and below the horizon when imbalance is south heavy for a top heavy weight effect compass located on the equator.

In cases (*a*) and (*b*) the permanent error may be eliminated by removing the cause, i.e. the compass must be rebalanced and the suspension checked for unwanted twist, or the error may be eliminated, in a given latitude, by resetting the lubber line on the master compass if the facility exists for doing so. If the error is low the lubber line must be moved the appropriate amount to starboard, and vice versa. In the case of (*c*) although no azimuthal displacement occurs the unwanted tilt might cause an axial load on one of the spin bearings: again, the compass ballistic frame should be rebalanced.

7.2.3 *Course, latitude and speed error*

It is implicit that, apart from the effects of damping error, a gravity controlled gyro-compass will settle with the northern end of the spin axis pointing to 'the port beam of the resultant motion' caused by the ship's own motion towards her destination and the eastward translation due to the earth's rotation.

Figure 17.23 shows the ship's own motion at V knots in a direction θ east of north, and this may be interpreted as an angular motion $\omega' = V/R$ where R is the earth's radius. The total angular motion about an axis parallel to the earth's surface will be the resultant of ω' and $\omega \cos \lambda$. Clearly, the gyro spin axis will settle in the direction of this

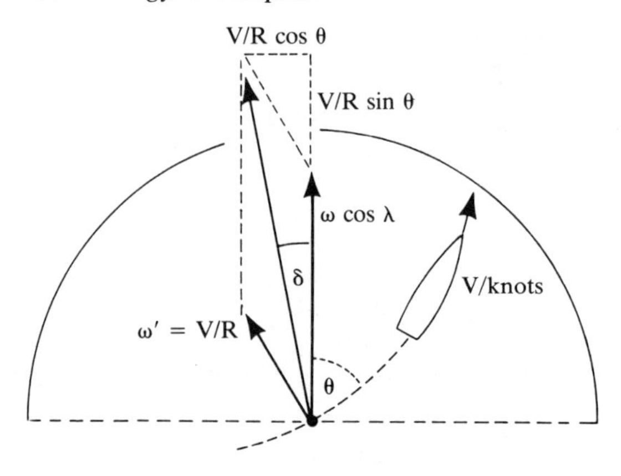

Figure 17.23

resultant which then defines compass north displaced by an error δ from true north. Figure 17.23 shows the superimposed angular motion caused by the ship's own movement to be resolved into components about a north/south axis (V/R sin θ) coaxial with the axis of spin and about an east/west axis (V/R cos θ) perpendicular to the spin axis of the gyro rotor. Expressed in knots instead of angular motion the component about the north/south axis is no other than 'departure' per hour, while the component about the east/west axis is the 'd.lat' per hour. Considering the value of δ to be small:

$$\delta^c = \frac{\omega' \cos\theta}{\omega\cos\lambda + \omega'\sin\theta} = \frac{(V/R)\cos\theta}{\omega\cos\lambda + (V/R)\sin\theta} = \frac{V\cos\theta}{R\omega\cos\lambda + V\sin\theta} \quad \textbf{[17.1]}$$

$$\delta^\circ = \frac{V\cos\theta}{900\cos\lambda + V\sin\theta} \times \frac{360}{2\pi} \quad \textbf{[17.1a]}$$

$$\delta^\circ = \frac{V\cos\theta}{5\pi\cos\lambda}\text{(very nearly)} \quad \textbf{[17.2]}$$

From this discussion it is evident that the north/south component of the ship's motion (V cos θ in equation [**17.2**] causes a tilting of the gyro which it is unable to distinguish from the tilting due to the earth's rotation. Furthermore, an examination of equation [**17.2**], which gives the magnitude of the error, shows how the variables affect it:

(*a*) The error is *directly proportional* to the *speed* of ship, i.e. if the speed is increased from 10 to 20 knots the error is twice as much.

(*b*) The error is *directly proportional* to the *cosine of the course*. It is therefore maximum on 000° and 180°; it is zero when steering 090° and 270°. On all northerly courses the error is westerly; on all southerly courses it is easterly.

(*c*) The error is inversely proportional to the cosine of the latitude; or, more easily, *directly proportional* to the *secant of the latitude*. This means that any error which appears when the ship is on the equator would become twice as large in latitude 60° (secant 60° = [**17.2**]) and would become larger still if the ship proceeded to further higher latitudes.

The denominator of equation [**17.1**] indicates that in very high latitudes the cosine becomes so small that the error increases to unmanageable proportions, and the compass becomes useless as 90° is approached. Fortunately merchant ships cannot navigate in extremely high latitudes and this contingency does not arise. If in the future merchant submarines are to undertake transpolar voyages under the ice then other forms of compass indication will have to be used. High latitude navigation adds a further embarrassment in this connection because the east/west component of the ship's motion ($V \sin \theta$) becomes sensibly large with respect to 900 $\cos \lambda$ under such conditions. The magnitude of $V \sin \theta$ may then no longer be considered insignificant and the course, latitude, and speed error ($\delta°$) must be computed from the more exact formula:

$$\cos \delta = \frac{903.55 \cos \lambda \sec \theta}{V} + \tan \theta$$

Although it is unlikely to occur for some years the fact remains that a combination of *high speeds*, particularly in *high latitudes*, creates the greatest values of the speed error to which gravity controlled gyro-compasses are liable, unless of course, measures are taken to eliminate it in all but the highest latitudes near the pole.

.2.4 *The correction of the course, latitude and speed error*

(i) In the first instance the error, which is known from equation [**17.2**] above, may be calculated and applied. Since the error is independent of compass characteristics it can be easily tabulated and this is sometimes done so that the table covers a wide range of latitude (to 70 degrees or more), speeds up to 30 knots and each of the possible courses. The table will indicate the sense of applying the error.

(ii) Sometimes a portable correction calculator is used where the extent of the damping error and the course and speed error combined are interpreted on the calculator by a white line indicating the course to steer. This corresponds to a red line set against the required true course. If the compass incorporates a transmission system to distant repeaters a similar correction calculator is made to realign the repeater and thereby eliminate the error.

(iii) The introduction of an artificial error of equal amount and of opposite sign is effected by moving the lubber ring on the master compass. When the error is HIGH the lubber ring is moved to PORT; when the error is LOW the lubber ring is moved to STARBOARD. Both the damping error and the course, latitude and speed error are eliminated simultaneously by two manual settings for latitude, one for speed and an automatic mechanism for course. The transmitter in this case is located on the lubber ring so that this movement is automatically communicated to distant reading repeaters by reorientating the bearing plates to the true meridian.

(iv) (*a*) The methods of dealing with the speed error outlined in (ii) and (iii) above constitute legitimate artifices whereby the master compass assumes some displaced position from the true meridian but is made to *indicate* an apparent absence of this displacement. No attempt is made to restore the spin axis to the true meridian. Other methods of dealing with the speed error are however possible in modern compasses in which the cause of the error is removed at source. Since the term $V \cos \theta$ predominates in equation [**17.2**] above the spin axis tilts at this rate and in the case of the Arma Brown compass, by means of a potentiometer setting, a torque about the vertical axis is created in order to precess the spin axis about the horizontal axis at the rate $-V \cos \theta$ and thereby removes the cause of the error by reducing this term to zero. It is proposed to adopt a similar procedure in the Italian Microtecnica compass.

(*b*) This unique innovation is also used in the Sperry Mk 20 compass where not only damping error, but course and speed error are eliminated by a motor which causes a torque about the vertical axis sufficient to produce a rate of precession which will eliminate (i) unwanted tilt due to the NS component of ship's motion (ii) unwanted tilt due to the rate of damping precession. The control system will then assume a tilt appropriate to the latitude in order to eliminate drift due to the earth's rotation.

17.2.5 Change of speed error

Since the course and speed error is independent of compass constants it follows that when a ship alters course and/or speed the spin axis of a compass, where no steps have been taken to reduce $V \cos \theta$ to zero, will finally settle in the compass meridian appropriate to the new course and speed. The difference between the error before and after a manoeuvre is called the *change of speed error*.

If the error is retained in circular measure (δ^c) and the E/W component of the ship's motion ($V \sin \theta$) considered negligible the change of speed error is given by:

$$\delta_1{}^c \sim \delta_2{}^c = \frac{V_1 \cos \theta_1 \sim V_2 \cos \theta_2}{R\omega \cos \lambda}$$ (equation [**17.1**], p. 288)

This is quite likely to be significant if, for instance, a ship alters to a reciprocal course at full speed, particularly in a high latitude. Left to itself it might be thought that the compass remains unsteady while it seeks the new error ($\delta_2{}^c$) from north on the new course. Certainly on the basis of a compass with a damped period of oscillation of nearly an hour and a half this unsteadiness might be expected while it seeks its new equilibrium position.

17.2.6 Ballistic deflection

Fortuitously such a compass experiences a horizontal acceleration during the manoeuvre which displaces liquid in the control unit in such a way that the compass is precessed towards the new error meridian.

For example, consider a ship steering 000° (T) which increases speed. The change of speed error is clearly westerly because the course and speed error which is westerly to start with becomes more so as the ship settles to the new increased speed. *During* the manoeuvre a horizontal acceleration takes place and this causes the vertical ring assembly (Figure 17.24) to respond to a false vertical and the liquid in the control system to associate itself with the instantaneous false or dynamical horizontal. In the particular case chosen, the additional south heaviness causes westerly precession

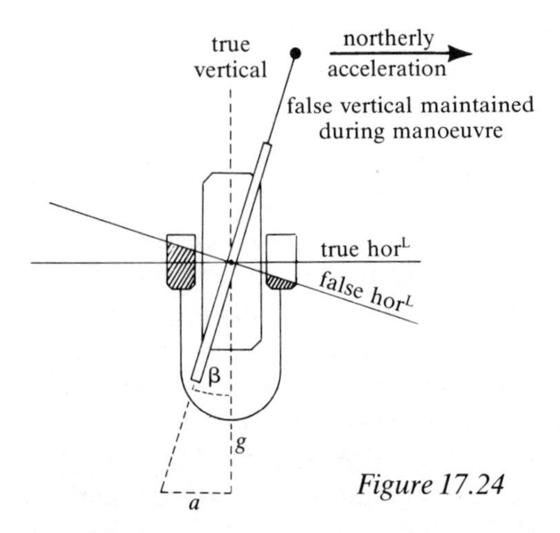

Figure 17.24

which is fortuitously in the *same direction as the change of speed error* mentioned above. Furthermore, this precession of the north end of the spin axis which is called *ballistic deflection* occurs only *during* the manoeuvre, i.e. it commences when the speed increases and it ends when the new speed is attained.

It is important to realise that for a given manoeuvre the *magnitude* of the ballistic deflection is the same regardless of the *duration* of the manoeuvre. If the increase of speed in the above example is effected slowly then β in Figure 17.24 is smaller than it would be if the increase were made more quickly. The displaced liquid in the former case would be of smaller amount and exert its influence for a longer time, than in the latter case where the displacement of liquid in the control system would be more pronounced but exert its influence for a shorter period of time. The total ballistic deflection in each case is the same and is independent of the duration of the manoeuvre.

In general the horizontal acceleration a experienced during a manoeuvre is given by:

$$a = \frac{V_1 \cos \theta_1 \sim V_2 \cos \theta_2}{t}$$

Figure 17.24 shows that provided β is small, which it invariably is, then,

$$\beta^c = \frac{a}{g}$$

where g is the acceleration directed to the earth's centre by gravity.

Since the rate of precession produced by the control system is $B\beta/H$ it follows that if t is the duration of the manoeuvre the *ballistic deflection* experienced is given by,

$$\frac{B\beta t}{H} = \frac{Bat}{gH} = B\,\frac{\left(\dfrac{V_1\cos\theta_1 \sim V_2\cos\theta_2}{t}\,t\right)}{gH}$$

In order to ensure that the gyro-compass spin axis is deflected to the new settling meridian by the end of the manoeuvre it is essential that not only must the sense of ballistic deflection be the same as the change of speed error, which has already been established, but the magnitude of the two must be the same also; otherwise, the compass spin axis would either overshoot or fail to reach the new settling meridian. This ideal condition may be stated as follows:

ballistic deflection = change of speed error

$$B\,\frac{\left(\dfrac{V_1\cos\theta_1 \sim V_2\cos\theta_2}{t}\right)t}{gH} = \frac{(V_1\cos\theta_1 \sim V_2\cos\theta_2)}{R\omega\cos\lambda}$$

i.e. $$\frac{B}{gH} = \frac{1}{R\omega\cos\lambda}$$

or, $$\frac{R}{g} = \frac{H}{B\omega\cos\lambda}$$

or $$2\pi\sqrt{\frac{R}{g}} = 2\pi\sqrt{\left(\frac{H}{B\omega\cos\lambda}\right)}$$

Pendulum — Gyro-compass

This expression is of the form $T = 2\pi\sqrt{l/g}$ which is the period of oscillation of a simple gravity pendulum where l is the length of the string. The length in this case, R, is the radius of the earth and the period of oscillation is the maximum obtainable from a simple gravity pendulum. Its value is 5060 secs or just under 84½ *minutes*.

This expression may be shown to be the undamped period of a gyro-compass acting under the control of gravity. For the numerical identity to be secured between ballistic deflection and change of speed error it is therefore necessary that the undamped period of the compass should also be 84½ *minutes*. This unique property is sometimes referred to as Schuler Tuning after the investigations made by the discoverer.

17.2.7 *To secure a steady compass in any latitude*

This discussion shows that since the undamped period of oscillation is to be 84½ minutes, on compasses where no steps are taken to eliminate the course and speed error at source, it can only be achieved in a specified latitude. Note the term cos λ in the denominator. In commercial gyro-compasses it is sufficient to establish these conditions for latitude 45° without any practical disadvantages occurring after executing a manoeuvre in some other remote latitude.

In certain naval compasses where closer azimuth indication is required, especially after a series of high speed manoeuvres, it is essential that the 84½-minute period is maintained for any latitude in which such manoeuvres may occur. In principle there are several possibilities open but in practice only three have been used. In the expression for the undamped period ($H/B\omega \cos \lambda$), the term under the radical sign indicates clearly that either H (which is the angular momentum of the gyro) must be made to alter proportionally with cosine latitude, or B (which defines the control characteristics) must be made to change proportionally with secant latitude. The three technical variants are discussed briefly as follows:

(i) The angular momentum (H) which is the product of the moment of inertia of the rotor and the rate at which it spins provides an easy means of maintaining the 84½ minute period by controlling the rate of rotation against a cosine latitude scale. This is done in the Anschütz compass and was also a feature of the former American Arma compass.

(ii) The gravity control characteristics signified by $B = 2Al^2p$ (see page 000) permit a further two possibilities of maintaining the 84½ minute period over a range of latitude in compasses which use a liquid system of control. Although in principle a variable density of the liquid used would satisfy the purpose it is awkward to achieve and is therefore not considered. It is much easier to arrange a suitable mechanism, suggested in Figure 17.25, whereby the square of the north/south distance between the liquid containers is made to change proportionately with the secant of the latitude. When the setting is made appropriate to the latitude the 84½ minute undamped period of the compass is secured and no further wander occurs after the speed or course is altered. This device has been used in former Sperry compasses.

(iii) The second alternative designed to change the value of B lies in the extent of the area of free liquid surface A. If this can be made

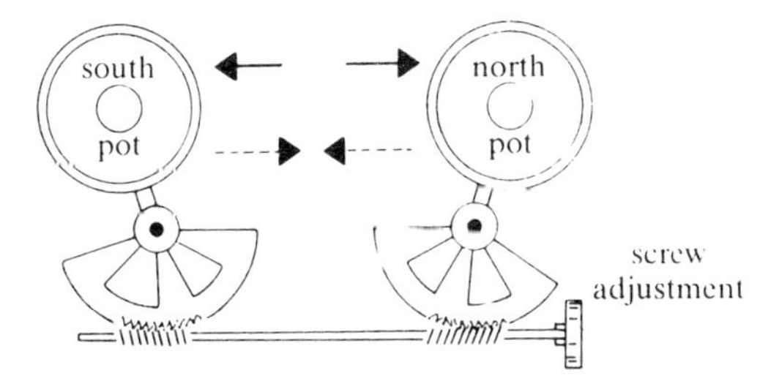

Figure 17.25

variable in sympathy with the secant of the latitude the steadiness of the compass, after executing a manoeuvre, can be ensured over a wide range of latitude.

Figure 17.26 shows how shapes cut to the form of a 'secant dome' can be immersed in the liquid of the ballistic containers in such a way that by either raising or lowering them the extent of the liquid free surface can be changed. Compasses which make use of this refinement have two north containers and two south containers fitted to the four corners of a square frame. The secant domes in the NE and SW containers are fitted 'point downward' while those in the NW and SE containers are fitted 'point upward'. The purpose of this alternate arrangement is to ensure that the overall level of liquid remains undisturbed and that when the height of the domes is set, after the latitude has been appreciably changed, the compass does not become excessively or deficiently top heavy. The vertical

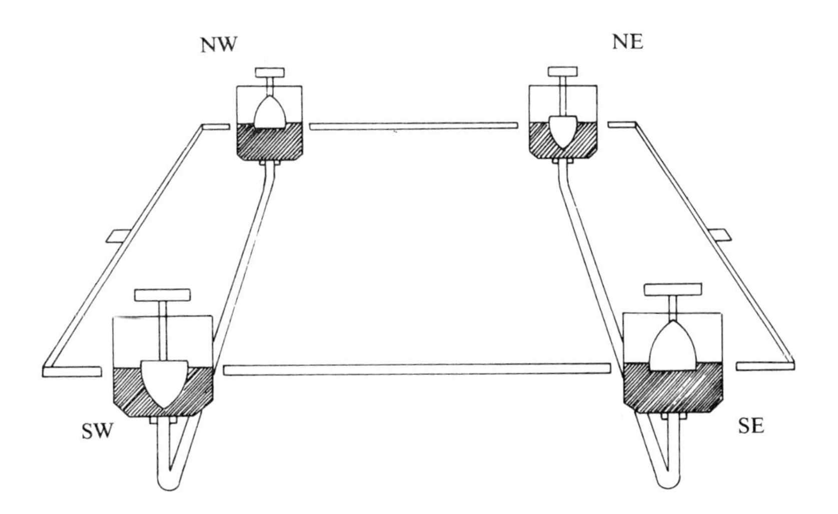

Figure 17.26

balance 'top against bottom' would otherwise be seriously upset if, for instance, each of the four domes were fitted point downwards. In such a case immersion would create excessive top heaviness.

17.2.8 *Ballistic tilt effect*

The fact that the control system acts about a false vertical during a manoeuvre has been shown to be the means by which the compass is deflected to the new settling meridian and remains steady. This is true provided that the torque required for damping purposes acts about the horizontal axis. In the event of a couple acting about the vertical axis to produce damping in 'tilt' it follows that this eccentricity will create ballistic tilt in sympathy with ballistic deflection so that although the compass will be precessed to the new settling meridian it will arrive there with a tilt which is slightly different from that which is appropriate to the latitude in which the manoeuvre is executed. This unwanted tilt is, of course, detected by the compass and is eliminated by successive circuits of a small damped spiral.

Figure 17.27 illustrates the case of westerly ballistic deflection ($B\beta/H$), due to an increase of the northerly component of motion (or a decrease in the southerly component), accompanied by a slight north end downward tilt due to damping torque about the vertical axis. After the manoeuvre is completed the spin axis, which has been deflected to the new settling meridian immediately departs from it on the initial stages of the damped spiral. Since the period of

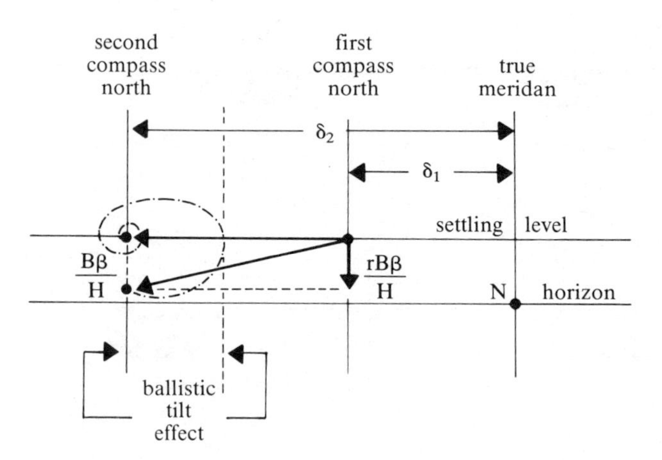

Figure 17.27

the damped spiral is of the order of 87 minutes it follows that a maximum wander from the new settling meridian will occur at the end of the first quarter period, i.e. about 22 minutes after the completion of the manoeuvre. Fortunately this wander is not serious and seldom exceeds more than 1½° at most. During the second quarter period this disappears and ballistic tilt effect causes no further concern.

Since ballistic deflection of the correct amount occurs during the manoeuvre regardless of whether the manoeuvre was executed quickly or slowly it may be said that ballistic deflection is therefore independent of the duration of the manoeuvre and that ballistic tilt is an effect which follows it. Nevertheless it is necessary to point out that in the unlikely event of a protracted manoeuvre the quarter period of some twenty minutes would preclude ballistic deflection ever reaching its required magnitude on account of unwanted ballistic tilt.

17.2.9 *Ballistic deflection on compasses where course and speed error is eliminated at source*

On page 291 the means were described whereby the course and speed error may be eliminated with a torque about the vertical axis to cause precession about the horizontal axis equal and opposite to the rate of tilting due to the north/south component of ship's motion. Under these conditions there is, of course, no change of speed error and the fortuitous circumstance of ballistic deflection instead of being an advantage becomes an embarrassment. Every reasonable effort is made to keep ballistic deflection at a minimum. Since there is no necessity to preserve the 84 minutes period a reduction in ballistic deflection is achieved by an increase of the damped period which approaches 120 minutes in some cases. The period cannot, however, be extended without limit because wander from the equilibrium settling position would tend to persist too long. The gravity sensing device, if suitably damped against short-term horizontal accelerations provides a second means of reducing ballistic deflection in cases where this can be done. Finally, the amplitude of the sensing device itself can be restricted within certain limits though these limits must be carefully chosen in design or saturation will take place and the performance of the compass be impaired.

It has also been suggested that in compasses corrected at source for course and speed error the speed potentiometer setting be set to zero if damping about the horizontal axis is used. The intention is that ballistic tilt effect should thereby offer a pronounced reduction of ballistic deflection. This is unlikely to be of any advantage unless manoeuvres were executed with extreme slowness; in practice this would not occur so that the settings are best made immediately a manoeuvre is completed and are appropriate to the new conditions.

17.2.10 *Rolling error*

It has been mentioned that both high speeds and high latitudes are anathema to gyro-compasses: high seas are no less disturbing because horizontal accelerations, which cause the compass to swing to and fro in the gimbal mountings, combine with the gravitational acceleration and tend to destroy the true vertical. One of the basic properties of a gravity controlled compass becomes in jeopardy once this occurs. The defences against this happening are fundamentally twofold. Firstly, the compass may be further stabilized along preferred axes by additional gyros. Secondly, the compass might be encouraged to disregard horizontal accelerations by reducing these to a minimum through placing the compass close to the dynamical centre about which the ship rolls, although it must be pointed out that this position itself experiences horizontal accelerations. Clearly, this would be very difficult to achieve if for no other reasons than that the ship not only rolls but also pitches and frequently does both simultaneously; in addition the ship must sail in both a light and loaded condition. It is therefore impracticable to place the compass in an ideal location because none exists to satisfy all conditions.

The detail of rolling errors must be examined so that some form of compromise can be made regarding their elimination. Broadly, rolling errors fall into one of two kinds. Each kind deals with the compass swinging in its gimbals. The mode of swinging can be either north/south, east/west or a combination of the two, i.e. quadrantally. The reader is cautioned against associating the mode of oscillation with the direction of the ship's head (i.e. the course). This is often done but is quite unjustified. For instance, the gyro-compass may swing east/west simply because the ship may be rolling on a north or south heading; the same mode of swing might however, have been caused by the ship pitching heavily on an east or west course. Indeed, with a cross-sea it is quite conceivable that the compass could swing east/west without the ship heading on any of the cardinal headings. For similar reasons the reader should guard against a quadrantal mode of swing being directly associated with some intercardinal heading. In fact, in the following discussions of what are called intercardinal rolling errors it must be understood that the *mode of swinging* is the criterion for determining the errors incurred and that this is only incidentally related to the ship's heading, and certainly never without considering the state of the sea. The reader is discouraged from attaching too much attention to the term *rolling error*; also, the term 'intercardinal' which refers primarily to the mode of oscillation, and not necessarily to the ship's heading.

7.2.11 *Quadrantal error no. 1*

The pendant compass does not possess asymmetrical distribution of its mass about a vertical axis in all horizontal directions. It obviously predominates along the axis which coincides with the plane of the vertical ring/rotor case/and rotor. A corresponding deficiency exists along the axis at right angles to the former. When the compass swings in its gimbals this asymmetry causes the 'long axis of weight' to tend to turn towards the plane of swing in such a way that a reactionary torque about the vertical axis is caused at the point of suspension, and so precesses the compass about a horizontal's axis leading in turn to unwanted tilt of the spin axis and subsequent wander in azimuth.

The precise reasons for the long axis of weight to turn in the fashion described are complex and beyond the scope of this book. The dynamical considerations involve both radial and circumferential accelerations during swinging and need distinctions to be made between two sets of forced oscillations. Sometimes explanations are offered which use the idea of a steel bar, suspended horizontally from its centre, set into a swinging mode which makes a small angle with the axis of the bar. Various results can be obtained by imposing different periods of oscillation. Simple examples of this kind are best left alone because the tendency is to oversimplify the phenomenon.

The gyro-compass is a compound type of pendulum and it is sufficient to state:

(*a*) with the plane of swing coincident with the long (east/west) axis of weight no tendency to turn exists

(*b*) with the plane of swing perpendicular to the long (east/west) axis of weight no tendency to turn exists
(*c*) with the plane of swing orientated at 45° to the long (east/west) axis of weight the *tendency* to turn reaches a maximum
(*d*) the *tendency* to turn is proportional to a function of twice the angle between the meridian and the plane of swing
(*e*) the long axis of weight of the vertical ring and the rotor case, although coincident behave differently in compasses where only the vertical ring responds to the false verticals while the rotor case is stabilized in the true vertical about its east/west axis. The predominating adverse effects lie in the behaviour of the vertical ring
(*f*) the rotor within the case does not contribute to these effects
(*g*) these effects do not occur if the vertical ring contains the vertical axis of the rotor case and is thereby stabilized in the true vertical.

Figure 17.28 shows an approximate plan view of a conventional vertical ring and rotor case assembly having its long axis of weight *AB* and its short axis of weight *CD*. The centre of gravity of the mass eastward of the line *CD* is shown at W_1 and the centre of gravity of the mass northward of the line *AB* is shown at W_2. The distances from their respective axes are shown as d_1 and d_2. A

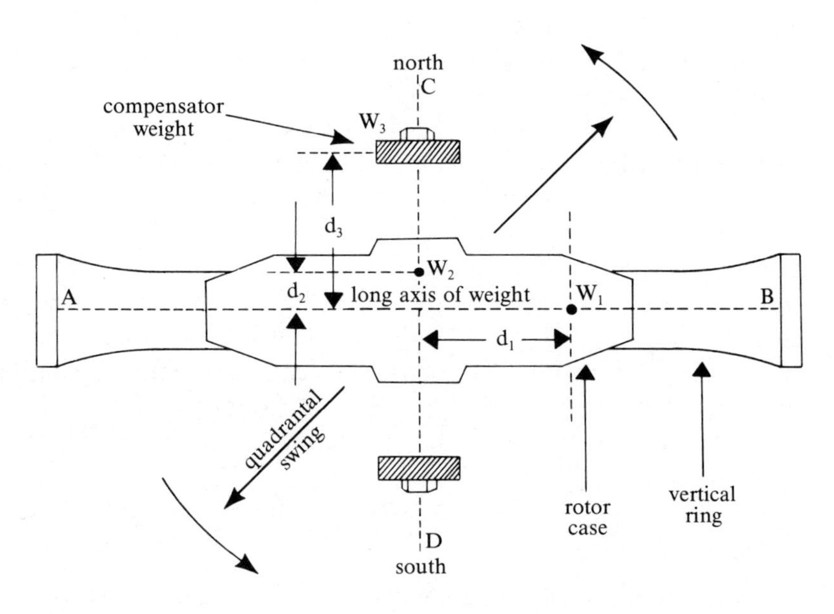

Figure 17.28

glance shows that W_1d_1 is a larger moment than W_2d_2. When the compass swings quadrantally as shown the moment of inertia about the axis *CD* exceeds the moment of inertia about the axis *AB* and this inequality tends to turn the assembly anticlockwise. To remedy this *compensator weights* are fitted to brackets, which extend perpendicularly from the plane of the vertical ring, at distances sufficient to make the *moments of inertia* about the two axes *AB* and *CD* equal. This is done by making $W_3d_3^2 \simeq W_1d_1^2 - W_2d_2^2$. In this way the moments of inertia about the vertical axis in all horizontal directions are made the same. The long axis of weight is effectively destroyed and this cause of error eliminated. If the compass were not fitted with the compensator weights then with the compass swinging quadrantally NE/SW as shown in Figure 17.28 with an amplitude of 15°, and a period of oscillation 3½ seconds in latitude 50° the rolling error would amount to something greater than 12°W.

7.2.12 *Quadrantal error no. 2*

The false vertical produced by horizontal accelerations tends to produce a second type of rolling error which is best explained by considering the effects when, in the first instance, the compass swings east-west in the plane of the vertical ring.

(i) *Compass swinging east–west*
As an example Figure 17.29 shows the outline of a Sperry type compass viewed from the south side. The offset connection between the ballistic frame and the rotor case is shown at *O* and occupies its designed position just east of the true vertical through the rotor case whenever the swinging compass crosses its central position. The position of this connection becomes displaced still further to the east of the true vertical when the compass swings eastward; it is likewise displaced to the west of its designed position when the compass swings westward. The effect of this is to create alternating torques about the vertical axis on opposite swings of the compass, the sense of each depending upon the natural north or south heaviness appropriate to the southern or northern hemisphere respectively. Because these torques are of opposite sense on opposite swings the precessions about the horizontal axis are also alternate with the result that no accumulated north or south heaviness occurs and the net effect over a series of east-west swings is zero.

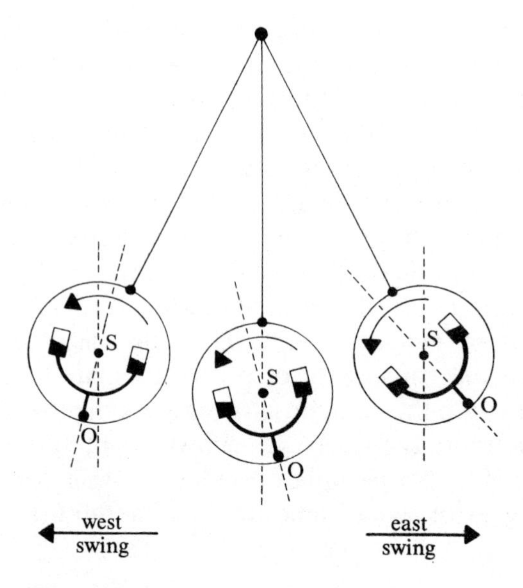

Figure 17.29

(ii) *Compass swinging north-south*

When the compass swings in the north-south plane the eccentric connection referred to in (i) above remains in its designed position just east of both true and false verticals, but Figure 17.30 shows that although gyroscopic inertia maintains the rotor case and ballistic

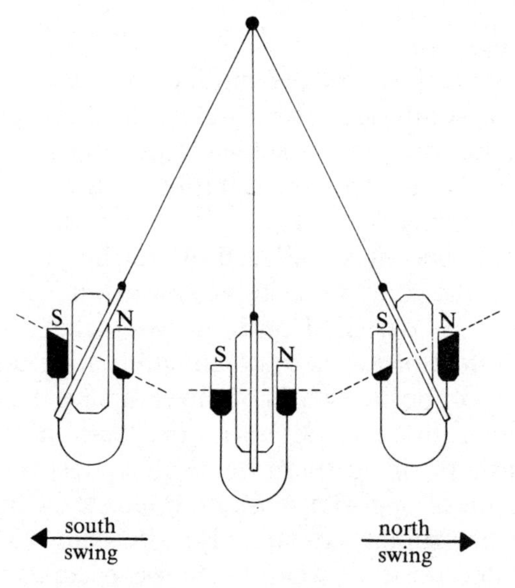

Figure 17.30

control upright the vertical ring and the instantaneous flow of liquid associate themselves with the false vertical and false horizontal at each alternate swing of the compass. Since the eccentric connection has remained undisturbed this means that alternate surges of liquid produce alternate torques, about the horizontal axis, which in turn precess the compass east-west in sympathy with north and south surges respectively. Again, the effect is of opposite sense on opposite swings and the net effect over a series of north-south swings is zero.

(iii) *Compass swinging quadrantally*

If instantaneous liquid flow is assumed (simulating top heaviness) then a maximum surge between north and south pots occurs simultaneously with the maximum displacement of the eccentric connection, for instance (Figure 17.31) on a NE swing excessive north heaviness due to the northerly component is associated with the maximum easterly displacement of the eccentric connection due to the easterly component. A clockwise torque is created about the vertical axis resulting in north end upward precession. On the SW half of the swing south heaviness is associated with maximum westerly displacement of the connection which again imparts a clockwise torque producing north upward tilt. If this sequence is allowed to persist liquid accumulates in the south pots and a

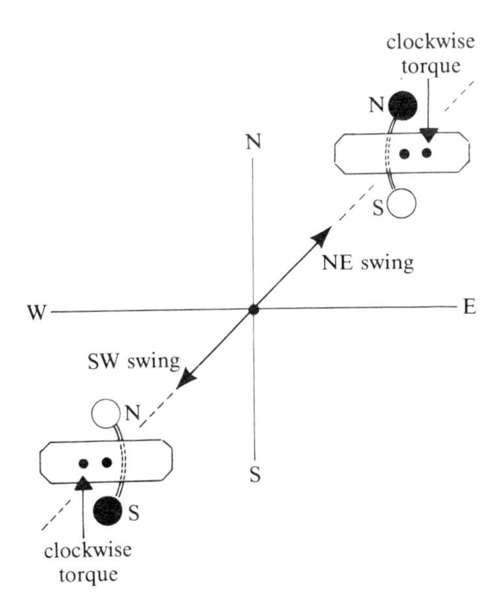

Figure 17.31

westerly wander of the north end occurs to give rolling error. When the compass swings quadrantally in the sense NW-SE an anticlockwise torque occurs, about the vertical axis which precesses the south end upward to give an easterly wander.

17.2.13 *Methods of eliminating rolling error no. 2*

East-West stabilization

Since this type of error is due to torques being applied about axes which are no longer truly horizontal or vertical, on account of transverse accelerations, one way to eliminate it is to maintain either the north-south or east-west axis truly horizontal and so prevent the cumulative effects when the compass swings quadrantally. Since the compass must have freedom to tilt this rules out stabilization about the east-west axis. Some manufacturers make every effort to prevent oscillations about the north-south axis, east-west stabilization being achieved with one or more additional gyros specially arranged for the purpose.

17.2.14 *Delayed action control system*

The synchronous effects of liquid surge and displacement of the eccentric connection described above assumed a free and instantaneous response of the liquid to lateral accelerations. This points to further means by which the error may be eliminated. A 90° phase difference can be introduced between the swinging of the compass and the liquid oscillations. By this means the maximum surge of liquid occurs in sympathy with the instants when the eccentric connection occupies its designed position, i.e. is not, for the moment, displaced either east or west of its correct position. The arguments used for north-south and east-west swinging are now applicable once more in the sense that torques of opposite direction occur on opposite quadrantal swings so that the ultimate precessions eliminate themselves and rolling error disappears.

At this point it should be mentioned that for this artifice to remain effective the delayed flow of liquid, which is created by restricting the bore of the liquid pipes, must always be equal to a quarter period of the swing of the compass in its gimbal mountings. This raises the question as to what this period is and the factors which control it. The compass itself has a natural period of oscillation of the order of one second when it is displaced by hand and permitted to swing as a pendulum. If $T = 2\pi\sqrt{(l/32)}$ this would support an effective length of the pendulum of the order of one foot which is reasonable. In practice however, it is the rolling of the ship which initiates the horizontal accelerations and maintains the oscillation. Clearly, this is governed by the period of roll of the ship and to some extent the period of pitch while both are associated with the period of the sea waves with which there may occasionally be some synchronous swings of larger than normal amplitude. Furthermore, the type, size and state of loading of the ship all add difficulties to estimating the final period of compass swing.

Before considering a range of possible periods let the most likely be, say, 10 seconds. If the liquid can be made to flow with 2½ seconds delay then rolling error is eliminated precisely when, and only when, the compass swings in its gimbals to and fro in 10 seconds. This would correspond to a simple pendulum approaching 100 feet long. Figure 17.32 shows that the graph of the rolling error would then be zero. If however, the period of swing were less than 10 seconds rolling error would occur in the sense that the 2½ second delay is excessive and would show the characteristics of a compass deficient in topheaviness, i.e. easterly wander would occur when the compass swings north-east south-west (cf. page 302). If on the other hand, the period of swing were greater than 10 seconds rolling error would occur in the same sense as that described on page 303 and the compass would then display the effect of excessive topheaviness.

17.2.15 *Range of oscillation period*

This discussion leads to the possibility of reducing the error and rendering any remainder of consistent sign over a *range* of likely periods of swing by eliminating the majority of the error by the addition of actual top-heaviness in the form of small weights. On the basis that ships (and therefore compasses) are unlikely to roll with periods in excess of 18 seconds this figure is chosen in certain Sperry type compasses. Figure 17.32 shows that at this period the error is zero and a 4½ second delay in the liquid flow has been incorporated. Between periods of zero and 18 seconds the error is consistent in the sense of the compass lacking topheaviness. Sufficient weight is now added to the top of the liquid containers to eliminate the rolling error shown by the shaded area and covering all periods of swing which are most likely, say, between 4 and 14 seconds. The unshaded

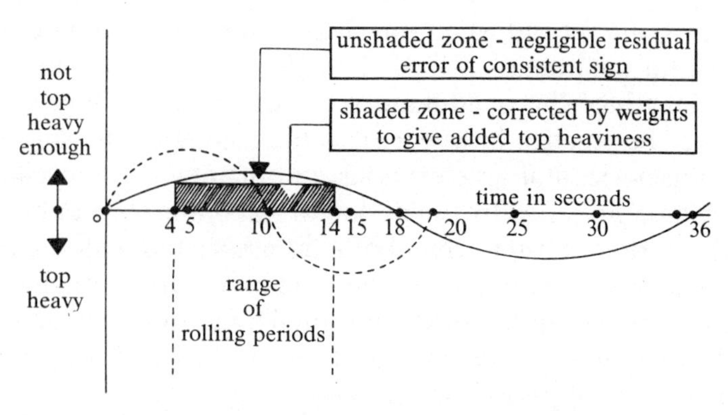

Figure 17.32

area remains uncorrected but is negligible. It is clear that this uncorrected area could be further reduced by adding further weight but only at the expense of a smaller, and therefore not so useful, range of rolling periods. The most useful compromise is therefore adopted as shown in Figure 17.32.

17.2.16 Short-period oscillation

The short-period swinging of the compass referred to in section 17.1.14 above is the only remaining embarrassment. To reduce this as much as possible dashpots are fitted to the fore and aft gimbal axis to damp part of the athwartship swing while friction brakes can be adjusted during heavy weather to reduce the fore and aft swinging of the compass when the ship pitches heavily.

17.3 Details of construction

The description of specific proprietary equipment has generally been avoided within this book. The preference being for general system descriptions which cannot so readily become obsolescent. In the case of gyro-compasses this technique is not appropriate and so two long established compasses are described in detail.

Sperry gyro-compass Mark E.14

The construction details of the Sperry Mk E.14 compass are conventionally listed under five headings which deal with the five main components which constitute the master compass.

17.3.1 *The sensitive element*

This consists of the *gyro rotor* within its case, the *vertical ring* and the *suspension wire*. The gyro wheel, which forms the rotor of a squirrel cage induction motor, is some 10 inches in diameter, weighs between 53 and 55 lb. (some models differ slightly according to whether the compass is made in England or the USA) and is made to rotate at 6000 revolutions per minute in air with a tolerance of 300 revolutions per minute or less. The sense of rotation is anticlockwise when viewed from the south, i.e. the rotor spins contra-earthwise as required by the top heavy weight effect of the control system. The south face of the rotor features a spiral line, which can be viewed through a window in the case, and appears to move radially outwards when the rotor is spinning counter-clockwise.

The rotor case forms the stator of the motor and contains the bearing housings and the oil for lubricating the bearings. Oil well windows are fitted to indicate the precise level of oil. The rotor case is provided with two studs, one on the east side of the case and the other on the west side, which define the horizontal axis. The studs are located in ball races in the vertical ring whose plane is normally east-west and therefore permit the rotor case to tilt about this east-west axis. The small amounts of tilt to which the rotor case is subject when the compass is operating are indicated by a *spirit level* which is fitted to the case. The graduated divisions each represent a tilt of 2 minutes of arc. On the north-east side of the rotor case three terminals are located to receive a three-phase, 50 volt, 210 cycle a.c. supply to drive the gyro rotor.

The vertical ring surrounds the rotor case and supports it along the east-west axis as mentioned above. A second ring, called the phantom ring, q.v., surrounds the vertical ring and is provided with guide bearings into which are located studs fitted to the vertical ring along the line of the vertical axis. This arrangement effectively defines the vertical axis of the sensitive element about which the

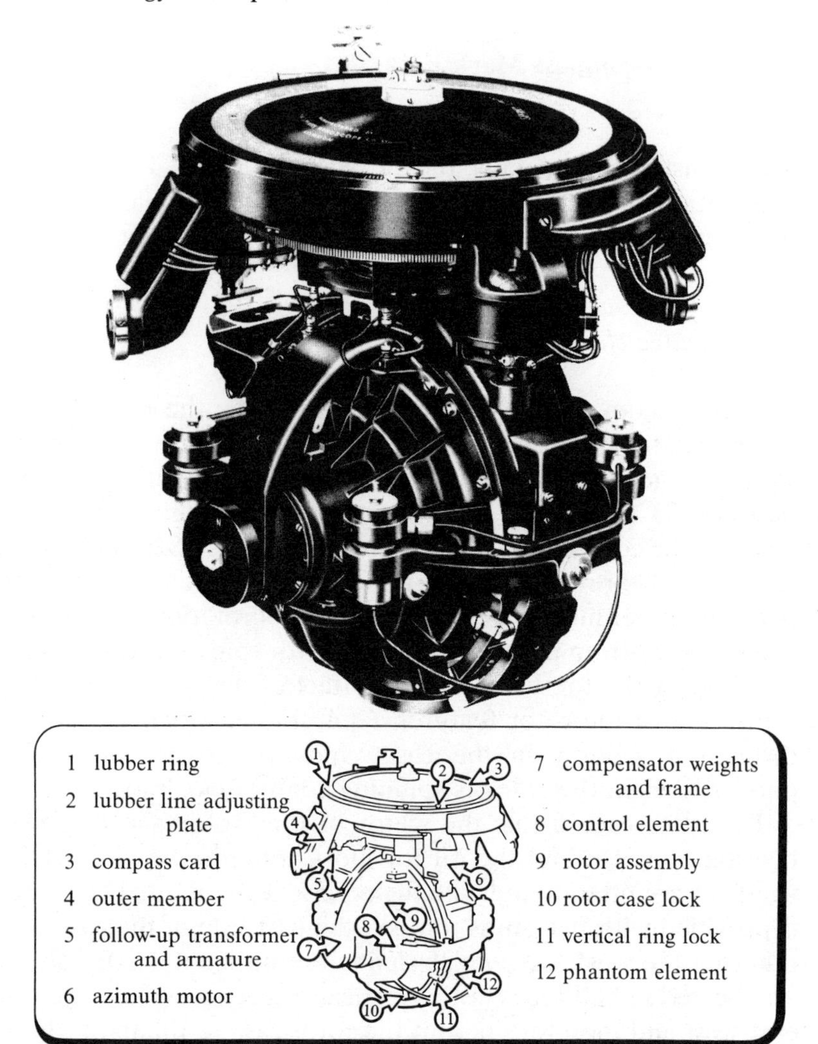

Figure 17.33 Sperry Mark E.14 gyro-compass.

ship turns in azimuth. From the top of the vertical ring two substantial brackets arch to the north and south and support *compensator weights* designed to eliminate one of the causes of rolling error by equalizing the moments of inertia of the system in all horizontal directions about the vertical axis (see page 301). On the east side of the vertical ring there is a twin armed bracket, encircling the phantom ring referred to, which supports the *armature* of the electric sensing device. Three leads carry the three-phase current down the top eastern quadrant of the vertical ring to a position close

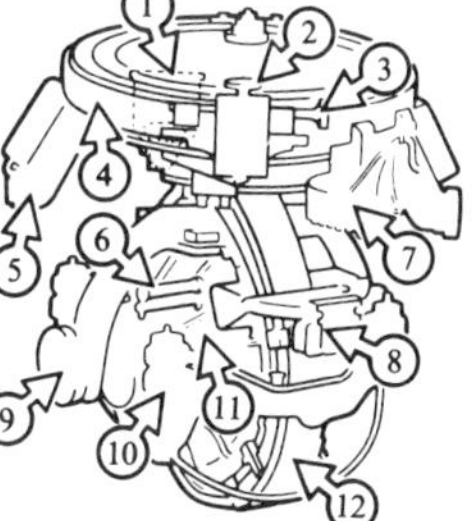

Figure 17.34 Sperry Mark E.14 gyro-compass.

to the horizontal axis bearing and end at three terminals. Special and extremely flexible leads connect the two sets of terminals to convey the current from the vertical ring to the rotor case. Care should be taken to ensure that these leads do not become foul or cause any unwanted torque on the rotor case.

7.3.2 *The phantom element*

The *phantom element* is a very distinctive feature of the Sperry design. Basically it consists of a secondary ring surrounding and

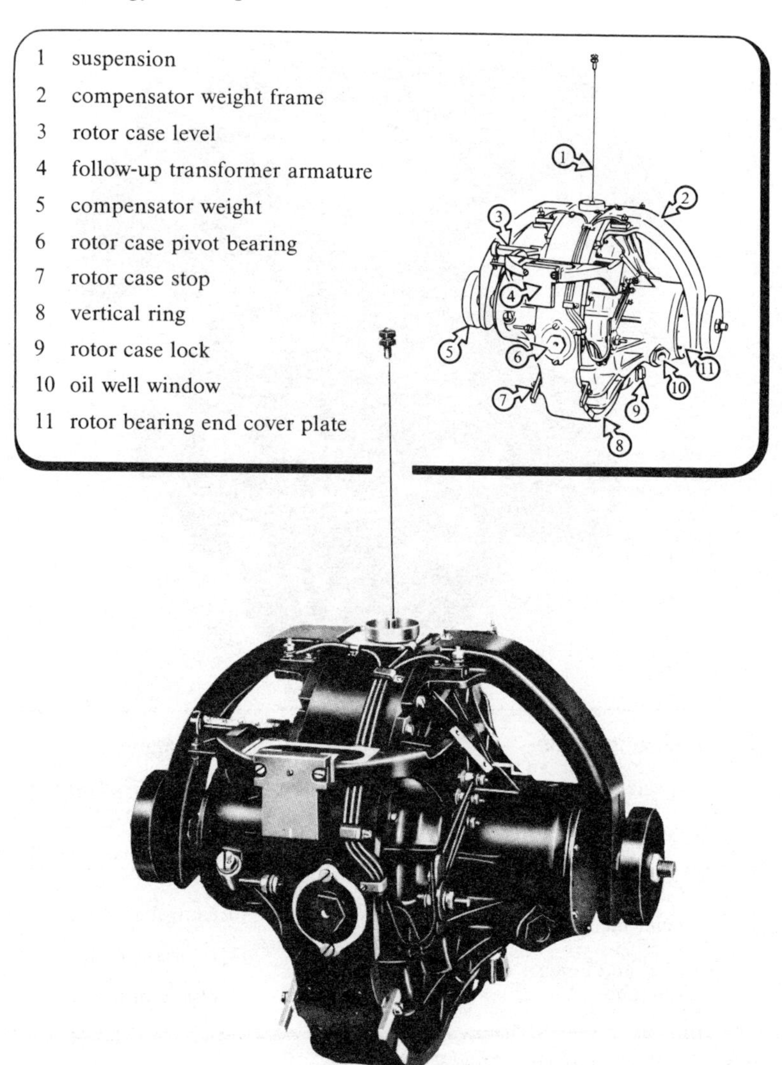

Figure 17.35 Sperry Mark E.14 gyro-compass.

located in the same plane as the vertical ring. The vertical ring is located within the phantom guide bearings. The phantom ring has a projecting stem at its uppermost point and is supported in the spinder element (q.v.) by two stem ball type bearings. The stem is hollow and through it passes the *suspension* which consists of a group of eighteen special steel wires. The top of the wire suspension is anchored to the top of the phantom stem by a support stud. This stud is clamped in a suspension adjuster which can be set in azimuth

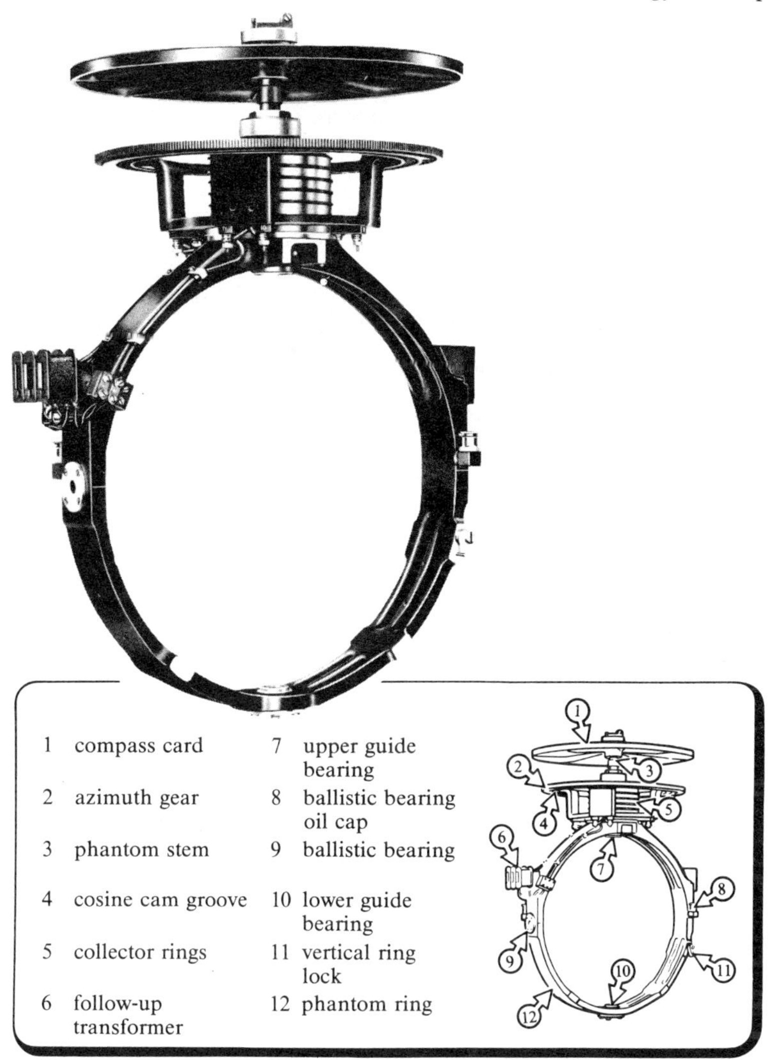

Figure 17.36 Sperry Mark E.14 gyro-compass.

through a small angle. The lower end of the suspension wire is attached to the vertical ring. The suspension wire serves to support the full weight of the sensitive element while the guide bearings simply locate the vertical ring within the phantom ring. The phantom element, as the name suggests, is made to follow every relative movement in azimuth of the sensitive element within it. The east side of the phantom ring therefore carries the second part of the sensing device which consists of an *E-shaped transformer* which

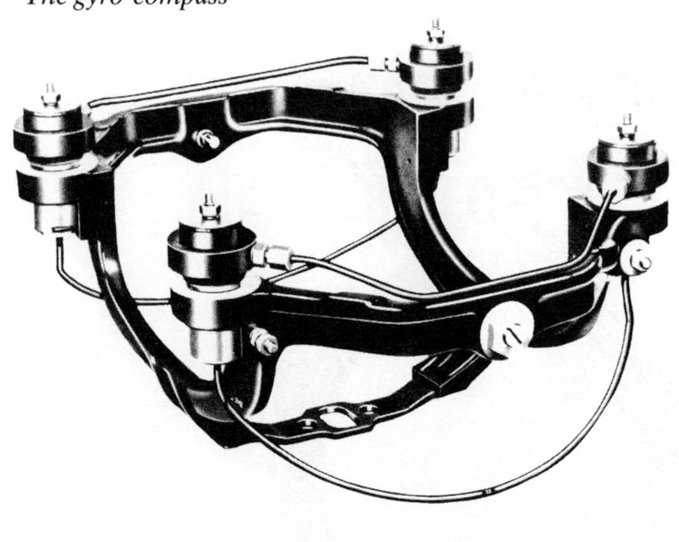

1 air tubes

2 bearing stud

3 mercury pot

4 mercury tubes

5 bearing stud locking nut

6 link arm

Figure 17.37 Sperry Mark E.14 gyro-compass.

detects every minor displacement between it and the armature carried by the vertical ring. Signal voltages, after amplification, energize an azimuth motor armature which serves to turn the whole phantom element into perfect alignment with the vertical ring. For this purpose the upper part of the phantom stem carries a large *azimuth gear* with which a gear train and the pinion of the azimuth motor engage. The azimuth motor is attached to the spider element.

This accurate sensing and follow-up system ensures that the suspension wire cannot be twisted. It constitutes Sperry's particular method of providing a near frictionless vertical axis for the sensitive element. Since the phantom identifies the azimuth orientation of the sensitive element it is used to support the *compass card* which is

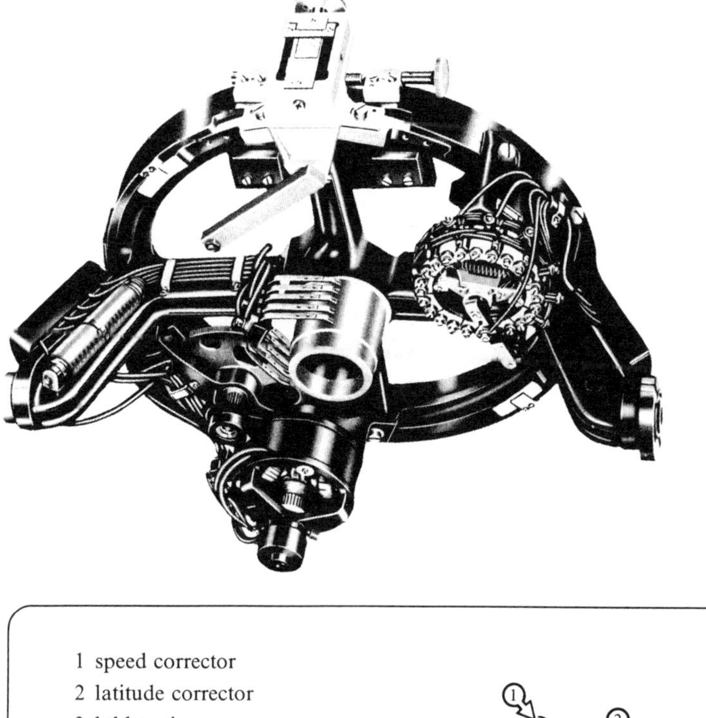

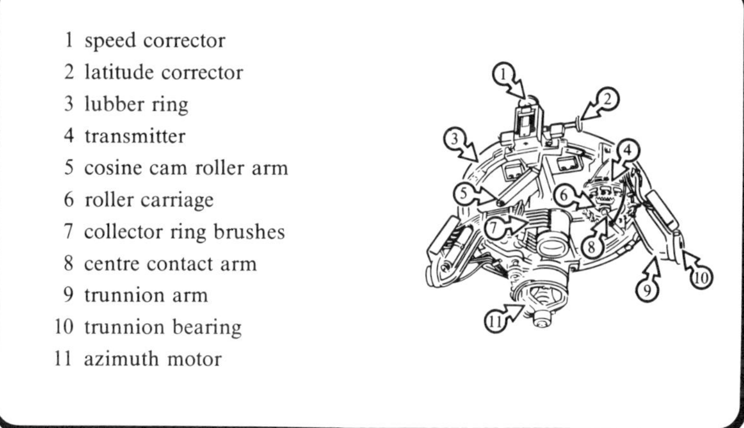

Figure 17.38 Sperry Mark E.14 gyro-compass.

located near to the upper end of the stem. The card, which is a steel ring, is engraved with degree markings 0 to 360. Encircling the phantom stem, below the azimuth gear, are five *collector rings* with brushes secured to the spider element. Three of these carry current for the gyro which is led onto the vertical ring by three flexible leads similar to those mentioned above. The two remaining rings convey the output signals from the induction sensing device.

The phantom ring has two recessed bearings on its outer surface

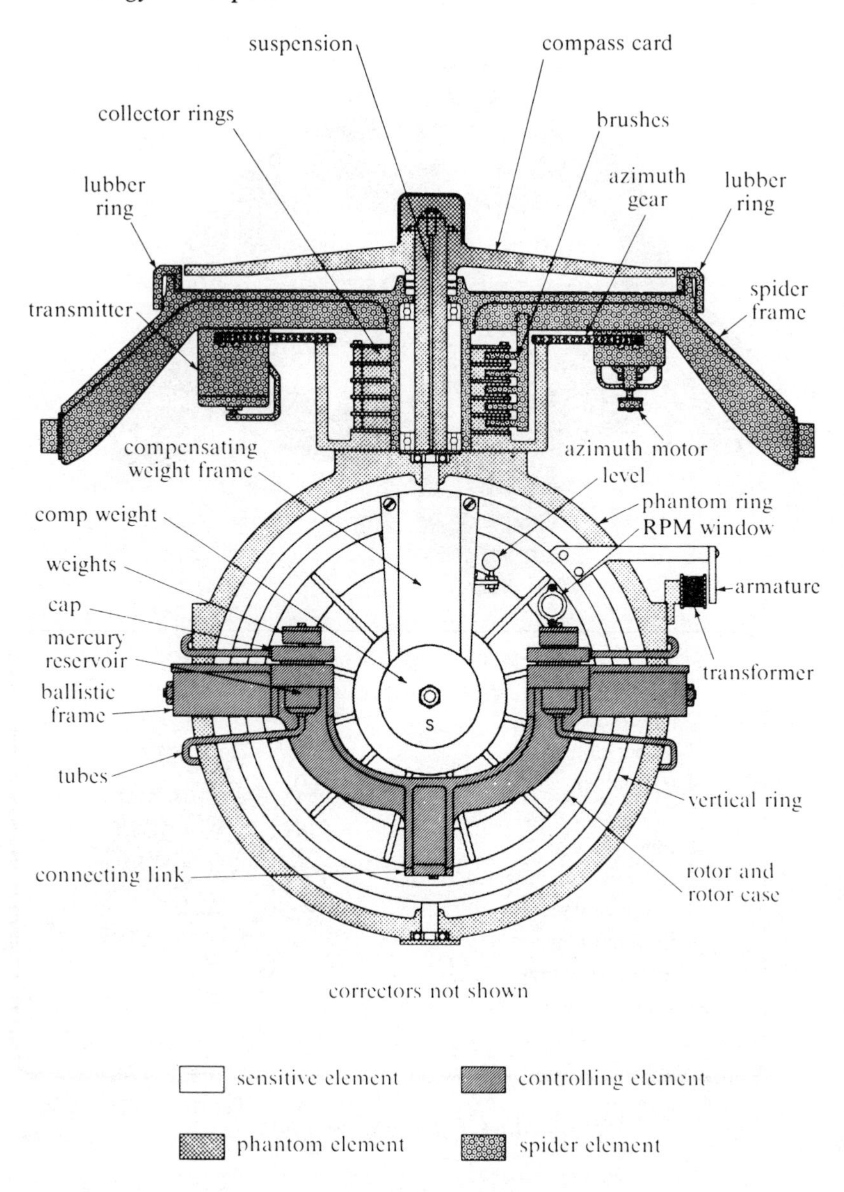

Figure 17.39 Four elements of the Sperry Mark E.14 gyro-compass.

in line with the east-west horizontal axis. These are located to receive bearing studs appropriately placed on the ballistic frame and provide support for the mercury ballistic (q.v.). An oil cup is provided for lubrication.

17.3.3 The mercury ballistic

This is the *control element* which simulates the effect of top heaviness. It consists of essentially a square frame at the four corners of which are located a pot or reservoir which contains 4 oz. of mercury. Studs are located at the centres of the east and west sides of the frame and these are received by the support bearings on the phantom ring referred to above. The mercury pots are connected in pairs, i.e. the two north pots are connected separately to the two south pots by stainless steel tubes through which mercury can flow whenever the system tilts. To prevent oxidation and evaporation of the mercury the tops of the pots are sealed with screw covers. The two north pots and the two south pots are similarly but separately connected by air tubes so that each pair forms an enclosed air/mercury circuit. Projecting below and forming part of the ballistic frame there is a central link arm (see Figure 17.37) by means of which connection is made with the rotor case. This ball connection is located 0.156 inches to the east of the true vertical through the rotor case. By this means the natural downhill flow of mercury is made to act eccentrically on the rotor case and so provide simultaneous torques about the horizontal and vertical axes. The former which may be thought to act at the bottom dead centre, or 6 o'clock, position causes precession about the vertical axis to eliminate drift, while the latter which may be thought to act at 3 o'clock (or, 9 o'clock) position causes precession about the horizontal axis for damping purposes.

The significance of this arrangement is very important because it shows that without the control element being supported by the phantom element there would be no means of creating a component of torque about the vertical axis for damping the compass oscillations. It is imperative to have a phantom element to *support the ballistic control* quite apart from its two additional functions of creating a *frictionless vertical axis* and to provide the means of *driving distant repeaters*.

The mercury ballistic is designed to be neither top nor bottom heavy and is balanced vertically by means of graduated balance weights on the frame and on the top of the mercury pots. The natural gravity flow from the high to the low reservoirs provides the effect of top heaviness.

17.3.4 *The spider element*

The part of the compass which is directly attached to the ship within the binnacle mounting is called the *spider element* because it vaguely resembles the insect form though imagination is required to establish any distinct identity. It consists of a ring structure reinforced by cross members. Two of these project downwards and contain trunnion bearings which are located in the athwartship axis of the main gimbal system. At the hub of the spider element the two ball bearings and the roller thrust bearing attached to the stem of the phantom are held securely in position. These support the weight of the compass and simultaneously ensure that the stem of the phantom remains substantially in the true vertical.

The spider element is important in the sense that it provides the final connection between the compass and the ship. It also supports several other necessary parts of the equipment, notably the *azimuth motor*, the *transmitter* and the course, latitude and speed *corrector box*. The brushes which make contact with the slip rings on the stem of the phantom are also located on the spider element so that power can be fed to the compass from the remote generator and output signals from the sensing device can be conveyed to the amplifier and subsequently returned to the azimuth motor.

The azimuth motor which is secured to the spider element drives the phantom through a gear train which ends with the main azimuth gear on the phantom. The rim of the spider frame, called the *cardan ring*, supports a fitted but sliding *lubber ring*. A small steel plate with an engraved lubber mark is fixed to the lubber ring by screws set in slotted holes so that any small permanent errors can be eliminated by an adjustment of the position of the lubber mark. The step-by-step transmitter is bolted to the lubber ring and its pinion engages with the main azimuth gear. The reason for this is that the transmitter responds to any intentional displacement of the lubber ring initiated by the corrector system. By this means any such corrections are immediately forwarded to the distant repeaters controlled by the transmitter.

17.3.5 *The course, latitude and speed corrector*

Bolted to the aft side of the spider is the corrector mechanism which is shown in Figures 17.40 (a) and (b). The purpose of the corrector is to eliminate both the *damping error* and the *course, latitude and*

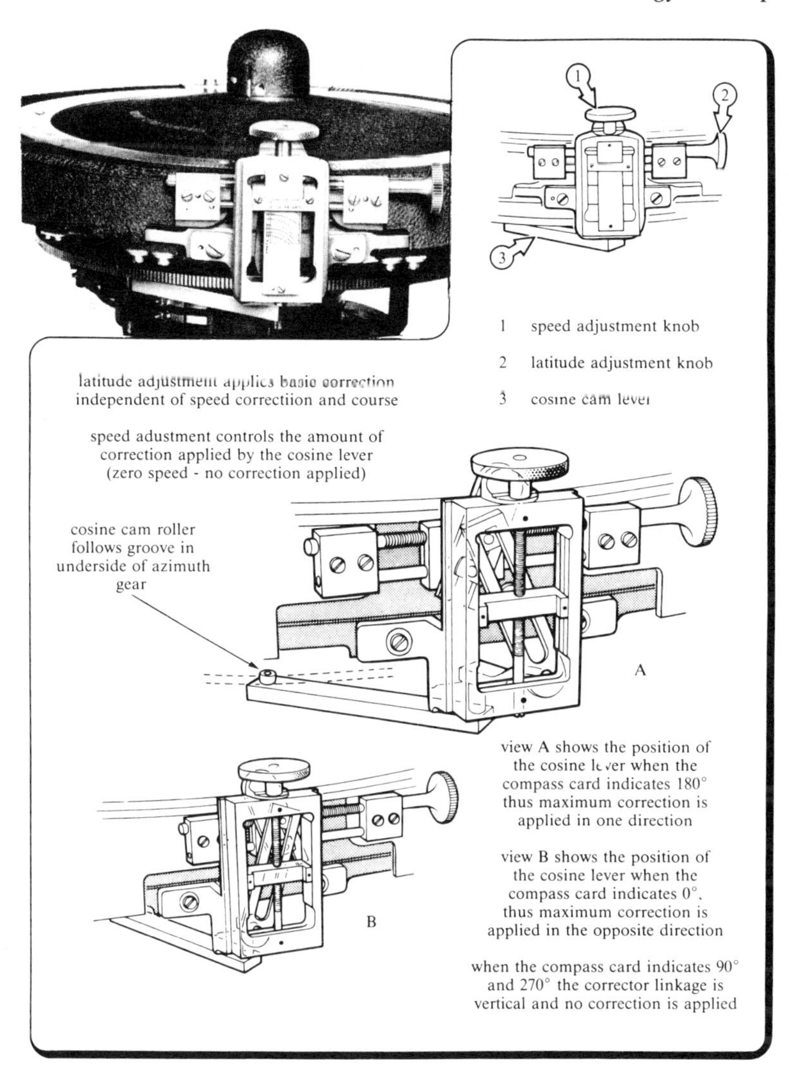

1 speed adjustment knob
2 latitude adjustment knob
3 cosine cam lever
latitude adjustment applies basic correction independent of speed correctiion and course
speed adustment controls the amount of correction applied by the cosine lever (zero speed - no correction applied)
cosine cam roller follows groove in underside of azimuth gear
A
view A shows the position of the cosine lever when the compass card indicates 180° thus maximum correction is applied in one direction
view B shows the position of the cosine lever when the compass card indicates 0°, thus maximum correction is applied in the opposite direction
B
when the compass card indicates 90° and 270° the corrector linkage is vertical and no correction is applied

Figure 17.40 Sperry Mark E.14 Corrector System.

speed error by displacing the lubber ring through an angle which is the sum of the combined errors. The lubber ring is displaced in the *same* scnse as the deflection of the gyro spin axis from north. The reader will recognize that this is a case of introducing an apparent *A* coefficient of opposite sign to that of the existing combined error.

There are two main parts of the corrector system. The *auxiliary latitude corrector* secured to the lubber ring and the *speed and latitude corrector* which is attached to the spider. The two correctors are interconnected by a moveable block and pivot which in turn moves the lubber ring the required amount depending upon the settings made. It will be remembered that the damping error is proportional to the tangent of the latitude, while the ship's motion error is proportional to the secant of the latitude, the speed of the ship and the cosine of the course. Each of these variables is contained in the corrector mechanism. For instance, the cosine cam roller rides in a cosine groove which is cut into the underside of the azimuth gear on the phantom. When the ship alters course the bell crank with the roller located in the groove moves the lower lever about the adjustable pivot in the crosshead. This movement is communicated in turn to the common pivot above. The upper arm moves the adjustable block and pivot, which is attached to the lubber ring, through the auxiliary corrector screw and the two fixed blocks. A close study of the diagram shows that the amplitude of the movement of the lubber ring, for a given movement of the cosine cam arm, is determined by the position of the crosshead. The position of the crosshead is therefore a function of the ship's speed and the latitude. Each of these is preset by means of a latitude scale engraved on a horizontal strip attached to the crosshead which moves over a plate engraved with speed curves. The pivot which ultimately moves the lubber ring is mounted on a block which is translated by the auxiliary setting knob in accordance with an engraved scale provided. Since this effectively introduces the damping error of opposite sign it is sometimes referred to as the *tangent latitude corrector*. The selection and preselection for the throw of the levers are made by hand while the cosine of the course is introduced automatically. The combination moves the lubber ring and the transmitter reorientates the repeaters in sympathy.

17.3.6 Transmission system

Distant reading compass indication is necessary because the master compass is rarely positioned for direct navigational use. The *transmitter* on the master compass and individual *motors in each of the repeaters* constitute the transmission system. One set of these is illustrated in Figure 17.41. Lead 4 carries the positive side of a 70 volt d.c. supply to the rollers of the rotary switch in the transmitter. The twelve commutator segments are arranged in four groups each having consecutive segments numbered 1,2 and 3. Segments 1 of each group are connected to each other and, via the cable, to the pair of coils numbered 1 in the repeater motor. The remaining

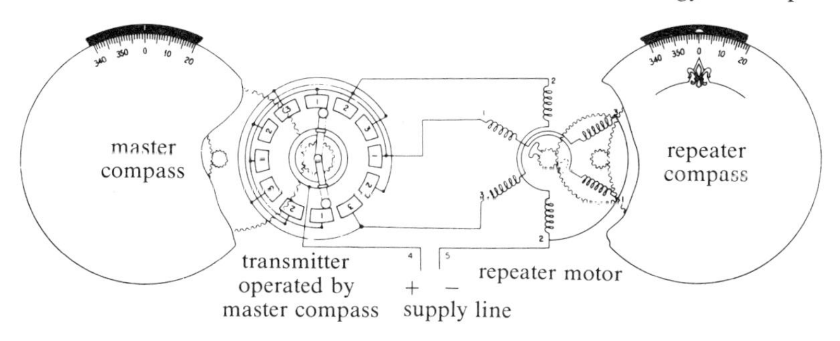

Figure 17.41 Transmission circuit of Sperry Mark E.14.

segments are connected in a similar manner. The negative lead is connected to the remaining terminals of the repeater motor. Figure 17.41 shows the rollers in contact with no. 1 segments so that the armature of the repeater motor is aligned with no. 1 coils. If the roller carriage rotates clockwise the upper roller will move to No. 2 segment while the lower roller remains on no. 1 segment. Nos 1 and 2 motor coils will both be energized and the armature will rotate to a position midway between them. A further slight rotation will bring the rollers to no. 2 segments and the armature will align itself to no. 2 coils. By this means the armature makes twelve steps per revolution but owing to the 180:1 gear reduction ratio at the transmitter each step represents $^1/_6$*th degree* movement of the card.

17.3.7 *The binnacle*

The binnacle houses and supports the compass. The spider element, as referred to above, is located in the inner main gimbal ring along the athwartship axis. The outer ring provides the fore and aft axis. This arrangement permits the compass to remain vertical within the limits of 60° of roll and 25° pitch. Twin cylinder adjustable *roll dampers* are fitted and consist of oil dashpots. The *pitch damper* is a friction brake acting on a cam fitted to the spider element. The suspension system includes shock absorbers which suppress unwanted vibrations from the ship's hull.

17.3.8 *The Arma Brown gyro-compass*

The compass described here was a radical development in British gyro-compasses, dispensing with a liquid control system and making use of the gyrosphere configuration.

Figure 17.42 Arma Brown gyro-compass.

It is interesting to note the unique methods used in the Arma Brown design to solve the common requirements of all gravity controlled systems. The construction details are listed as follows under conventional headings.

17.3.9 The gyro ball

Hitherto most gyro-compasses have been described initially by dealing with the sensitive element. Since in the Arma Brown compass the *sensitive element* comprises the *gyro, its spherical case,* a *primary gimbal* and a *follow-up tank* it is necessary to deal with each in turn. The gyro ball is not a complete sphere but one which has a deep annular groove in the east-west plane similar to the shape of a child's yo-yo. The two hemispheres are joined at the centre and the *stator* of the rotor is set in the bore of the bobbin. The *rotor* shaft is set through the centre, supported in bearings, and constitutes the rotor of a three-phase electric motor. A heavy rimmed *spinner* is attached to each end of the rotor shaft where it projects into each hemisphere (*end-bell*) of the case. The gyro rotor is made to rotate at 11,800 revolutions per minute in a clockwise direction when viewed from the south, i.e. *earthwise*. The two end-bells of the gyrosphere are made of copper; the unit is sealed and filled with helium. In line with the spin axis an *electromagnet* is secured both to the north and south inner faces of the end-bells. These form part of the sensing device whose function is described below.

.3.10 *Primary gimbal*

At this stage the design of the Arma Brown compass departs from other more conventional systems. A *primary gimbal* is located in the annular groove of the gyro-ball. It consists of a square light alloy frame with two sides vertical and the other two horizontal. The former are connected by two *horizontal torsion wires*, mounted in beryllium copper spring supports which bridge the annulus at its external circumference, and pass inwards to the gimbal to which they are mounted in glass insulated terminals. Two *vertical torsion wires* are secured to the horizontal sides of the gimbal and are attached at their top and bottom ends to the surrounding tank unit (q.v.). Covering each of the torsion wires is a flexible silver helix the purpose of which is to convey electric current to the gyro stator windings and the electro-magnets of the sensing unit.

.3.11 *The tank*

The *flotation tank* is an aluminium alloy forging with a cast aluminium cover mounted above it to form a container which is sealed at the top and bottom with covers and to the north and south with rotary damp adapters. The inside of the tank is machined to conform to the shape of the gyro-ball with sufficient clearance to give the ball freedom of movement. The space between the tank and the gyro-ball is filled with *Fluorolube*, a liquid whose SG of 1·914 is similar to that of the assembled gyro-ball and primary gimbal. The gyro-ball rests with *zero buoyancy* within the tank so that the torsion wires take zero weight and the gyro-ball remains immune to the effects of linear accelerations communicated to it by the wires.

The tank is mounted in the *tilt gimbal* by two trunnions fixed to the gimbal and fitted with ball races. The bearing housings incorporate a fluid damping mechanism designed to minimize oscillations of the tank about its north-south axis in response to rolling accelerations. *Stops* limit the swing of the tank to 60°. Along the same axis, and set in the wall of the tank, is a plate which carries two flat horizontal and vertical sets of *pick-off coils* which together with the electromagnets on the gyro-ball constitute the azimuth and tilt sensing device. The tank is made pendulously heavy, within the tilt gimbal so that it normally hangs vertically. The tank supports the *pendulum unit* (q.v.).

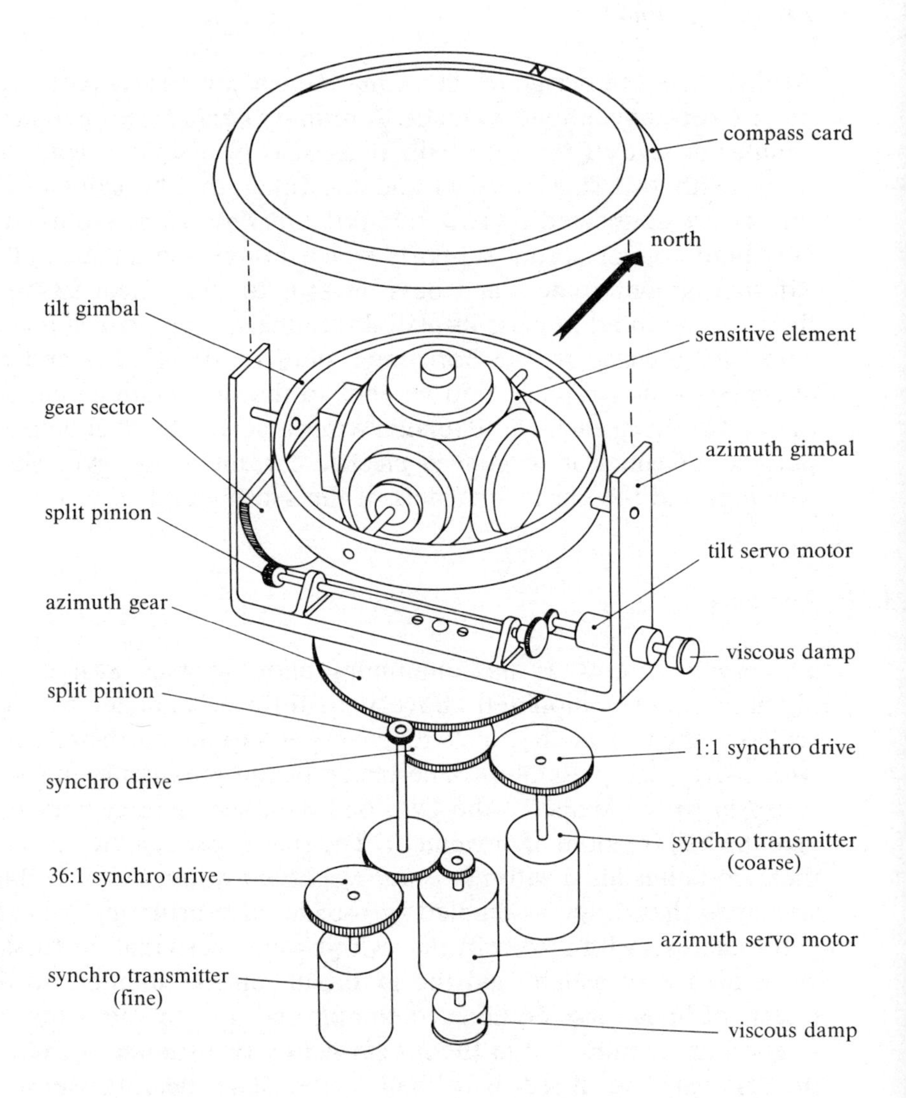

Figure 17.43 Schematic arrangements of Arma Brown gyro-compass.

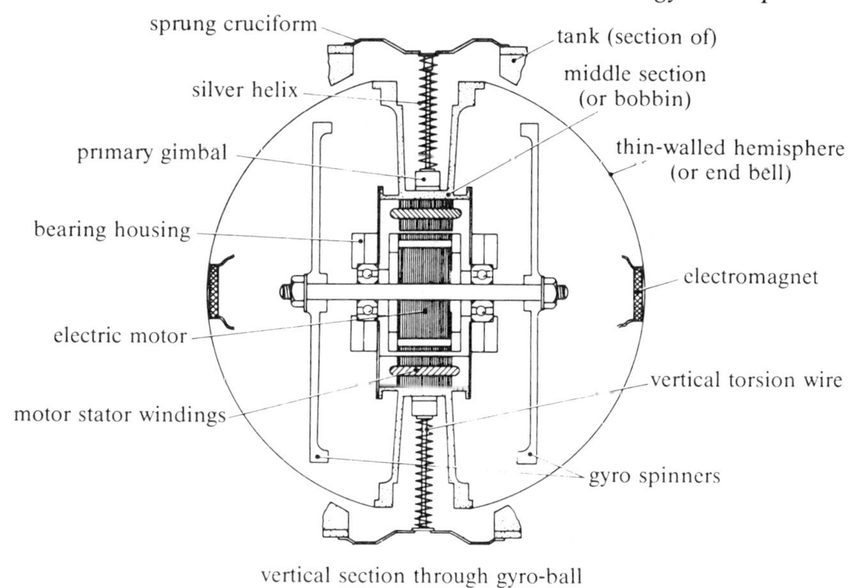

Figure 17.44 Arma Brown gyro-compass – arrangement of gyro-ball.

3.12 The pendulum unit

The *pendulum unit* is the gravity sensor and substitutes for the liquid ballistic in the conventional Brown compass or for the naturally pendulous arrangement used in most continental makes. It is an electrically operated mechanism containing a pendulum. It consists of an *armature* (the pendulum bob) and an *E-shaped transformer* fixed to the case. The armature is suspended by two flexible beryllium copper strips from the top of the unit. The transformer is E-shaped and carries the exciting and pick-off coils. The container is

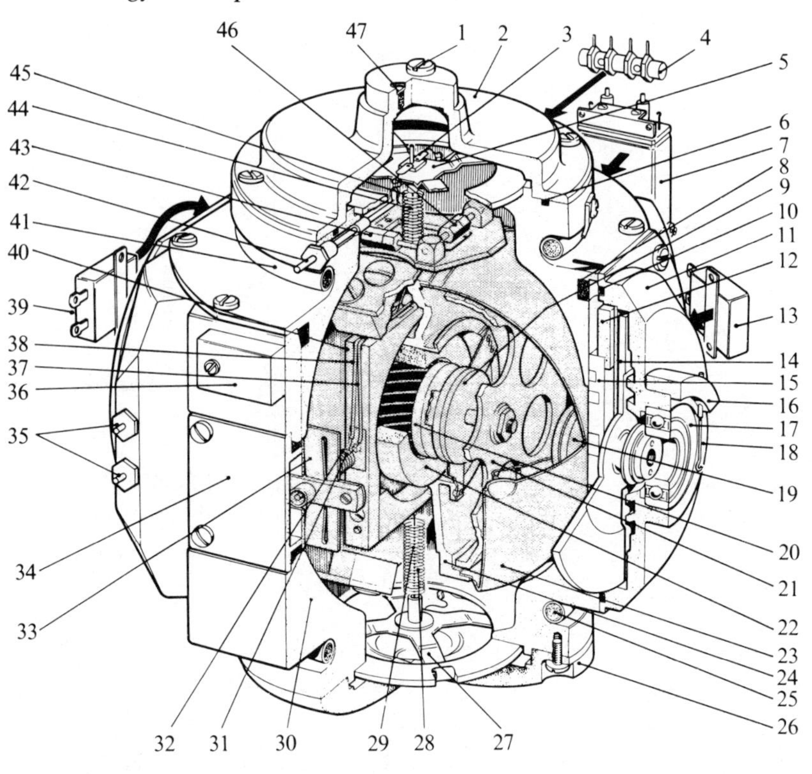

Figure 17.45 Arma Brown gyro-compass – general assembly of sensitive element.

1. Filler plug
2. Tank top cap
3. Flexible connection
4. Terminal strip for pendulum unit
5. Top cruciform
6. 'O' ring
7. Pendulum Unit
8. 'O' ring
9. Motor bearing housing
10. 'O' ring
11. North bearing housing
12. Pick-up coil retaining plate
13. Thermostatic switch (S5) (not fitted on Mk. 1C (cold) compass
14. Rotary damper disc
15. Pick-up coil
16. Brush block
17. North bearing
18. Brush
19. Electro-magnet
20. Electric motor (rotor)
21. Gyro wheel (1 of 2)
22. Motor stator
23. Thin-walled hemisphere or end-bell (1 of 2)
24. Middle section or bobbin
25. Bottom Calrod heater (not fitted on Mk. 1C (cold) compass)
26. Tank bottom cap
27. Bottom cruciform
28. Bottom vertical helix
29. Bottom vertical torsion wire
30. Tank body
31. East horizontal helix
32. East horizontal torsion wire
33. Torsion wire spring plate
34. Side cover plate
35. Pick-up coil connections
36. Balance weight
37. Torsion wire electrical connection
38. Helix strap electrical connection
39. Thermostatic switch (S4) (not fitted on Mk. 1C (cold) compass
40. Primary Gimbal
41. Top Calrod heater (not fitted on Mk. 1C (cold) compass)
42. Electrical connection (E1) to top cruciform
43. Ball N/S balancing nuts
44. Top vertical torsion wire
45. Top vertical helix
46. Ball E/W balancing nuts
47. Breather

filled with *viscous silicone fluid* so that the pendulum oscillation is damped to a period of between two and three minutes. The amplitude of the pendulum is restricted to 43 minutes of arc. The damped period prevents the pendulum from responding to short term horizontal accelerations caused by rolling and the restricted amplitude helps to reduce ballistic deflection. The pendulum unit is secured to the west side of the tank, level with the centre line and is orientated north-south to detect any tilt of the tank about the east-west axis of the tilt gimbal.

7.3.13 *The tilt and azimuth gimbals*

It is important to understand the gimbal arrangement of the Arma Brown compass because the tank which contains the gyro-ball is effectively the *follow-up unit*, not merely in azimuth but also in tilt. Whenever the gyro-ball tilts (due to the earth's rotation) or moves relatively in azimuth when the ship turns the tank must be made to repeat these movements. Consequently each part of the gimbal arrangement which supports the tank must be driven by motors.

The *tilt gimbal* supports the tank along its north-south axis as mentioned above. It is in turn supported along an east-west axis in the *azimuth gimbal* by ball races located in housings. The east pivot is hollow to take electric leads from the azimuth to the tilt gimbal. Fitted to the west side of the tilt gimbal is an aluminium *gear sector* which meshes with the ultimate pinion of the tilt servo motor gear train. Also fitted to the west side of the tilt gimbal is a *spirit level* which has a sensitivity of 8 minutes of arc per 5 graduations. The north trunnion holds a commutator assembly which decouples, or transfers, part of the azimuth sensing signal into the tilt servo-loop whenever the tank inclines about the north-south axis by more than 25°.

The *azimuth gimbal* is an elaborate supporting framework which has two support arms in which are located the bearings supporting the east-west pivots of the tilt gimbal. The arms curve inwards towards the base of the frame which is fitted with a vertical shaft at its centre about which the whole of the assembly previously described can turn in azimuth. The two support arms also project upwards and carry the compass card. Integral with the shaft is a set of twenty slip rings ('hot' compass), or ten slip rings ('cold' compass), through which electrical connections are made from the static base of the compass to the moving parts within. The base or *chassis* supports the *azimuth servo motor* which engages with an *azimuth gear* so that the azimuth gimbal and all parts of the compass may be driven into alignment with the gyro-ball. Special concentric gearing about the shaft also drives the *synchro transmitters* which actuate the distant reading *repeaters*.

17.3.14 *The binnacle*

This is the outer case of the compass and contains the chassis which is the base plate supporting the moving parts of the compass described above. Below the chassis is a light alloy casting forming the *control box* which contains the tilt and azimuth transistorized amplifiers, power packs and controls. The control knobs and graduated scales are located on the front panel of the control box.

17.3.15 *Sensing and control*

The construction details outlined above are insufficient to appreciate how the compass works without some further description of the several sequences.

The electromagnets on the gyro-ball and the pick-off coils on the tank sense every displacement both in tilt and azimuth between these two parts. Since the tilt and azimuth servo motors receive amplified signals from the sensors it is clear that the orientation of the tank and the gyro-ball is maintained. At this point the tank and the gyro-ball would act simply as a free gyro and would not constitute a gyro-compass. It is likewise clear that if the tank is intentionally misaligned with respect to the gyro-ball then torques will be created by twists in the torsion wires and the result of these will be rates of precession which correspond to the amounts of twist. If the horizontal torsion wires are twisted the torque about the horizontal axis will cause precession about the vertical axis; similarly, if the vertical torsion wires are twisted precession will occur about the horizontal axis. With these effects in mind it can be understood that the gravity sensor (pendulum unit) is used to feed a bias signal into the *tilt servo loop* to precess the compass about the vertical *axis* to initiate an elliptical oscillation, and another bias signal into the *azimuth servo loop* to precess the compass about the *horizontal axis* to damp out the former oscillation and produce a settled compass. This arrangement shows that the Arma Brown compass is controlled by a simulated *bottom heaviness* and is *damped in tilt* by a torque about the vertical axis. In loop systems of this kind it is not difficult to inject electric signals at any stage or point required and this leads to a consideration of the correction system.

7.3.16 *Correction system*

Since the compass is damped by a torque about the vertical axis a damping or *latitude error* will be caused. Furthermore, the north-south component of the ship's motion causes the *speed error*. As with the other gyro-compasses which incorporate a correction system the Arma Brown compass is corrected for each of these errors. Taking the second error first, a bias signal is fed into the *azimuth servo loop* which is proportional to the *cosine of the course*, and the *speed of the ship*. The cosine function is derived from the voltages in the synchro stator windings because these voltages are proportional to the cosine of the angle between the axes of these windings and the rotor winding axis. The signal strength proportional to the speed of the ship is made by a manual setting of the *speed potentiometer* on the control panel. On receipt of the combined signal the azimuth servo motor precesses the compass at a rate equal to the north-south component of ship's motion but in the opposite direction, so that the speed error is eliminated at source. The damping error on the other hand is eliminated by setting a *latitude potentiometer* which injects a signal, proportional to the sine of the latitude, into the tilt servo loop and thereby precesses the compass about the vertical axis at a rate equal and opposite to the rate of drift for the latitude concerned caused by the earth's rotation. This means that the gyro can run substantially *horizontal* though the tank will be slightly tilted and the damping error is eliminated because nothing remains to cause it.

7.3.17 *Presetting and use as a directional gyro*

Apart from the latitude and speed potentiometer settings the only other controls consist of the main operating control and the control used for presetting and erecting the compass. The *start* sector causes the gyro to assume its normal running speed in about five minutes. Having checked that the *slew rate* switch on the left of the panel is vertical the main control is set to the *free slew* sector. The purpose of this is to enable the gyro to be preset to the required heading and to remove the tilt if any. A suitable slow rate of precessional slew is selected and at the touch of the appropriate push-button the compass precesses to the required heading and removes any existing tilt as indicated by the spirit level. The delayed action of the pendulum sensing unit may require the operator to re-level after two or three minutes. The main control is now set to *settle* in which position the compass seeks the meridian vigorously by a *heavily damped spiral* trace, and after some 25 minutes is set to the final sector marked *run*. The compass now maintains the meridian with a reduced damping factor. If the main control is retained in the *free slew* mode the pendulum signal is disconnected from the tilt amplifier but is still connected to the azimuth amplifier. The instrument no longer acts as a north seeking compass but assumes the character of a *directional gyro*. Used in this way some limit must be placed on the initial azimuth chosen. In high latitudes a preference may be exercised but in low latitudes an azimuth fairly close to the meridian should be chosen. Used in this mode the instrument should be *checked periodically* by sun or star azimuths in case there is any wander after which the gyro should be reslewed to the correct required azimuth. The need for a directional gyro aboard merchant ships may be questioned but it certainly provides stabilized azimuth indication in very high latitudes, should it be required, when conventional gyro or magnetic compasses cannot function properly.

18 Radar

Peter Yarwood

18.1 The basic principle of radar

18.1.1 *Electromagnetic reflection*

The principle of electromagnetic propagation is dealt with in another section of this book. A source of radio frequency power known as the transmitter excites the antenna which, acting as a transducer, launches the RF electromagnetic wave into the atmosphere. Figure 18.1.1 illustrates a portion of the EM wavefront travelling along a selected line of propagation away from the antenna and carrying with it an amount of the transmitted signal power in the form of alternating electric and magnetic fields (the E and H fields). Regardless of the transmission frequency, the velocity of the wavefront along the propagation line is essentially that of light and can be taken to be $3 \times 10^8\ \mathrm{ms}^{-1}$. This value is allocated the symbol 'c' and is the velocity of light in free space. In air the value is very slightly less but not significantly so.

When such a wavefront encounters an abrupt change in the boundaries of the medium through which it is passing, for example a solid body, a portion of the wavefront's RF power is re-radiated and

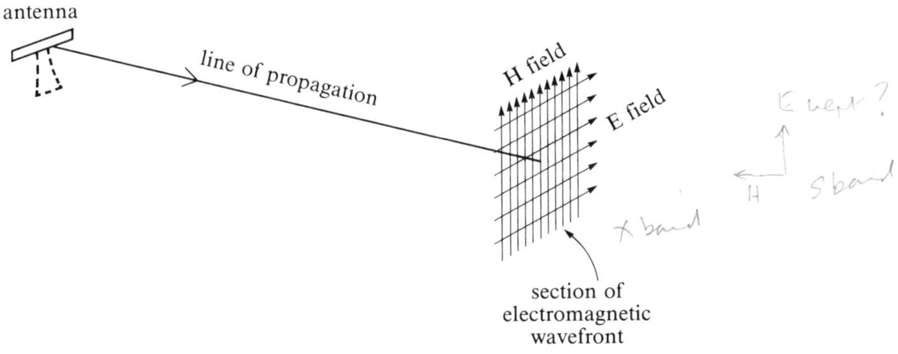

Figure 18.1.1

scattered in various directions. Some of the reflected power returns along the line of propagation at the velocity of light, forming a reflected wave which has the same frequency as the transmitted wave. This is termed the signal *echo* formed by the reflection. After a short but measurable time interval the echo will have reached the site of the transmitting antenna which will now have been switched to act as a receiving antenna. The total echo power reaching that site is usually only a minute fraction of the total transmitted power, depending upon many factors such as the distance travelled to and from the reflecting object known as the *target*, the nature of the target, transmitted power and antenna characteristics and the effect of the medium through which the wave propagates.

A sensitive radio receiver connected to the receiving antenna amplifies the echo and when used in conjunction with high-speed time measuring circuits, the exact elapsed time can be measured between the instant that the transmitted wave leaves the antenna and the instant the echo reaches the antenna. Since the velocity of the wave is known to be constant, the distance along the line of propagation and reflection is related to velocity and elapsed time as follows:

$$\text{total distance} = \text{velocity} \times \text{elapsed time} \qquad \textbf{[18.1]}$$

18.1.1.1 Range

Evidently if the direction of the line of propagation is known then the distance between antenna and target, known as the target *range* can be found since:

$$\text{range (m)} = \frac{3 \times 10^8 \times \text{elapsed time (s)}}{2} \qquad \textbf{[18.2]}$$

see Figure 18.1.2.

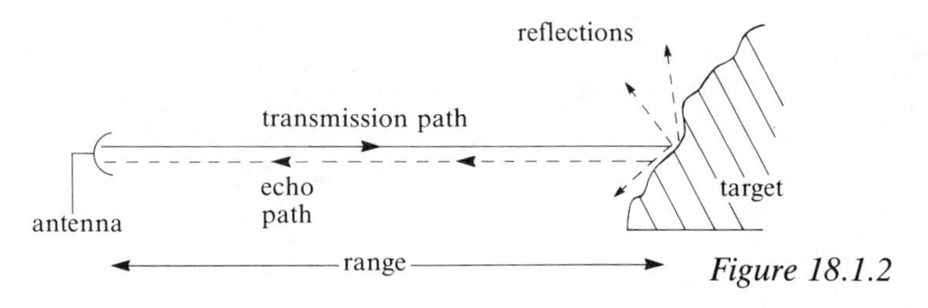

Figure 18.1.2

In practice the direction can accurately be measured and in mobile marine systems the direction of the target is generally related to an azimuth bearing with reference to a selected datum.

The process which has been outlined forms the basis of *radio detection and ranging* from which the acronym *radar* emerged.

18.1.2 *Pulse transmission*

Continuous transmission of RF power as described will indicate the presence of a target, and certain types of radar use this CW method of transmission. For simple radars CW radar techniques are impracticable for basic range measurement since, following the initial return of the echo wavefront, all time references relative to transmitted and returned signal will be lost in the continuous transmission and reception process. The timing difficulty is obviated by transmitting the RF signal in the form of very short bursts of a sine wave carrier known as the transmitted *pulse*. The pulses are repeated at accurately timed intervals. Figure 18.1.3 illustrates the principle.

A rectangular high voltage pulse is generated periodically and is known as the *modulating* pulse. The time for which the pulse exists determines the time for which the transmitter is switched on and hence the length of time for which the RF transmitted power will be radiated from the antenna. This time is variously known as the *pulse width*, *pulse length* and *pulse duration*. Modern mobile marine civil radars employ pulse widths which are typically fractions of a microsecond.

1.2.1 *Pulse repetition frequency and pulse repetition period*

The pulse repetition frequency (PRF) or pulse repetition rate (PRR) is simply the frequency at which transmitted pulses occur in time. Measured in pulses per second (pps) a PRF of 1000 pps will, for example, have 1000 pulses generated in one second of time.

It follows that since the pulses are periodic then the interval between two consecutive pulses is given by:

$$\text{period between pulses (s)} = \frac{1}{\text{PRF}} \qquad \textbf{[18.3]}$$

Again, Figure 18.1.3 indicates that the PRF is 1000 pps giving a pulse repetition period (PRP) of 0.001 seconds, the PRP being the interval of time measured between identical points on two consecutive pulses.

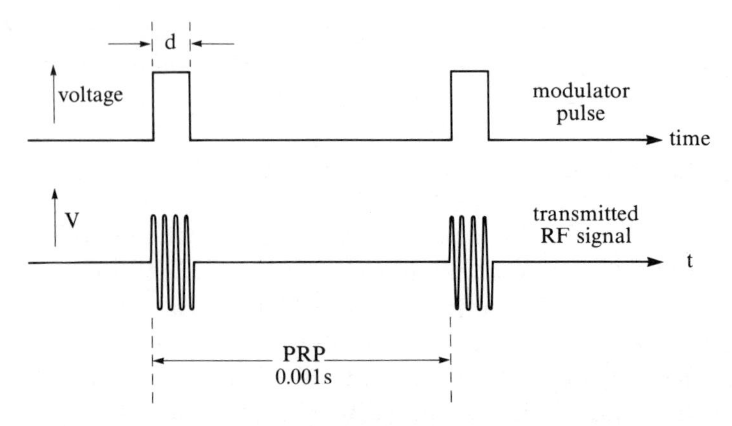

Figure 18.1.3 Showing modulator pulses and resultant bursts of radio frequency signal transmitted. The pulse repetition frequency is 1000 pps. The pulse width 'd' could be 0.05 microseconds. (Not drawn to scale.)

18.1.3 A basic transmitter

The essential sections of a simple pulsed radar transmitter are shown in Figure 18.1.4. Circuits in the PRF waveform generator periodically produce pulses to control the rate at which the modulator generates a high voltage rectangular waveform. The modulator electronics determine the pulse width which can be altered in modern radars. This generated waveform is then used to switch on the transmitter stage for the selected duration of the pulse and for the remainder of the pulse repetition period the transmitter remains off. An electromagnetic wave passes from the transmitter via a waveguide to the antenna and is propagated.

A more detailed explanation of the antenna appears in section 18.3. For the moment it is sufficient to appreciate that the antenna is engineered to form the radiated power into a highly directional beam. Most of the power lies within a volume of space clearly defined by angles subtended at the antenna and taken in both the horizontal and vertical planes. Taken in the horizontal plane the *beamwidth* is of the order of one degree, a narrow horizontal beam which when directed at a target within range of the equipment will irradiate that target and produce reflected echo signals. Simple measurement of the beam direction relative to a chosen direction will indicate the relative bearing at which the detected echoes are being returned. Figure 18.1.5 shows the principle.

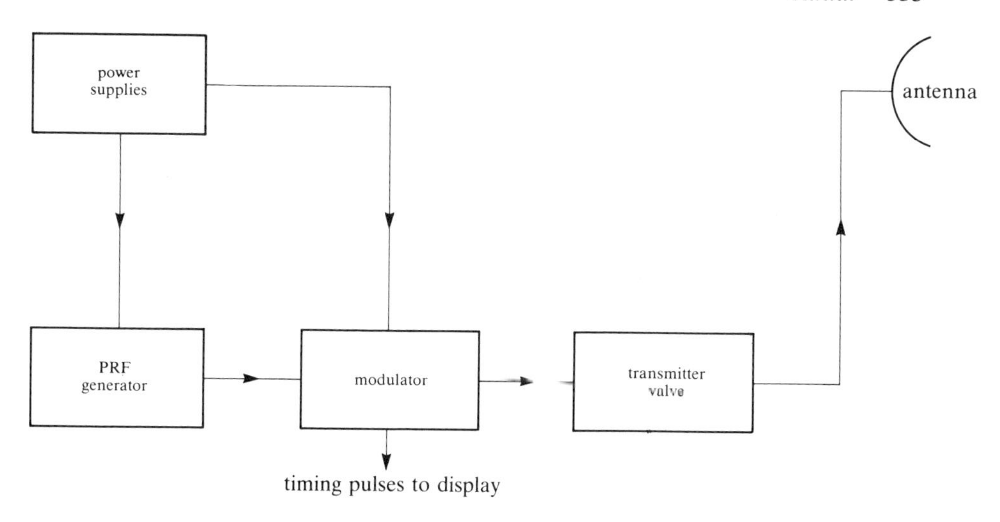

Figure 18.1.4 Essential sections of a simple pulsed radar transmitter.

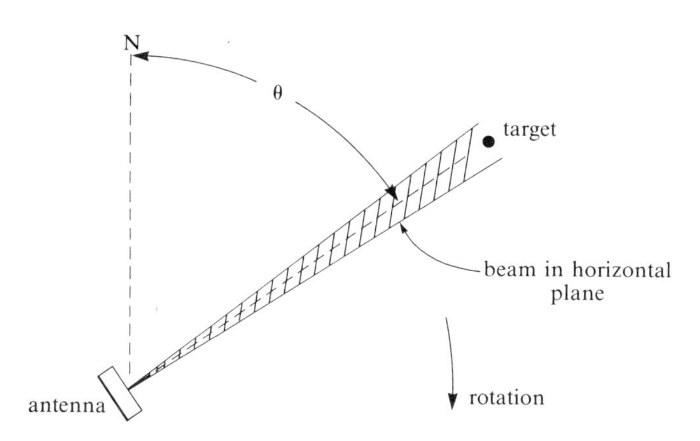

Figure 18.1.5 A narrow beam is produced in the horizontal plane by the antenna. θ is the beam bearing relative to N in this case.

18.1.4 *Pulsed range measurement principle*

Pulse modulation of the transmitted wave provides a period of time during the PRP when the transmitter is quiescent; sometimes termed the *resting time* or *dead time* of the transmitter. During that period of time useful target echoes can return to the receiver and since the antenna is no longer being used by the transmitter it is rapidly switched to direct the received echo signals to the radar receiver section. This action takes place immediately upon cessation of the transmitted pulse. Such echo signals will be of the same pulse length as the transmitted pulse but of diminished power and they will arrive at the receiving antenna having a time delay relative to the transmitted pulse which generated the echoes. This time delay can be measured accurately and the range of the target computed from the relationship given by expression [**18.2**].

Consider a transmitted pulse of width 0.5 μs having a PRF of 1000 pps. Figure 18.1.6 shows three such consecutive pulses. Evidently the transmitter is switched *on* for the pulse duration and *off* for a period given by:

$$\begin{aligned}\text{resting period} &= \text{PRP} - \text{pulse duration (s)}\\ &= 1000 - 0.5\ \mu\text{s}\\ &= 999.5\ \mu\text{s}\end{aligned} \qquad [\mathbf{18.4}]$$

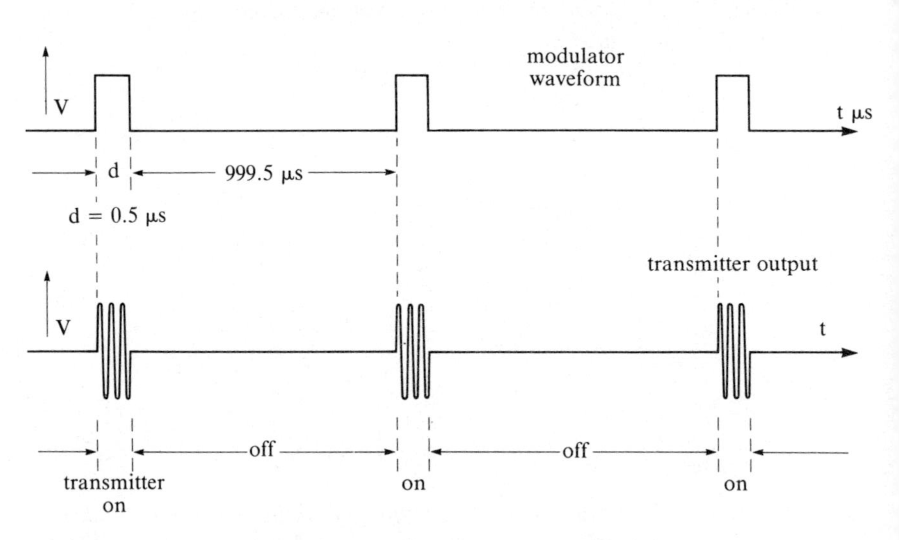

Figure 18.1.6. Timing of modulator pulses and transmitted waveform for pulse width of 0.5 μs and PRF 1000 pps (not to scale).

This value is a theoretical maximum for the resting period since in practice the transmitter cannot switch on and off instantaneously. It is evident that for the majority of its operational time the radar set will be in its receiving mode. Note that whilst the antenna is used for transmitting, the receiver has to be made inoperative and range measurements cannot be made during the transmit period. The significance of this action is detailed in section 18.2.

In marine radar systems the time interval between a transmitted pulse and any echo returned from that pulse is directly translated into a measure of distance or target range by means of a cathode ray tube display and the display control circuits which are synchronized in time to the transmitted pulse.

The simplest type of display centres the cathode ray tube (CRT) electron beam along the tube axis so that the electron beam comes to a sharp focus at the centre of the tube's flat circular screen (see section 18.4). At the instant the RF energy starts to leave the antenna, the electron beam comes under the deflecting influence of a linearly increasing magnetic force. The deflecting force is produced by the action of a sawtooth current waveform produced by a circuit known as the *timebase generator* which is synchronized to start the deflecting process at the onset of each transmitted pulse.

The electron beam current can be adjusted by means of a *brightness* control, so that the tube phosphor glows at the point of impact of the electrons. The linear deflecting force is arranged to *sweep* the glowing point from the centre to the tube periphery and then very rapidly return the beam to its central position ready for the next transmitted pulse and timebase sweep. The process is known as *scanning* the tube phosphor and one timebase scan or sweep will produce a luminous radial line on the tube face. During the period when the beam is returned to its central position the beam current is suppressed, the period being known as the *flyback time*. Figure 18.1.7 shows idealized relationships between the transmitted pulse, timebase waveform and the resultant timebase trace.

In a practical situation the tube brightness is adjusted so that the path traced by the beam is below the threshold of visibility. Echo returns which occur during the timebase waveform period are processed by the receiver and display circuits and emerge from these circuits as rectangular voltage pulses which are representative of the echoes. They are known as *video pulses*. The video pulses are then applied to the cathode ray tube and serve to lift the brightness of the timebase trace above the threshold of visibility for the brief period of an echo's duration so that the target echo appears on that particular trace as a bright luminous spot (known as target *paint*).

Since the radial trace is linear time related to the transmitted pulse, the distance along that line from the scan origin at the screen centre to the target paint is a scaled representation of the target's range from the antenna. A simple relationship between transmitted pulse, timebase waveform, returned echo and the resultant paint of that echo is illustrated by Figure 18.1.8.

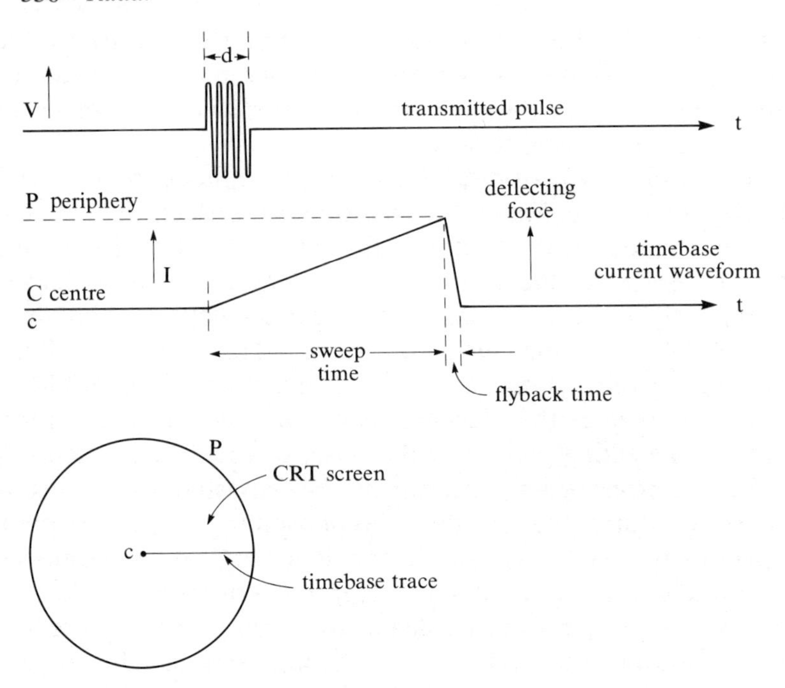

Figure 18.1.7 Idealized relationship between sawtooth timebase current waveform, transmitted pulse and radial trace.

18.1.5 Nautical mile, statute mile, radar mile

Using expression [**18.2**] and expressing velocity of light in yards per second, target range can be deduced as

$$\text{range in yards} = \frac{327.36 \times 10^6 \times \text{elapsed time (s)}}{2}$$

or range = 164 yards per μs of elapsed time (approx.). [**18.5**]

Since one nautical mile = 6080 feet = 2027 yards (approx.), one nm of range is represented by an elapsed time of 12.35 μs. A statute mile of 1760 yards is represented by an elapsed time of 10.7 μs. A radar mile of 2000 yards is sometimes used and has an elapsed time of 12.2 μs. Figure 18.1.8 shows that the timebase waveform depicted rises linearly to its maximum value in 148.2 μs during which time a radical scan is completed at the display. Screen radius directly represents a linearly scaled maximum measured range of 12 nautical miles. For a screen diameter of 16 inches each radial inch is therefore equivalent to 12 ÷ 8 nautical miles, assuming that the full radius is used.

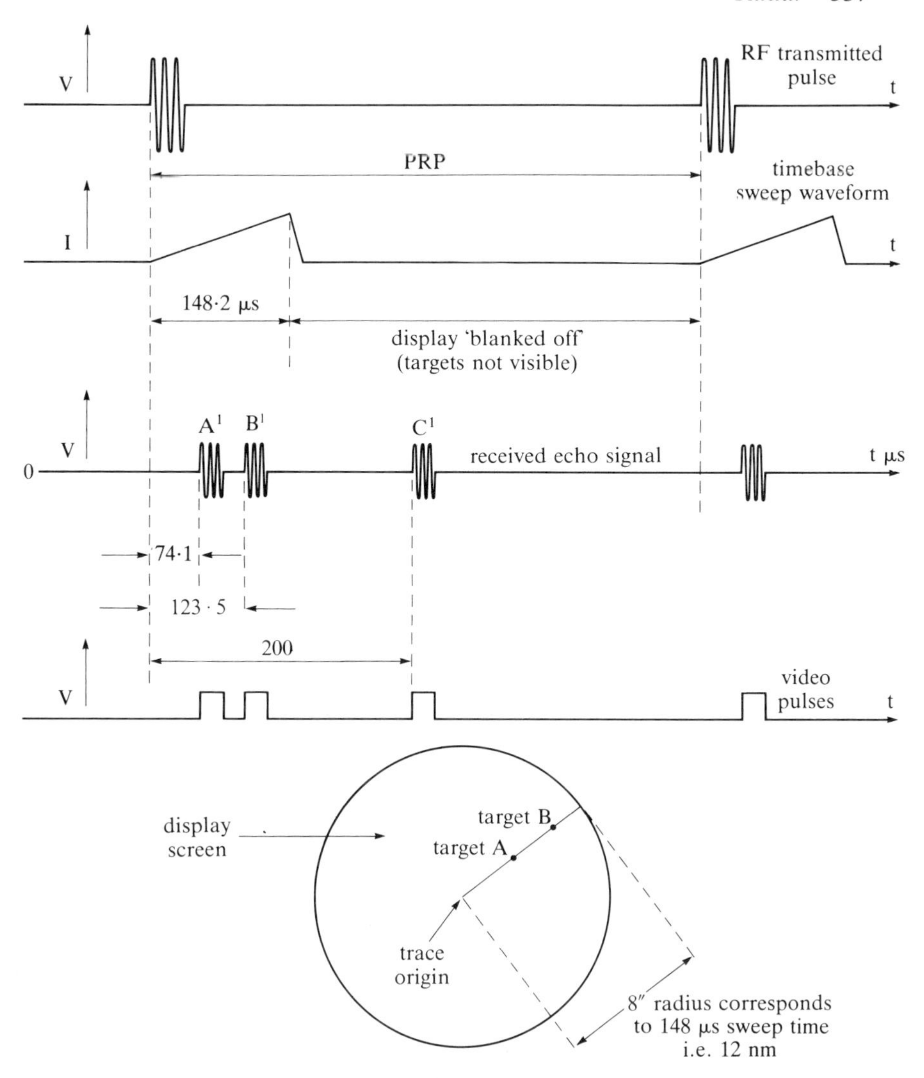

Figure 18.1.8 Target paint A and B produced from echoes A^1 and B^1. C^1 is not painted because it lies beyond the selected maximum displayed range.

A target paint at *A* corresponds to an echo received from target *A* 74.1 μs after the transmitted pulse. Target paint *B* appears on the same bearing and therefore on the same trace after a longer time delay of 123.5 μs being due to return from target B^1. The scale distances may now be measured directly from the tube face to give A^1 at 6 nm and B^1 at 10 nm. Bearing information may be obtained from the known relative direction of the aerial main beam. More distant targets than B^1 may return echo signals to the receiver as at C^1. Such signals will be processed by the receiver and display circuits but cannot appear anywhere on the display since the beam current is virtually suppressed during flyback and for the remainder of the resting period. Range measurement is made with the aid of electronically generated *range rings* and *variable range marker*. (See section 18.4.)

18.1.6 Antenna rotation: generating a PPI picture

The antenna is free to rotate in azimuth, RF signals being coupled to and from the antenna via a rotating joint in the waveguide feed. An electric motor drives the antenna through a reduction gearbox so that the rotational speed is sensibly constant. As rotation occurs through one complete revolution, so the pulsing main beam of the antenna sweeps through 360 degrees irradiating targets within its range. Bearing information regarding the direction in which the main beam points is continually electrically transmitted to the radar display. At the display the magnetic field which produces radial electron beam deflection is caused to rotate concentrically with the tube axis and in synchronism with the antenna. This action is brought about by a *bearing transmitter* at the antenna drive unit and a *bearing receiver* at the display. When the main beam position aligns with the ship's heading, a clearly visible radial line is generated on the display which indicates to the observer the direction in which the ship is heading relative to other displayed information. The line is variously known as the *heading marker, heading flash* or *heading line* and also enables the observer to measure bearings of displayed targets relative to the ship's head. In the simpler displays the heading line is arranged to appear at the top of the display screen usually aligned with 000 degrees on a bearing scale arranged around the display periphery. This is the *ship's head up* presentation. Those units comprising the antenna drive, antenna, bearing transmitter, heading switch and other relevant ancillaries are collectively known as the *scanner assembly*, its action being to scan an area around the vessel. A rather loose usage of the term scanner is often applied to the antenna.

Antenna speeds vary somewhat, a common nominal value being 20 revolutions per minute, corresponding to one revolution every three seconds. Using these values it is seen that a transmitter with a PRF of 1000 pps will transmit 3000 pulses during one complete revolution of the antenna. Since the radial scan lines on the display rotate in synchronism with the antenna and are generated in synchronism with the transmitted pulses, 3000 such lines will be generated in the same time. Each trace will advance on the previous trace by a small angular increment in a clockwise direction. From the given example, each degree of angular rotation contains 3000/360 = 8 timebase traces (approx.). Each adjacent pair of lines will subtend an angle of one-eighth of a degree at the scan origin, i.e. the scan centre, effectively covering the entire screen area during one antenna revolution.

Radar display CRTs have phosphor coatings on the inner face of the screen which when energized by an electron beam of sufficient intensity will emit light which persists for a time after the energizing beam has been removed. These are *high persistence* tubes and for a short time store on the screen the image of any target paint.

As the CRT radial trace rotates, each individual swept area of the screen will display any returned target information and build up a picture of the surrounding targets within the range selected. The tube persistence is so arranged that target paint decays to a low level of intensity during the period of one antenna rotation, the paint being renewed with each revolution by the rotating timebase trace. The type of picture presented by such a display shows the scan origin as the site of the radar antenna and that point on the display is known as *own ship*. It is known as a *plan position indicator* or PPI presentation since target information appears on the screen as it would if a map was drawn of all detectable echoes within the range selected for display. Figure 18.1.9 illustrates the generation of a PPI picture.

18.1.7 *Essential sections of a basic receiver and display*

Figure 18.1.10 shows the essential sections of a simple radar receiver and PPI display. The transmitter, modulator, PRF generator and power supplies have been omitted (see Figure 18.1.4).

A brief outline of the basic function of each section will be given here.

Antenna drive unit; rotates the antenna at constant speed
Heading marker switch; closes when main beam is in line with ship's head and causes the PPI heading mark to appear
Bearing transmitter; usually a small machine driven by the antenna which transmits electrically the antenna beam bearing information

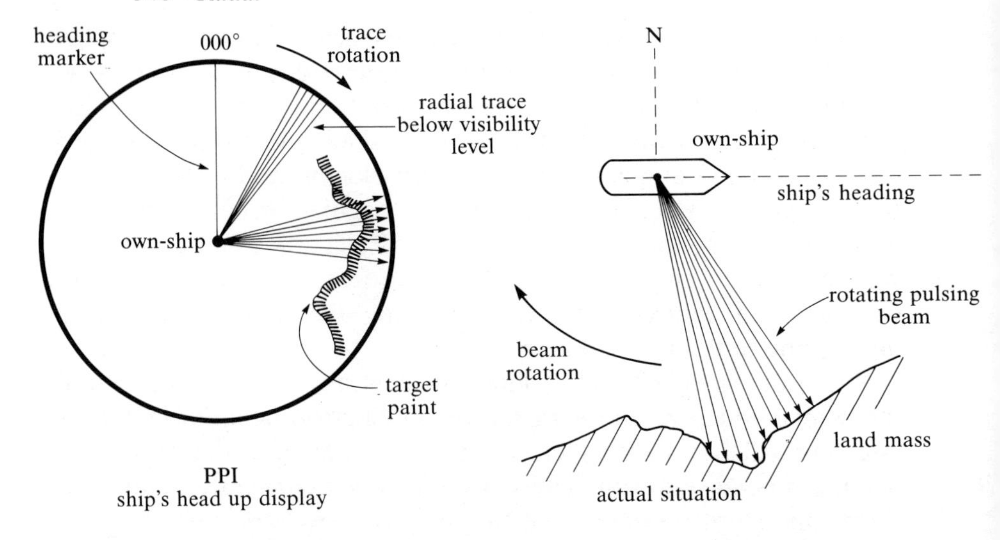

Figure 18.1.9 How a PPI picture is built from successive sweeps and the relative target echoes.

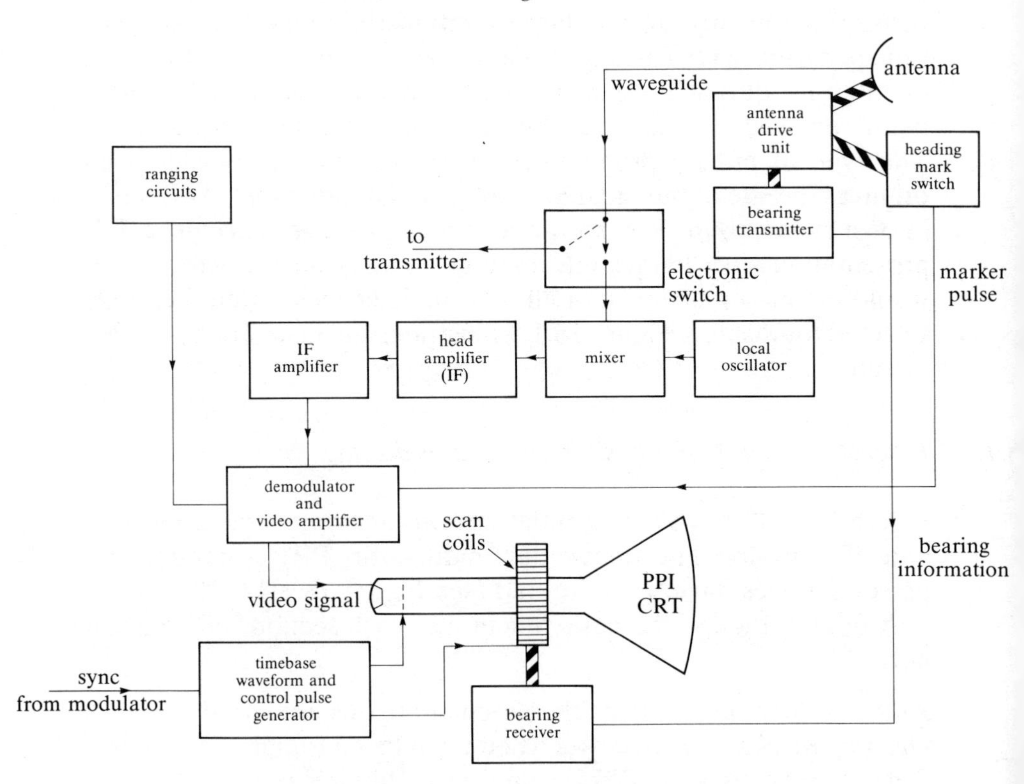

Figure 18.1.10 The essential sections of a basic radar receiver and PPI display.

Electronic switch; known as a transmit/receive cell (T/R), switches at high speed between transmit and receive modes: essentially a receiver protection device

Mixer; an electronic circuit which converts the incoming echo signal at the transmission frequency to a much lower value known as the intermediate frequency (IF); the lower frequency is easier to process in the succeeding stages

Local oscillator; provides a frequency stable output signal having a value of frequency either higher or lower than the transmission frequency by a value equal to the IF. Simultaneous application of the local oscillator output and the echo signal to the mixer will produce a difference frequency which is the IF. The IF signal contains the same information as the incoming echo signal

IF amplifier; provides amplification for the IF signal

Demodulator; produces video pulses from the IF signal pulses

Video amplifer; amplifies and processes the video pulses to a level adequate to intensity modulate the PPI CRT beam current

Timebase waveform and control waveform generator; generates the timebase sawtooth sweep waveform and other rectangular waveforms used to control the display of targets during the sweep time only: the circuits are synchronised to the transmitted pulse

Bearing receiver; for the case illustrated is a small machine which receives antenna bearing information and applies mechanical drive to rotating scan coils. The coils rotate in synchronism with the antenna. There are other methods of producing a rotating scan at the PPI

Ranging circuits; two separate circuits one of which produces periodic short pulses to display accurately spaced concentric rings on the tube face, the other circuit produces a variable radius ring linked to an accurate range scale. Thus echoes coincident with such rings may be electronically ranged with considerable precision.

18.2 Basic radar parameters

18.2.1 Frequency bands

Civil maritime radar operating frequencies are allocated within the following bands of frequencies:

(i) the S band from 2000 to 4000 MHz

(ii) the X band from 8000 to 12 500 MHz

these correspond to a range of wavelengths of:

(i) S band 0.075m to 0.15m
(ii) X band 0.024m to 0.0375m.

There are very small differences between the extremities of the British and American S and X bands.

Typical values used by all vessels lie within these two bands and are assigned by the International Telecommunications Union. For example, a vessel might transmit on 9445 ± 35 MHz in the X band and 3050 ± 10 MHz in the S band. Such values are, in terms of wavelength, loosely referred to as 3cm and 10cm radar wavelengths.

These (and higher) frequencies have been reached due to the continued research and development since the advent of the second world war when a need arose for small, high-powered radars having improved range and bearing resolution. Scientists fully realized that such improvements could be achieved at microwave frequencies with the added bonus of antenna structures of reduced size and weight. The significance of employing centimetric wavelengths is emphasized by examining the *radar range equation*.

Other types of radar exist, mainly for military or scientific applications, which will not be considered in this text.

18.2.2 *Antenna beamwidth and gain*

In order to achieve bearing resolution on a target the antenna radiates RF power in the form of a highly directional beam. Maritime installations are designed specifically to detect targets which are lying virtually in the horizontal plane with the radar equipment on an unstable platform. The antenna is therefore engineered to propagate a fan-shaped beam, narrow in the horizontal plane and relatively wide in the vertical plane. Since the antenna has directivity it is said to have a *power gain* in that direction. Antenna gain is an important radar parameter and power gain in particular is considered in the radar equation.

Beamwidth is another of the important criteria since it specifies boundaries within the antenna radiation pattern which are considered to be the limit of useful radiation (or reception). Figure 18.2.1 illustrates the concept of beamwidth and shows that because the beam shape of a marine radar antenna is not conical with the cone apex at the antenna, there exist two important beamwidth figures; one is taken in the horizontal plane, known as the *horizontal beamwidth* (HBW) and the other is taken in the vertical plane, being known as the *vertical beamwidth* (VBW).

The HBW tends to assume more importance than the VBW because of its effect on the radar's bearing discrimination and hence the navigational information interpreted at the display. The VBW in a marine system is largely a compromise between a requirement to illuminate a target whilst own-ship is rolling and also minimize unwanted echoes from the surface of the sea whilst optimizing the power gain characteristics of the antenna.

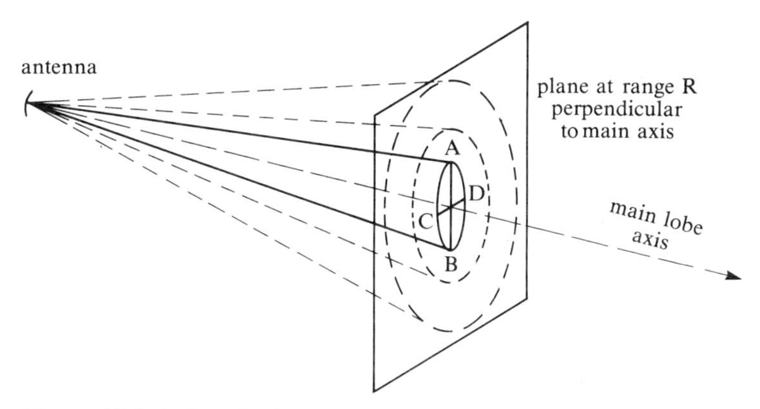

Figure 18.2.1 Simple three dimensional concept of radar antenna radiation.

Figure 18.2.1 attempts to show that the portion of the radiated beam which illuminates a target at distance R units and shown in solid outline is considered to be the useful *main lobe* of the radiation pattern. Power measured at A, B, C and D is one-half the measured power at the main lobe axis along which maximum radiated power acts. These (and other points lying on the ellipse $ABCD$), are known as the *half-power points* within the beam.

18.2.2.1 Beamwidth defined; the decibel; minor lobes

The horizontal or vertical beamwidth is then conveniently defined as the angle subtended by the selected half-power points at the antenna (see Figure 18.2.2). The half-power points are also known as the *minus three decibel* points or *three decibels down* points, written -3dB and meaning 3 decibels lower than the maximum power measured at the main lobe axis at range R. The decibel and its application is explained in appendix A. Beamwidth can also be

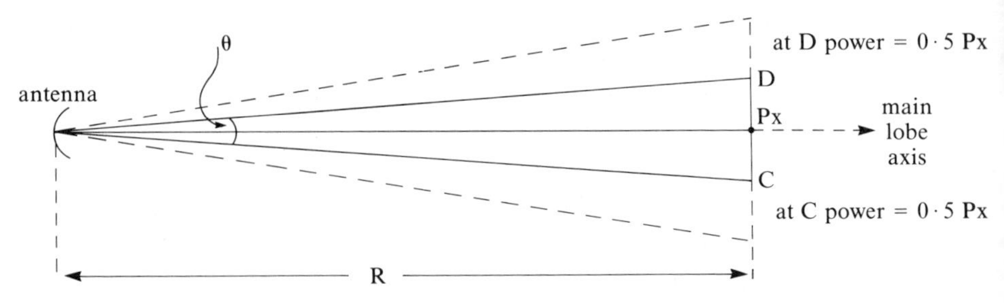

Figure 18.2.2 θ = HBW subtended from C and D which are half-power points relative to Px the maximum power at range R.

defined in terms of the electric field strength, being the angle subtended at the antenna between points in space where the RMS EMF in the electric field is 0.707 of the maximum RMS EMF at the main lobe axis.

Vertical beamwidths of the order of 22–25 degrees and horizontal beamwidths of 0.8–1.5 degrees are common.

It is important to realize that the main transmitted lobe or *major lobe* does not solely contain all the transmitted power. That would be an ideal condition, not realizable in practical antenna design. Figure 18.2.1 indicates that other minor radiation lobes exist, shown in dotted outline, but the power in those lobes may be reduced by at least as much as 30dB upon the main lobe power. Such a power reduction represents one-thousandth Px and in normal circumstances contributes a negligible echo from a distant target. These *minor lobes* are known as *sidelobes* and modern slotted waveguide antennas are designed to minimize such lobes which can cause confusing display echoes, particularly from targets at short ranges.

Earth proximity and its effect has thus far been ignored. The presence of the earth will have an effect on the radiation pattern as outlined in sections 18.3 and 18.4.

18.2.2.2 Antenna power gain

Antenna *power gain* is defined as the ratio of the power density produced at a given distance from the directional antenna in the direction of maximum radiation, to the power density produced at that site from an *isotropic radiator* fed with the same power. The power gain is quoted in dB.An isotropic radiator is a convenient fictitious point source of radiation which radiates uniformly in all directions. In practice a short uniform dipole is used as the reference antenna aligned such that its omnidirectional polar diagram lies in

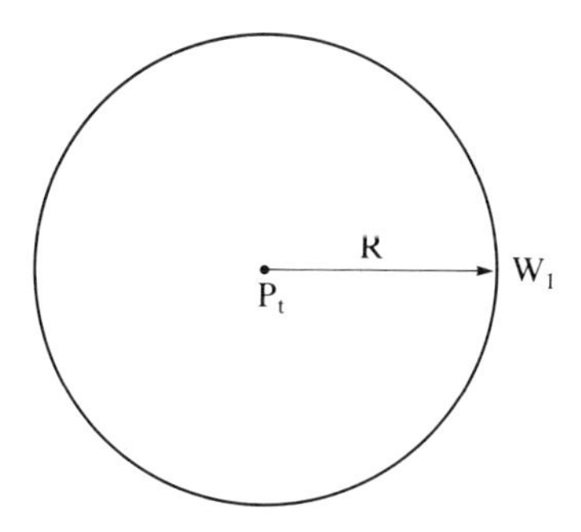

Figure 18.2.3 (a) Shows the omnidirectional polar plot (horizontal) of the reference antenna. At range R power density is W_1 $W.m^{-1}$.

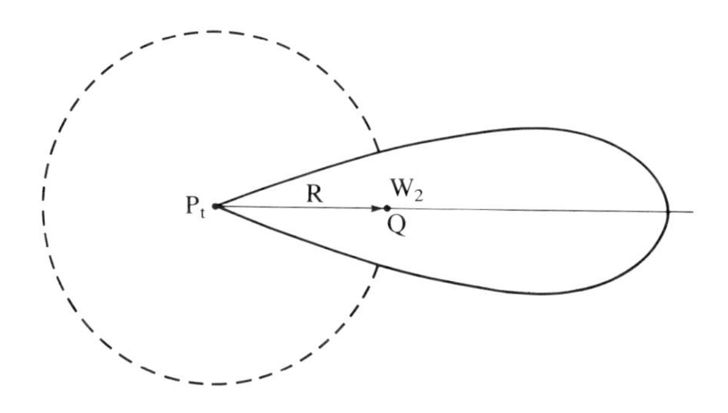

Figure 18.2.3 (b) The power flux density at range R, point Q, in the direction of maximum radiation is W_2 for the directional antenna.

the requisite plane of measurement. Figure 18.2.3 illustrates the principle. Suppose the reference antenna is supplied with power P_t and produces a power density W_1 $W\text{m}^{-2}$ at Q, distant R units. The reference antenna is now replaced with the directional antenna under test which is fed with the same power P_t assuming identical radiation resistances for the two conditions. The power density at Q in the direction of maximum power radiation is now found to be W_2 $W.\text{m}^{-2}$. The antenna power gain G_a is given by:

$$G_a = \frac{W_2}{W_1}$$ and expressed in decibels is

$$G_a = 10\log_{10}\frac{W_2}{W_1}\ \text{dB}. \qquad \textbf{[18.6]}$$

Table 18.2.1 shows practical relationships between HBW and VBW for slotted waveguide antennas. Also shown is the relative sidelobe power level relative to the main lobe axis at ±10 degrees from that axis.

Antenna size	HBW degrees	VBW degrees	Sidelobes ± 10 degrees
12′ S band	1.85	22	−28 dB
12′ X band	0.65	22	−30 dB
9′ X band	0.85	22	−29 dB

Table 18.2.1 Some practical values of antenna size, horizontal and vertical beamwidth and relative sidelobe power level.

18.2.2.3 Antenna aperture or effective area

This parameter is determined from the dimensions of an area perpendicular to the direction of radiation (or reception), over which the antenna can be assumed to be radiating or receiving. The principle is expanded slightly in section 18.3. The effective area is directly related to the *linear* measure of a slotted waveguide antenna. In general, for a given wavelength increasing the aperture will *increase* the power gain and *decrease* the horizontal beamwidth. See Table 18.2.1 and compare values of beamwidth and antenna length.

18.2.2.4 Antenna reciprocity

The antenna reciprocity theorem simply states that an antenna used for transmission (for given operational conditions) may be used for reception. The results obtained when used as a receiving antenna are reciprocal with those of the antenna when used for transmission. According to the *Rayleigh-Carson reciprocity theorem*, the gain of a given directive antenna is the same whether used for transmission or reception; therefore receive and transmit *polar diagrams* are identical. It follows that the antenna has the same aperture whether transmitting or collecting energy. An equation relating the effective receiving area A_e of an antenna to the power gain G_a and wavelength λ in use, is:

$$G_a = \frac{4\pi A_e}{\lambda^2} \qquad [\mathbf{18.7}]$$

Expression [**18.7**] is the *per-unit* gain, not the gain expressed in dB.

18.2.2.5 Polar diagrams; radiation pattern

Figure 18.2.4 illustrates the concept of the polar diagram. A brief study of section 18.1 and sections 18.2.2 to 18.2.2.4 indicates that the directive properties of an antenna and means of measuring these are important in radar systems.

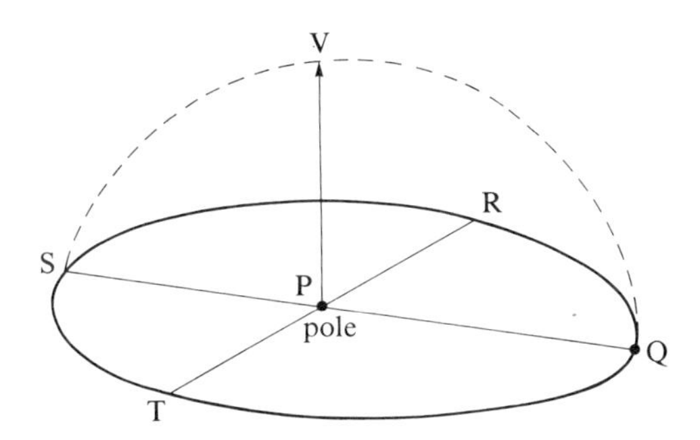

Figure 18.2.4 Any plane swept by PQ in a vertical direction such as Q, V, S, etc., is the vertical plane. Any plane swept by PQ in a horizontal direction such as Q, R, S, T, etc., is the horizontal plane.

The standard approach is to place the antenna at the centre or *pole* of two intersecting sets of planes. One is the *horizontal* or equatorial plane; any plane *normal* to it is therefore the *vertical* or meridian plane. A receiver may be imagined to move about radii such as *PQ* or PV and produce relative electric field readings in each plane. From the values obtained a relative plot of the radiation intensities can be produced. A cartesian plot for one plane as in Figure 18.2.5(a) is known as a *radiation pattern*. A *polar plot* obtained as in Figure 18.2.5(b) shows relative horizontal or vertical radiation patterns and sidelobes from a different view point. It is a matter of convenience which representation is chosen.

Although polar plots may be drawn to relative scales and values regarding gain (or loss), it is usual to indicate such values on radiation patterns.

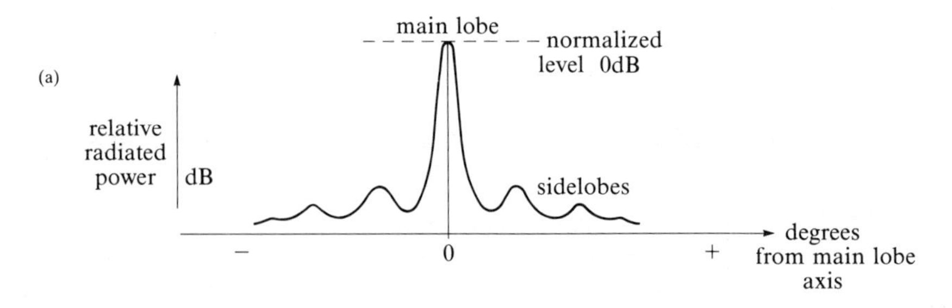

Figure 18.2.5 (a) Cartesian radiation diagram (pattern).

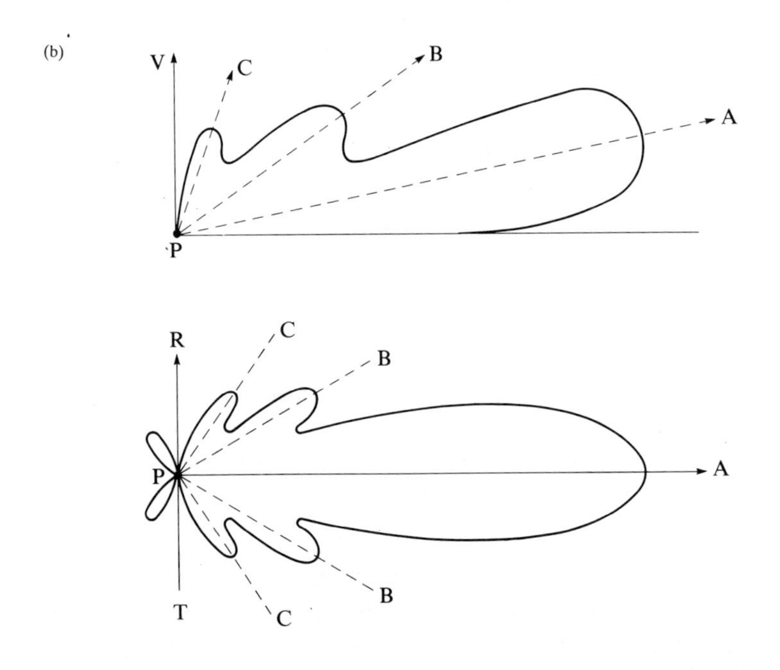

Figure 18.2.5 (b) Polar plot of radiation pattern in vertical plane (above) *and horizontal plane* (below). *A is major lobe axis, B and C are minor lobes.*

18.2.2.6 Target effective echo area

The echoing area of a target depends upon many factors beyond the influence of the radar system. At any particular instant the area can be considered to be the area of a perfect reflecting plane perpendicular to the incident transmitted wave which produces the same effect at the receiver as the signal actually received from the target.

18.2.3 *The radar range equation*

The radar range equation shows the relationship between radar system parameters, target and range. It is useful for determining the theoretical maximum detectable range of targets and forms the basis for an understanding of the system operation, indicating why certain design decisions are made with regard to the transmitter, antenna and receiver.

A simple form of the radar range equation is given as:

$$R_{max} = \left[\frac{P_o . G_a . A_e . A_t}{(4\pi)^2 . P_{r\ min}} \right]^{1/4} \text{ metres} \qquad \textbf{[18.8]}$$

The symbols have the following meaning:

R_{max} is the maximum distance at which a target can be detected in metres
G_a is the power gain of the antenna
A_t is the target effective echo area (m^2)
A_e is the effective receiving area of the radar antenna (m^2)
P_o is the output power of the radar transmitter (W)
P_r is the minimum detectable received signal power (W).

Appendix B provides the basic derivation of the expression together with cautionary notes.

It should be emphasised that expression **[18.8]** has been produced assuming free-space conditions. No provision is made for effects of earth proximity and propagation effects in the earth's atmosphere or for changing conditions of effective target area. These effects and others, reduce the practical maximum range to a value much less than is indicated by the equation. Assuming a constant effective antenna area and substituting **[18.7]** for G_a in **[18.8]** gives:

$$R_{max} = \left[\frac{P_o . 4\pi \ A_e^2 \ A_t}{(4\pi)^2 . \ \lambda^2 . \ P_{r\ min}} \right]^{1/4} \qquad \textbf{[18.9]}$$

Expression **[18.9]** although simplistic, indicates the following important points:

(i) R_{max} is proportional to $\sqrt[4]{\dfrac{P_o}{P_{r\ min}}}$

(ii) R_{max} is proportional to $\frac{1}{\sqrt{\lambda}}$

(iii) R_{max} is proportional to $\sqrt{A_e}$

(i) Suggests that the maximum range is proportional to the fourth root of transmitted power. A sixteen-fold increase would be necessary in the value of P_o in order to double R_{max} (assuming such an increase was possible). It also seems that R_{max} is proportional to $(P_{r\ min})^{-4}$ indicating that the receiver would need to detect signals sixteen times less in power in order to double range, other factors in the equation remaining constant. In practice the receiver sensitivity is ultimately limited by the acceptable signal-to-noise ratio.
(ii) Implies that as the wavelength is reduced, the maximum range increases. A relationship supported by the fact that for a given antenna aperture a narrower beam is formed if λ is reduced thus increasing the concentration of available propagated power in a given direction.
(iii) Is a factor which indicates that R_{max} is proportional to the linear dimensions of the antenna, since A_e is a measure of area. The length of a slotted waveguide antenna will have an effect on the beamwidth for a given value of λ. For similar operating conditions a twelve-foot SWG antenna will have a higher gain than a six-foot antenna.

For the two radar bands given, three-centimetre radar is used for relatively high-bearing definition and good echo return for low to medium peak output power (25–45 kW). Target re-radiation is enhanced as the ratio λ/A_t decreases. In free space, energy loss (attenuation) is negligible. Normal energy dispersion occurs in accordance with the inverse square law. In the atmosphere gas molecules absorb some energy, but three-centimetre waves can experience severe attenuation and reflection in rain and dense fog. Wavelengths of ten centimetres are less affected by rain, fog and snow and also permit returns from large targets beyond the normal radar horizon of the shorter wavelength. A higher power output is generally transmitted at the longer wavelength, typically 60 kW. Many modern radar installations permit transceiver switching for either three-centimetre or ten-centimetre operation according to navigational requirements. An option of choosing circularly polarized propagation is also generally available at the longer wavelength. A significant number of S band radars use vertical polarization.

18.2.4 *Pulse repetition frequency; pulse length*

Choice of PRF is influenced by several design factors. One of those being the anticipated maximum detectable range of a target. Evidently, successive pulses must be separated by a period of time which permits the echo from a transmitted pulse to return and be displayed before the onset of the next transmitter pulse. Hence, as a minimum theoretical value, the PRP must have a length which is equivalent to twice the one-way trip time from antenna to target. In practice the PRP is greater than the theoretical minimum. Typical values in use vary between about 500 PPS to 4000 PPS and a modern radar will have two or three PRFs giving a choice of PRP to suit the ranging conditions. In addition, the pulse length will be switched to suit those conditions.

Table 18.2.2 shows some typical values of PRF and appropriate pulse lengths. The values are automatically selected when the radar operator selects a desired viewing range but an added switching facility allows the operator to override short pulse selection on certain ranges, replacing it with one of the available longer pulse lengths should his experience deem it necessary. Selecting a longer pulse in lieu of a shorter one will always degrade range discrimination.

adar	Range nm	¼	¾	1.5	3	6	12	24	60
	Pulse length (μs)	0.05	0.05	0.05	0.05	1.0	1.0	1.0	1.0
	PRF	3200	3200	3200	3200	800	800	800	800

adar	Range nm	¼	1½	3	6	12	24	48	64
	Pulse length (μs)	0.06	0.06	0.06	0.5	0.5	1.0	1.0	1.0
	PRF	3600	3600	3600	1800	1800	900	900	900

Table 18.2.2 Some practical values of PRF, pulse length and displayed range.

18.2.4.1 Target ambiguity; second trace echoes

If the designed PRF is made too high a possibility arises for target echoes due to one transmitted pulse to be displayed on the timebase trace of the next successive pulse. Figure 18.2.6 illustrates the effect. Approximate values are used in calculations.

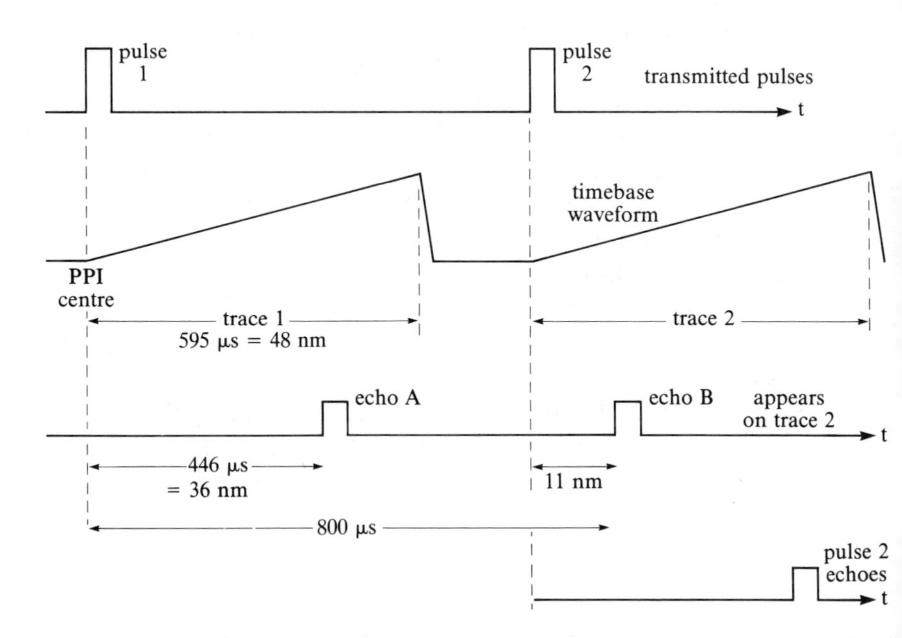

Figure 18.2.6 Showing target range ambiguity due to second trace recpetion of echo B from previous transmitted pulse.

Here, transmitted pulse *1* has sufficient energy to return a discernible signal echo from targets *A* and *B*. The radar user is able to change the range scaling at the display to suit particular navigation conditions. Table 18.2.2 indicates the value of maximum displayed range which might exist for given PRFs and pulse lengths. In Figure 18.2.6 the selected maximum range is 48 nm, producing a timebase sweep of $12.35 \times 48 \simeq 595.0$ μs. Target *A* returns an echo 446.0 μs after timebase waveform initialization, indicating that the target lies at 36 nm range on the *first transmission trace*. A PRF of 1500 pps is chosen giving a prp of 666 μs. Target *B* returns its echo due to pulse *1* after 800 μs and is due to the target lying at range 64 nm. The echo due to target *B* therefore returns 134 μs after the start of the timebase trace due to pulse *2*. The paint due to echo *B* then appears on timebase trace *2*, producing an ambiguous echo at an apparent range of 11 nm. All unambiguous echoes will, of course, be displayed at normal ranges.

Second trace echoes, as described, can cause confusing range measurement. In a practical situation it is usual for such echoes to be displayed only when the pulse transmitted has sufficient energy to cause a discernible echo return from a large target area such as a land mass beyond the visible horizon but within the radar horizon, or under circumstances where a propagation anomaly known as *super refraction* is occurring. A displayed coastal outline will be compressed towards the central area of the CRT screen, and an experienced navigator will in any event be suspicious of such unexpected targets and aware of the possibility of super refraction in the atmosphere. Second trace echoes can be resolved by changing pulse repetition frequencies as, for example, will occur by switching to a lower range. Since most radars employ at least two PRFs, these echoes present no problem to the radar observer.

18.2.4.2 Duty cycle; magnetron principle

Pulse length and PRF are linked by a transmitter parameter known as its *duty cycle*. In centrimetric radars the transmitter stage is a *magnetron valve*; the type used to provide high-peak transmitted powers of the order of 75 kW is the *resonant cavity* magnetron. In principle the magnetron is engineered so that the interaction between a powerful magnetic field and a high potential gradient acting perpendicular to that field influences the trajectory of electron charges which are produced by heating a material having a low work function (the cathode). A high-voltage rectangular pulse applied to the magnetron from the modulator acts as the valve's momentary d.c. anode voltage. During that application the electrons from the cathode move in orbits in a space between the cathode and anode to excite and sustain microwave oscillations in a series of cavities surrounding and concentric with the cathode. This process occurs in a vacuum and the microwave energy comprising the RF radiated pulse is coupled via an inductive loop and glass vacuum seal from the cavities into a waveguide and thence to the antenna. A supply pulse for a high-power cavity magnetron may be of the order of 20 kV.

The transmitter is active only during the pulse and is inactive for the rest of the PRP. The ratio pulse length/PRP is called the *duty cycle* of the transmitter, i.e. the ratio of the time for which the magnetron is *on* to the total time between successive pulses. Figure 18.2.7 illustrates duty cycle, and using the shortest two values given

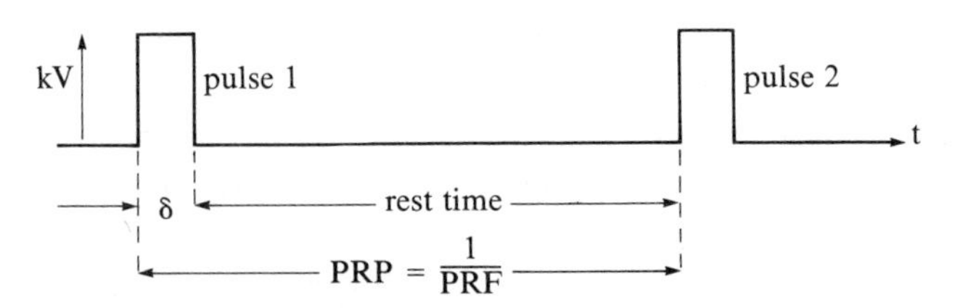

Figure 18.2.7 Illustrating duty cycle:

$$D = \frac{\delta}{PRP} = \delta \times PRF.$$

in Table 18.2.2 we have:

(i) $D = 0.06 \times 10^{-6} \times 3600 = 0.0002$ (approx.)
(ii) $D = 0.05 \times 10^{-6} \times 1800 = 0.0009$ (approx.).

These values are sometimes expressed as a percentage. A continuously transmitting CW radar would have a duty cycle of one, or 100 per cent.

Duty cycle also relates the pulse length and PRF to the peak and average power from the transmitter. For the very brief period a magnetron is on it will radiate RF power between about 25 kW to 70 kW, the latter value commonly appearing in ten-centimetre radars.

18.2.4.3 Peak and average power

The power output P_o watts used in the radar range equation is considered to be the peak transmitted power. This value is not the peak instantaneous power of the sine-wave carrier as is commonly supposed. It is, in fact, the average power taken over one carrier frequency cycle occurring at the maximum value of the pulse of power output. When so defined, peak transmitter power approximates to half the maximum instantaneous power.

Average power (P_{avg}) is taken as the average power transmitted during the entire pulse repetition period.

Using these definitions the peak power can be regarded as a substantially rectangular pulse of width shown in Figure 18.2.8: it is seen that the *energy* in a pulse is given by $P_o \times \delta$ Ws, and is represented by the area shaded under the pulse shape. That same shaded area when averaged over the whole PRP produces an average pulse power of:

$$\frac{P_o \times \delta}{PRP} \text{ watts}$$

from which:

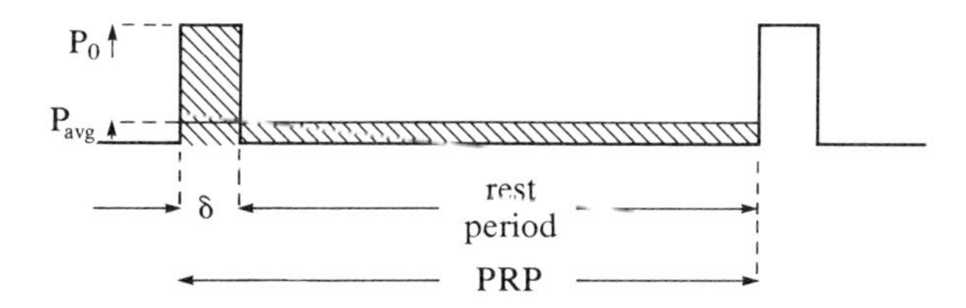

Figure 18.2.8 Showing the total energy $P_o \times \delta$, in the pulse can be averaged over the PRP.

$$P_{avg} = P_o \times \delta \times \text{PRF watts} \quad \textbf{[18.10]}$$

this can be written as:

$$P_{avg} = P_o \times \text{duty cycle watts} \quad \textbf{[18.11]}$$

Duty cycle is an important radar parameter. Manufacturers stipulate the maximum duty cycle possible for a given magnetron together with its average power capability. Exceeding that capability can quickly damage the device.

As an example the following practical values may be applied for a modern system:

ximum displayed ge (nm)	¼,	½,	¾,	1½,	3,	6,	12,	24,	48,	64
ıge number	1	2	3	4	5	6	7	8	9	10
se length (μs)	.06	.06	.06	.06	.06	.5	.5	1.0	1.0	1.0
F (pps)	3600					1800		900		

Nominal peak output power on all ranges is 25 kW
Average power capability of magnetron is 80W.
Hence for ranges 1 to 5: $P_{avg} = P_o \times \delta$ PRF = 5.4W
for ranges 6 and 7: P_{avg} = 22.5W
and for ranges 8 to 10: P_{avg} = 22.5W.

The results show that operation is well within the capability of the magnetron. Note that magnetron valves are not very efficient and they are likely to dissipate (as heat) more energy than they actually output. Note also that the shorter ranges in the example given can use lower transmitted energy in the pulse than the higher ranges since echo energy is returning from targets at shorter ranges.

As the measured range increases so the equipment is arranged to increase the pulse length, thus providing higher pulse energy and an improvement in echo level. These comments assume that the peak power output remains substantially constant. Pulse length also affects other important parameters namely *minimum detectable range* of a target and *range discrimination*.

18.2.4.4 Minimum detectable range

Figure 18.2.9 illustrates a simplistic but useful method of modelling the ranging behaviour of a radar pulse. Since an electromagnetic wave travels away from the antenna at the velocity of light, Figure 18.2.9(a) shows that the RF radiation can be considered as a packet of energy which, for a 1.0 μs pulse duration, will occupy a length in air of 300m. During the 1 μs that the pulse is being formed at the transmitter the RF waveguide path to the receiver is switched off by the quick acting electronic transmit/receiver switch. This precaution is necessary to prevent the high-power output from the transmitter damaging the sensitive receiver circuits. The leading edge of this spatial pulse is at A and the trailing edge is at B. Immediately the pulse ceases and the trailing edge B has just left the antenna, the transmit/receive switch will open the waveguide path to the receiver. However, during that 1.0 μs of transmission there may be a target, at for example 100m range, returning an echo. The sequence of events which occurs as the echo energy returns is shown in Figure 18.2.9 (b), (c) and (d). It is apparent that the pulse length is such that the leading edge A' of the echo is lost since it cannot be received and processed for range measurement. An echo at 150m will have its leading edge A' just reach the antenna as the receiver switches on, as shown in Figure 18.2.10. Thus, 150m is the *theoretical* minimum range at which a target is detectable and the value of that minimum range depends on the pulse length. In practice the receiver cannot be switched on at the instant the transmitted pulse ceases because of the *recovery time* necessary for the T/R switch to function and for the receiver circuits to become active. The delay usually amounts to a small fraction of a microsecond and it extends the theoretical minimum range. Manufacturers quote minimum ranges as '20m' or 'better than 15m'.

Consider a practical minimum pulse width of 0.05 μs used on the lower ranges of a radar set; the minimum theoretical range is given by $0.05 \times 300/2 = 7.5$m. The manufacturers of that particular equipment quote the minimum detectable range as 'better than 15m'. Such quoted values do not indicate how much better than 15m the minimum range is. This is because T/R switches can vary slightly in switching times and are subject in use to an ageing process which

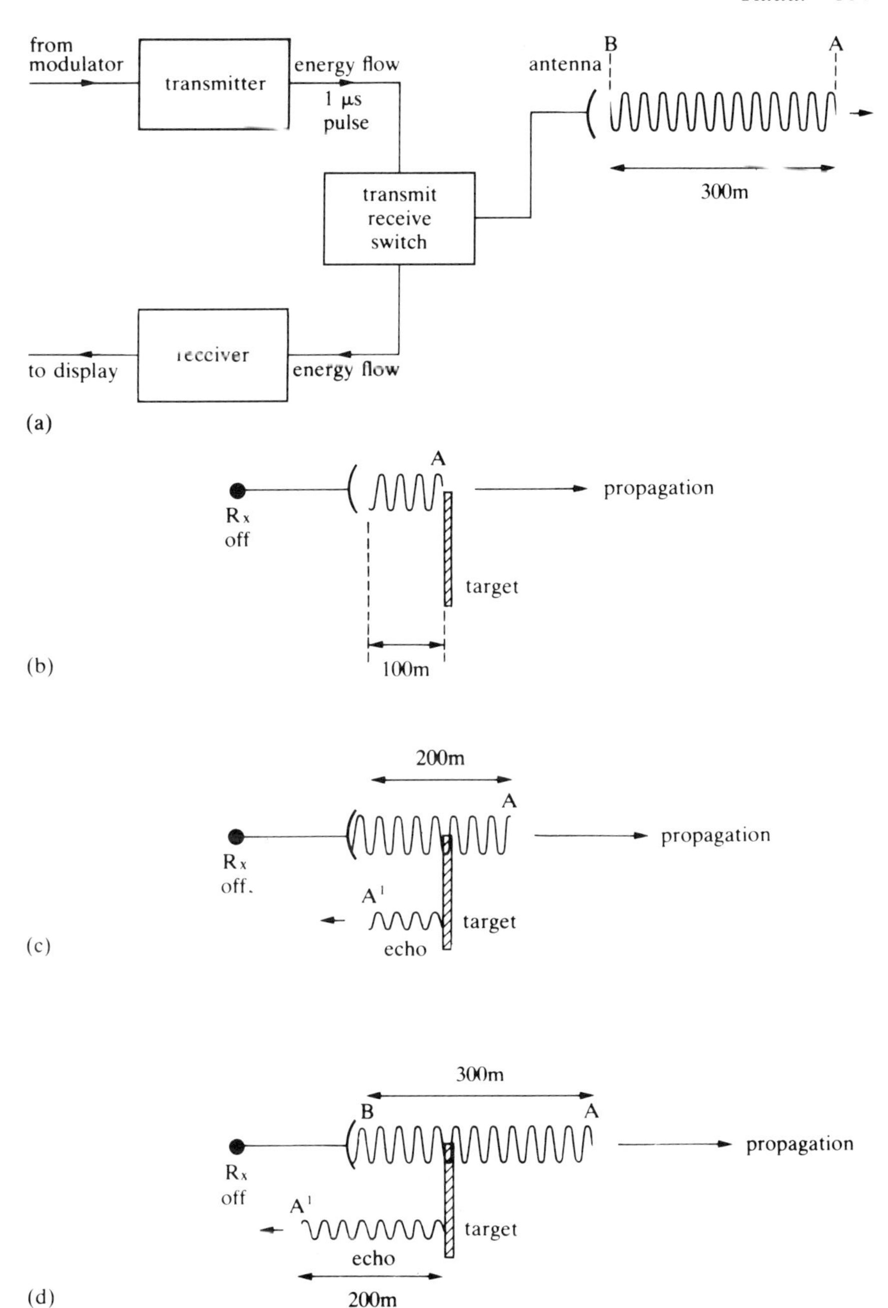

Figure 18.2.9 (a) The pulse considered as a packet of energy in space
(b) Situation 0·33 μs into a 1·0 μs pulse
(c) 0·66 μs into the pulse
(d) After 1·0 μs trailing edge B is just leaving antenna

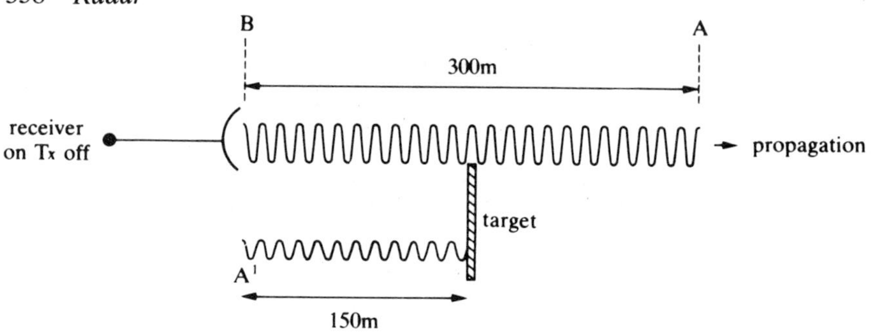

Figure 18.2.10. Theoretical minimum range is 0·5 × pulse length in space. A^1 reaches the antenna as the receiver turns on.

tends to increase that time. As a general rule, the practical minimum range is taken to be twice the theoretical minimum. Short pulse duration improves minimum detectable range.

18.2.4.5 Range discrimination

Range discrimination is defined as the ability of a radar to detect two or more targets lying close together on the same bearing relative to the antenna and at differing ranges and display those targets as separate echoes. Figure 18.2.11 depicts a situation where the antenna beam has rotated to irradiate two targets *C* and *D*, each lying on the same bearing and separated by 150m range. For clarity it is assumed that the antenna is momentarily stationary. The transmitted pulse length is 1.0 μs and occupies 300 metres length in air along the direction of transmission. The Figure shows that two targets on the same bearing must be separated by a distance *greater than* half the length in air of the transmitted pulse if they are to be detected as two separate targets, that is:

$$\text{target separation} > \frac{\text{pulse duration } (\mu s) \times 300}{2} \text{ m}$$

If this basic requirement is not obtained then two such targets will be displayed as one continuous paint on the display having the displayed characteristic of a return due to a target larger than either individual target.

Manufacturers of equipment often refer to range discrimination as *range resolution*. Typically, it may be given in equipment specifications as ‘better than 12 metres’, or in another case ‘18 metres or 1 per cent, whichever is greater’. In the latter case, the one per cent refers to the maximum value of the display’s selected

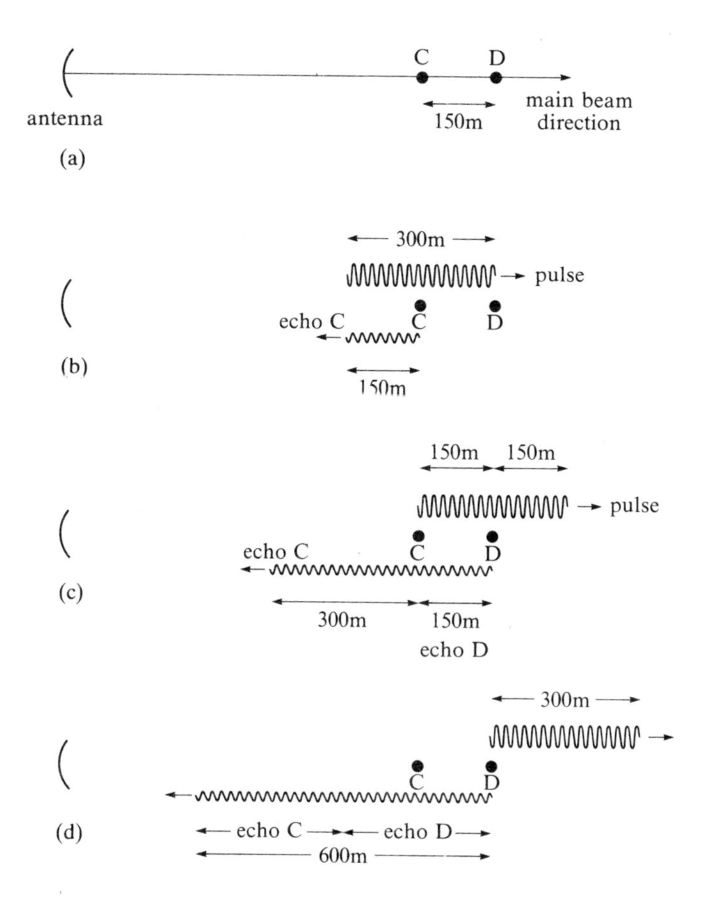

Figure 18.2.11 (a) Targets at 150m separation; (b) pulse returning echo from C; (c) pulse leaving C, echoes from C and D merging; (d) pulse leaving D: effective echo pulse is 600m length since both pulses have merged.

range on which the targets are observed and plus or minus 1 per cent is implied. Very often values are given in yards. It is seen that range discrimination will be degraded as pulse length increases. In a set having three pulse lengths (see Table 18.2.2) of 0.06 μs, 0.5 μs and 1.0 μs, the range discrimination cannot be better than 9m, 75m and 150m respectively.

18.2.4.6 Bearing discrimination

Bearing discrimination or *bearing resolution* is a measure of the radar set's ability to separate targets which lie close together at the same range from the antenna but on different bearings. Figure

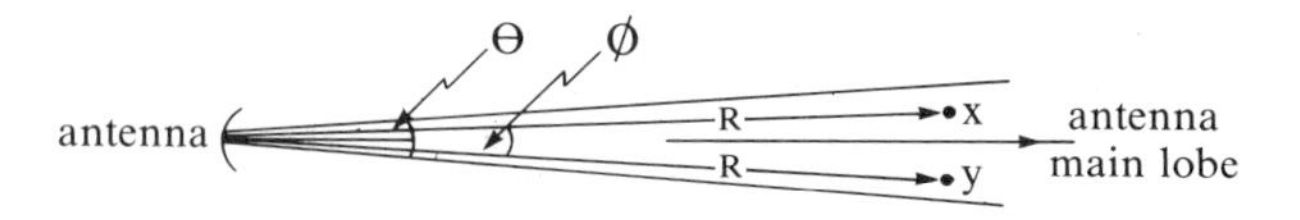

Figure 18.2.12 Showing that if Ø < Θ at range R, X and Y cannot be separately displayed.

18.2.12 shows two targets *X* and *Y* lying at range *R* metres. Using an idealized plan of the antenna beam it is seen that when the angle ϕ subtended by the two targets at the antenna is less than the horizontal beamwidth, the two targets will be irradiated simultaneously for a fraction of the beam rotation period and they will appear as one displayed target. It follows that whatever the value of *R*, the subtended target angle must be greater than the horizontal beamwidth ϕ for the two targets to be displayed as separate echoes. The appearance of such merged target echoes on the display is similar to that which would occur from a larger target having its length distributed in the horizontal plane.

Horizontal beamwidth is governed by the relationship between antenna aperture and wavelength in use. The larger the aperture for a given wavelength, the narrower is the HBW. For a slotted waveguide, antenna aperture is approximately proportional to length. Table 18.2.3 gives practical values.

	X band antennas				S band
Antenna Length	4′	6′	9′	12′	12′
Horizontal Beamwidth	1·7°	1·5°	0·8°	0·65°	1·85°

18.2.4.7 Relationship between antenna rotation speed, PRF, target and horizontal beamwidth

Assume the following values: PRF of 1000 pps; HBW of one degree and scanner rotation speed of 20 revolutions per minute. Consider a *point* target *A* at range *R* from the antenna. The point target is a convenient fiction, but for practical purposes it can be imagined as a target which subtends an angle to the antenna which is much less than θ the HBW and yet returns a detectable echo. Figure 18.2.13 shows these conditions. As the antenna rotates through one degree the time taken will be 60/(20 × 360) seconds, or 1/120s. Since the PRF is 1000 pps the number of pulses striking the point target is 1000/120 = approx. 8 pulses.

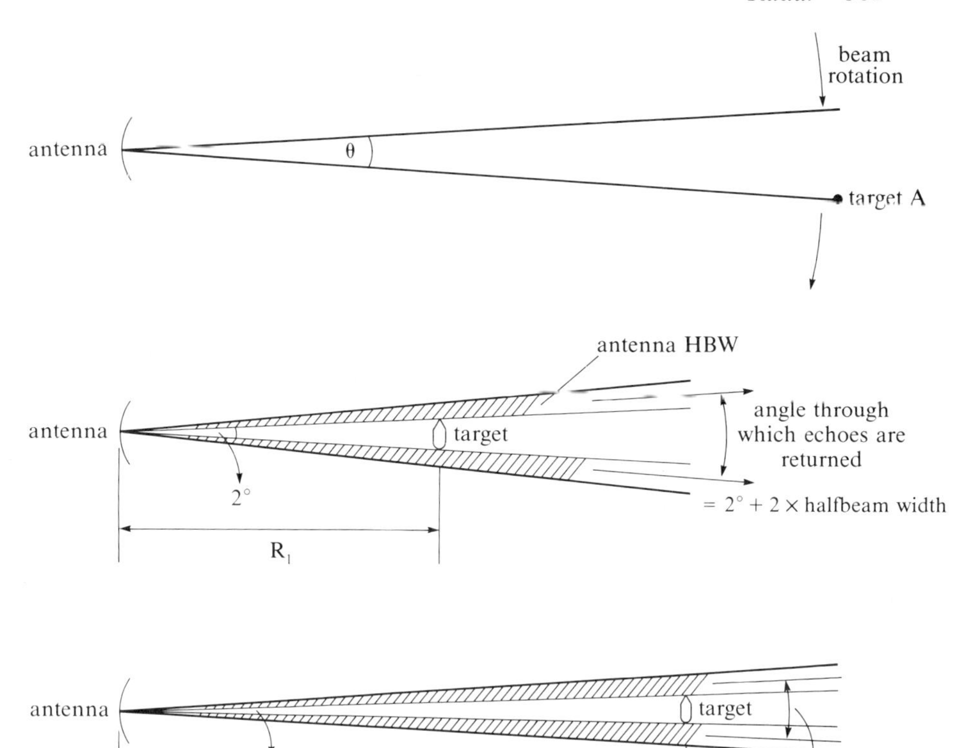

Figure 18.2.13 Showing effect of range on return angle. If $\theta = 1$ degree, PRF = 100 pps and antenna speed = 20 RPM, approx. 8 pulses will strike target A as beam passes A.

The number of strikes on a theoretical point target is therefore calculable from:

$$\text{strikes} = \frac{\phi \times \text{PRF}}{6N}$$

where ϕ is the horizontal beamwidth in degrees and N is the speed of antenna rotation in revolutions per minute.

If all of those *strikes* produced detectable echoes then each successive timebase trace related to those echoes will energize the tube phosphor and build up a bright paint on the CRT. The tube phosphor action *integrates* the individual echo paints to produce easily visible target paint.

In practical cases the target is not a point and will itself subtend an angle with the antenna depending upon target size, aspect and range. These are variables outside the influence of the radar system. Suppose a stationary target subtends an angle of two degrees with the antenna at range R_1 and an angle of 0.5 degrees at a greater range R_2. This condition is shown in Figure 18.2.13. For a beamwidth of one degree, at range R_1 the target will theoretically return echoes as the antenna turns through three degrees, the echoes returned for a rotational speed of 20 RPM and PRF of 1000 pps will be $3000/360 \times 3 = 25$ pulses. At the greater range R_2 the target will return echoes as the antenna turns through 1.5 degrees, giving $3000/360 \times 1.5 = 12$ pulses.

Evidently the same target, having a similar aspect, will produce different painting characteristics at the display when viewed at different ranges. The closer range tends to display a more compact brighter paint from a larger number of returns. Video pulses produced from the longer range are less in number and at the display the individual paints are spread over a wider arc (being further from the tube centre), which reduces paint brightness for that target. Various design solutions can be applied to counter the latter effect.

18.2.4.8 Target glint

The radar equation and results obtained above are difficult to apply in practice because some of the parameters outside the influence of the radar designer are unpredictable or *statistical* in nature. One of those factors concerns the changing aspect of the target, for example a ship which is rolling and pitching, as indeed is own-ship. Another very important receiver parameter outlined in section 18.4, is the *signal-to-noise* ratio at the receiver necessary to provide adequate detection of target echoes despite the presence of other degrading and unwanted noise impulses. The signal strength at the receiver can also be affected by propagation conditions along the transmit and receive path. Statistics show that of the 25 pulses reaching the target (section 18.2.4.7) approximately 10–15 detectable pulses may return. Successive pulse transmissions will, however, build up a bright area of paint on the display which appears at the range of the centre of echo returns. Changes of target aspect with respect to the transmitting antenna which produce changes in the apparent centre of reflection and also in the reflected signal's plane of polarization and strength at the receiver produce an effect known as *target glint* where the target may appear during one antenna scan and disappear during the next.

18.2.4.9 Summary

Although the radar range equation in its simple form is not precise due to the assumptions made, it is nevertheless a useful tool and highlights a number of the more important parameters which affect the performance of a radar installation. Some of the requirements conflict. In order to produce a good signal return from a distant target the energy in the transmitted pulse should be high. For a given peak power this suggests a long pulse duration. However, lengthening the pulse will degrade range discrimination. As is the case with most engineering problems, a compromise is reached where short pulse lengths giving good range discrimination are chosen for displaying targets at the shorter ranges and where good discrimination is an essential navigational requirement. Sufficient energy is contained in the pulse for such close returns. Good minimum detectable range is also ensured by having a short pulse length.

At longer displayed ranges, where range discrimination is not so critical, the pulse energy is increased by choosing a longer transmitted pulse. The PRF and the pulse length determine the average transmitter power for a given peak output power. Also, the PRF determines the maximum displayed range and is preferably chosen so that second trace effect is avoided. Again a compromise is accepted to provide adequate average power and not violate the duty cycle requirements.

Evidently reducing the antenna speed for a given PRF will provide more paints per target; it is not feasible to increase antenna horizontal beamwidth since this would reduce antenna gain and degrade bearing discrimination. But a conflict exists between CRT phosphor persistence and antenna rotation speed. The persistence tends to determine the minimum speed of rotation, so that a compromise is reached regarding antenna rotation, paints statistically expected, horizontal beamwidth, persistence of tube, PRF and duty cycle. Since those factors affecting the antenna radiated frequency and CRT are not normally variable, the PRF and pulse width are generally made variable being switched to a desired value when the operator of the equipment changes the displayed range. In this way adequate compromises are made between the various conflicting requirements.

Most modern marine systems provide both three-centimetre and ten-centimetre transceivers between which the display circuits can be switched in order to provide the facility of enhanced range and detection afforded by ten-centimetre systems in conditions of high precipitation and sea clutter. Finally the minimum detectable power returned, depends statistically upon variable quantities beyond the control of the user; the ability of the receiver to determine useful navigational echoes within externally and internally generated random noise voltages is dependent upon the signal-to-noise ratio at the receiver's demodulator and is the limiting factor in the detection of weak signals.

18.2.4.10 Super- and sub-refraction

The path of an electromagnetic wave is bent slightly as it passes through the atmosphere even under what might be considered to be normal atmospheric conditions for that region. This refraction is a result of the index of refraction of the atmosphere being a function of height above the surface of the earth. Refraction causes the 'radar horizon' to appear at a distance greater than if the radar waves travelled in a straight line.

For *visible* waves the index of refraction is dependent only on temperature and pressure in the atmosphere, whilst for radar waves it depends in a more complex way on temperature, pressure and water vapour content. For this reason, microwave radar returns echo signals from distances greater than the visual horizon.

Many radar observations have been recorded for target returns extending to ranges which are well in excess of the normal radar horizon. The majority of these have been paths over the sea. Such extended ranges are caused by propagation through a 'non-standard' atmosphere and is generally caused by a *ducting* phenomonena. Microwaves do propagate beyond the normal horizon due to earth *diffraction* (see appendix B.3) but this is a consistent propagation factor for which an allowance is made in the application of the *earth radius factor*. Atmospheric ducts are caused by rapid decrease of refractive index with altitude which can itself occur due to an increase in temperature and/or a decrease in humidity. It is the latter humidity gradients which are recorded as producing the most pronounced changes in refractive index. Enhanced propagation can occur by ducting if the radar antenna or the target is near the water which places the radar antenna and the target for most maritime applications within the duct. When evaporation occurs from the sea into a still atmosphere a layer of moist air is produced, extending perhaps 50 to 100 feet and having a vapour content which decreases rapidly with height. The condition causes partial trapping of radar energy and is further accentuated if there exists at the same time a *temperature inversion*, i.e. temperature reduces less rapidly with increase of height. Such conditions can cause radar energy to reach and be returned from targets many times the normal radar horizon and is known as *super-refraction*. Figure 18.2.14 shows the effect. Echo returns are generally from targets having a large radar reflecting area such as high land masses. Due to their extended detection range such echoes return on the next consecutive PPI trace causing *second trace effect* (see section 18.2.4.1). In general warm dry air settling over a relatively cold sea produces conditions ideal for super-refraction.

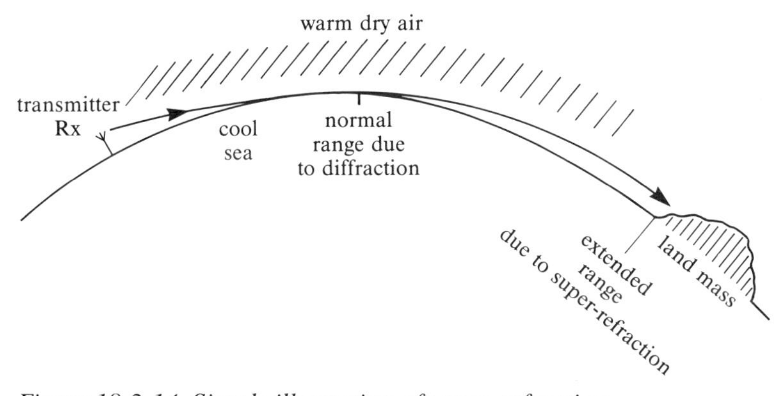

Figure 18.2.14 Simple illustration of super-refraction.

A potentially more hazardous anomaly is that of *sub-refraction* which causes a reduced radar detection range. Figure 18.2.15 illustrates this effect. Sub-refraction occurs where there is a very rapid reduction of temperature with height and/or increase in humidity with height. These conditions can arise when a very cold air layer settles over a relatively warmer sea. This is less frequently encountered than super-refraction, probably due to the fact that those regions of the world giving rise to the effect are in or toward the polar regions. Detection ranges are reduced to line-of-sight distances or in more severe cases reduced below this where instances can arise of targets being visible to the eye and return no echo on the display. Large masses of ice drifting in relatively warm sea currents can produce the necessary temperature lapse rate: these being conditions where the navigator needs reliable radar returns. In addition such conditions commonly generate fog.

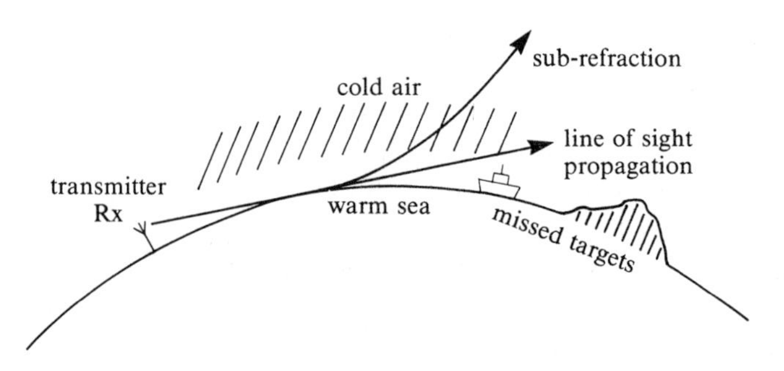

Figure 18.2.15 Simple illustration of sub-refraction.

18.3 Antennas and ancillary devices

18.3.1 Slotted waveguide antenna

Reference to Appendix C shows that an electromagnetic wave propagated down a length of waveguide in the dominant mode produces an RF current flow directly across the narrow 'b' face of the guide. At regular half-wavelength intervals along the guide, the phase of the current reverses. If a number of slots are cut across that face, the current will energize the slots which then act as an array of slot antennas producing a narrow radiated beam in the horizontal plane if the guide is horizontal. A horizontally polarized EM wave is launched, having in practice a beamwidth of the order of 1 degree in the horizontal plane for a guide length of about 2 metres.

Slots are cut every half-wavelength but since a phase reversal occurs in the RF current for each half-wavelength, the slots are slightly inclined alternately in opposite directions, producing the effect of phase reversal in the excitation of each slot – thus all slots radiate in phase. Figure 18.3.1 shows the basic arrangement, indicating that the antenna is fed at one end from a waveguide of similar cross-section. Such antennas used for 10cm radiation may be fed from a coaxial cable.

Due to the inclination of the slots in the array the electromagnetic emanation has a vertical component of radiation. This component can be eliminated by placing a number of short conductive plates, interspersed between the slots, which act as a virtual short circuit to the vertical radiation component, so producing an orthogonally polarized (horizontal) radiated beam. Manufacturers strive to

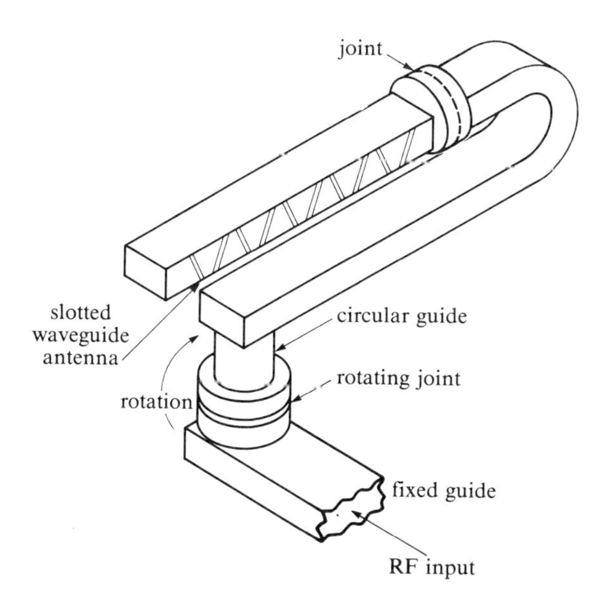

Figure 18.3.1 Showing basic arrangement for rotating slotted waveguide antenna. S band antennas may be end-fed by suitable low-loss coaxial cable via a rotating coaxial joint.

reduce the vertical component to effect minimum sidelobes. Typical horizontal and vertical beamwidths for practical antennas are given in Table 18.2.1.

18.3.1.1 Squint angle

Manufacturers of SWG antennas endeavour to ensure that RF power is radiated uniformly along the length of the guide to produce a good transformation (matching) with the atmosphere. Various methods are employed with different degrees of success. These are beyond the scope of this text. Slight non-uniformities in emission along the guide introduce a slewing of the main beam axis so that it is not truly perpendicular to the length of the antenna. The small angle so produced in the horizontal plane is called the *squint angle*. Evidently, if the antenna was laid at right angles to the fore-aft line of the ship, the main beam squint angle would introduce an unacceptable bearing error of all targets should the heading marker on the display be aligned to the antenna/ship geometry and not the main beam axis. Any existing squint is compensated during installation and check procedures which are outlined later in this section.

18.3.1.2 Radiation pattern

Figure 18.3.2 shows a radiation diagram typical of a slotted waveguide antenna. Note the well-defined horizontal beam and very low level of the sidelobes in that pattern. The sidelobes cannot entirely be eliminated, but to avoid signal returns via the sidelobes which will produce false displayed echoes they are maintained at about 23dB to 30dB down on the power in the main lobe. A radiation pattern in the vertical plane may also be obtained and would show a vertical beamwidth of between twenty-two to twenty-five degrees.

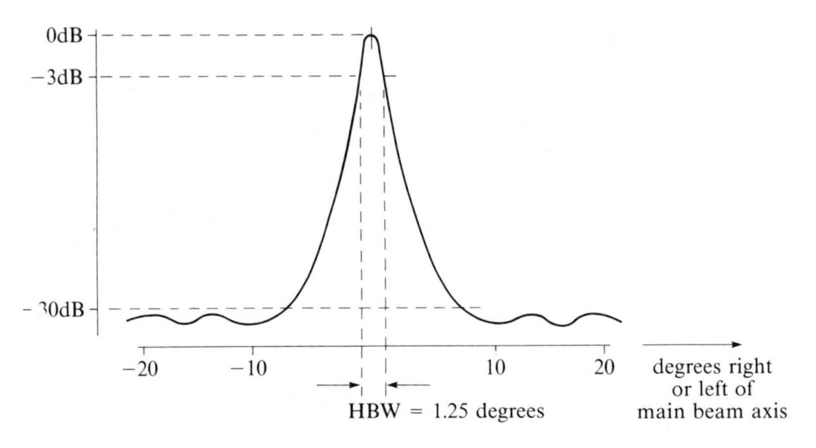

Figure 18.3.2 Horizontal radiation pattern typical of a six-foot slotted waveguide antenna X band.

18.3.1.3 Effective aperture of an antenna

The effective aperture or effective area of an antenna is a useful concept and can be defined as the area over which the antenna is assumed to collect energy from a distant transmission. If the slotted waveguide antenna is receiving a signal from a plane wavefront having a power density of W watts per metre squared and the antenna receives P_r watts, the antenna has an effective aperture of P_r/W metres squared. The gain of an antenna is related to the effective area and the wavelength of the wave being propagated as follows:

$$\text{Gain} = \frac{4\pi \times \text{Effective area}}{\lambda^2} \qquad \textbf{[18.12]}$$

The antenna is assumed to be lossless. In simple terms, expression [**18.12**] shows that, for a given effective area, high gain can be obtained by operating at very short wavelengths. Increasing the effective area for a given transmission frequency will also increase antenna gain. Any such changes bring about a narrowing of the beamwidth, thus increasing the directivity and gain. In the case of the slotted waveguide the aperture can be increased by increasing its length. It follows from this and the antenna reciprocity that the same effective aperture exists for both reception and transmission in a marine radar system.

18.3.2 *The scanner*

An exploded view of the single-purpose scanner assembly is depicted in Figure 18.3.3. The SWG antenna is protected by a rigid weatherproof housing, being aerodynamically shaped to reduce wind resistance and having a strong glass-fibre window running the entire length of the slot array. The window is transparent to electromagnetic waves and as a precautionary measure the general practice is never to paint its surface since even a thin layer of paint can considerably attenuate weak echo signals. For similar reasons antennas situated close to funnels can accumulate exhaust deposits and should regularly be inspected and if necessary, cleaned for optimum performance.

Turning power is derived from a relatively small electric motor which drives the antenna via a reduction gear of some kind, providing torque necessary for constant antenna speeds in winds of up to 100 knots. Various methods are adopted; toothed reinforced flexible belts and fibre gear trains being in common use, the main objective being to provide a positive drive having no play in the system which would adversely affect the display bearing accuracy. In Figure 18.3.3, those sections of the scanner assembly shown within the dotted line are generally housed in a substantial weatherproof housing, access being gained via removable plates in the casing. In the generalized scanner unit shown, a safety switch is provided for personnel working at or in the vicinity of the antenna site to disable the antenna drive supply and the transmitter. This prevents accidental starting or RF radiation occurring. As a cautionary note; the RF power distribution in the field directly in front of an antenna can be as high as 100W/m^2 for a magnetron nominal peak output of about 25 kW at very short distances (say 0.25m). This power density can cause permanent damage to the eyes and great care should be exercised in the proximity of

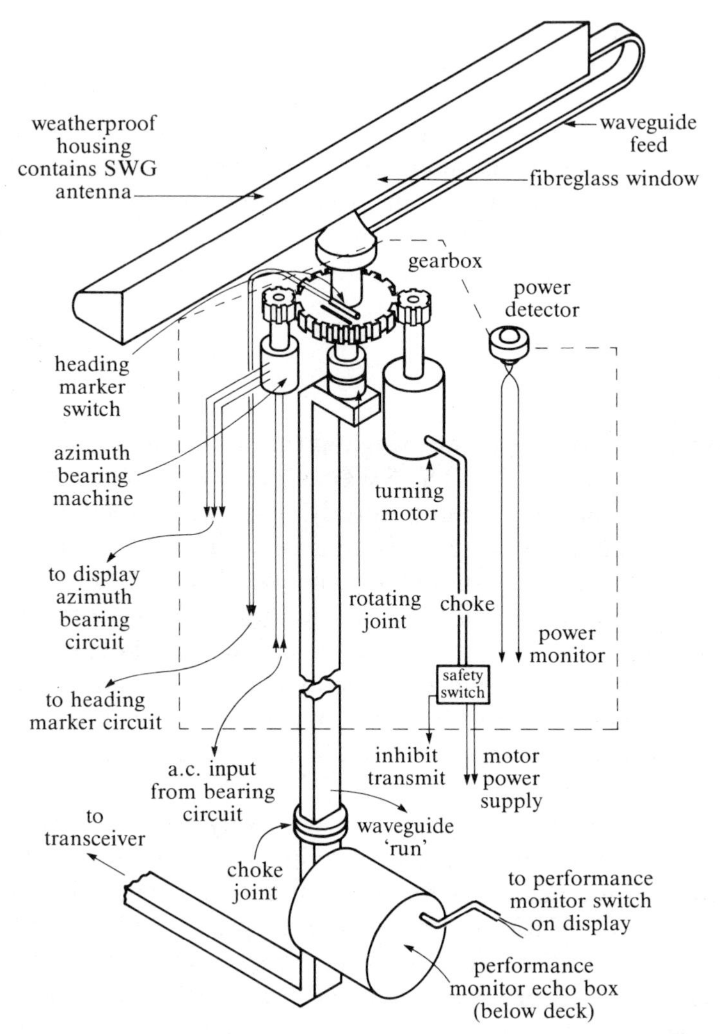

Figure 18.3.3 Exploded view of typical single-purpose scanner assembly. Units shown within dotted line are housed in weatherproof casing. Several choke joints and bends or corners can appear in the waveguide run, usually limited in length according to radar design.

stationary antennas. Some antenna sitings are close to exhausts which produce noxious fumes and personnel should take appropriate precautions to avoid breathing such fumes.

The drive housing has a utility power outlet separate from the isolated supply for the connection of low power instruments and electrical tools. Many installations have intercom facilities to the transceiver site.

18.3.2.1 The power monitor

A variety of circuits and devices are used to provide power monitoring facilities, the main purpose being to give at the display an indication that power is being radiated in the main beam. In the simplest types a neon tube situated in the scanner housing is exposed to the main beam with each revolution of the antenna. Ionisation of the tube varies with power irradiating it and the subsequent change of tube resistance can be used as a direct indication of the power in the beam. Usually the detector is incorporated in a circuit which drives an indicator lamp at the display, the lamp brightness indicating relative power being transmitted. More sensitive and accurate methods employ crystal detectors which can be calibrated to indicate fall-off in relative performance of the transmitter compared to some optimum level attained at installation of the equipment. Figure 18.3.4 shows the general arrangement.

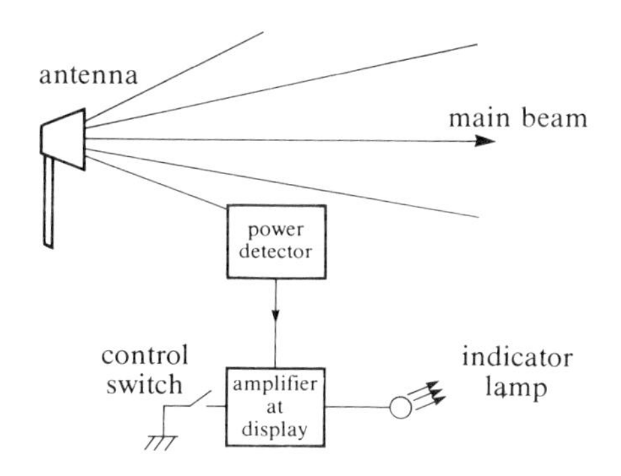

Figure 18.3.4 Basic power monitor arrangement.

18.3.2.2 The performance monitor

More accurately known as the 'overall performance monitor', its function is to provide an indication, when brought into circuit, of the overall performance of the transceiver unit. It comprises a resonant cavity, essentially a box-like structure made of low loss copper or brass and having a high Q factor and dimensions which allow it to resonate within the marine band. Radio frequency energy due to a transmitted pulse is fed into the cavity via an aperture causing the cavity to resonate; and since the cavity has a high Q factor, the resonant oscillations will persist in the cavity for some time after the pulse has terminated. During the resting period of the transmitter (receive period) the cavity couples energy *out of its aperture* forming a return echo signal for the duration of the cavity oscillations.

This signal is processed by the receiver and display circuits and appears on the display, brightening the timebase traces from the scan origin and extending for a measurable distance radially from that origin. Since the energy radiated from the performance monitor resonant cavity is directly proportional to the power injected into it (losses being constant) the length of the brightened trace is a measure of the overall transmitted and received power. Should the transmitter, the receiver or both be operating at less than some measured optimum value, then the length of the displayed trace will indicate a fall-off in the *overall* performance.

Performance monitor cavities are sometimes referred to as 'echo boxes'. Figure 18.3.3 shows one type as an integral part of the waveguide section close to the transceiver, being aperture coupled to the waveguide. Generally these devices have an internal conductive plate or paddle which is vibrated or rotated by a small external motor when the performance monitor is brought into use from the display by means of a biased switch. Releasing pressure on the switch will incapacitate the monitor. In action, the paddle sweeps the tuning range of the cavity at regular intervals through a resonant value corresponding to the magnetron frequency and permits the cavity to be used on a range of magnetron frequency values without the necessity of fine cavity tuning.

18.3.2.3 Plume and sunburst patterns

A monitor of the type shown in Figure 18.3.3 is coupled to the main waveguide run close to the transceiver. With the monitor switch depressed, the cavity re-radiates energy directly into the waveguide during the magnetron quiescent period and, due to the sweep action involved, produces a sunburst pattern as shown in Figure 18.3.5. The length of the major 'spokes' in the pattern occur when the echo box is resonating at magnetron frequency, measurement of this length from the scan origin is a direct representation of the power received.

Some installations use separate echo boxes installed above deck, usually sited on part of the ship structure in a shadow area, or an azimuthal sector offering echoes of low navigational priority. Energy enters the box via an aperture whenever the main beam of the antenna sweeps over it, and the box, when switched on, re-radiates energy as previously described. In this case the returned signal displays a plume on the indicator approximately twice the

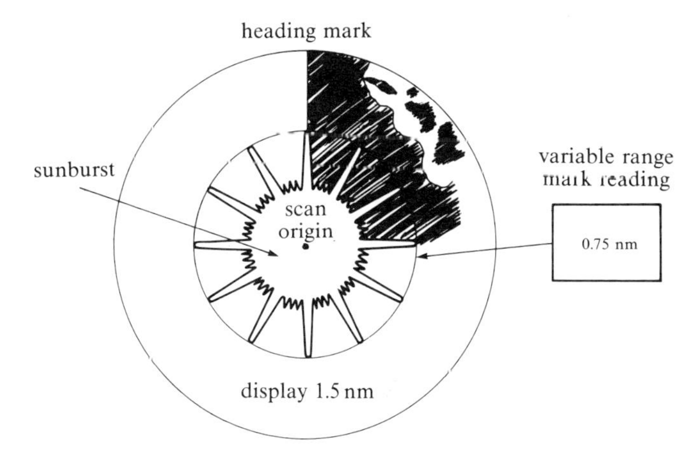

Figure 18.3.5 Sunburst PM pattern. The VRM is adjusted to the peaks of the displayed pattern.

width of the horizontal antenna beam. Figures 18.3.6 and 18.3.7 illustrate the arrangement and the displayed plume.

Whilst performance monitor echoes are being displayed, some target echoes may be obliterated, so use of the monitor is therefore limited to occasional performance checks and for optimum tuning procedures.

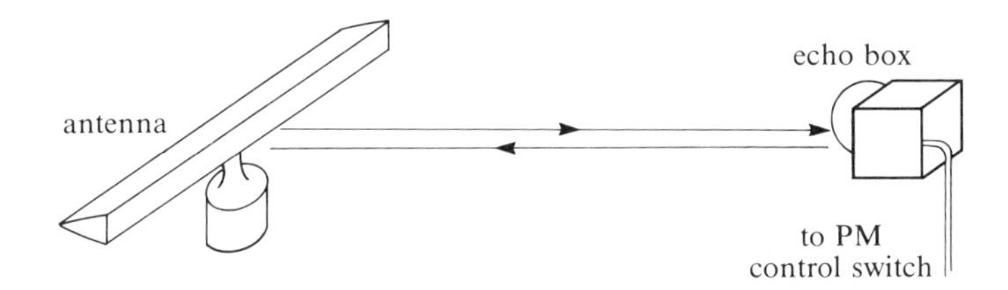

Figure 18.3.6 Showing arrangement of separate echo box.

18.3.2.4 Using the performance monitor

During use, controls such as the anti-sea clutter or anti-rain clutter control are set to a minimum so that they have no effect on the displayed PM plume, although this process is automatically carried out in modern sets when switched to performance monitor operation. Gain and brightness are adjusted as detailed in section 18.4. In some circumstances sea clutter may tend to obscure the PM display, in which case discrete use of the anti-clutter control may be necessary.

The same pulse length is chosen each time the performance

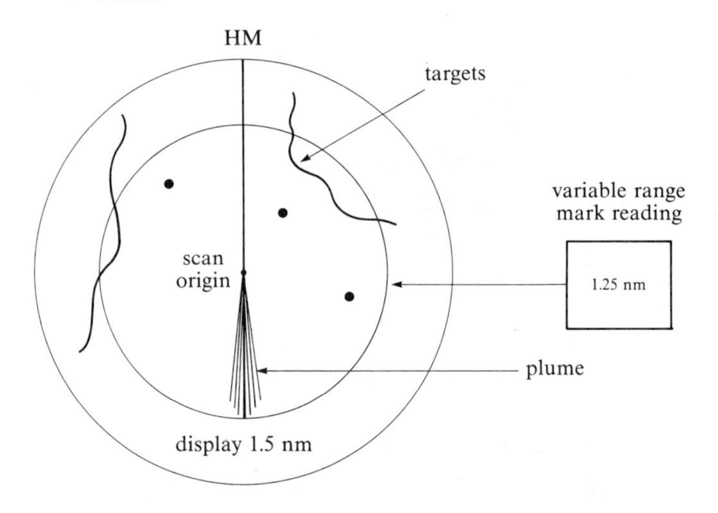

Figure 18.3.7 Plume pattern.

monitor is used. A low range is selected, such as the 1.5 nm range, and the calibrated VRM strobe is adjusted to lie as close as can be visually estimated to the extremities of the pattern as in Figures 18.3.5 and 18.3.7. Correct tuning of the radar receiver will produce optimum observed length. When the radar is first commissioned the extent of the PM pattern is entered in the radar log, and comparison can be made with this initial value obtained when the set is considered to be at its best performance level. The radar observer quickly becomes familiar with a particular radar set and any shrinkage of pattern in subsequent checks indicates reduction in overall performance of the set. This could be due to a variety of causes; for example ageing of the crystal, TR cell or magnetron, receiver detuned, water in the waveguide, etc. Output power reduction should be indicated by the power monitor and skilful use of both monitors can isolate the cause of performance reduction to a section of the installation. The PM is used when a radar set is first switched on and is particularly useful for spot checks on overall performance and tuning in conditions when there may be no visible returns on the display to immediately indicate the set is correctly tuned. Because of the proportionality between transmitted and returned power, a reduction of the performance monitor plume to, say, one half of its length would indicate a fall off of 3dB in overall signal performance. Under these circumstances the possibility of losing many of the weaker target returns is high.

If, in given circumstances, the radar set cannot be tuned to extend the plume to its optimum value, the reduction in dB can be found relative to the normalized optimum by:

$$10 \log \frac{R}{R_0} \text{ dB} \qquad [18.13]$$

where R_0 is the optimum range of the plume and R is the measured reduced value.

The performance monitor is generally of simple robust construction and gives little trouble; but should a reduced plume be displayed in the presence of good normal echo returns then the monitor is suspect. The more common causes of overall performance reduction are:

(1) incorrect receiver tuning
(2) defective receiver crystal(s)
(3) obscured antenna window
(4) ageing T/R cell (active types of cell)
(5) magnetron ageing
(6) water in waveguide
(7) damaged waveguide, loose choke joints.

Items (1) to (4) produce reduction in PM plume, leaving the power monitor indicator normal (depending on degree of obscuring in (3)). Items (5) and (6) reduce power monitor output and performance monitor plume; (6) can also produce 'hotspots' in the waveguide where a quantity of water might gather to waste transmitted energy as heat – received signals are severely attenuated. Item (7) can produce flashover in the guide, reducing output power as indicated by the power monitor: it can also cause erratic display action. Loose joints can produce intermittent effects, usually most noticeable on received signals and with consequent reduction in PM plume length.

18.3.2.5 Waveguide run and rotating choke joint

Sections of waveguide connect the transceiver below deck to the scanner unit. The sections will necessarily contain a variety of bends and elbows and are joined by machined sections known as *choke joints*. The joints are constructed so that they electrically reflect an impedance which makes the inner waveguide surface appear to be electrically continuous despite the obvious mechanical discontinuities introduced when sections are joined. Leakage of transmitted power is also prevented by the choke action. Each joined surface is sealed against ingress of moisture by a neoprene ring.

Where vibration is a problem sections of flexible waveguide may be inserted comprising essentially a continuous choke joined helix of thin copper strip weather sealed by an exterior rubber sheath. A maximum value for waveguide run is invariably stipulated (typically 20m).

The rotating choke joint (Figure 18.3.3), is arranged so that the rectangular waveguide section meets a transformer producing a transition without loss or reflection from fixed rectangular to circular and back to rectangular waveguide. The circular section supports a concentric field pattern and the tube wall can be rotated without disturbing the pattern. The rotating choke joint splits the short circular waveguide section and provides electrical continuity without RF leakage, at the same time permitting the rotating antenna section of the guide to rotate coaxially with the fixed section.

18.3.2.6 Heading mark switch

Arrangements vary in different equipments. Figure 18.3.3 shows a magnetically-operated reed switch, the contacts of which are operated whenever a small permanent magnet mounted on the antenna drive passes over the switch. The switch is invariably mounted on a small adjustable base plate carrying a scale graduated in degrees of azimuth. An overall possible adjustment of ±5° is adequate, to a required accuracy of at least 0.5°. Switch action and magnet are mutually arranged to coincide when the main lobe of the antenna beam is pointing along the ship's heading. This action produces a bright radial line at the display with own ship at the scan origin by brightening a few consecutive traces, to show the ship's heading on the indicator. Type tested radars display heading lines whose thickness is 0.5° or less at the tube periphery. The maximum error permitted of the heading marker line is ±1° and most installations have an accuracy better than this.

18.3.2.7 Adjustment of heading mark

An acceptable method is to place the display in an unstabilized condition (relative motion ship's head up), centre the scan origin to lie in the exact centre of the indicator – this being facilitated by the mechanical bearing cursor. Where appropriate, the PPI display is rotated by the display controls to align the heading mark with zero degrees on the bearing scale. A small target is selected on which visual bearings may be taken, its range being such that the radar will display the target echo as a distinct paint lying near the periphery of the PPI on one of the shorter displayed ranges (say 1.5 nm). Own-ship is then aligned and a visual bearing on the target is taken, preferably from the antenna site. When the ship's head and the visual bearing coincide, any error is noted in the heading marker bearing at the PPI which should indicate zero degrees. Error is compensated by adjusting the heading mark contacts in the scanner housing and the visual and radar bearings once more checked. In some equipment the heading marker can be adjusted by rotating the stator housing of the synchro transmitter.

It is advisable to double check bearings of easily distinguishable radar and visual targets on other bearings relative to the ship's head in order to average bearing errors. An ideal situation may arise when own-ship is swinging at anchor in an area relatively uncluttered by radar returns and which affords adequate targets with slowly changing bearings and aspects. The procedure of alignment automatically compensates for any squint angle introduced by the antenna; although some antennas carry a sighting plate which is basically a groove used to sight along the fore-aft line of the vessel, these should not be regarded as accurate devices.

Heading marker accuracy should be checked at regular intervals and should adjustment arise but not immediately be possible, a notice advising radar users of any error should be shown at the display and an entry made in the radar log. Contacts should be adjusted when feasible. Modern magnetrons are pre-plumbed devices and are manufactured tuned to a nominal frequency within the radar band. Slight changes in the transmission frequency which can, for example, occur when a magnetron is replaced, can alter the squint angle of a slotted waveguide antenna. It is advisable to carry out heading mark check procedure following such replacement. Coincidence of the heading mark and the electronic bearing marker in the unstabilized condition should also periodically be ascertained for correct alignment.

18.3.2.8 Azimuth bearing transmitter and receiver

Figure 18.3.3 shows a small machine driven via a gear train from the antenna drive unit. The machine is the electrical *bearing transmitter* and generally is either a *synchro* or *resolver*. To the radar user both machines produce the same effect at the display, transmitting the bearing information from the antenna to the display but they differ electrically. The synchro principle is shown in Figure 18.3.8 where the bearing transmitter is shown as a machine having three stator windings arranged symmetrically. Inside these a rotor winding on a ferromagnetic core is rotated by a drive shaft through a step-up gear train from the antenna drive. The rotor is energized via slip rings from an a.c. supply and induces alternating EMFs into the stator windings which are themselves arranged so that the relative induced EMFs are 120 degrees out of phase with each other thus forming a simple three-phase generator at the transmitter end.

Each transmitter stator winding is connected to a corresponding stator winding at the distant *bearing receiver* which is an electrically similar machine situated at the display unit. The rotor of the receiver is energized from the same a.c. as that of the transmitter. EMFs induced in the stator windings of the transmitter, due to the rotating alternating field of its rotor, now cause currents to flow in the stator windings at the receiver and produce an alternating receiver field rotating in synchronism with the transmitter field. Hence the name 'synchro'. The receiver rotor R_2 (Figure 18.3.8), being fed by the same a.c. supply as the transmitter rotor R_1, will automatically align itself to the stator field and rotate with it. The drive shaft of the rotor can be used to drive via a reduction gearbox, the rotating scan coils of the display cathode ray tube (section 18.4).

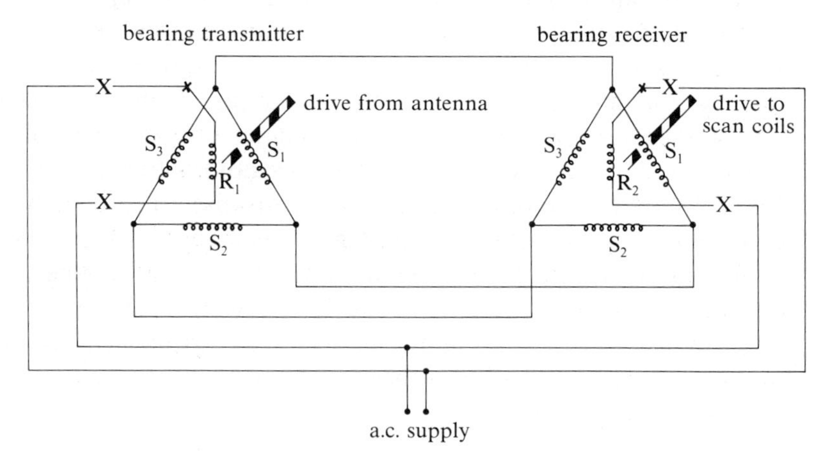

Figure 18.3.8 A basic synchro system: XX denotes slip rings; S_1, S_2, S_3, stator windings; R_1, R_2, rotor windings.

It is usual to have some form of step-up gearing driving the transmitter synchro and an appropriate reduction (step-down) gearing at the scan coil mechanism to improve machine resolution and provide the necessary torque. When the set is switched off, synchronism between transmitter and receiver is lost. On restart the displayed heading line and picture may assume one of 'N' possible positions on the screen, where N is the gear ratio involved. A self-aligning circuit or manual picture position control is provided. In the latter case rotation of the control causes the displayed picture and heading mark to rotate so that the heading marker can be brought to zero degrees for unstabilized relative motion display. Once aligned in this manner the heading marker will follow correctly the course of own-ship when the display is compass stabilized. Other methods of bearing transmission and reception are dealt with in section 18.5.

18.3.3 *Antenna siting*

The antenna should be placed in a position which avoids or minimizes obstacles presented by the ship's structure in the path of the radiated beam. Such obstacles produce shadow sectors and blind areas which can hide targets of navigational importance and give rise to false echoes appearing on the PPI. Antenna height above sea level is also of importance, since it has an effect on the *radar horizon*; in principle, radar range improves with height. A practical limit is reached when the incident angle of the vertical beam lobe extremities becomes sufficiently acute to return strong echoes from the sea surface which increases sea clutter at the display and can obscure targets at close range. There is also a practical limitation placed on the length of the waveguide run (18.3.2.5). All installations need to have ready access for safe maintenance and should not be placed where they adversely influence magnetic compasses.

18.3.3.1 Antenna height; target height

If the electromagnetic energy radiated from a radar antenna travelled in a straight line, ray optics theory could be applied to the main beam geometry and the horizon 'seen' from the antenna would extend to points where the beam was tangential to the earth's surface, which is assumed to be spherical. In fact, such a glancing wave path causes an effect known as *diffraction* by introducing slight differences in the velocity components at different parts of the wavefront. Diffraction causes the path of the wave to follow the earth's curvature for a distance determined by such factors as frequency, surface conductivity and atmospheric permittivity. The diffraction effect is greater for lower propagation frequencies and ten-centimetre wavelengths will bend to follow the earth's surface for a greater distance than will three-centimetre wavelengths, other factors being equal. The overall effect is illustrated by Figure 18.3.9(a) and Figure 18.3.9(b) showing that increased antenna

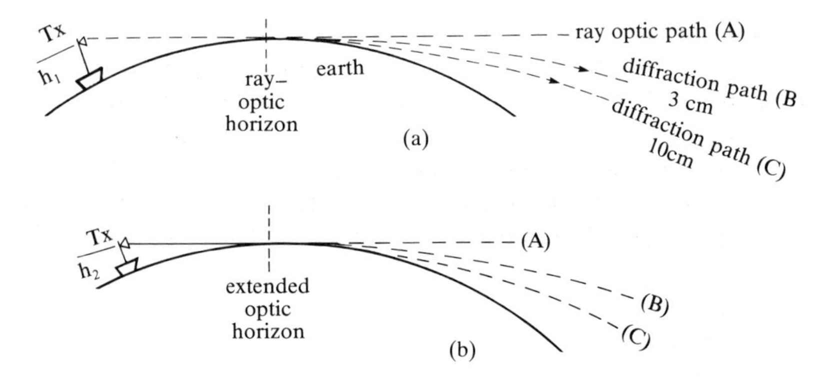

Figure 18.3.9

height above sea level will extend the radar horizon. Evidently, the height of a target above sea level will also determine whether the target is irradiated by energy following paths *B* or *C* (Figure 18.3.9) and therefore has an influence on the detectable range of a target.

Appendix B contains a simple approach to determine the effective range between a radar transmitter and target assuming ray-optic techniques are applicable. If a target return is detectable at a radar set which has its antenna at height h_1, with the distant target at height h_2, the approximate optical range is given by:

$$\text{Range} = \sqrt{2R}\,(\sqrt{h_1} + \sqrt{h_2})\ \text{km} \qquad \textbf{[18.14]}$$

where *R* is the radius of the earth, considered to be 6370km. Values of *h* are also in km. From this it is seen that the range separating a target of height 300m and an antenna of height 15m can be:

$$\sqrt{2} \times 6370\,(\sqrt{0.3} + \sqrt{0.015})\ \text{km}$$

$$= 75.5\ \text{km or } 40.37\ \text{nm} \qquad \textbf{[18.15]}$$

Due to diffraction, the optical range is increased as if the earth had a larger radius. The result of [**18.15**] is then increased by the *earth radius factor* which, for a standard atmosphere, is 4/3. Therefore, in a standard atmosphere, the range will be approximately:

$$= 40.37 \times 4/3\ \text{nm} = 53\ \text{nm} \qquad \textbf{[18.16]}$$

The theoretical figure is modified by the characteristics of the target and the actual propagation conditions which may exist over the propagation path.

Very small targets close to the ship which might otherwise be easily discriminated may not adequately be irradiated by the main beam if the antenna height is too great; this and the increased sea clutter return can cause such target return to be lost on the display. Selection of antenna height is a compromise between these conflicting factors.

18.3.3.2 Shadow sectors and blind sectors

Any part of the ship's structure which forms an obstruction to the main beam of the radar antenna can cause a *shadow sector* or *shadow zone* on the PPI. A plan view of such zones might appear as in Figure 18.3.10 but it should not be forgotten, particularly in larger vessels, that extended shadow sectors exist in the vertical plane as depicted by Figure 18.3.11. Small but important targets can lie hidden in those sectors the angular width of which depends largely on the aspect of the obstruction relative to the beam.

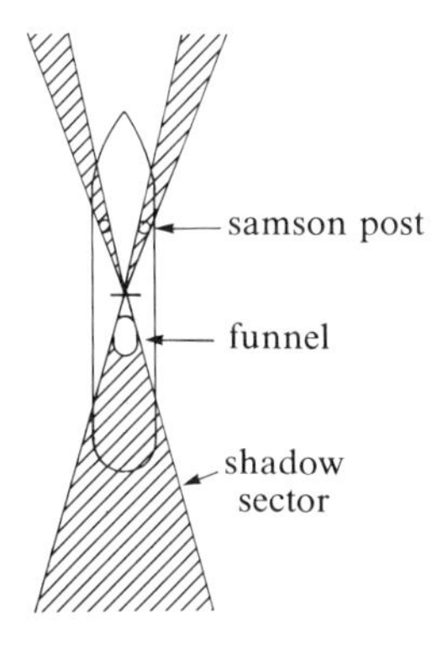

Figure 18.3.10 Horizontal shadow sectors.

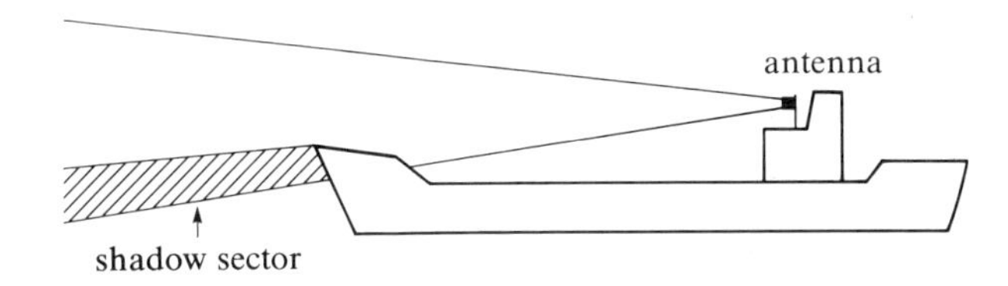

Figure 18.3.11 Vertical shadow sector.

An antenna would be ideally situated if it lay on the centre line of the vessel at optimum height for the length of the waveguide run and simultaneously avoided all obstructions in the beam path. Usually the best compromise is reached to reduce the shadow sectors to the minimum number and width, and if possible to place these where targets of high priority are not likely to appear. In practice, some diffraction of the electromagnetic wave occurs and the sectors are not as well defined as shown in Figures 18.3.10 and 18.3.11; but typically there is a core of the shadow sector within which there exists a total *blind sector*.

18.3.3.3 Shadow sector records

The angular width and the bearings of all shadow sectors are found from observation of the PPI, and details concerning these are kept at the display (or displays) of each radar set. Installations of more than one set have the separate antennas erected at different heights and each display may show different shadow sector size and bearing. The separate details should be carried near the appropriate display. Shadow sectors may be measured by selecting the echo from a small fixed object, such as a buoy, adjusting the display for optimum performance in minimum clutter conditions and observing the bearings on which the echo enters the shadow regions as own-ship circles slowly at a short distance from the buoy. The sectors may also be estimated by observation of the PPI in conditions of weak sea clutter and preferably in the absence of all other targets (open sea): such sectors then show as dark areas free of light sea clutter. Heavy clutter tends to mask the edges of the zone and zonal estimates should not be made in such conditions. Figure 18.3.12 illustrates the appearance of shadow zones in conditions of light clutter. It should be apparent from Figures 18.3.10 and 18.3.11 that differences in trim of the vessel can alter the sectors, as can the carriage of large items of deck cargo.

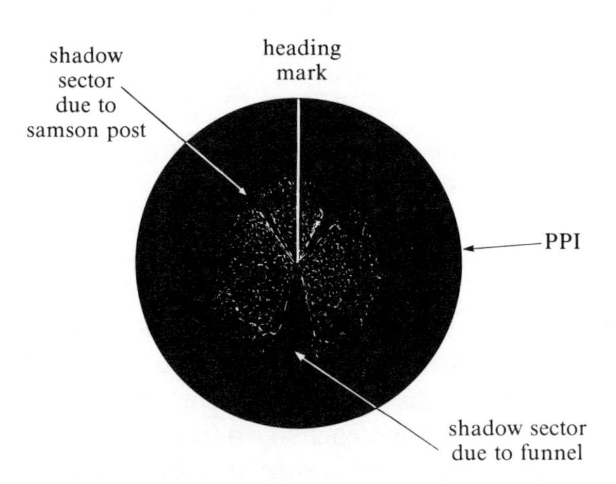

Figure 18.3.12 Observation of shadow zones in sea clutter.

18.3.3.4 False echoes (due to ship obstruction)

False echoes can appear on the display due to reflection of echo energy from an obstruction on the ship which is causing a shadow sector. Usually such false echoes are manifest in the shadow sector and appear at virtually the same range as the true target echo. They may also appear at slightly reduced brilliance in radar sets which display raw video signals. Figure 18.3.13 illustrates the situation and the appearance of true and false echo on the display for one such condition. In general the echoes are produced by fairly large echoing areas at close range; a change in course of own-ship causing the false echo to disappear or a new false echo from a different target to appear in the blind sector.

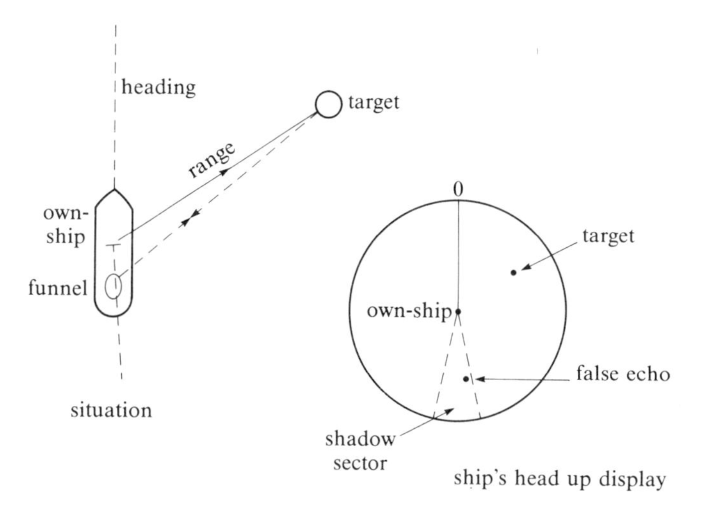

Figure 18.3.13 Showing a falsely displayed echo due to indirect transmission and reception via ship's funnel.

18.3.3.5 False echoes between own-ship and target

Such echoes are produced due to the transmitted energy bouncing between a target at close range and own-ship. The effect is to produce from one transmitted pulse a series of echoes which appear on the display equally spaced at multiples of the target range and extending beyond the true target echo with gradually diminishing paint. Figure 18.3.14 shows this condition.

False echoes appear on the same bearing as the true echo and are usually easily recognized by the radar observer due to their regular pattern and distinctive characteristic. As own-ship changes attitude relative to these targets the echoes tend to change or disappear.

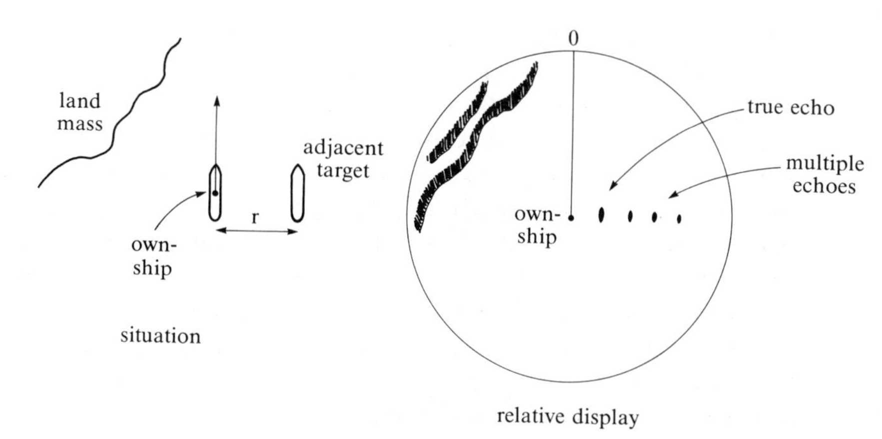

Figure 18.3.14 Showing typical appearance of echoes between own-ship and target at multiples of r on same bearing.

18.3.3.6 Sidelobe echoes

Modern slotted waveguide antennas produce very low output power in the sidelobes compared to the main lobe. For targets which are close to own-ship, however, the energy can be transmitted and received not only via the main lobe but also via the sidelobes. The effect is to severely degrade bearing discrimination of adjacent targets and to produce on the display bright arcs of paint at the target range and close to the scan origin. Sidelobe echoes diminish rapidly with range. Figure 18.3.15 illustrates the effect observed on the display.

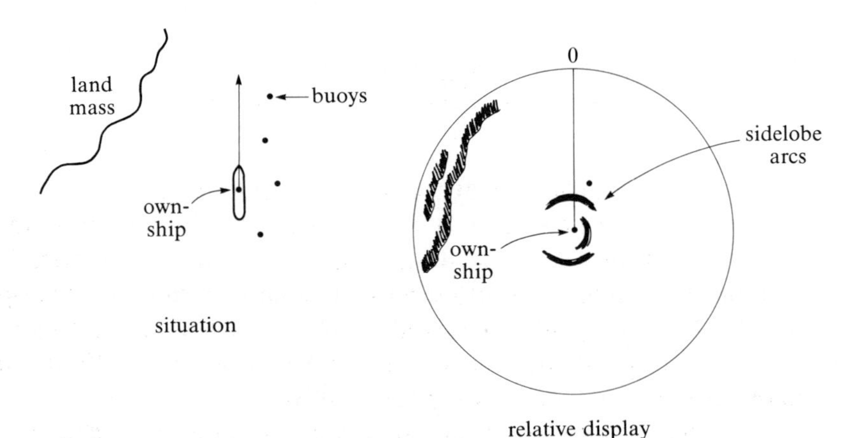

Figure 18.3.15 Showing arcs painted close to scan origin due to energy transmitted and received via main and sidelobes.

8.3.4 Circular polarized transmission

In certain operating conditions, particularly in the presence of rain, the clutter effect produced on the PPI by returned rain echoes can be minimized by transmitting and receiving a circularly polarized electromagnetic wave.

Precipitation echoes are influenced by such factors as transmission frequency, wave polarization, transmitted power and pulse width. At frequencies up to X band the degree of signal attenuation in rain and snow is low but the overall effect of the random returned echoes from precipitation can clutter the PPI with unwanted returned echo energy which can severely limit the radar's ability to separate low energy target echoes lying in the clutter. Conventional methods used to minimize the so-called *rain clutter* resort to differentiation of the video pulses, a technique outlined in section 18.4.3.3.

18.3.4.1 Wavelength

If the wavelength of the electromagnetic wave is large compared to the effective diameter of the reflecting particle (raindrop), the radar cross-section is proportional to $1/\lambda^4$ or proportional to f^4 where λ is the wavelength and f the frequency of propagation respectively. Reduction of frequency (increase of wavelength) significantly reduces echo energy returned from precipitation and where such echoes occur the radar wavelength may be switched from 3cm to 10cm to provide enhanced target returns in comparison to the rain clutter returns. Such an arrangement requires the installation of a dual radar system comprising a ten-centimetre transceiver and antenna and a separate three-centimetre transceiver and antenna, either of which may be switched to a common indicator or have separate displays. Figure 18.3.9 indicates that ten-centimetre operation enhances the radar horizon for a given target.

18.3.4.2 Effect of polarization on echo return

Raindrops can be considered as small spherical targets whilst ships and land masses are relatively complex reflecting targets. Various reflecting surfaces modify the polarization of returned echo wavefronts in different ways. Horizontal polarization is generally used in civil marine systems giving good overall reflection from targets lying in the horizontal plane.

Use of a circularly polarized wave will improve the ratio between target signal and clutter because a spherical target will reflect a circular polarized echo signal but *reverse* the wavefront rotation relative to the duplexed antenna. The antenna in its receive mode will then reject such echoes. A ship target, on the other hand, reflects energy which, statistically, is evenly divided in the two senses of rotation so that the antenna always receives some portion of the complex target echo energy even though the signal power reaching the receiver is reduced compared to a plane polarized echo. At the receiver the effect is to improve the ratio of target echo to clutter echo.

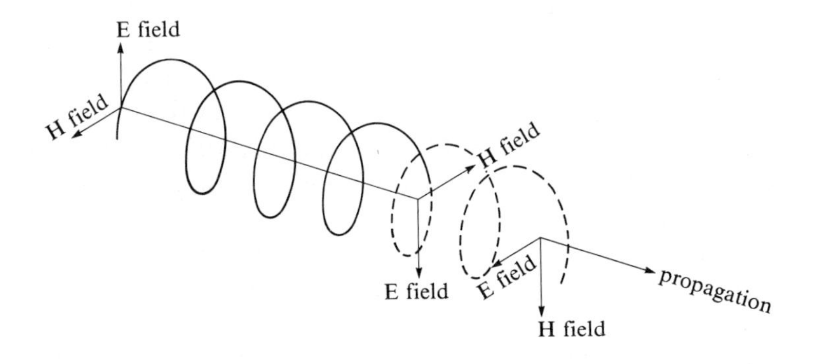

Figure 18.3.16 Concept of circular polarization. As the EM wave travels away from the antenna towards the target the wavefront not only alternates but rotates its direction of polarization.

In practical circumstances where very heavy rainstorms are encountered, the raindrops are not truly spherical and some rain clutter will be experienced but the circularly polarized radar will still give better clutter rejection than a horizontally polarized system. The concept of a circular polarized wave is shown in Figure 18.3.16, generated by combining a horizontally polarized wave with a vertically polarized wave of the same frequency and amplitude but having a phase difference of $\pi/2$ radians. The resultant wavefront has a direction of polarization which rotates as it travels along the line of propagation, in the manner of a helix. One method of producing a circular polarized wave from a horizontally polarized slotted waveguide antenna is to place a polarizing filter or grid in the path of the wave emanating from the antenna. The grid comprises a large number of thin conductive plates placed at 45 degrees to the waveguide and covering the entire length of the antenna window. The basic idea is shown in Figure 18.3.17. The action of the RF EMFs induced in the plates produces circulating currents which

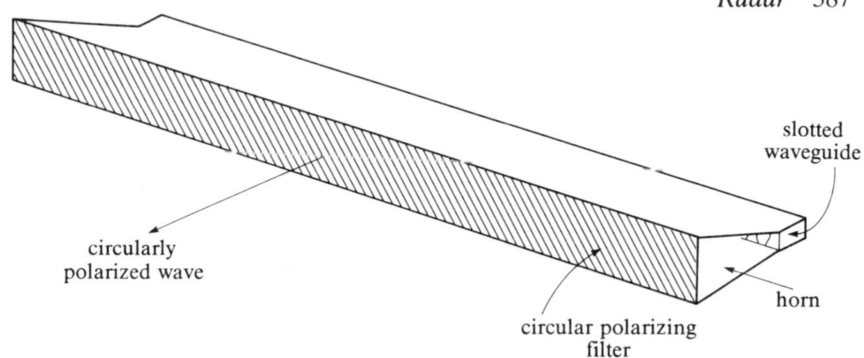

Figure 18.3.17 Showing basic arrangement of three-centimetre horizontally polarized slotted waveguide fitted with circular polarizer.

generate both a vertical and horizontal radiated component. The plate inclination ensures that equal amplitudes are radiated so producing a circularly polarized electromagnetic wave. Such an arrangement will accept return echoes which are similarly polarized, i.e. have the correct rotation.

A marine radar can optionally be provided with either conventional plane polarized three-centimetre emission or by operating a switch at the display can change to circular polarization mode. The arrangement requires that two slotted waveguide antennas are housed back-to-back inside the same weather housing; one of these is fitted with the circular polarizer. A single waveguide leads through the rotating joint to a solenoid or motor-operated waveguide changeover switch in the antenna housing. The RF energy can then be directed via the switch to either antenna at will. Since the two antennas point in opposite directions it is necessary to effect a changeover in heading marker contacts at the time antenna switching occurs. The popular option in today's installations is that of dual three-centimetre and ten-centimetre sets to provide the advantages of each system, the single system with circular polarized capability is relatively rare.

18.4 Radar configurations

18.4.1 Noise

Noise is a term used for any impulsive voltage source which is undesirable in any circuit and which cannot be separated from a desired signal. In marine radar sets, sources of noise due to natural phenomena have negligible effect and emphasis is placed on noise voltage sources *within* the electronic equipment which can seriously limit the overall performance of the receiver, namely *thermal noise*. Thermal energy absorbed by conductors, semiconductors and resistive devices produces random, free electron motion which introduces minute voltage fluctuations within the circuit. At normal temperatures (nominally 20°C) noise is always being generated. The average power due to noise is proportional to absolute temperature and bandwidth, available noise power being:

$$P_n = kTB \text{ watts}$$

P_n is the average available noise power in watts
T is the conductor temperature in degrees kelvin
B is the bandwidth of noise spectrum in hertz
k is Boltzmann's constant $= 1.38 \times 10^{-23}$ joules per degree kelvin.

It can be shown that the average noise voltage squared is given by:

$$v_n^2 = 4\,kTBR \qquad \textbf{[18.17]}$$

in any circuit of resistance R ohms. The significance of this expression in a radar receiver lies in the fact that the average noise voltage is proportional to the square root of bandwidth; wideband circuits admit more noise than narrow band circuits, other factors being equal and reference to this fact is made later in the chapter.

A radar contains many electronic devices each contributing noise, some of which will ultimately appear at the display cathode ray tube together with the desired target video information. Amplitude versus time-related random noise in which target echoes are embedded is shown in Figure 18.4.1. If noise amplitude approaches that of the signal it is difficult to separate the two and display the target on the screen. Excessive noise reaching the PPI causes a random brightening or speckling effect over the entire screen area hiding the paint from small targets and reducing the overall contrast between the normally highlighted target paint and the darker background. Normal and abnormal noise appearing on the PPI is shown in Figure 18.4.2. Maladjustment of the receiver gain control shown in Figure 18.4.3 can produce excessive displayed noise and is to be avoided.

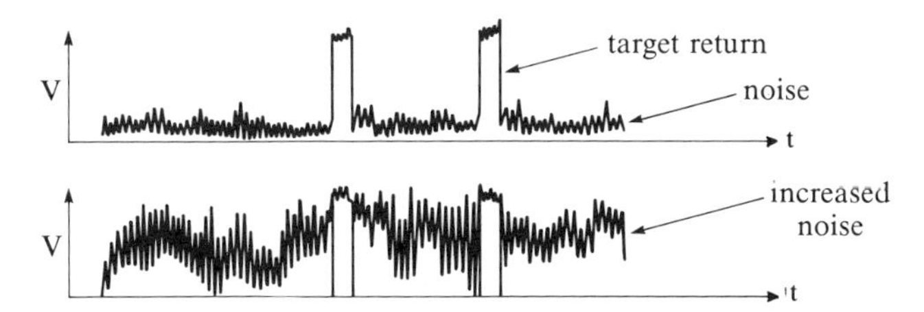

Figure 18.4.1 (a) Shows signal easily distinguishable from noise voltage; (b) signal hidden in noise. Any attempt to display the signal will also display the noise.

18.4.1.1 Noise factor and noise figure

The presence of noise in the radar receiver is an important limiting feature in the performance of the receiver since a point is eventually reached where target signals of very low level will be lost in the noise. Any attempt to increase the amplification of the signal (by increasing the receiver gain) will equally amplify noise – particularly that generated at the first stages of the receiver. These first stages are known as the 'front end' and comprise a special low-noise amplifier immediately following a crystal mixer. This is shown as the *head amplifier* in Figure 18.4.3.

The ratio of signal input power to noise power at the receiver input is

$$\frac{P_{Si}}{P_{Ni}}$$

and at the output the ratio is $\dfrac{P_{So}}{P_{No}}$

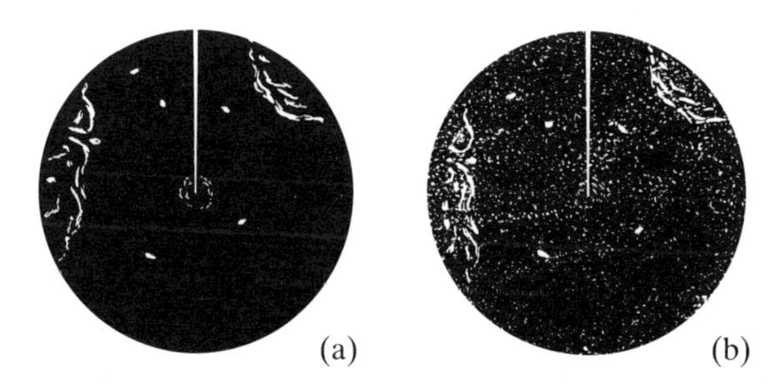

Figure 18.4.2 (a) PPI picture with noise barely visible. Receiver gain adjusted so that noise is just visible then backed off; (b) excessive noise in picture. Gain too high.

The ratio of these two values is known as the *noise factor F*:

$$F = \frac{P_{Si}}{P_{Ni}} \times \frac{P_{No}}{P_{So}} \qquad \textbf{[18.18]}$$

Noise factor *F* is often quoted in dB as the *noise figure*:

$$F\text{dB} = 10 \log_{10} F \qquad \textbf{[18.19]}$$

and is a value often given by manufacturers.

For a three-centimetre radar the manufacturer may quote a noise figure of 9.5dB nominal for the receiver overall which includes the IF and video stages, generally using selected mixer crystals to provide the best performance figures. A noise figure of 9.5dB equates to a noise factor of 8.9 indicating as a figure of merit that the signal-to-noise input ratio is nearly nine times greater than the signal-to-noise output ratio. Modern receivers have overall noise figures between about 9.5dB to better than 11dB.

18.4.2 *The radio frequency head (RF head), three-centimetre radar*

The RF head comprises sections of waveguide and stripline components together with the crystal mixer, local oscillator and transmit receive cell. It will also carry ancillary passive devices such as attenuators. They are arranged in one unit, the T/R cell and receiver crystal being replaceable whilst the modern local oscillator is an integral part of the head. The head can be considered to have two branches or 'arms', one connecting the antenna to the magnetron transmitter during transmission, the other connecting the antenna to the receiver at all other times. It is sometimes known as the RF *duplexing* head since it allows a single antenna to perform the dual function of transmission and reception.

18.4.2.1 T/R cell

The cell prevents magnetron high-level power from entering the receiver arm and protects the sensitive receiver crystals from damage when the transmitted pulse is present. During magnetron rest periods the cell allows low-level received signal power to reach the receiver crystals. The cell is a chamber filled with inert gas and sealed at each end by a window transparent to RF. The chamber acts as a wideband 1:1 transformer in the receiver arm (see Figure 18.4.3) and admits relatively weak received signals to the receiver section. When the magnetron fires, the gas very rapidly ionizes producing a switching action which directs the RF power to the antenna and away from the receiver. At magnetron switch-off, the cell very rapidly de-ionizes to allow received signals at short ranges to be processed by the receiver. Note that any excessive delay in cell de-ionizing will cause loss of returned target echoes from very short ranges since the cell will still be blocking the receive arm. This act will increase the minimum detectable range of targets.

Modern T/R cells are of the passive type utilizing a radioactive isotope of tritium as the gas. Since the T/R cell is ionized by the transmitter pulse it is inevitable that a spike of energy will leak through the T/R cell and possibly damage the crystals. Excessive RF energy reaching the receiver amplifiers can also cause receiver paralysis increasing the receiver recovery time beyond acceptable limits. An arrangement of a T/R limiter combination is used where the T/R cell is supported by a varactor diode limiter. The varactor reduces leakage power to a very low acceptable level. Some types of T/R cell use a 'keep-alive' electrode maintained at about 1 kV to place the gas on the verge of ionization, application of the transmitted pulse then rapidly ionizes the gas. The passive T/R limiter is the more reliable arrangement, has a longer life than the keep-alive method and introduces less signal degradation with age.

8.4.3 *The mixer*

Receiver crystals are in fact semiconductor diodes capable of rectifying radio frequency currents in the SHF band and are arranged in a *mixer* circuit to operate in conjunction with the *local oscillator*. The arrangement converts all incoming RF signals at super-high frequency down to a much lower radio frequency known as the intermediate frequency or IF where the signal can be amplified and processed using conventional RF circuit design. The conversion is achieved by superheterodyning the received signals with a locally generated stable signal produced by a local oscillator circuit operating at a frequency (f_t + IF) or (f_t − IF) where f_t is the magnetron transmission frequency.

A Gunn diode is the preferred choice of local oscillator since it produces adequate output power (about 5 mW), operates at low voltage levels and has lower noise output than its predecessor, the klystron. The local oscillator is tuneable over a range of a few megahertz by means of an electronic tuning control (Figure 18.4.3). A balanced mixer circuit tends to be used rather than single mixer systems due to inherent noise rejection characteristics, the crystals and the local oscillator commonly being deployed in a hybrid ring arrangement. Strip line techniques have reduced this assembly to a fairly simple and economical production exercise. Appendix C (Figures C16 and C17) contain further information on stripline and hybrid rings. The target signals now converted to the IF are first amplified by the *head amplifier* which is a cascode arrangement providing immediate amplification whilst minimizing noise in the stage. Quite often the head amplifier is situated at the RF head and connects to the rest of the IF amplifier via a short length of coaxial cable. The advent of compact printed circuit and integrated circuit applications has led to the entire IF amplifier strip being placed at the RF head.

Because the crystal diodes have a non-linear characteristic the effect of applying the target return signal at f_t and the local oscillator signal at $(f_t + \text{IF})$ is to produce sum and difference frequencies at the mixer output. Thus the output contains $f_t + (f_t \pm \text{IF})$ and $f_t - (f_t \pm \text{IF})$ together with the original frequencies. For example, if $f_t = 9400$ MHz and f oscillator equals 9460 MHz OR 9340 MHz, the frequencies produced are 18 860 MHz OR 18 740 MHz (being the sum values) together with 60 MHz, the difference frequency. The sum and original frequency components are decoupled at the mixer output and the difference frequency is the desired IF. Values of IF lie typically between 45 and 60 MHz. Raw echo signals converted to the IF will retain the same pulse length as that of the transmitted and received SHF signal.

18.4.3.1 Local oscillator tuning

Although the Gunn diode is structured to operate at one particular frequency, it can be 'tuned' by placing it in a resonant cavity and then tuning the cavity which is an integral part of the local oscillator. Cavities are tuned to a particular resonant value by means of a mechanical screw projecting into the cavity and fine electronic tuning is effected by use of a variable capacitance diode placed in the cavity and having its capacitance varied by means of bias applied from the manual tuning control at the display position. The fine tuning facility enables the user to produce the best possible display of target returns. Installations have a tuning override control, sometimes known as the *local* control, which enables the receiver to be tuned at the transceiver, usually some distance from the display. A small microammeter is provided at the transceiver which monitors the crystal current and is a useful facility to indicate correct local oscillator tuning.

18.4.3.2 The IF amplifier

The purpose of the IF amplifier is to amplify the low-level IF signals produced by the mixer to a level sufficient to operate the video detector (usually a few volts). At the same time the noise must be kept to an acceptable level. A gain of the order of 137dB will occur between the head amplifier and the video detector. The first section (the head amplifier) is generally a cascode arrangement providing about 20dB gain whilst minimizing noise, and succeeding IF amplifier stages progressively increase the signal level.

A suppression pulse derived from the magnetron trigger is applied to bias off the head amplifier during the transmission of a pulse. The desensitizing action ensures that the transmitter pulse leakage is not handed on to the remaining IF stages where overloading could occur, particularly in the latter stages, resulting in paralysis of the circuit and possible component damage. Paralysis occurs where charge storage in semiconductors renders them inoperative as amplifiers for a period during which the charge leaks away. During that time no signals can be processed and the minimum detectable range is considerably extended. Figure 18.4.3 indicates the source of the pulse and Figure 18.4.4 shows its application.

18.4.3.3 Swept gain

A manually controlled swept gain circuit operates on the head amplifier of Figure 18.4.3. Waveforms demonstrating the principle of operation are shown in Figure 18.4.4. The purpose of the circuit is to suppress signal return from the sea (sea clutter), which tends to obscure the centre of the display. In conditions of high seas the clutter can extend several miles from the centre of the display. Figure 18.4.5 shows the displayed effect of excessive sea clutter. The swept gain control reduces the gain of the head amplifier at short range on each transmitted pulse and as the timebase generator sweeps the electron beam across the tube face the gain is progressively increased with range. Excessive anti-sea clutter control can cause loss of small short-range targets at the PPI, correct adjustment being subjective. Figure 18.4.6 indicates a reasonable compromise adjustment on the indicator screen.

The manual anti-sea clutter control acts on each timebase sweep, treating the received signal information as though sea clutter was

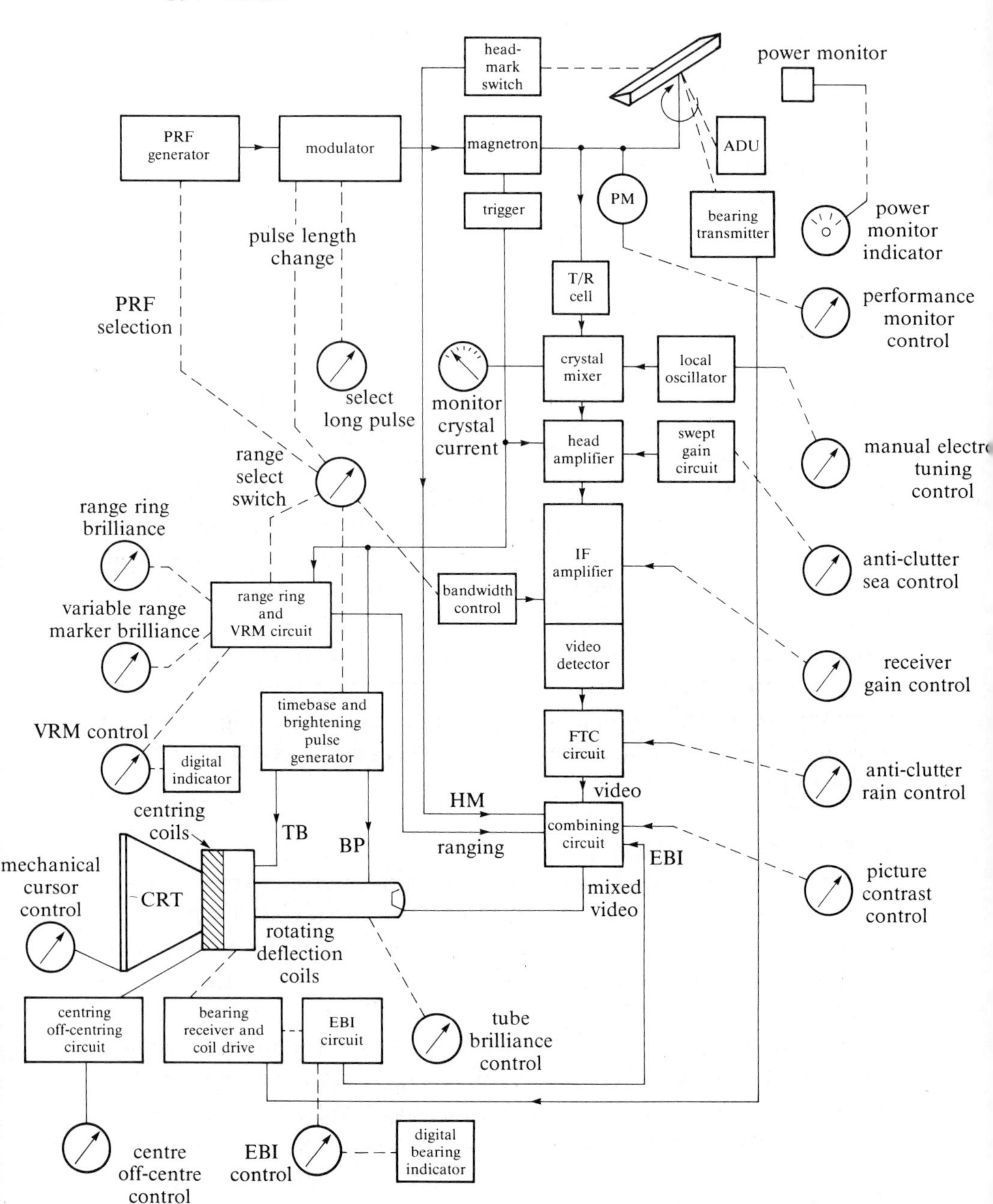

Figure 18.4.3 Block schematic of a basic radar transceiver and display circuits showing operating regions of essential user controls and their functions.

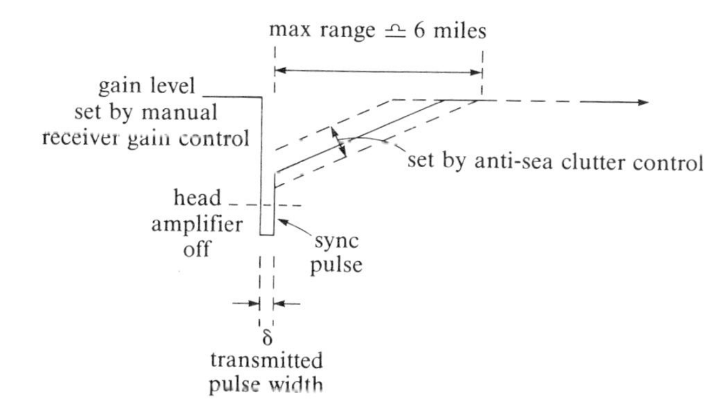

Figure 18.4.4 *Swept gain waveform showing the manner in which receiver head amplifier gain is swept from a reduced level to optimum value. Anti-sea clutter control determines the range at which optimum gain is reached.*

Figure 18.4.5 (a) *Targets at close range on 12 nm range virtually obliterated by sea clutter returns; (b) excessive use of anti-clutter sea control leaving a hole in picture centre; small targets have also been lost.*

equal through 360 degrees of azimuth. In practice, the sea returns vary as the scanner rotates and an automatic swept gain circuit can be employed which dynamically adjusts the correct level of sea clutter control by rapidly sampling signal returns from short ranges. It is a mandatory requirement in type tested radars that automatic anti-sea clutter control can be switched off by the user.

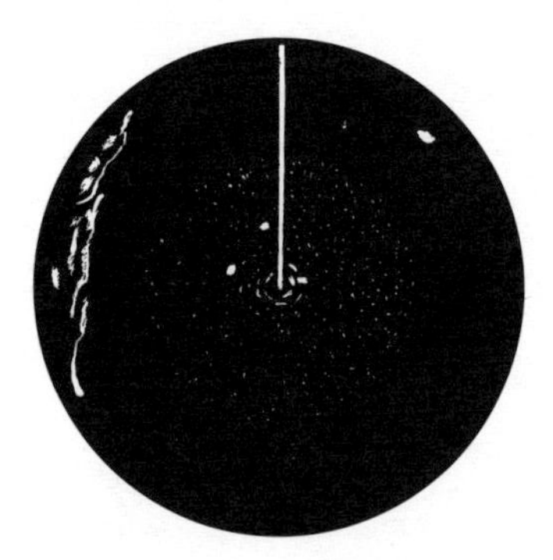

Figure 18.4.6 A compromise setting for the anti-sea clutter control. Targets at close range are discernible – some tolerable clutter is evident.

18.4.3.4 IF gain control – contrast and brilliance

Receiver gain control acts on the IF amplifier allowing the radar user to adjust the overall signal amplification. Excessive gain will cause background noise to intrude upon the target paint (see Figure 18.4.2). Acceptable adjustment is obtained by turning the gain up to a point where noise is just visible and then backing off the gain control until the noise just disappears.

Control of brilliance can be effected by altering the d.c. bias conditions of the cathode ray tube. Figure 18.4.3 implies that grid bias is varied whilst video signals are applied to the cathode. The brilliance control should be adjusted to suit the ambient light conditions in the vicinity of the PPI. Excessive brightness can cause defocusing of the screen image, an effect known as *blooming* where the target paint is saturated by electron bombardment causing a bright halo to form around the paint. Excess brilliance demands higher beam current than normal levels; it will ultimately reduce the tube life and can result in the tube phosphor being burned. To some extent tube brilliance control is interactive with the contrast control. The contrast control adjusts the video level at the cathode in Figure 18.4.3, and careful adjustment selects the best level to provide a sharp contrast between the paint and the darker unpainted background. Maladjustment of brilliance can destroy the contrast and negate the effect of the contrast control.

18.4.3.5 Receiver bandwidth

Analysis of a train of short pulse envelopes separated by a relatively long time interval shows that the pulse envelope occupies a band of the frequency spectrum having both odd and even harmonics of the periodic frequency of the pulses. The amplitude of the spectral components depends on the duty cycle and the harmonics extend over a wide frequency range. If the pulse is passed through a filter, such as the tuned circuit of the IF amplifier, it is essential that the filter shall pass the most important harmonics otherwise the pulse envelope becomes distorted and loses its rectangular shape. For a radar set the squareness of the target signal pulse is essential to providing a clear paint and good target discrimination.

It can be shown that very short pulses occupy more of the spectrum than longer pulses, thus when pulse length is changed in a radar set it is usual to switch the bandwidth of the IF amplifier. The bandwidth defines the useful band of frequencies over which the amplifier will operate without severely attenuating or distorting the input signal. Typical bandwidths for a 45 MHz IF are 24 MHz with 0.05 μs pulse, 4 MHz with 0.5 μs pulse and 700 kHz for 1 μs pulse. Some radars are designed with a constant IF bandwidth and the video amplifier bandwidth is altered. For the radar shown in Figure 18.4.3 the former method is used where operation of the range switch will automatically select the pulse length appropriate to a given range and at the same time change the receiver bandwidth to suit the desired response. Radars are provided with a control to override the short transmitted pulse and substitute a longer pulse where the user feels the need for higher energy target returns. Whether or not the bandwidth is changed depends on the particular set's design since a bandwidth suitable for a short pulse will also serve the longer pulse without distortion. It follows from section 18.4.1 that switching to a wider band will introduce noise which is not present on the narrower bands. It may be necessary on some sets to adjust receiver gain and/or brilliance and contrast when bandwidth switching occurs.

An IF amplifier may have *linear* or *logarithmic* characteristics, some manufacturers provide a choice in their sets. A linear amplifier will operate over a small range of outputs in truly linear fashion before saturation takes place. This will produce a limiting effect on signal amplitude and can cause loss of large target signal echoes. The swept gain circuit action prevents sea clutter signals from saturating the amplifier. A logarithmic amplifier has a gain which at any particular level of input is inversely proportional to the input level and has the advantage of reducing sea and rain clutter effects at the display.

The video detector is a rectifier circuit sometimes known as the *video demodulator*. Its purpose is to retrieve the envelope of the IF signal applied at its input and is driven by the output from the final stage of the IF amplifier (Figure 18.4.3). After detection and filtering the video pulse is applied to the wideband video amplifier which amplifies the rectangular video pulse with negligible distortion. Band switching may be applied at the video amplifier instead of the IF amplifier, choosing a narrow band for long pulse and a wideband for short pulses whilst maintaining the IF bandwidth constant.

18.4.3.6 Anti-clutter rain control (FTC)

The effect of echo returns from raindrops is considered in section 3 where circular wave polarization is outlined. The echo energy due to an individual raindrop is very small but the combined effect of returns from rainstorms causes *rain clutter* which is displayed on the indicator as a bright indistinct area of paint. Targets can be hidden in the clutter. In linear IF and video amplifiers the returns can saturate the amplifier and the display can be improved by use of logarithmic amplification. In addition, the video signal can be passed through a *differentiating* or *fast time-constant* circuit (Figure 18.4.3). The FTC circuit is a high pass filter, generally a combination of capacitance and resistance having a time-constant much shorter than the returned video pulse length. Such pulses will therefore be differentiated when passed through the filter.

Rain clutter builds up an RMS voltage due to the volume of the rainstorm and behaves as if the rain was returning a much longer pulse than that transmitted. The act of passing all video signals through such a circuit is to considerably reduce the effect of the rain on the display whilst having little effect on a discrete target return which echoes a consistent pulse. Figure 18.4.7 shows that only the leading edge of the video pulse is displayed showing a bright leading edge with a contrasting shadow behind it. The degree of control may be fixed or variable; variable controls must be used with discretion since loss of signal can occur due to maladjustment.

The video signal after amplification is used to intensity modulate the cathode ray tube. Figure 18.4.3 shows the video signals applied to a combining circuit where pulses from heading mark, ranging circuits and electronic bearing indicator are combined to produce a composite video input to the CRT.

18.4.3.7 Range ring generator and control

Fixed range rings are displayed on the PPI when the range ring brilliance control is turned from its 'rings off' position. Adjustment of the range ring brilliance control should lift the rings to a visible

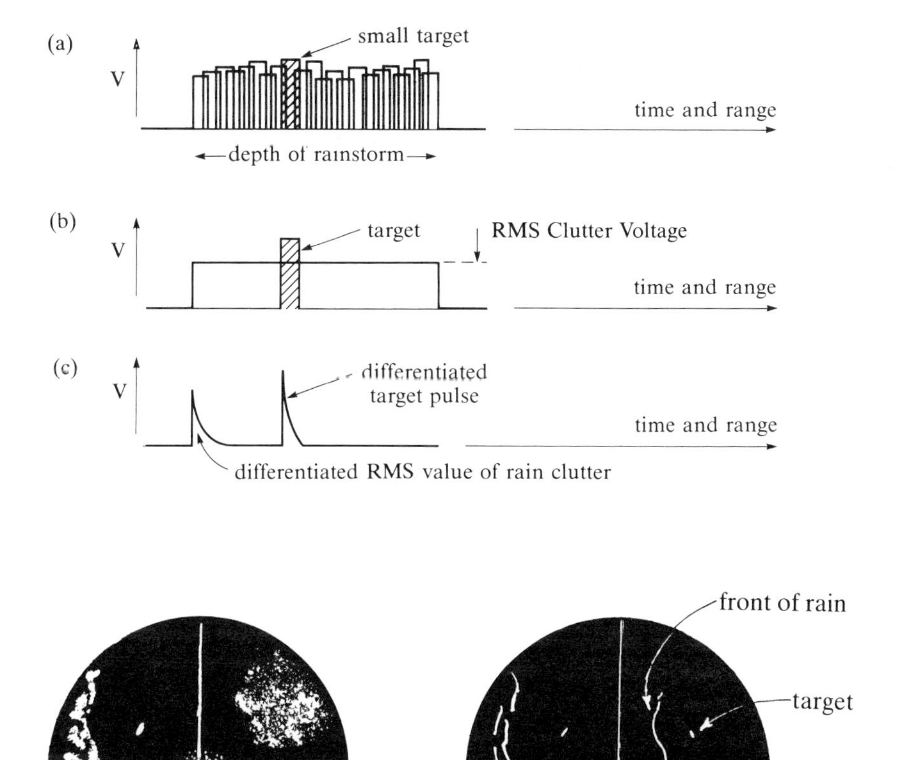

Figure 18.4.7 (a) Shows rain clutter with small target embedded and (d) is effect on PPI. (b) Effective values of target and rain clutter (c) differentiated signals and (e) effect of differentiating on display.

level without excessive brightness. Range ring generation is a mandatory requirement in type tested radars. The range scales provided for such sets have as a minimum requirement range scales of 1.5, 3, 6, 12 and 24 nm, and one range scale of not less than 0.5 and not greater than 0.8 nm. Manufacturers are able to add scales of less than 0.5 nm or greater than 24 nm. On the 0.5 to 0.8 nm scale at least two range rings are specified and on the other mandatory ranges six range rings must be provided. Rings are generated from the output of a stable oscillator which is in some cases crystal controlled. The waveform generated is shaped to give short duration pulses at the generator frequency and synchronized to the timebase sweep. These pulses or *ranging pips* are superimposed on the video information in the combining circuit (Figure 18.4.3) and appear on the screen at equal intervals of time and therefore range, along the timebase sweep as bright dots of light. Rotation of the sweep produces a series of concentric rings from which the range of a target may be interpolated. Figure 18.4.8 illustrates the relationship between ranging pips and timebase. Where off-set of own-ship is provided, the timebase generator of the range in use must overscan and range rings must be displayed to the full extent of the off-set display. The range ring generator is switched to the appropriate frequency by the action of the range selector switch which also controls receiver bandwidth, pulse length and PRF selection for the particular range in use (Figure 18.4.3).

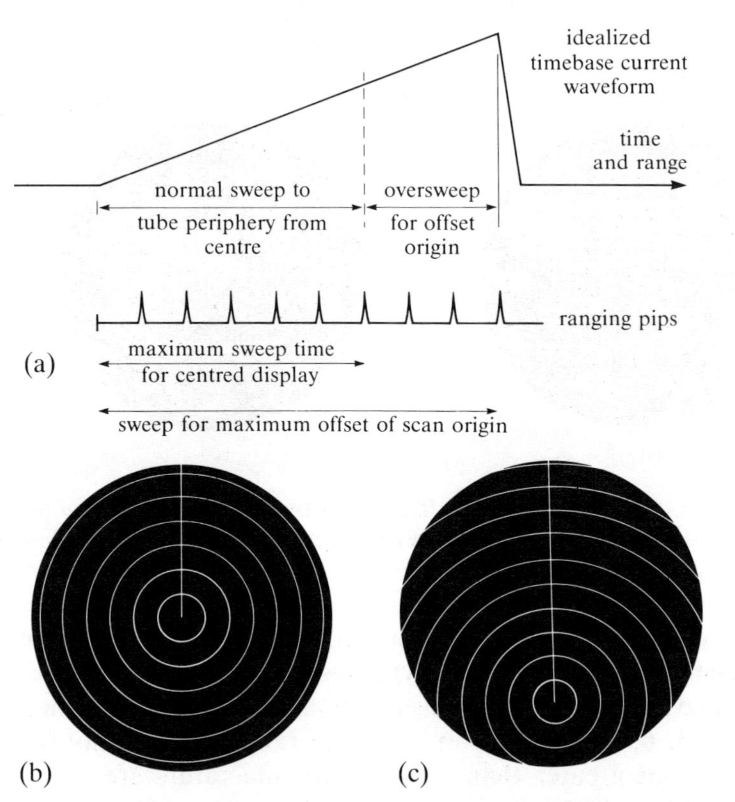

Figure 18.4.8 (a) Relationship between sweep time and ranging pips; (b) appearance of concentric range rings on 1·5, 3, 6, 12 and 24 nm ranges; (c) range rings with overscan and centre off-set by 60 per cent of radius.

18.4.3.8 Variable range marker

A circuit is provided which generates a marker pip synchronized to the timebase sweep. A manual control varies the delay between timebase initiation and the time at which the VRM pulse is produced. The time relationship is converted directly onto range measurement from the scan origin by means of a digital indicator. Because the sweep is rotated the range marker pip will produce a bright circle on the PPI having a variable radius. The VRM control is manipulated so that the ring lies on the desired target and its range is then read directly from the digital indicator.

Use of the VRM allows accurate interpolation of targets which lie between range rings. The range scale of the VRM should coincide with the calibrated range rings value when the VRM is overlaid with each ring on the PPI. Fixed range rings and variable range markers found in type tested equipments must enable the range of a target to be measured with an error not greater than 1.5 per cent of the maximum range of the scale in use or 70 metres, whichever is greater. Variable ring and calibrated range rings must be capable of removal from the PPI, usually achieved by turning the appropriate brilliance control fully anticlockwise.

18.4.3.9 Electronic bearing indicator

The electronic bearing indicator circuit generates a brightened radial line, sometimes broken into dashes on the PPI and emanating from the sweep origin. The EBI is variously known as the electronic bearing line and electronic bearing marker. Its purpose is to enable the bearing of a target to be measured regardless of the position of the scan origin on the screen, bearings being taken directly from a digital indicator controlled by the EBI circuit. Because the bearing line is generated at the tube surface there is no parallax error introduced into the measurement as can occur when using the mechanical bearing cursor. Many EBI circuits generate the bearing line only once per antenna revolution which can make the application of the indicator a slow process, since in moving the line the operator has to wait one antenna revolution to ascertain its new position. Modern radar sets generate the bearing line during the *interscan* period which is the time between the end of one timebase sweep and the beginning of the next sweep. Using this method the EBI is continuously displayed on the screen at all periods of an antenna rotation, facilitating more rapid application of the line for bearing measurement. The EBI is most useful where the display is off-set from the geometric centre of the screen, as in Figure

18.4.8(c). There are certain mandatory requirements for the EBI in type tested equipments; it must have controllable brilliance, be free to rotate clockwise and anticlockwise continuously through 360 degrees, the direction of turning being the same as that of the control, have a maximum thickness not exceeding 0.5° measured at the edge of the display, be updated at least once per antenna revolution and be clearly distinguishable from the ship's heading marker. An accuracy of ± 1° in bearing measurement is required for a target having an echo lying at the edge of the display.

18.4.3.10 Off-centring

The scan origin can be off-set from the geometric centre of the PPI by manipulating the off-centre control. A navigator may wish to adjust the display to provide maximum radar coverage ahead of the ship, as in Figure 18.4.8(c). Off-centring is achieved by adjustment of current passing through a set of fixed deflection coils on the PPI tube neck. Figure 18.4.3 shows the off-centring coils separate from the timebase deflection coils which are in this case, rotating deflection coils. In systems which use fixed deflection coils these will perform the dual function of timebase deflection coils and off-centring coils.

18.4.3.11 Timebase generator

The timebase generator has briefly been touched upon in section 1 in order to describe the principles of radar. It is the circuit where a sawtooth current waveform is generated which when passed through the PPI deflection coils causes linear deflection of the electron beam and linear movement of the luminous spot on the PPI. When the range switch is changed, the timebase is also changed to provide the necessary beam deflection. At short ranges the deflecting current amplitude necessary to deflect the spot from the scan origin to the edge of the screen has to be reached in a very much shorter time than on one of the longer ranges. Assuming a centred display for the 0.5 mile range it will be 6.2 μs and for the 48 mile range it will be 595.2 μs. If the tube radius is eight inches, the luminous spot on the screen is travelling at a velocity 96 times greater on the shorter range than on the longer range and would produce a very faint trace if the same brightness level was used in both cases. A brightening waveform known as the *brightening pulse* is generated and applied to the grid of the CRT (Figure 18.4.3). This pulse is a rectangular waveform acting in synchronism with the timebase waveform and serves to lift the trace to the threshold of visibility.

18.4.3.12 Mechanical bearing cursor

Type tested radar displays have as a choice an electronic bearing indicator or a mechanical cursor and are mandatorily provided with a bearing scale around the periphery of the display which is centred on the geometric centre of the PPI. Modern displays provide both methods of bearing measurement.

The mechanical cursor is a circular sheet of transparent material usually orange tinted to act as a filter of light emission from the inner phosphor coating of the PPI. It is mounted above and parallel to the tube face and is free to rotate through 360 degrees, being driven by a manual friction drive. The PPI display is observed through the cursor. A fine line is inscribed on the cursor diameter whose centre must always remain above the display centre when the cursor is rotated through 360 degrees. Drawn parallel to that diameter are a number of additional lines whose equal spacing coincides with that of the range rings on the various displayed ranges. It is usual to provide inscribed lines coincident with each other on opposite faces of the cursor thus providing a means of minimizing optical parallax error.

The mechanics of the system require an accuracy of alignment with the bearing scale such that when the diametric bearing line indicates 000° its opposite end is within ±0.5° of 180°, and when indicating 090° the diametrically opposite reading is within ±0.5° of the 270° mark. The general appearance of the cursor on the display is shown in Figure 18.5.22. Because of their appearance the parallel lines are sometimes referred to as *parallel index lines*. In a centred display the diametric cursor is aligned to bisect a target whose bearing may then be obtained from the peripheral scale. In off-centred displays the cursor can be adjusted for an appropriate parallel index line to bisect the echo.

18.4.3.13 Orthogonal PPI deflection system

Previous PPI generation examples have described the electro-mechanical rotating deflection coil method commonly used to effect rotating deflection forces on the CRT beam. A number of disadvantages are evident in these systems, mechanical wear, noise, conveying deflection currents via moving slip rings and lack of flexibility in beam manipulation. Set against this is relative cheapness and availability of a well tested product. Where the simplest types of relative and stabilized displays are required with no off-centring facility a basic rotating coil system is adequate.

When the need arises in such displays for off-centring or true-motion, an additional set of fixed deflection coils is needed, details are given in section 18.5.1.4. The type of deflection assembly described there is known as the *fixed deflection yoke* to include the orthogonally mounted coils and the permeable core which surround the CRT neck.

Such deflection coils can be designed to carry not only the d.c. off-centring currents and the relatively slowly changing true-motion deflection currents but also the rapidly changing sweep currents from the timebase generator circuits. One set of fixed coils is then able to provide all necessary functions, reduces complexity of the deflection system, improves reliability and provides a psychological bonus – a noiseless display.

18.4.4 *Deflection system principle*

Manufacturers of equipment have adopted a variety of methods aimed at providing a radial rotating scan using fixed X and Y deflection coils. All meet certain basic requirements, these are resolution of antenna azimuth bearing into X and Y coordinates and the generation of a deflection current ramp waveform from those values for each individual PPI sweep. A microprocessor-based system is used where bearing data arriving from the antenna is related to the time of a particular transmitted pulse. Antenna information may be inputed in coded form directly from a digitally encoded disc but is more likely in modern marine radars to be derived from a synchro or resolver. The resolver principle is shown in Figure 18.4.9. Antenna bearing measurements are calculated relative to some datum, usually the heading mark pulse. Figure 18.4.10 shows a simple block arrangement and indicates that compass stabilizing information is easily processed in digital form, the microprocessor algorithm being devised to add or subtract compass and antenna bearing data. One method of generating the scan coil waveforms is shown in simplified form by Figure 18.4.11. Here the antenna bearing information is resolved from *polar* form $V_{ref} \angle\theta$, where V_{ref} is a regulated reference voltage and θ is the main antenna lobe angle relative to the head mark, into its 'vertical' and 'horizontal' components or *cartesian* values of $V_{ref} \cos \theta$ and $V_{ref} \sin \theta$ respectively. Each of these voltages must generate a ramp of current, the sin θ current passes through the vertical coil to produce horizontal beam-motion, the Cos θ current passes through the horizontal coil to produce vertical beam-motion. The resultant

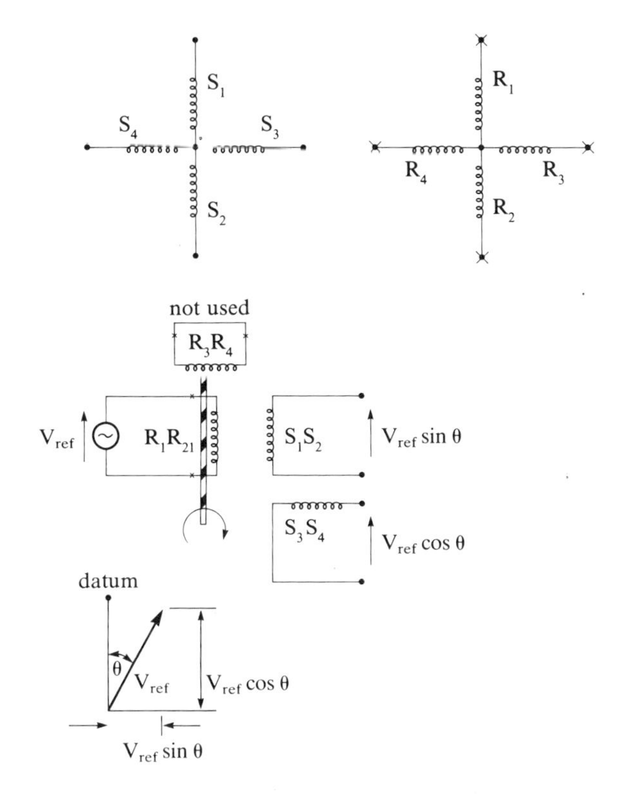

Figure 18.4.9 Basic principle of a resolver. Above: stator and rotor windings on resolver. 'X' denotes a slip ring connection. The machine has two pairs of rotor windings and a corresponding set of orthogonal stator windings. These afford a variety of configurations for different purposes. Centre: resolver polar to cartesian conversion. V_{ref} *is exciting voltage equal to the modulus of a polar co-ordinate. Shaft angle is argument* θ. *Due to stator physical displacement two outputs appear;* V_{ref} *Cos* θ *and* V_{ref} *Sin* θ. *The cartesian co-ordinates of* $V_{ref}\angle\theta$.

direction in which the CRT beam is deflected is determined by the resultant magnetic field of the current components.

The processor calculates the desired values and outputs them in digital form to two digital to analogue converters each producing from the digital information the required voltage ramp. Two drive amplifiers convert these into current ramps necessary to drive the deflection coils. As the antenna rotates, new bearing data are input, ramp values are updated and on the next consecutive trigger pulse

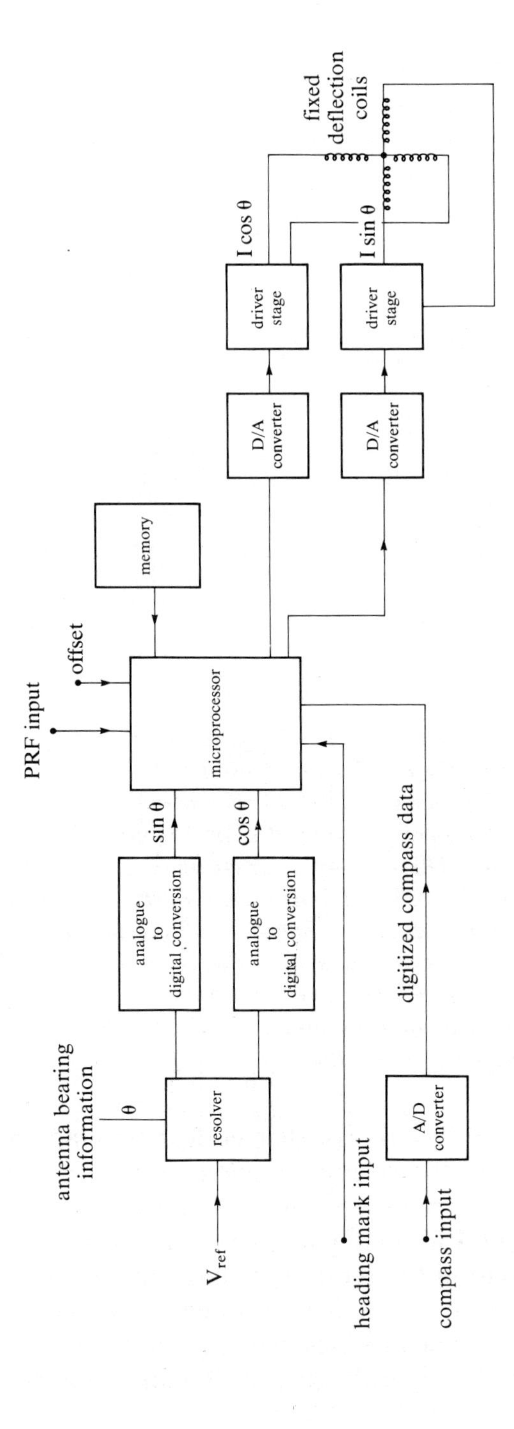

Figure 18.4.10 Digital principal of generation of rotating sweep for fixed deflection coils. Antenna bearing information may be directly digitally encoded by optical disc encoder.

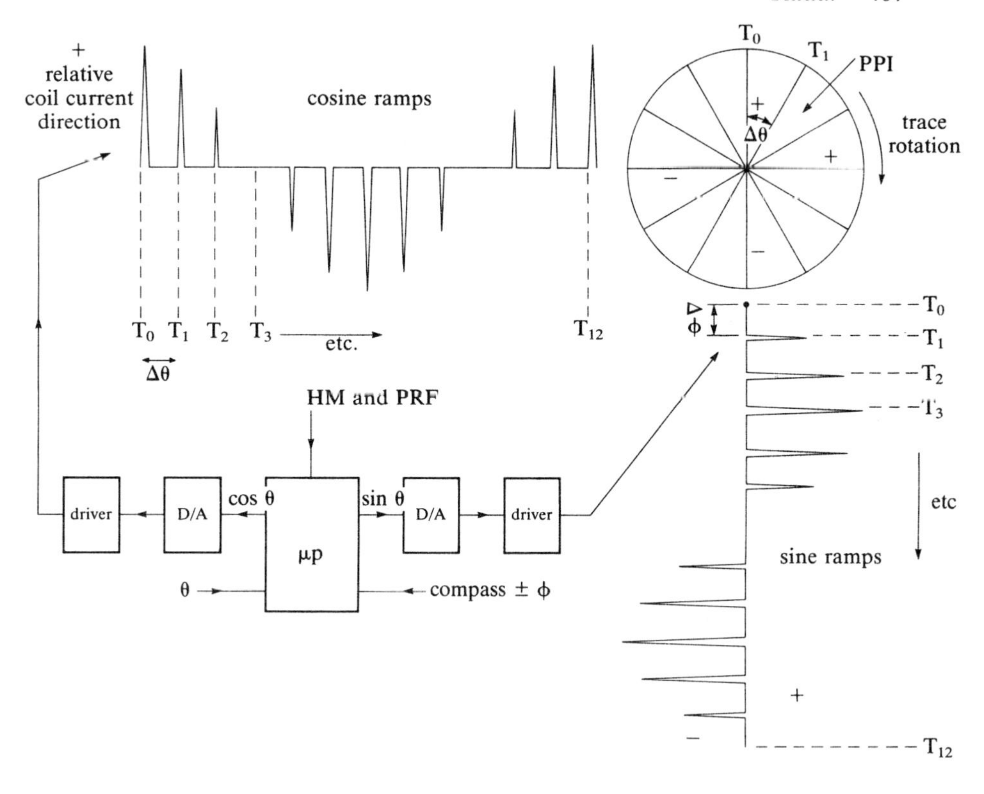

Figure 18.4.11 Diagrammatic representation showing rotating sweep generation from synchronized sine and cosine deflection current ramps. Twelve widely separated sweeps are shown for clarity. If stabilization is needed the compass input $\pm\ \phi$ is added to T_0.

these are sent to the D to A converter to produce the next trace, and so on. In this manner a rotating trace is displayed on the PPI, the current being reversed through individual coils as the trace passes through each quadrant.

In most microprocessor-controlled displays the processor will be performing other tasks in apparent simultaneity, time-sharing the deflection generation time with those other jobs. Other more sophisticated display circuits designed to process and make decisions upon a large volume of incoming data, increase their processing speed and power and provide a certain amount of redundancy to improve reliability by using two, three or more processors. Such arrangements are found in automatic radar plotting displays.

18.4.4.1 Burn protection

Should the deflection waveforms fail, the electron beam comes to rest and expends its energy on one small spot on the phosphor surface. Prolonged periods in this condition at normal brilliance level will irretrievably damage the phosphor. Most displays now incorporate a sensing circuit which in the event of sweep failure immediately biases off the cathode ray tube current. It should be evident that if a display is left with the scan origin at the PPI centre even in normal operation, the phosphor at the centre is almost constantly energized and a burned area of phosphor tends to appear over a period of time. Where an area of sea clutter has been displayed for extended periods (many weeks) a darkened area of phosphor is often observed. Targets subsequently displayed in these damaged regions of the tube face suffer from reduced brightness. Many users become accustomed to this display idiosyncrasy and if possible off-set the display to a clear area. Whilst presenting a minor irritation on the longer ranges, neglect of the tube condition can lead to serious small target loss at short range. The only remedy is to replace the CRT.

18.4.4.2 Overscan or oversweep

In displays where the scan origin is able to be off-centred either manually or in true-motion mode, it is necessary to provide a time-base sweep waveform of sufficient amplitude to deflect the beam to the furthest extremity of the tube periphery. Off-setting the origin from the tube centre by 70 per cent of the radius requires that the sweep be increased proportionately. Figure 18.4.8 indicates the oversweep or overscan. Unfortunately, this amount of off-set will also overscan the tube face on the *shortest* radial sweep by 70 per cent and overscan is taking place to a lesser extent around the tube periphery. The worst case condition has been described for most applications.

During the overscanned display periods the electron beam for a large proportion of each antenna revolution is falling on the aquadag coating of the tube and very high beam currents ensue. In addition, the deflection amplifiers are generating maximum deflection current at each sweep, regardless of possible reduced sweep requirements. Using microprocessor control the deflection limits of the screen periphery are always known, that is they can be stored in memory. When the scan origin is moved, its coordinates within the geometry of the tube face can then be calculated and related to the ship's head and tube periphery.

Since each sweep is digitally generated the individual amplitudes at all angles through the 360 degree sweep can be 'tailored' to meet the beam deflection demand and to fit within the screen boundary. The net result is a reduction in beam power, reduction in sweep drive amplifier power, lower rated supplies to those areas of circuit, cooler operation and increased reliability.

18.4.4.3 Sweep stealing

Placing the display sweep generator circuits under microprocessor control increases the flexibility of the display enormously compared to analogue methods. Some of these benefits will be discussed in later sections. A relatively simple example of sweep manipulation involves the *stolen sweep*. In a fairly standard type of marine display one picture scan occupies about three seconds, and for a PRF of 1000 pps target information is presented on about 3000 sweeps. The absence of several of those sweeps represents a negligible loss of picture information and is unnoticeable if they are 'stolen' at distributed intervals in the picture. During the interval of the stolen sweep other information can be displayed. The stolen sweep method is widely employed in marine radars to display the electronic bearing line (EBL) or electronic bearing indicator (EBI). One method is to keep a count on the number of sweeps; sweep 32 is suppressed and the EBL information, say a dotted radial line, is substituted. The count is repeated over the next consecutive sweeps again stealing the thirty-second and substituting EBL display.

Each time the stolen sweep 'time slot' is encountered, the processor inputs EBL bearing information from the front panel manual control, resolves this into its sin and cos components, and displays the EBL at the correct bearing. The advantage of this method is that EBL display is rapidly updated and, instead of having to wait for a complete scan to make fine EBL adjustment, as in the older mechanical systems, very quick and precise alignment is possible.

Other methods of EBL generation under microprocessor control use the *interscan period* (outlined in later sections).

18.4.4.4 Delayed synch

Waveforms in Figures 18.1.7, 18.1.8, 18.2.6 show that the timebase generator waveform, brightening pulse, etc., at the display are synchronized to the transmitted pulse. In order to accurately measure range the timebase sweep must begin at the instant which represents zero-range video signals arriving at the display.

Due to waveguide propagation delays there is a finite time lapse between the transmitted pulse reaching the antenna and the received RF echo signal reaching the receiver. In addition there will be propagation delays through the receiver and in the triggering circuits of the timebase.

A delay is therefore introduced into the synch pulse path between transmitter and timebase generator to ensure that zero-range video signals reach the CRT at the instant that the time-base sweep starts. A tapped delay line is provided for this purpose which is adjusted when the set is installed. Solid state gates can also be used to introduce the required delay.

Where very short waveguide lengths are encountered the delay in the trigger and timebase circuits can exceed the signal delay, in which case a delay is introduced into the video signal path.

18.4.5 *The cathode ray tube*

Cathode ray tubes are used exclusively as the indicators in marine radar sets. They provide an intermediate storage and display unit where, due to very low electron beam inertia and correct selection of screen phosphors, information can be written to the screen at very high speeds providing a luminous output which decays relatively slowly to be refreshed with each revolution of the antenna.

A basic CRT is shown in Figure 18.4.12 and indicates the essential sections of a CRT used in a radar display: *a* is the evacuated glass envelope – shown dotted; *b* is the heater – an element which raises the temperature of the cathode *c* to about 750°C and causes *c* to emit a cloud of negatively charged electrons. These are accelerated by a relatively positive potential applied to *f*, the accelerator anode which is cylindrical and concentric with the tube axis. Accelerated electrons pass into the influence of zone *A* where focusing of the electron beam is achieved; two basic focusing methods are used:

(i) magnetic focus
(ii) electrostatic focus.

In (i) a magnetic field acts from the outside of the tube causing the electron beam *e* to converge at a point of focus–arranged to be at the CRT phosphor screen surface *j*. The magnetic field may be generated by passing a direct current through focusing coils or by the use of permanent magnets. In the former case, adjustment of

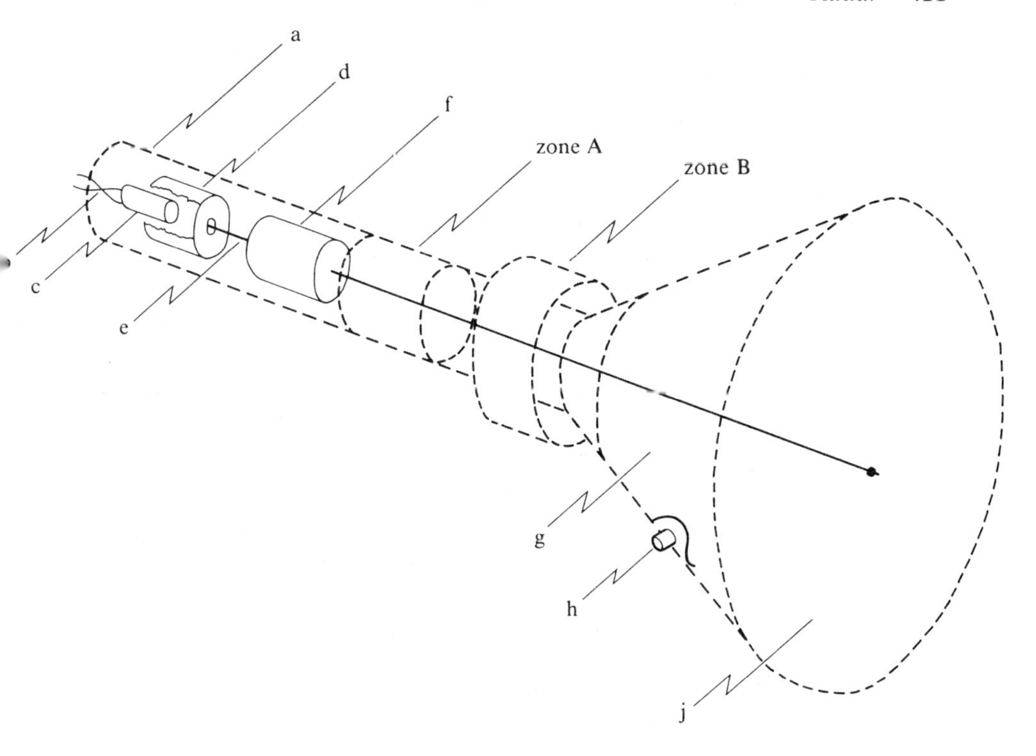

Figure 18.4.12 Essential structure of a PPI cathode ray tube.

the current brings about correct focus; in the latter, the position of the magnet is manually adjusted.

In (ii) a series of concentric conducting cylinders are arranged inside the tube to have differing electric potentials. This creates an electrostatic field having an effect on the electron beam similar to that of a converging lens on a light beam. Adjustment of relative potentials on the cylinders brings about correct focus at the screen.

There is an electrically conducting coating of graphite deposited on the inner surface of the CRT bulb (*g*). A connection is made to *g* from an extra high tension voltage supply via an external electrode *h*, the EHT connector. The coating, known as the *aquadag electrode*, acts as the final accelerating anode for the electrons in the beam and has a potential of about 20 kV for a sixteen-inch diameter tube. Electrons in the beam reach a high velocity, gaining kinetic energy and upon impact with the screen *j* can cause it to fluoresce. The point of impact of the beam then becomes visible as a luminous spot. The display screen (*j*) is itself a complex piece of engineering. In the simpler tubes it comprises at least two layers of phosphorescent material to form a double coating on the inner surface of the CRT flat area. Such construction is known as a *cascaded phosphor*.

In action the inner phosphor reacts quickly to the impacting electrons and emits light. This phosphor has a low persistence and its afterglow fades rapidly. It serves, however, to activate the outer phosphor layer which has a much longer persistence and a slowly fading afterglow which is still discernible after a complete radar picture will have been painted on the screen. The majority of screen phosphors are classified according to their persistence and light emitting properties and a group of tubes exist which have multiple phosphor layers each capable of emitting a different light colour. These are *penetration tubes* and they operate on an entirely different principle to the shadow-mask colour tubes used in colour video screens. Zone *B* comprises the beam deflection system. In civil marine radars the method used for beam deflection is electromagnetic. There are two basic methods adopted to produce a rotating CRT trace to rotate in synchronism with the antenna main beam:

(i) rotating deflection coils
(ii) fixed deflection coils

to produce a varying, rotating magnetic field and hence rotating trace. Figure 18.4.12 shows the principle of (i). In all cases the electron beam sets up a *beam current* comprising an electron flow from the cathode to the screen. The circuit is completed via the EHT supply since the aquadag collects all electron charge from the screen itself. The beam current can be considered to flow conventionally from screen to cathode in a conductor of virtually zero mass.

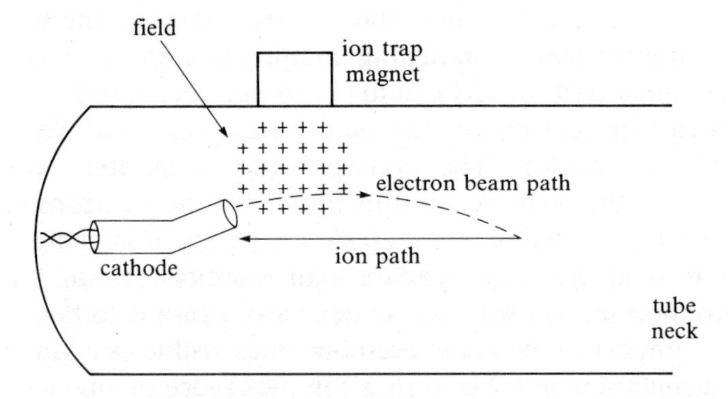

Figure 18.4.13 Principle of ion trap.

18.4.5.1 The ion trap; aluminizing

It is impossible to attain a perfect vacuum within the tube envelope; in any case as the tube operates there are minute quantities of occluded atmospheric gases trapped within the electrode structure which are subsequently released. As electrons in the beam collide with atoms of the (rarefied) gases the action produces positive gas ions. These have a much greater mass than the electrons and move toward the negative cathode, gaining kinetic energy in the process. If the massive ions are allowed to bombard the emitting surface of the cathode it will rapidly be destroyed and lose its emitting properties. The tube life is prolonged by preventing ions from bombarding the cathode surface if an *ion trap magnet* is used. In conjunction with the magnet, the axis of the cathode is tilted from the tube axis as shown in Figure 18.4.13. The ion trap magnet is supported on the outer surface of the tube neck in the proximity of the cathode, its field penetrating into the tube. The field deflects the relatively low-mass electrons back along the tube axis where they form the beam current. The more massive positive ions, however, travelling in the opposite direction along the axis are not deflected by the ion trap magnet and they impact harmlessly on the tubular cathode structure – not its emitting surface. The presence of the ion trap has an effect on the beam focus and due allowance is made for this in the tube adjustments on installation.

Modern tubes of all types are invariably *aluminized*, a term applied to the process of depositing a thin film of aluminium on the inner surface of the phosphor layers. The electron beam first meets this film and easily penetrates it into the phosphors which are activated to emit light. Normally, some of this light would not only travel through the PPI screen to the observer's eye but also a large percentage of it would travel *back into the tube*. The aluminium film, however, reflects the majority of this backward travelling light to provide a brighter picture to the viewer.

18.4.5.2 Colour tubes

Colour cathode ray tubes allow information to be displayed in more than one colour, thereby allowing circuits to be devised which can highlight features of the picture presented. It is possible to display for example specially generated marker symbols, alphanumeric information or target information of special significance in a colour different to that of the rest of the display. Visual data is then more easily discriminated. The shadow-mask tube commonly used in colour television receivers depends on a triad of phosphor dots to produce colour output. The geometry of these limits the available screen resolution. Modern colour tubes have, however, been applied in radar displays to produce a true multicolour PPI display (see section 18.7). The basic principle of the colour tube is shown in Figure 18.4.14.

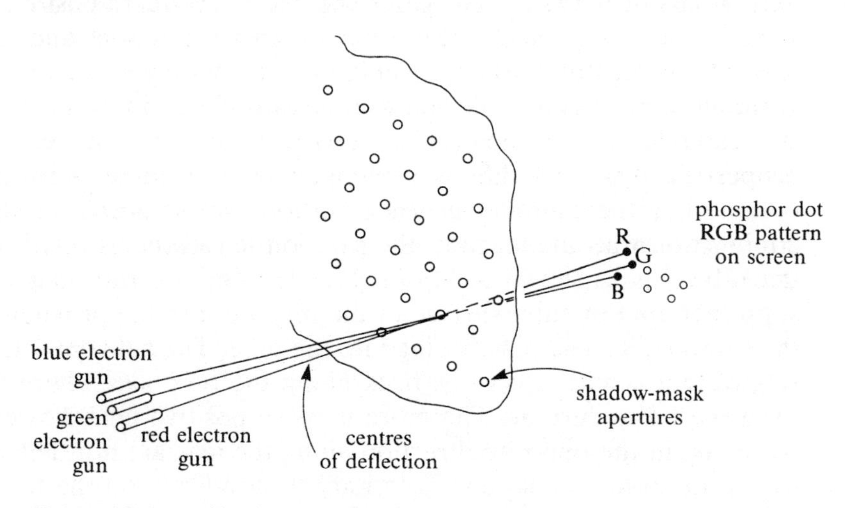

Figure 18.4.14 Shadow-mask colour tube principle.

Three electron guns are assembled at 120 degree intervals about the CRT axis, inclined to converge the undeflected electron beams at a common point on the screen. The beams must pass through a small aperture in a thin steel sheet known as the *shadow-mask* which is aligned about 10mm from and parallel to the screen. Triads of red, green and blue phosphor dots are arranged on the screen, one triad for each mask hole. Where the red beam meets the screen a red phosphor dot will always lie, and all such dots will lie in the shadow of the mask from the aspect of the blue and green beams which cannot then energize red dots. Blue and green phosphor dots are similarly energized only by blue and green beams. In a medium resolution CRT, as used in the Racal Decca digital scan conversion display, separation between triad centres is about 0.42mm. Control of individual beam currents permits control of the RGB emitted light and hence of the colour components at each picture point. The eye averages the emission so that any colour can be obtained under such control.

Various civil marine displays employ the colour *penetration tube*. It operates on the principle that the active penetration of the electron beam into a phosphor layer is a function of the kinetic energy imparted to the primary electrons. Variation of the potential on a colour control anode will vary the extent of the beam penetration. Two or more layers of phosphor, each separated by a thin insulation layer which is transparent to light are deposited on the inner surface of the tube, each emitting light of a different colour. Adjustment of the control anode potential can be binary, thus allowing one or the other of two colours to be selected. If the electron acceleration is arranged to penetrate the first layer and partially penetrate the second then the eye will average the two according to the intensity of the second colour to produce some intermediate colour. One of the main drawbacks of colour penetration tubes is that colour switching needs about 4 kV potential change. More than three layers produces deflection sensitivity and defocusing problems.

18.4.5.3 Front-end amplifiers

The past decade has seen the majority of innovations in marine radar design appear in the display section where signal processing and display techniques have been transformed by the availability of cheap reliable digital processing devices. During the same period transceivers have undergone relatively little change with the exception of the introduction of solid state devices in lieu of thermionic devices. The vast majority of receivers use mixers as the first stage in a superheterodyne and we see from section 18.4.1.1 that the noise figure of the receiver is governed mainly by the mixer performance. Interest in satellite transmission reception has produced relatively cheap, small pre-amplifier circuits for X band operation and radar manufacturers have recently been turning their attention to their application in marine radar receivers.

Use of a front-end amplifier provides signal gain before the mixer and a low-noise pre-amplifier improves the overall receiver noise figure and receiver sensitivity. A typical device produced by JRC is the NJT1906 which comprises a low noise RF amplifier, double balanced mixer and local oscillator. The assembly is pre-plumbed into a small section of waveguide. The pre-amplifier yields a 3dB improvement in noise figure; effectively a 5 kW peak magnetron is equivalent in performance to a 10 kW magnetron. Several advantages arise from the use of pre-amplifiers as follows:

(i) increased receiver performance as described permits use of lower transmitter powers to achieve the same displayed results

(ii) reduced transmitter power increases life of magnetron overall and increases the reliability

(iii) increased reliability reduces service time and makes masthead fitted transceivers a more attractive proposition

(iv) masthead mounted systems need no waveguide run, are less lossy and less expensive in installation

(v) reduced power demands place less stress on power supplies, switching circuits etc., thus introducing another enhanced reliability factor.

Set against these advantages is the present extra cost of the GaAsFET amplifier together with its protective circuitry, PIN diodes commonly being used in that role. Lower output powers allow ferrite circulators to be used instead of T/R cells (see Appendix C), to provide about 20–25dB isolation between transmitter and receiver, but at increased cost. Inclusion of a front end stage whilst improving noise figure may need some receiver IF stage gain adjustment since the overall receiver dynamic range may be exceeded. Masthead fittings are notoriously difficult to access for servicing. Figure 18.4.15 illustrates a typical low-power high performance masthead transceiver arrangement.

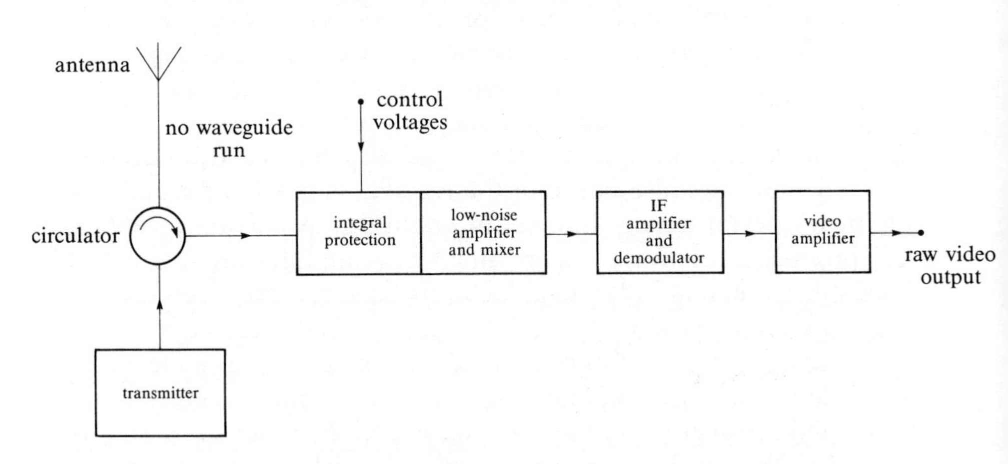

Figure 18.4.15 Basic arrangement of a low-power masthead transceiver using front-end low-noise amplifier.

From the manufacturers' point of view there are design bonuses in other directions which arise when front-end amplification is considered. Consider as an example the antenna installation and its drive unit. Conventional marine radars generally achieve narrow beamwidths and high forward gain by use of long slotted waveguide structures approaching 4m length, these are required to operate reliably in winds up to 100 knots. A shorter antenna of 1.5m will meet the desired beamwidth specifications, is less expensive to manufacture and will require lower drive power. All these factors yield lower unit costs. However, if the antenna aperture is halved there needs to be a 3dB increase in system power budget achievable by use of the front-end amplifier as outlined.

18.5 Displays, methods used, presentation of information

8.5.1 *Relative motion display (unstabilized)*

The simplest form of PPI display in terms of equipment requirement is the *relative motion display* where the PPI shows 'own-ship' as the scan origin situated at the centre of the display. Own-ship's heading, indicated by the heading marker is usually adjusted to point at 000° on a bearing scale surrounding the indicator. Bearings of target echoes can be measured by means of an electronic or mechanical bearing cursor which is effectively a radial line passing through the scan origin and reaching the periphery of the tube face.

The cursor is adjusted by the observer to lie upon a displayed target echo whose bearing relative to the ship's head can then be determined from the bearing scale; the target range can be measured using the ranging circuits provided (see section 18.4.3.4, 18.4.3.5). Before such bearing and range information can be laid off on a navigation chart the observer must account for the ship's true heading. Since the 'top' of the PPI is generally arranged to display the heading mark, it represents 000 degrees relative and the resultant displayed information is known as 'ship's head up' (SHU) presentation. It is also known as an 'unstabilized' display since no compass bearing information is fed into the display. Some relative motion display consoles utilize repeater motors driven from the ship's gyro-compass to drive a compass stabilized bearing scale which surrounds the indicator. This simplifies conversion from relative to true target bearings. Other systems provide a readout of the compass bearing from a small repeater-driven indicator or a digital indicator. For various navigational manoeuvres the navigator may wish to offset the scan origin from the centre of the indicator and panel controls are usually provided for such manual off-setting.

A relative motion display will show all targets moving in their correct scaled speeds and directions in a manner similar to that which would be seen by an observer situated at the antenna site.

Hence target echoes on the display are seen to move in scaled motion relative to the scan origin (own-ship), which is itself moving through the water or over the ground at own-ship's speed and heading. It is the task of the navigator to derive information concerning the movement of targets with a view to predicting possible navigational hazards and hence take the necessary avoiding action well in advance of such a hazard being imminent. To do this he uses a *plotting screen* which is an optical device placed over the PPI face to correct for the convex surface of the PPI tube. The principle is shown in Figure 18.5.1.

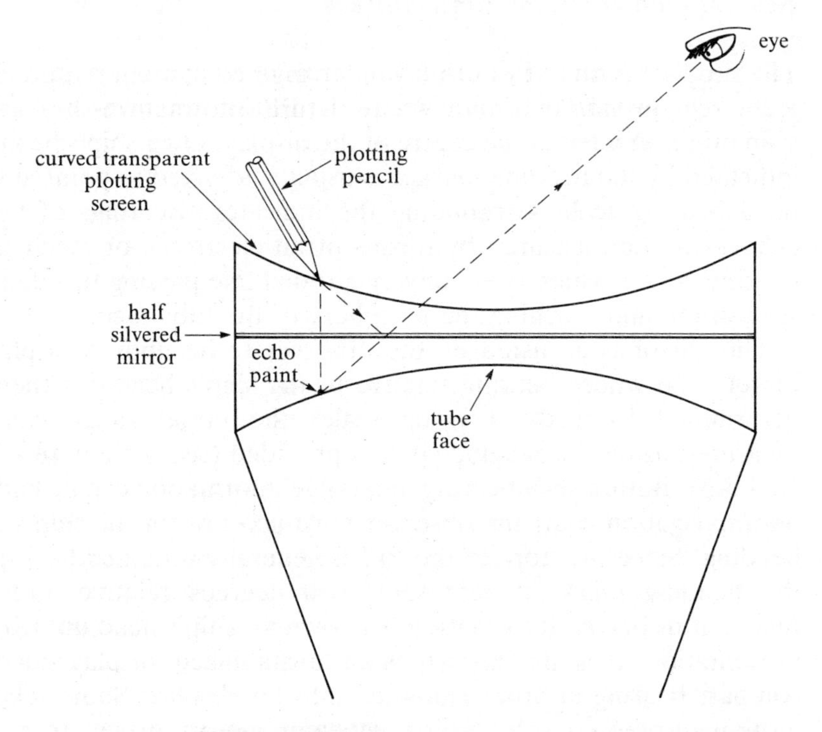

Figure 18.5.1 Principle of the reflection plotter. Curvatures have been exaggerated. The device operates on ray optics theory, regardless of the eye angle relative to the echo. The point of the plotting pencil covers the target when its image is seen to be directly over the echo.

A target is selected for observation and its position is marked on the plotter using a marker pencil. After a measured elapsed time the new position of the target is marked. Using triangulation methods it is possible to allow for own-ship velocity and to plot the motion of the target relative to own-ship motion.

Such methods are adequate for a situation where few targets need to be 'tracked'. It is evident, however, that the navigator is always attempting to make decisions based on the recent history of all recorded target information and the plotting process becomes very complex as the number of possibly hazardous targets increases, the frequency of maneouvres increase and the frequency of plots rises.

From the point of view at the display, relative motion has the added disadvantage that all fixed targets are seen to move on the display at the scaled own-ship speed on a contrary course. A vessel heading due north will, for example, display a coastline apparently moving south at the same scaled speed. This produces a blurring of fixed target detail as each antenna revolution refreshes the picture information. If the vessel makes a change of course, the heading mark remains at 000 degrees but picture information appears to rotate in a direction contrary to the course change, producing a smearing of target detail while the change of course is in progress. Similarly, target smear occurs when own-ship is yawing in bad weather conditions. Figure 18.5.2 illustrates.

The block schematic diagrams of Figures 18.1.10 and 18.4.3 show the arrangement of the various sub-units for a relative motion radar set. Figure 18.3.8 and section 18.3.2.8 explain the principle of bearing transmission from antenna to display for such an arrangement.

18.5.1.1 Stabilized display

The presentation is variously known as *compass stabilized display*, *north up display*, *chart plan display* or *azimuth stabilized display*. Its introduction was a step towards making the presentation of relative motion target information less confusing. Such information is more easily transferred from indicator to navigation chart by representing the 'top' of the screen at 000 degrees as true north, achieved by feeding the ship's heading from the gyro-compass to the display. As own-ship changes course the displayed heading mark swings to the new course whilst other targets are shown on their correct scaled relative courses referred to the top of the indicator. The change of say, collision risk, brought about by the course change is readily apparent during the manoeuvre. Actually, although north stabilizing assists matters when own-ship's course is close to north, it can produce an awkward display if its course is, for example, due south. In that case the heading line of own-ship points constantly to the bottom of the PPI and for this reason modern radars provide a facility for *course up* display; in this example it would be *south up*. A variety of display controls can be manipulated to provide the

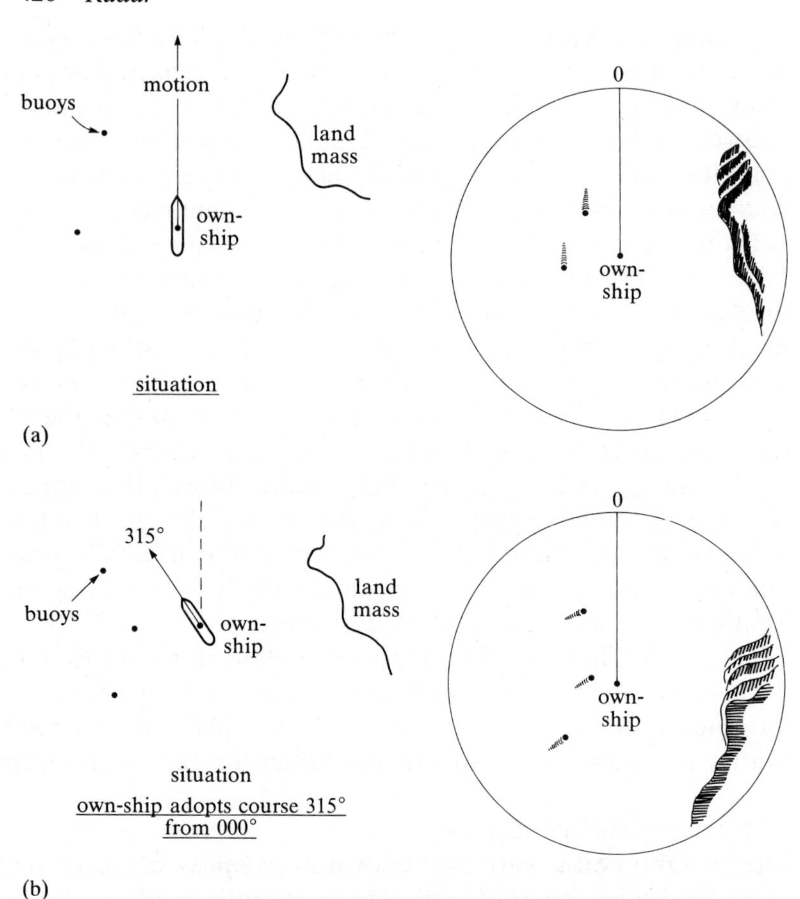

Figure 18.5.2 Relative display (unstabilized): (a) fixed targets appear to move contrary to own-ship motion. Smearing occurs: (b) fixed targets smear and rotate contrary to course change.

facility. When in the course up mode a gyro-driven bearing scale reduces the possibility of erroneous true and relative bearings of targets being taken. In all cases great care has to be exercised when translating bearing information from the display and clear directions should be left at the display console that the display is in course up mode.

18.5.1.2 Differential rotary transformer

A common method of effecting compass stabilization makes use of an electrical machine known as the *differential rotary transformer* or, sometimes, a *differential synchro*. A schematic arrangement is shown in Figure 18.5.3. Essentially, the transformer is inserted between the transmitter synchro and receiver synchro shown in Figure 18.3.8, the transmitter being situated at the antenna site and the receiver at the PPI. Referring to Figure 18.5.3, it is seen that the differential machine has stator windings similar to those of the other two machines but in addition its rotor is also three-phase, being fed via three separate slip rings.

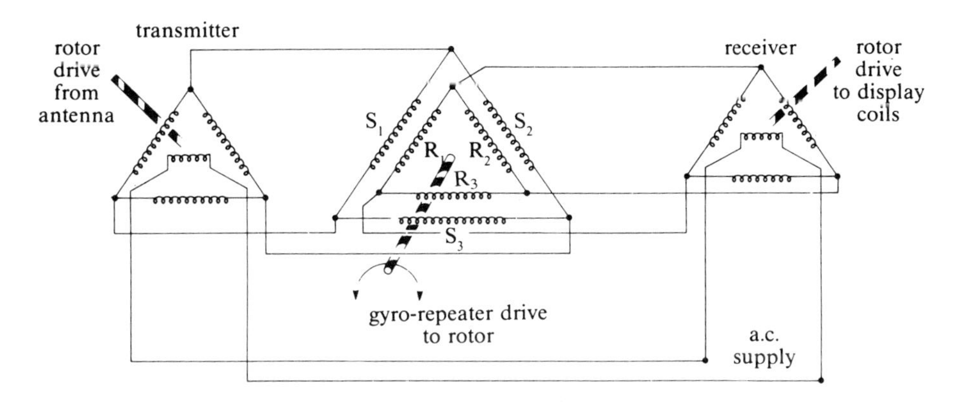

Figure 18.5.3 Schematic arrangement showing principle of compass stabilization using a differential transformer between receive and transmit synchros.

The transformer rotor is driven mechanically by a gyro-compass repeater motor, the rotor assumes a position determined by the ship's true heading. When the stator windings S_1, S_2 and S_3 are exactly aligned with the respective rotor windings R_1, R_2 and R_3, the bearing information from the transmitter synchro to the receiver synchro suffers no angular displacement. If the rotor is re-aligned due to gyro-repeater motion, the shaft angle turned through appears as a displacement between transmitted and received bearing information. Thus all displayed information is displaced on the CRT by the magnitude of the gyro-compass repeater angle relative to some datum (say north). The displayed heading mark will appear on the screen showing the true heading of the vessel and target bearings can be measured relative to north or own-ship heading as desired. Target presentation appears much as it would on a chart having north at the top of the chart. If own-ship alters course, the heading line will swing to the new heading whilst targets remain in the same relative positions on the CRT having north at the top of the display. Figure 18.5.4 depicts a north up stabilized relative motion display.

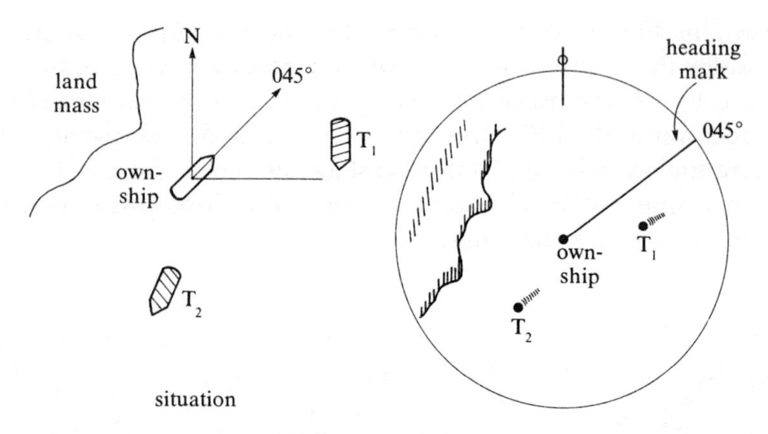

Figure 18.5.4 Relative motion NU stabilized display. Heading mark indicates own-ship course; other targets displayed in relative motion.

18.5.1.3 True motion principles; true motion display

It is seen that north up presentation stabilizes the picture information relative to north; all targets, however, have relative motion with respect to own-ship and the radar observer has to carry out plotting on targets in a similar manner to that required for unstabilized displays. The main advantage lies in the more immediate perception of possible collision courses and the effect of own-ship manoeuvres.

Own-ship and targets can be displayed moving with *true motion* on the PPI, that is having correct scaled velocities by inclusion of a *true motion unit* to modify the presentation of target and ranging information on the screen. If the PPI scan origin is itself moved at own-ship's scaled velocity, all other moving targets will be displayed in real time moving at their scaled velocities. Fixed targets do not appear to move if one's own course and speed are relative to ground. The arrangement provides a display which is chart-like in presentation and from which plotting information is quickly obtained, usually in conjunction with a reflection plotter.

A compass stabilized display is converted relatively simply into true-motion mode by use of a set of orthogonal fixed deflection coils. In some types of display these may also serve as the sweep deflection coils. The true-motion unit provides the necessary coil deflection currents to move the scan origin on the PPI. Essentially, the unit comprises a small analogue computer, or in more recent developments a digital and analogue hybrid circuit. It simultaneously processes voltage analogues of own-ship's scaled speed and course input, together with scaled tide speed and direction, if appropriate, to derive own-ship's scaled velocity (i.e. speed in a given direction). The polar information, velocity = speed $\angle\theta$ m.s.$^{-1}$, is converted into corresponding X and Y (horizontal and vertical) deflection forces to act on the CRT beam via the orthogonal coils.

Deflection force is provided by current passing through the appropriate coil. The horizontal component of deflection and hence motion is derived from speed $\times$ sinθ analogue signal, the vertical component being obtained from the analogue of speed $\times$ cosθ. Two basic methods will be outlined using on the one hand an analogue approach and on the other a mainly digital approach.

18.5.1.4 A conventional analogue TMU

Refer to the block diagram of Figure 18.5.5 showing the essential sections of an analogue true-motion unit. Compass stabilized display with centre off-set facilities and necessary overscan is assumed (18.4.4.2). Scan origin is off-set from the tube centre by passing d.c. bias currents through the deflection coils. The radar observer will in most cases arrange the maximum area of the PPI to lie ahead of his proposed track thus affording a protracted view of targets ahead and on port and starboard quarters by effectively increasing the sweep and displayed range in that zone. Most targets of navigational interest will lie in this area. A typical true-motion display is shown in Figure 18.5.6 and a comparison can be made with the true situation. If the scan origin is moved at exactly scaled ship's velocity over the ground for the selected range, all targets will appear to move with true-motion. Due to tube persistence the afterglow from moving targets, including own-ship, leaves a gradually disintegrating luminous tail which indicates the target track and serves as a plotting aid to the observer. Plotting techniques will need to be applied to the true-motion display in order to estimate target speeds, CPA and TCPA (closest point of approach and time to closest point of approach). True-motion presentation facilitates the plotting procedure when the display is correctly interpreted.

Analogue true-motion unit circuits incorporate *operational amplifiers* designed to perform arithmetic-like operations on analogue signals – hence the name. In this control context the term 'signal' embraces d.c. levels of input. The amplifier circuit can be configured to provide various mathematical operations on the input so that its output may be the analogue of the algebraic sum of two or more inputs, or it may be arranged to produce an output which is the

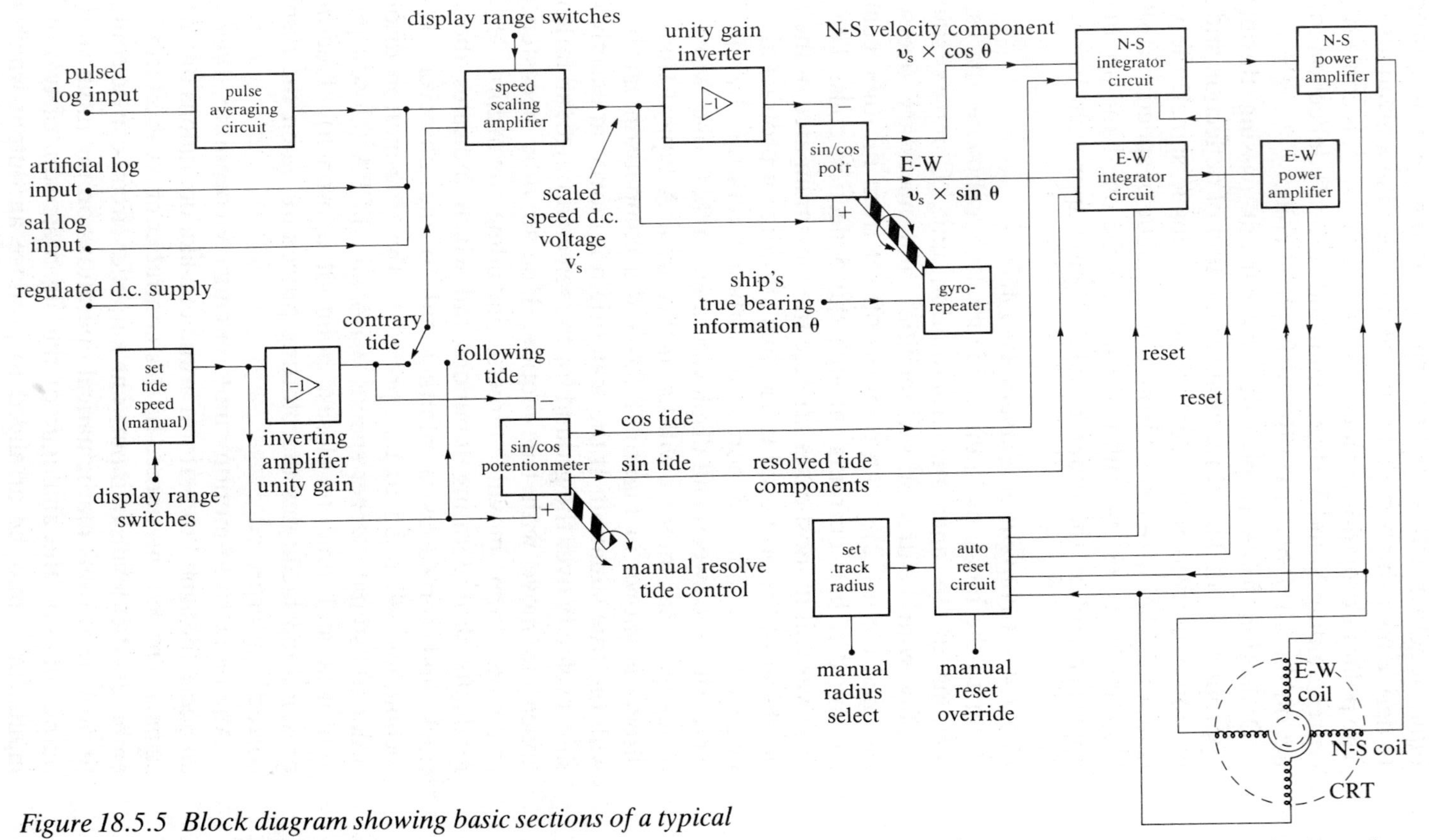

Figure 18.5.5 Block diagram showing basic sections of a typical true-motion unit.

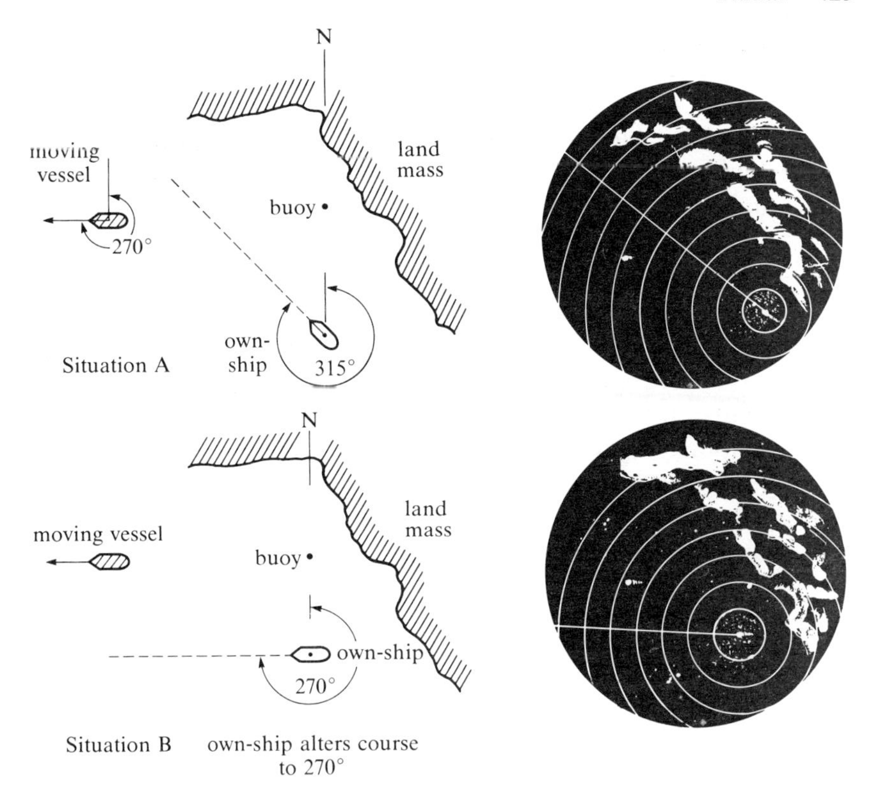

Figure 18.5.6 True motion displays with situations. Situation A; true-motion north-up display, ground stabilized. Situation B; true-motion north-up display, ground stabilized.

integral of the input with respect to time. These are just two of the possible applications; a brief summary of the more common operational amplifier functions in TMUs is given in Appendix C, and Figure C.13.

18.5.1.5 The sine-cosine potentiometer

Two basic types are used, the *flat card potentiometer* and the *contoured card*. These are high-precision wire-wound potentiometers to which a dual pole regulated voltage is applied, $+V$ and $-V$, with respect to OV. Two outputs are available, $V.\cos\theta$ and $V.\sin\theta$, θ being the angular displacement of the potentiometer shaft from some datum position (usually contrived to represent true north). The operating principle is contained in Appendix C, Figures C.14 and C.15. These devices are used as *resolving elements* in analogue circuits to convert own-ship's course and speed and tide direction and speed from polar form into their sine and cosine components.

18.5.1.6 Speed sources

An input signal representing the ship's speed parameter is necessary and is derived from a *speed-log*. Various options are generally available; a pulsed input from the ship's log typically providing 200 pulses per nautical mile, doppler logs, pressure tube logs, etc.

Unless a bottom-locked four-beam doppler log is used or correct resolved tidal information is fed in, the display is said to be *sea stabilized*. This mode of presentation has certain advantages since own-ship trails and target trails shown on the PPI indicate the target headings regardless of tidal forces. The *artificial log* (Figure 18.5.5) generates an input signal representing ship's speed selected by the radar operator, usually when it is impracticable to obtain ship's log input. Simulated values up to 30 or 50 knots are common.

18.5.1.7 Speed scaling

Output from the selected log source is applied as a representative d.c. voltage to the speed scaling amplifier. The speed amplifier acts in conjunction with the display's range selection switch adjusting the circuit parameters to provide a correctly scaled amplifier gain for each of the ranges offering true-motion facilities.

The circuit is necessary because the same PPI is used to represent ranges from say, ¼ nm to 24 nm in true-motion mode. If a 406 mm tube is used it is apparent that when own-ship has a speed of 12 knots and the three-mile range is chosen, the scan origin must traverse 203 mm of the display in 15 minutes. If the 24 nm range is chosen the 203 mm movement must take two hours. Correct scaling of the input speed voltage is in this case achieved by switching the voltage gain of the speed amplifier to be eight times greater on the shorter range than on the longer range. Similar reasoning applies to other true-motion ranges. Here we have assumed that total tube radius is used. In practice it will be less than this value.

18.5.1.8 Tide speed; resolved tide speed

When a tidal or river current is following or contrary to ship's motion an appropriate analogue can be input to the scaling amplifier at a *summing junction*. The resultant analogue output is the scaled algebraic sum of ship and tide speed. For tidal forces other than in-line following or contrary, it is necessary to resolve the tide speed information.

A sin-cos potentiometer acting on the analogue of tide speed (as shown in Figure 18.5.5) is manually adjusted to provide the resolved N–S and E–W components. The two resolved values are then input to the N–S and E–W integrator circuits via a summing junction. Circuit action integrates the algebraic sum of the ship's speed and tide speed at each integrator. Evidently, misuse of tide speed and resolved tide controls can introduce errors in displayed target information, particularly in open sea situations. These controls are used to best advantage when the display can be adjusted to display a prominent fixed echo as a stationary target producing no paint smear, i.e. a *ground stabilized* display. Observation of apparent drift of fixed targets on the PPI gives the navigator an estimate of tide speed; the direction of drift indicates the required setting of the sin-cos resolver contrary to that drift.

It is a mandatory requirement for type tested equipments that the manual tide resolver control is calibrated in degrees or similarly shows the cardinal compass points and in operation indicates the direction of the water motion. The tide speed control must provide analogue inputs from zero to *at least* 9.9 knots continuously variable or in steps not greater than 0.2 knots.

18.5.1.9 Ship's course resolver

Course information is obtained from a gyro-compass repeater mounted within the true-motion unit. It drives a sin-cos potentiometer shaft, the angle of which when taken relative to some datum position represents own-ship's true bearing. When the vessel changes course the repeater transmits the change to alter the shaft angle turned through. Let this angle be θ. Due to the manner of connection (Appendix C), two outputs are derived from the potentiometer, these being $v_s. \cos\theta$ and $v_s. \sin\theta$, where v_s is the voltage analogue of scaled ship's speed. The two resolved components are thus voltage analogues of the N–S and E–W components of ship's velocity respectively. Translated to the PPI they represent scan origin motion in a vertical or horizontal direction and are in turn applied to the N–S and E–W integrators.

If own-ship changes speed or course these will also change to new resolved values and the rate at which the changes occur will depend on the manoeuvring ability of the vessel.

18.5.1.10 Integrator circuits and power amplifiers

Two identical integrating amplifiers act on the resolved N–S and E–W signals over the period for which the inputs are applied to produce an output voltage which is the integral of the input with respect to time. If the input remains at a steady voltage level (positive or negative polarity), the output will appear as a linear ramp (rising or falling). Figure 18.5.7 illustrates one set of conditions.

Each integrator has at its input a summing junction where the resolved ship's course signal can be algebraically summed with any resolved tide signal, i.e. the two signals are combined there. Output from the N–S integrator is therefore the integral of the combined input signal due to ship and tide vertical motion. Similar reasoning applies to the E–W integrator and ship and tide horizontal motion.

A power amplifier associated with each integrator then provides a rising or falling ramp of drive current to the N–S and E–W orthogonal deflection coils causing the scan origin to move on true course at scaled speed.

18.5.1.11 Set radius and reset circuit

Type tested radars operating in true-motion mode must be capable of resetting the scan origin. The origin is not permitted to move outside a circle centred in the PPI which has a radius 75 per cent that of the display. Manual or automatic reset can be provided and in the former case an audible alarm is necessary to indicate the prescribed limit of traverse. Automatic reset circuits are provided with a radius setting control which is manually initialized to provide automatic origin reset within the prescribed radius limits. The trace origin may be manually reset at any time by pressing a *manual reset* button.

In practice the operator will adjust the scan origin for maximum view ahead and to port and starboard quarters. This is generally at an off-set radius diametrically opposed to the ship's track and effected by an *off-set radius* and *direction* control. These controls position the origin at its start of track and select a particular voltage level to represent the limits of track. When the outputs from the integrators agree with the set level the integrators are *reset* automatically. This action causes the scan origin to rapidly return along the ship track to its original point on the screen. Since the display ahead is newly painted in a different screen position a short time lapse and picture blanking are devised so that a confused presentation does not occur. The integrators then proceed from their initialized condition as described. The effect at the PPI seen by the observer is that of own-ship traversing the screen with targets moving in true motion, then at the preset radius the picture is momentarily blanked and own-ship appears at the original setting point on the screen. All targets previously displayed in the range ahead will have moved correspondingly and preserve their displayed true-motion characteristics. New targets which lay ahead and are now within display range will appear on the display. The origin continues to move. If own-ship makes a change of course the displayed effect is that of the heading marker swinging to the new course with some smearing of the heading mark trace as this occurs. All other moving targets are shown at correct relative bearings and range with their afterglow giving an indication of target headings, no target smearing occurs.

18.5.2 *True-motion achieved digitally*

Sections of this book dealing with video processing and automatic radar plotting techniques outline the increased application of digital electronic engineering to the radar display circuits. Digitally-controlled circuits applied to the generation of a true-motion picture are able to provide deflection accuracies better than analogue counterparts, have a lower component cost overall and reduced power consumption. In addition, certain problems encountered with analogue circuit design concerning thermal effects and integrator drift which often lead to complex circuits and high cost, are largely avoided using digital methods. Figure 18.5.8 shows the true-motion unit employing such control. This has necessarily been simplified and would form only a small part of the display's digital system. The microprocessor will usually perform more than the relatively simple task of moving the scan origin on the PPI.

18.5.2.1 Interface logic

A variety of arrangements are possible but in the arrangement shown the interface logic can be considered to comprise several registers acting as temporary storage for the digitized information from the several sources shown. It will also contain logic for address decoding and for the generation of control signals.

18.5.2.2 Inputs

Tide speed and tide direction inputs act directly on digitizing circuits which output a binary pattern whose 'code' represents these parameters. Instead of the more familiar rotating manual control associated with these functions it is possible to have push-buttons which are held as the desired value is clocked into its appropriate

+
volts 0
start
V log-input to scaling amplifier
t_1
time

volts 0
−
V contrary tide input

+
volts 0
output of speed scaling amplifier
$v_s = V \log - V \text{ tide} \times$ scaling fact
t_1

volts 0
−
output of unity - gain inverter = $- v_s$

+
volts 0
output N-S integrator at t_1
for $\theta = 045°$
$v \text{ out} = \int v_s \times 0.707\, dt$
corresponds to deflection of origin vertically
t_1

+
volts 0
output E-W integrator at t_1
for $\theta = 045°$ this will equal N-S integrator output.
Corresponds to deflection of origin horizontally
t_1

+
amps 0
−
corresponds to horizontal line through PPI centre
N-S deflection coil current ramp
t_v
t_1

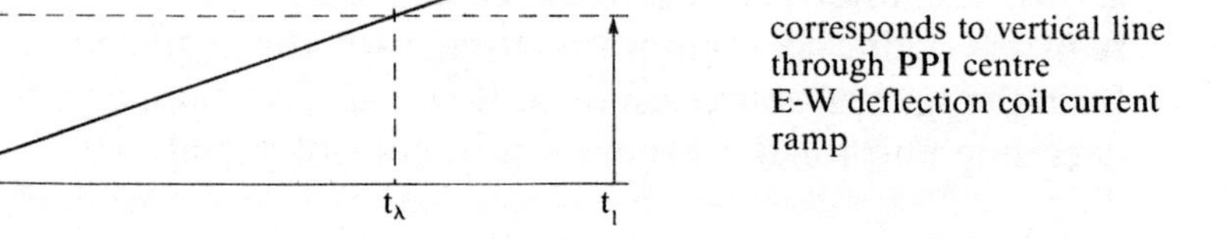

Figure 18.5.7 (a) Idealized true-motion unit signals. The value t_1 is dependent on ship speed. The track traversed on the PPI might appear, as in Figure 18.5.7 (b).

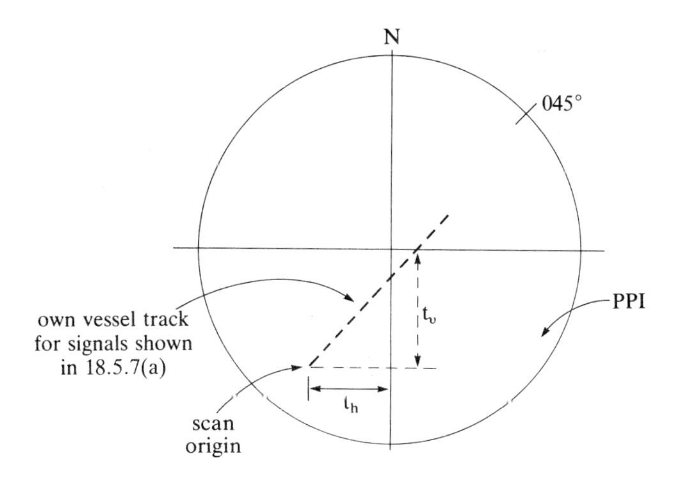

Figure 18.5.7 (b) Track of own-ship for conditions in Figure 18.5.7(a). Scan origin is offset by operator. Origin tracks automatically to set limit then resets to positions shown.

register. A numerical indicator shows that value. These comments apply also to the 'set radius and direction' control which has the same function as in the analogue system.

Log inputs are for the majority of travel time in a steady state condition, with the exception of pulsed or make and break log inputs. The former are converted from their analogue level to a digital code representing that level in an analogue to digital converter (A to D converter). The output code is then representative of speed. Since pulsed log inputs represent speed by the pulse rate it is necessary only to buffer that input and detect the presence or absence of a pulse relative to time; a requirement which is met very easily in the microprocessor program. Similar remarks apply to the range switches which are either open or closed. Compass bearing input is commonly converted to digital form in one of three ways. In the first of these the compass repeater drives a resolver, a machine which has orthogonal stator and rotor windings (see section 18.4.4). It produces an output which is the cartesian equivalent of the polar input. These two values are converted from their analogue form into digital (in A to D converters) and are then mathematically manipulated under program control. The second method is to employ a digitally encoded disc whose principle is outlined in Appendix C, the disc being driven by the compass repeater. Some manufacturers still derive the resolving function by use of a sine-cosine potentiometer following this with A to D conversion. There

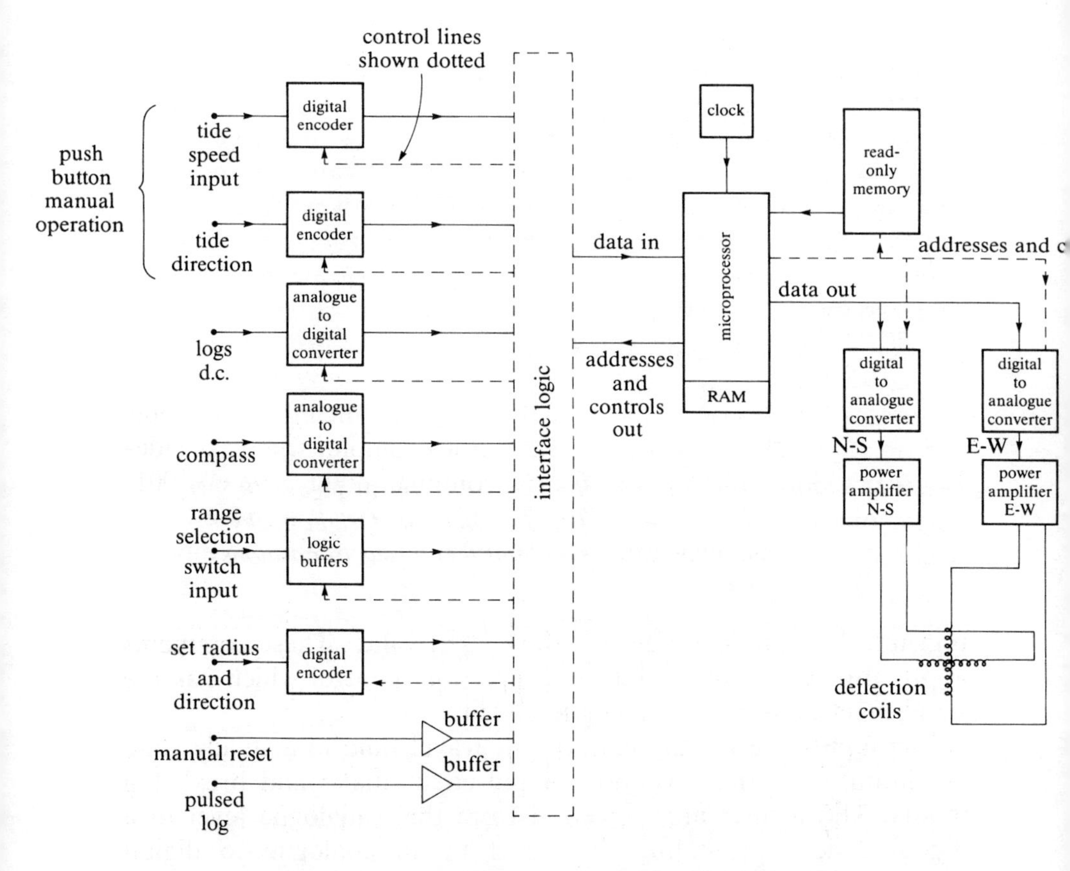

Figure 18.5.8 Basic principle of digitally-controlled TMU.

are also resolving methods used which incorporate three-phase synchros.

18.5.2.3 Processor: read only memory and output peripherals

The microprocessor shown uses an area of 'on chip' RAM containing the program sequence which under processor control manipulates input and output data. Since the random access memory is volatile its program is bootstrapped into the RAM from a non-volatile memory whenever the circuit is powered up. The ROM also contains program routines used in the various arithmetic processes for speed scaling and integration. Data are output directly into digital to analogue (D to A) converters which produce the relatively slowly changing N–S and E–W voltage ramps. Finally, analogue power amplifiers provide the necessary drive current to the CRT fixed deflection coils.

18.5.2.4 Principle of action

An external crystal controlled clock produces a square waveform from which the microprocessor derives its timing cycles. Each peripheral device has a unique address and as the program steps are executed each device will be addressed in a given sequence. An address decoder generates a 'chip select' control pulse from that address. During the *read* cycle of the microprocessor, data are transferred from the selected input device to the processor internal registers via the data bus. During this period all other data inputs are deselected, a process which effectively disconnects them from the data bus.

An appropriate program section identifies what the input represents and operates on the digital contents. For example, when the tide-speed encoder contents are input its contents might be 0000 1001, representing 09 decimal or 9 knots. This value is transferred to a RAM location for inclusion in calculations. The tide-speed input is then de-selected and becomes invisible to the processor and other devices. Tide direction input is then addressed and data are input in a similar manner, being saved in a unique RAM location. Each input is *polled* in the same way and saved. The program in RAM draws upon routines stored in ROM to calculate correct scaling for the range in use, summing of various inputs and the integration process.

The processor outputs data to the N-S and E-W digital to analogue converters during a *write* cycle, addressing those devices in turn, whose individual outputs then assume the analogue voltage level represented by the digitally-calculated inputs. Continuous updating of the resolved, scaled analogue deflection voltage occurs to produce output signals similar to those of Figure 18.5.7. These functions can be performed in a few tens of microseconds, depending on the length of program routine, which is far faster than is necessary for the comparatively slowly changing inputs and outputs. A great deal of the computing power and control capability of the processor will be redundant in such a simple application and most systems will have the microprocessor performing many other tasks – apparently simultaneously, in other sections of the display circuitry. The tasks outlined could well be performed using simpler devices such as programmable logic arrays (PLAs) but the microprocessor can still prove to be the most cost effective way to do the job.

An alternative method of programming is to operate the processor on *interrupt* which avoids the relatively time-wasting polling routine outlined. Here the peripheral requiring servicing by the processor pulls down a unique *interrupt request line* connected to the processor and all other peripherals. The processor is then forced into an interrupt routine where it recognizes the device, services it and then returns to the task which it had temporarily abandoned at interrupt. Where the processor is not likely to be heavily engaged in a task wherein the machine cycle timing is likely to be critical, interrupt operation is probably the most efficient method. For relatively slow uncritically timed tasks such as those few already mentioned, the polling method is adequate.

18.5.2.5 *Speed error*

A truly accurate speed-log does not exist; all are subject to error and it is *not always possible for the navigator to be aware of this error*. In the absence of locking the display to a prominent fixed target to obtain ground stabilization, the best compromise for civil marine use is the bottom-locked dual axis doppler log. Not all true-motion units are equipped to deal with input from such a log. Given that log input may contain some speed error the true motion unit must operate on that input to produce a scaled motion of the trace origin having an error not exceeding 5 per cent or 0.25 knot, whichever is greater. Although manufacturers produce artificial logs generating up to and beyond 75 knots, it is probably more realistic to assume a maximum speed of, say, 20 knots for our example. In this case the maximum error permitted becomes one knot which proportionately diminishes with reduced speed. These tolerances are relatively easily designed in analogue circuit form. In digital form we may represent the speed by means of a single byte (eight binary digits). This will give a resolution of 1/256 or, at our 20 knots, 0.078 knots error. Even at 75 knots the digital method produces only 0.293 knots resolution error, or 0.4 per cent of the input speed which comes very close to the desired value. Adding an extra bit to the word gives 0.146 knots resolution. The above approximations have assumed a one-bit accuracy. Supposing that such accuracy is required (or desirable), the digital method is superior in that circuit design is simpler, generally less costly and not subject to ageing and thermal effects which beset analogue circuits.

18.5.2.6 *Course error; drift error*

For type tested systems, the course error generated by the TMU circuitry must not exceed 3 degrees. Gross errors should initiate checks on heading mark accuracy, bearing transmission circuits and finally readjustment of analogue tracker circuits.

Achieving accurate alignment in analogue circuits is not trivial since the track depends not only on consistently accurate bearing resolution but also on long-term consistent performance of speed amplifiers, unity gain inverters, integrators and resolvers on each of the true-motion ranges. The integrators generally pose the greatest design problem, being sensitive to small impedance changes. They have typically been assembled in enclosed compartments containing a desiccator to offset the effects of humidity. This is the *dry box* structure. The appearance of linear integrated circuitry and the integrated circuit operational amplifier has considerably simplified design and alignment procedures but humidity problems remain. The integrator is also subject to *drift error*.

A typical analogue true-motion unit specification will quite likely quote *zero speed drift* in terms of the observed scan origin motion *when there is zero speed input*. As an example, it might be that drift will occur over not more than 5 per cent of one quarter of the tube diameter during a 30 minute period or 0.25 knots, whichever is the greater. For a 406 mm display this amounts to 5.0 mm. On the 24 nm range the drift becomes equivalent to 1.2 knots, a drift of 0.6 nm in 30 minutes. On the 1/4nm range the same distance represents 0.00625 nm or 0.0125 knots. If the 0.25 kn tolerance is applied to this shorter range the drift could be as great as one half the radius in one half-hour period.

To appreciate how an analogue integrator can drift, the simple circuit of Figure C.13(d) (Appendix C) will suffice. A feedback component is inserted between the output and the input terminal of the operational amplifier. A constant current component supplied by the output voltage needs to be drawn through that path in order to satisfy bias conditions at the input. If a d.c. feedback path is provided, as in the case of the scaling amplifier, a constant off-set voltage appears at the output. In the integrator circuit a capacitive feedback path is provided through which the op-amp action tends to force a steady current. The capacitor charge increases with time, affecting the input value and causing the output to drift from a steady value which should be maintained.

Drift can be minimized by careful adjustment of parameters, but over a long period the integrator will drift to saturation. On a true motion PPI the effect is to introduce apparent scan origin motion where none should exist. Since the integrator is usually reset at frequent intervals on short ranges, this action discharges the capacitor to produce zero integrator output voltage. The problem is not then crucial. Over longer tracking periods a track error will result due to drift which, being on the longer ranges, is seen from the previous example not to be critical for the observed errors.

18.5.2.7 Anti-collision markers

In true-motion mode one system offers an additional target supervisory feature in the form of anti-collision markers. These are displayed as short straight lines during the intersweep period and can be adjusted independently of the radar display. One end of the marker displays a bright spot which is placed over a target of interest. The marker extends from the target towards own-ship, maintains a fixed range and bearing from own-ship and moves across the screen with the scan origin. All targets move with true-motion whilst a target moves in relative motion with respect to markers. Figure 18.5.9 indicates that a target moving steadily down the length of its marker is on a collision course. Deviation from the marker track indicates a target passing ahead or astern. Simple plotting will indicate CPA.

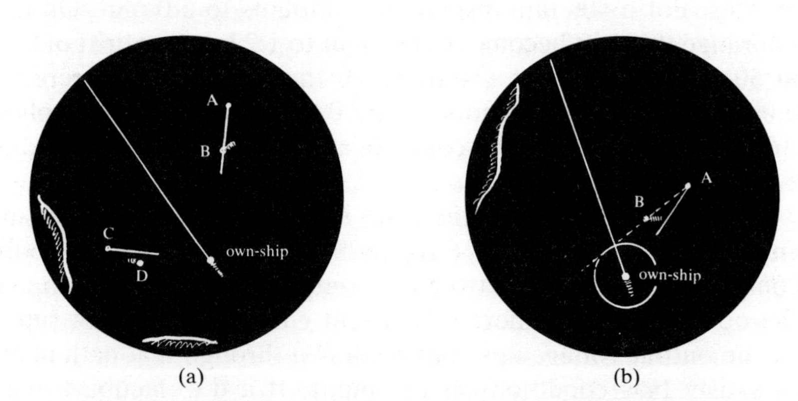

Figure 18.5.9 (a) Anti-collision marker placed on Target B at A. B has not altered course and is on a collision course. Target D, acquired at C, is going to pass astern. (b) Target B marked at A is moving across own-ship heading. The dotted line through B from A gives CPA for these maintained courses.

18.5.3 Digital displays – introduction

Considerable development has taken place during the last decade in the field of semiconductor technology and, in particular, in the area of large-scale and very large-scale integrated circuits. There is now available a very wide range of high-speed, low power, highly reliable and relatively cheap digital devices. Memory chips of all types have steadily increased in capacity, microprocessors have become faster and more versatile and high-volume production of analogue-to-digital and digital-to-analogue conversion chips with improved resolution has reduced their cost. In parallel with these developments there has been an increased demand for more sophisticated marine radar equipments of improved performance and operational capability and with improved reliability in a fairly hostile environment – all at a moderate cost to the shipowner.

Most of the technological and design modifications in civil marine radars in this period have been in the display section. With improved processing power signal processing, techniques previously found only in costly civil aviation or military systems are now common in marine radar sets. When compared to an analogue display system a simple digital display offers greater flexibility and is very cost competitive. Such displays are easily integrated into digital data processing systems whereas analogue displays are not. The principles underlying digital display of target information and methods used to present that information will be outlined in the following sections.

18.5.3.1 Signal conditioning

In an analogue display the 'raw' video signal is used to intensity modulate the beam current of the CRT. Opening paragraphs of section 18.4 emphasize the importance of target echo detection in the presence of unwanted noise components and the video output from a receiver will show a randomly varying signal amplitude with time. This contains target information, the desired signal, embedded in noise. The latter component depends upon a number of factors which, for simplicity, we can consider to be subdivided into two sections: system noise generated by the radar set and external noise which is generated by randomly varying aspects of target returns and includes interference from other marine radars. System noise can be minimized by careful design but never eliminated. Random returns from wave surfaces and rain create the so-called sea and rain clutter which is variable according to weather conditions and own-ship attitude.

Detection of a target return depends on its signal amplitude exceeding the level of the noise in an unambiguous manner and normally a *detection threshold* is set up by adjustment of receiver gain. Signals which exceed that threshold are recognized as targets, signals whose amplitudes are below the threshold are not recognized and will generally be lost. Manual adjustment of the threshold can reduce it to a level where weaker signals are salvaged but noise begins to intrude; the appearance on the PPI is of an increased background speckling which visually intrudes on the displayed information. As described in section 18.4.3.3, the adjustment of sea and rain clutter controls and receiver gain is a matter of some skill if small important targets are to be preserved. If the threshold is adjusted too high all lower amplitude returns are lost together with noise but so also will larger amplitude signals be suppressed and not appear on the display. An event where noise can be mistaken for a desired signal is known as a *false alarm*. Evidently a compromise is reached where the operator uses skill and judgement to adjust the threshold from time to time in an effort to minimize noise and maximize signal output, in other words, the threshold is adjusted to produce a desired probability of a false alarm. Unfortunately, slight changes in the threshold level can cause large changes in the false alarm probability and human control of that level is relatively slow. For example, sea clutter returns are not similar throughout a 360 degree azimuth scan being greater at close ranges to windward than to leeward – they change with the rolling and pitching of own ship and as the ship manoeuvres.

Modern displays which use digital control of video data and which may in addition provide computer-controlled tracking facilities generate a *constant false alarm rate* (CFAR) for all pointing conditions of the antenna, immediately and automatically. This provision is particularly important for automatic *track while scan* applications found in automatic radar plotting systems or the tracker will be overwhelmed by false alarms and be unable to function.

18.5.3.2 CFAR; principle

Various design approaches are possible. The objective of this text is to describe the general principle rather than particular circuit detail which varies considerably with different manufacturers. In order to derive a threshold level which adapts itself to changing signal levels from sweep to sweep, an *adaptive gain* circuit is commonly employed. The adaptive circuit samples the receiver output level over a sweep period and derives from that a dynamically adjustable threshold level. This level is then applied to the receiver circuits. Figure 18.5.10 shows the principle.

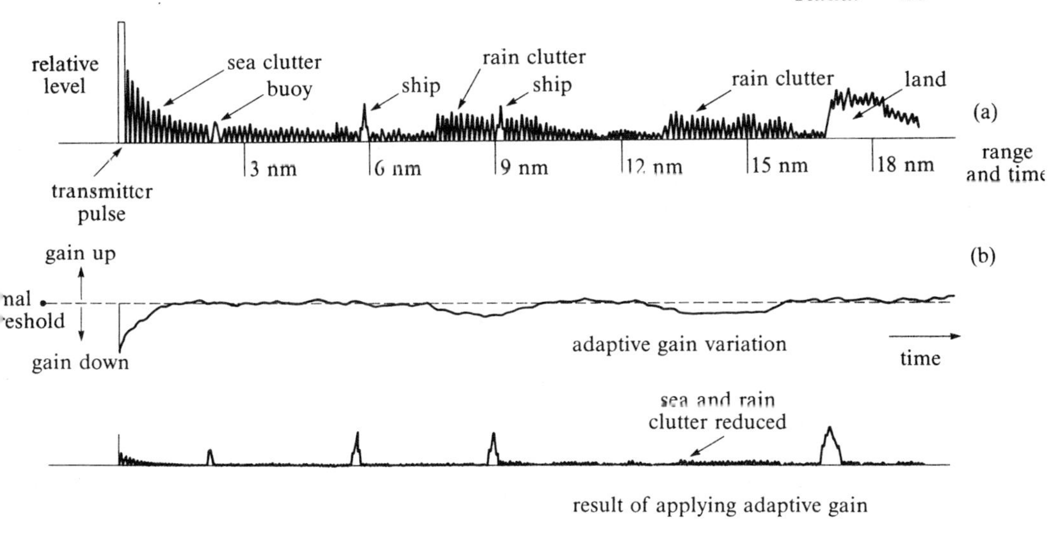

Figure 18.5.10 Effect of adaptive gain control on analogue signal.

Generation of the adaptive threshold level can be implemented using a variety of analogue techniques; most seem to favour a fairly direct method of integrating the signal. It will be appreciated that although the false alarm rate can be reduced to a very low and sensibly constant level, it cannot be reduced to zero. As it is, the circuit action will eliminate some very small target echoes from scan to scan. The number of false alarms which do occur will therefore produce a very light background speckling when the *manual* gain is increased.

Another more sophisticated method of adaptive gain control inspects the signal level in short 'time slots' immediately prior to and following a time slot in which a target may lie, and from the results a gain level is derived. Powerful computing and storage facilities are needed, these being readily available as single-board microcomputers.

In this way an immediate and automatic CFAR is obtainable. In practice the raw video is usually subjected to a controlled degree of differentiation to break up large blocks of clutter, and use of the technique in displays affects the appearance of Racon responses and other profiles of targets which appear along the length of the sweep, for example, coastlines. Echoes have the characteristic appearance of a sharp leading edge and lack of depth. Parameters of the circuit must be adjusted to provide limits to the dynamic range of automatic adjustment or the circuit may saturate the display on the one hand or suppress paint on the other, thus creating dark 'holes' in the picture.

There are digital displays used in the marine market which, when operated in relative motion, compass stabilized modes without additional automatic plotting facilities, do not employ CFAR circuitry. Should the display require conversion to ARPA mode a number of optional circuits and a tracking computer may be obtained for that purpose.

18.5.3.3 Digital video

Target returns shown on the display are stored and manipulated as digitized video. After reducing the rate of unwanted echoes a relatively noise-free video signal is produced by eliminating all signals below a certain level. In the simplest of circuits a binary condition is accepted, signal or no signal; thus all video appearing above the no-signal level has a logical '1' value irrespective of its actual analogue value. A logical '0' is associated with no-signal level. Some systems quantize the analogue levels, attaching binary codes to them. These codes, whatever their complexity, can be stored during ranging time and *replayed* or read from the memory in which they were stored, during the period between successive sweeps.

Several advantages immediately become apparent. All echoes can be displayed at full brightness by the simple expedient of brightening the trace optimally whenever a logical '1' is output and switching the beam current off during logical '0'. The full dynamic range of the tube phosphor can be utilized to obtain maximum contrast. Since the echoes are digitally stored the data can be read from memory and displayed on the CRT at a constant sweep rate *no matter what range may be selected*. This allows data to be written to the tube phosphor more slowly on the shorter ranges producing a bright display, and provides brightness compensation on the longer ranges. Because a high rate of change of sweep deflection current is avoided for the lower ranges, the analogue deflection circuit bandwidth is reduced which simplifies design. On the longer ranges the sweep period is reduced which reduces power dissipation in the deflection circuits. It is a simple matter to artificially lengthen a digitized echo to enlarge and brighten its paint at the PPI, a process known as *pulse enhancement* or *pulse stretching*. Finally, the stored digitized echoes from one sweep can be compared with those in the next sweep to retain consistent echoes and eliminate random signals. This process is known as *sweep to sweep correlation* or *two out of two* correlation and removes high level noise and interference from other radars. A similar *scan to scan* correlation can be made when target echoes are

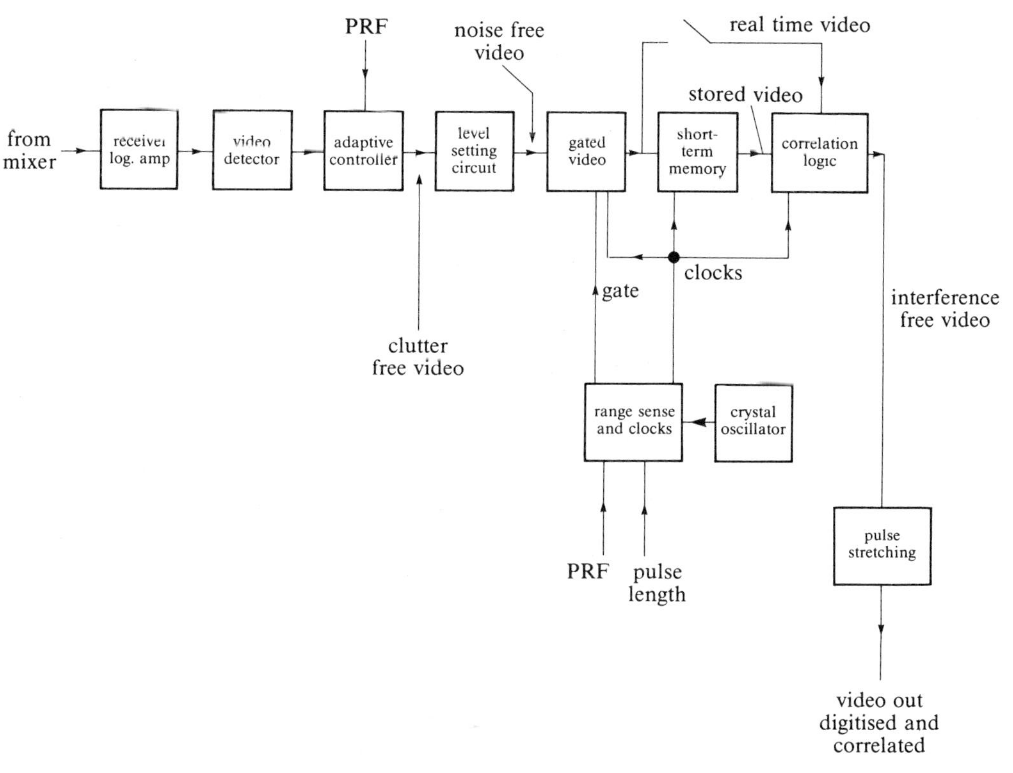

Figure 18.5.11 Basic arrangement for signal conditioning and processing.

being tracked in a tracking computer. Figure 18.5.11 illustrates the basic system requirement.

18.5.3.4 Digitized video processing principles

Processing principles are best explained with the aid of a few simple diagrams and examples, which is the approach adopted here. Some of the digital display methods will be better understood by a short recapitulation of the relationship of a displayed analogue signal to its pulse width and range discrimination; the reader is referred to section 18.2.4.5. If a transmitted pulse length of 0.05 μs is employed on a particular range (commonly between 1/4 and 3 nm, Table 18.2.2), target returns will generate a video pulse of the same length. A rectangular pulse envelope will be assumed. Suppose the effective PPI diameter is 366mm. In a centred condition one radial sweep will extend 183mm and on the 3nm range the sweep time is 3 × 12.3 μs = 37 μs. Since the video pulse occupies 0.05 μs it can be considered to occupy one of 37/0.05 *time slots* or *time cells* in the sweep time, that is one of 740 cells extending from the scan origin to the extremity of the trace. In a similar way the tube radius can be thought of as being subdivided into 741 cells, each of length 180/740mm, or 0.25mm.

Note also that a single paint from a point target would have only 50 nanoseconds painting time. If a 75 per cent oversweep is arranged for off centring, the longest sweep contains 1258 cells.

Selection of a long pulse of, say, 1 μs produces 63 cells with oversweep. Similar reasoning can be applied to the longer ranges where a range of 48 nm has a sweep time of 590.5 μs and using a pulse length of 1 μs each sweep (no off-centring) has about 590 cells. Evidently the cells are a function of sweep time and pulse length and differ on each range. Recollect that if a 1 μs pulse is used on the 3 mile range a return from one point target will paint over radial length of 5.0 mm, which is twenty times greater than the short pulse paint – but there is a corresponding decrease in range discrimination. The same pulse length used on the 18-mile range will paint over 0.3mm, a very small paint.

Because of these pulse and paint discrepancies and the high and low phosphor writing speeds on different ranges, the brightness and contrast in the PPI picture changes from range to range in an analogue display, despite efforts to compensate the brightness level automatically. Furthermore the writing speed is so high at very short ranges that it is not possible to obtain a bright *daylight viewing* picture with good contrast because of the limited dynamic range of the tube (about 10dB). A digitally-processed picture can, however, overcome these limitations and provide a picture bright enough to be viewed in daylight and with displayed pulse enhancement to emphasize the target paint when desired. The display cannot, of course, be comfortably viewed in direct sunlight. Range resolution is a function of cell size. Due to the binary nature of target storage and the discrete manner of strobing it is possible for digitized video to be 'shifted' in range by ± one-half cell. If the cells bear a one to one relationship to the 'bins' used for storage then range resolution of the digital signal falls within that desired for type tested radars. If the ratio of bins to range cells is increased, the ranging accuracy improves. The technical trade-off required is that clocking frequencies must rise which can raise design problems on very short ranges.

18.5.3.5 Video processing

During reception in real time, the conditioned radar signals are stored in a short-term memory. Numerous design solutions exist; the two most commonly applied are shift registers and random access memories. In application both produce the same manipulation of video data.

The principle of storage is illustrated in Figure 18.5.12. A transmitted pulse initiates a *ranging gate* which admits the returned echoes during a ranging period equivalent to the normal sweep time for the range in use. This is extended to about 1.7 of the range for those ranges using true-motion. Since the ranging time can be divided into picture ranging cells, the memory capacity is arranged as an equivalent number of one-bit memory locations commonly known as *bins* (binary storage unit). At the gate start a train of clock pulses are produced from a crystal-controlled source which continue for the period of the ranging gate. Timing of the clocks produces a strobe at the centre of each range cell period (at least), transferring the analogue condition of the range cell into its corresponding memory bin. At the end of the ranging gate the memory contains

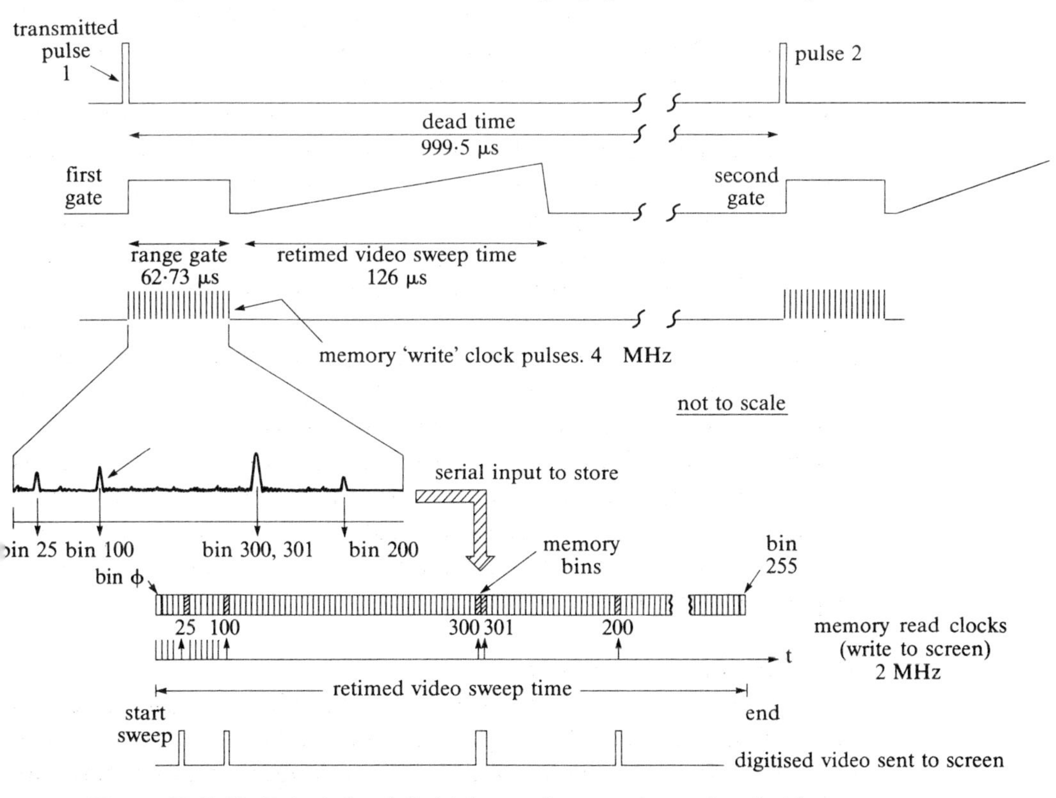

Figure 18.5.12 Principle of digitizing and generating retimed video – uncorrelated.

digital video information corresponding to recognisable signal levels and further returns from the receiver are then blocked.

Suppose a 0.25 μs pulse is transmitted on the 3nm range. The ranging gate has a duration 3 × 12.3 × 1.7 = 62.73 μs. The pulse length divides the radial scan into 251 range cells and we choose to have 256 bins (simply because of the easy binary arithmetic). The range cells will need strobing every 62.73/256 μs at the centre of each cell, the clock frequency then being at least 4 MHz. If the number of bins is increased to, say, 512 and the clock frequency to 8 MHz, one echo pulse fills two bins and there is an improvement in the digital discrimination. The latter arrangement ensures that an analogue target return cannot fall precisely between two strobe pulses and be missed.

Memory contents can now be written to the CRT at a slower speed than the real-time ranging gate, thus producing a much brighter paint. The 'replay' can take place at the end of the range gate when the sweep waveform is generated to produce a constant sweep time *for all displayed ranges*, typically 126 μs, which corresponds to the 6 nm range plus oversweep. During that time the contents of the 256 bins due to the 3 nm actual range are serially clocked or *read* from the memory and written to the CRT in the order in which they were filled using a lower clock rate of 2.0 MHz. The video input to the tube is known as *retimed video*. On each range the memory capacity is adjusted to accord with the desired number of range cells and appropriate clock rates selected to *write* the signals to the memory, storing them as video data.

For the example given, data stored on ranges below 6 nm will be retimed and written to the CRT at slower rates than real-time whilst on ranges above 6 nm data is sent to the display at higher rates. On those ranges above 24 nm where true-motion is not provided, there will be no provision for oversweep. Data on the 6 nm range is clocked into and out of memory at the same rate and intermediate storage may seem unnecessary. All data is, however, stored in this manner to facilitate other digital processing techniques such as target correlation.

18.5.3.6 Paint integration

A method of increasing the brightness of displayed digitized echoes on all ranges is that of *multiple painting*. Some manufacturers use this to balance the paint brightness for longer ranges where the video retiming process writes target information to the screen faster than it is initially stored. Once the video data has been stored it can be written to the CRT two or three times during the dead time. Figure 18.5.12 shows the basic timing for the 3.0 nm range using a PRF of 1000 pps. The ranging gate is 62.73 μs followed by a retimed video sweeptime of 126 μs, the deadtime being 999.75 μs. Allowing a few microseconds for flyback, etc., the whole process takes about 200 μs, leaving nearly 800 μs of unused dead time. The data may therefore be rewritten over the same strip of phosphor by generating a further sweep and memory readout. The integrating effect at the phosphor produces a very bright paint. There remains sufficient dead time to display other information in the form of *synthetic video* (section 18.5.3.9).

18.5.3.7 Basic correlation technique; interference rejection

Interference from other radars produces a distinctive pattern on the uncorrelated display or on an analogue display. Figure 18.5.13 shows the general appearance as a series of spurious paints in a spiral pattern appearing randomly from the centre of the PPI. Since they are due to high-level transmission from other radars, they may have the level and width to be digitized and painted as described in section 18.5.3.4. In normal radar operation as the antenna beam sweeps over a target it will register several hits at the same range on successive sweeps. Noise will appear randomly. Digitized video from two consecutive range gates is saved in two identical memories. The contents of the two sets of ranging bins are then compared, bin for bin. Where the contents of two identical range bins correlate a valid echo is assumed to exist at that range and it can be clocked to the CRT. Where correlation does not exist the

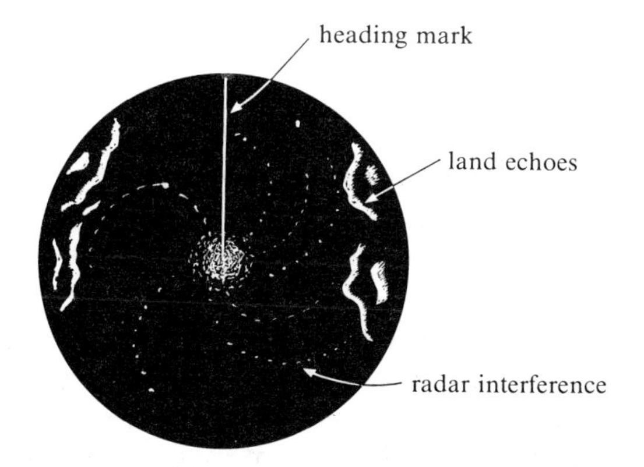

Figure 18.5.13 Diagram showing appearance of spurious radar interference.

contents of that bin are ignored. The latter case has a higher probability of occurring when interference from other radars is randomly being stored. In the sweep-by-sweep correlation method there have to be *two-out-of-two* identical range cells filled on consecutive sweeps before a target is accepted as existing at that range. Note that the first range gate contents are not displayed but this action has an insignificant effect on the display. Figures 18.5.14 and 18.5.15 illustrate the process.

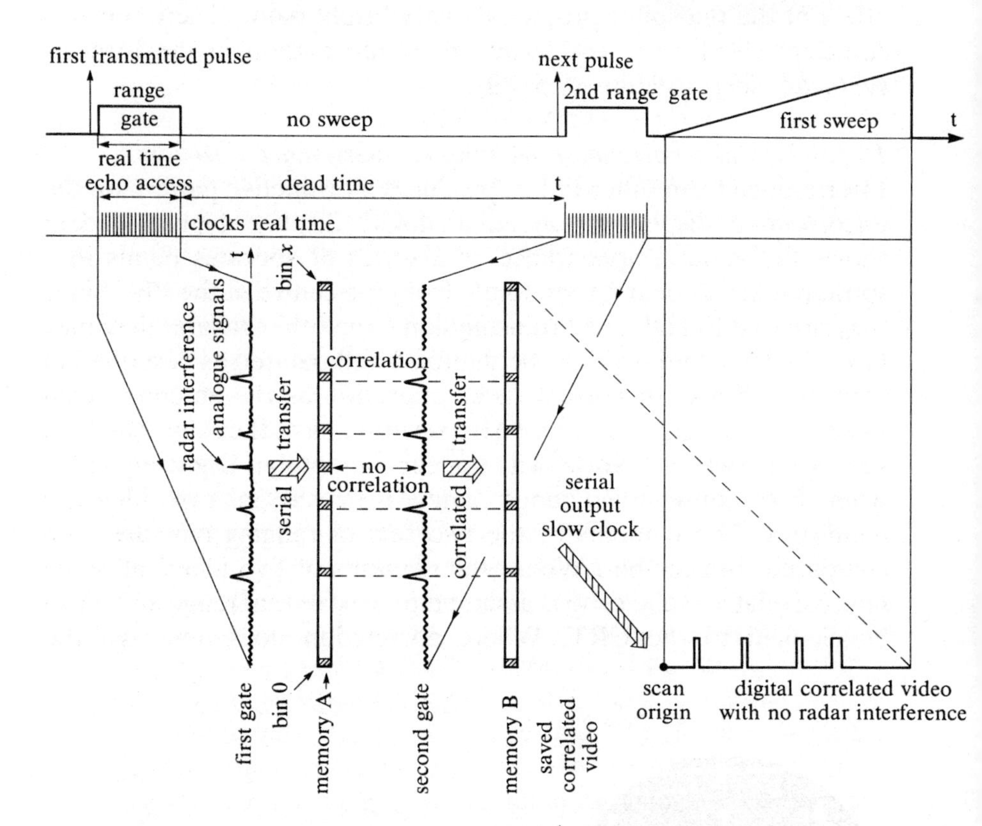

Figure 18.5.14 Correlation method. Memory A is used on 3rd range gate and contents correlated with B, and so on.

Occasionally noise will be correlated or very weak and intermittent target returns may be rejected but this is a relatively rare occurrence during each antenna scan. A more rigorous correlation can be exercised by increasing the correlation interval, achieved by performing a scan-to scan target correlation or by increasing the number of sweep-to-sweep correlations. Its application requires a

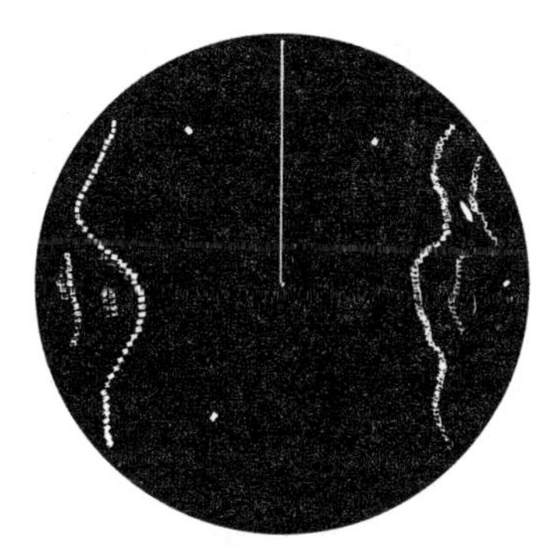

Figure 18.5.15 Clutter-free, interference free digital picture. The cell-like target structure has been exaggerated.

large memory capacity. In the scan-to-scan application the contents of each individual range gate are saved during an entire picture scan, the first not being displayed. On subsequent scans the memory contents are correlated as before but for the same relative sweep. The probability of noise occurring in the same bin on the same range gate is very low. Scan-to-scan correlation is the basis of *tracking* in ARPA systems.

18.5.3.8 Pulse elongation

This is also known as *pulse stretching* and *pulse enhancement*. A facility is provided to artificially lengthen the stored video signals by a fixed length of time. The effect at the display is to produce a more prominent paint of the digitized video. Pulse stretching is usually made selective in that it is only applied to pulses above a fixed duration usually on the 6–96 nm ranges and only to displayed video beyond about 3 nm. This precaution avoids excessive paint cluttering the picture centre. Since range is measured from the leading edge of a displayed echo, which is sensibly unaltered, accurate range measurement of stretched targets is obtained. Echoes having a duration less than that of the transmitted pulse are rejected and not elongated which eliminates many high-level noise impulses. In pulse correlated circuits only the correlated pulses are able to be elongated; the degree of stretching varies according to the transmitted pulse length and the range in use.

18.5.3.9 Synthetic video

By common usage the label *artificial* or *synthetic* is applied to any *system-generated* video signal producing an intentional display not derived from target returns. The simplest of such signals have been used as markers and indicators since the inception of the plan position indicator and include heading marker, range markers, electronic bearing line, etc., all being electronically created by the display circuitry. The availability of cheap digital processing and associated integrated circuits has led to the widespread development of computer-aided marine track-while-scan radars which use displayed synthetic symbols to interact with the machine's human operator. These take the form of distinctive shapes, not easily confused with the radar information, such as circles, ellipses, squares, triangles, hexagons, crosses and so on, limited only by the programmer's imagination. All PPI displayed alphanumeric symbols are included. Figure 18.5.16 illustrates that very simple waveforms acting independently on the orthogonal deflection system can easily form such symbols. Messages can be written to the screen to improve machine-man interaction since most modern digital displays carry sufficient computing power to generate the messages. Symbol generation must not interfere with the display of navigational information. For this reason, signal processing circuits are separated from synthetic video circuits.

Symbol generator circuits comprise microprocessor-controlled D to A converters which are multiplexed to the sweep deflection amplifiers and gated to produce output during unused portions of the dead time. Figure 18.5.17 shows a simple arrangement. Tube persistence is high and the refresh rate on each sweep is sufficient to provide a flicker-free presentation of symbols. When large amounts of symbol data need writing to the screen, the stolen sweep technique can be used (section 18.4.4.3). For most marine requirements this is unnecessary. Displays which employ penetration tubes (section 18.4.5.2) have a facility to display symbols in colour to further highlight visual information.

18.5.3.10 Built-in test equipment

As radars have evolved the shipboard system has become very complex. If the system is reduced to its various sub-systems and these further reduced to sub-sections and eventually to the basic building blocks of the digital and analogue circuits we find that circuit operation is simple and easy to follow. However, fault finding techniques which were applicable to discrete component circuit structures are applicable only to a very limited area in a modern radar. Familiar circuits such as video amplifiers are structured on a single chip. Similarities can be found throughout the system design. Additionally, the widespread use of several single-board micro-

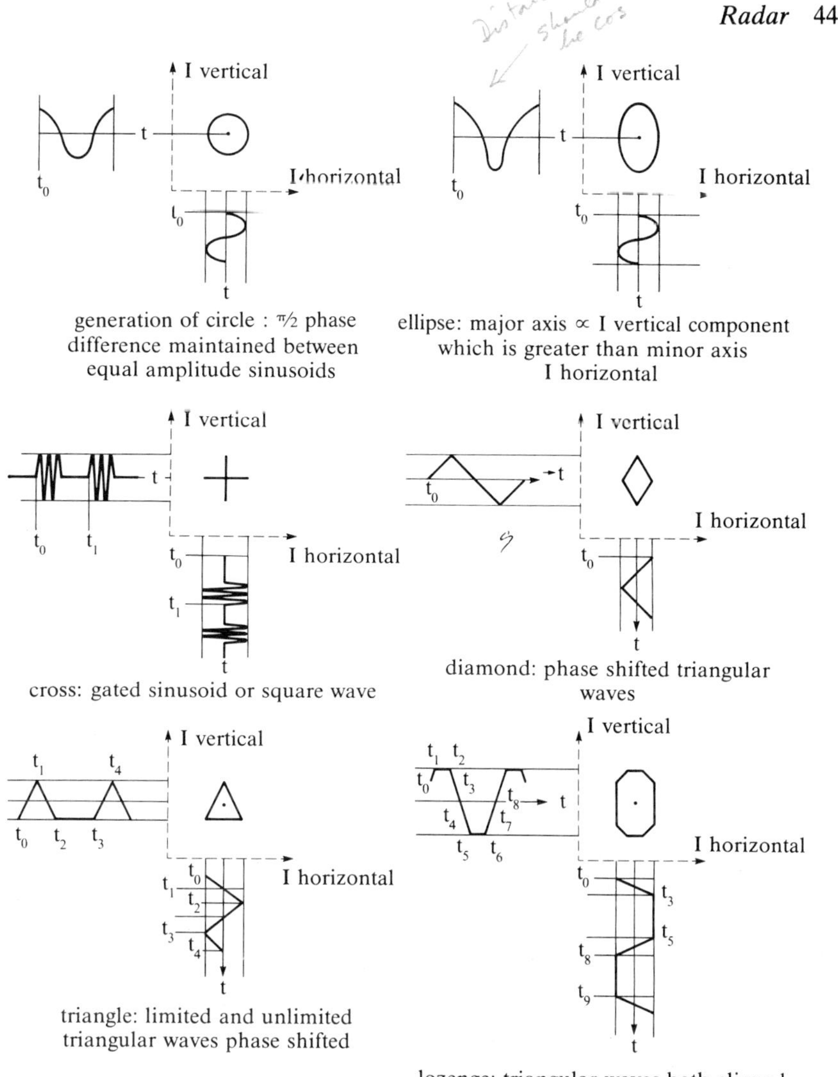

Figure 18.5.16 Common synthetic video symbols and basic method of generation.

computers in the more complex displays, each handshaking with the others and interdependent of control signals, data and addressing, renders the job of shipboard trouble-shooting a difficult task. Design trends now tend to have increased dependency on the proven high reliability of solid state fabrication. The overall intention is to reduce the skills previously considered necessary to maintain a radar in an operational condition and modern technology

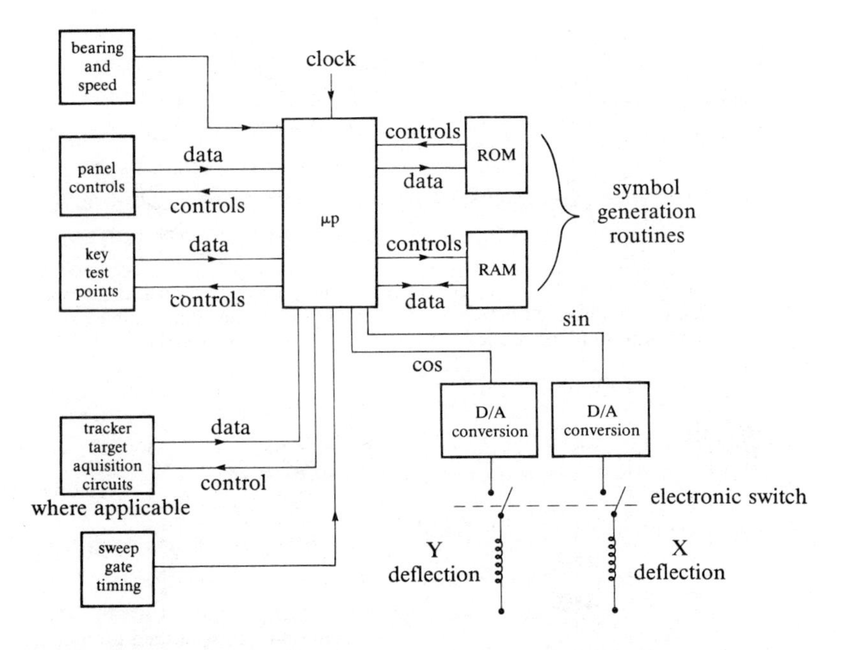

Figure 18.5.17 Basic arrangement of microprocessor control to generate symbols. ROM contains look-up tables and binary patterns, also software routines; RAM provides temporary storage and registers.

has moved the radar maintenance area from the realm of the skilled technician who worked down to component level to the relatively unskilled task of printed circuit board changing.

The simplest sorts of built-in test equipment comprise circuit 'tell tales'. These are generally light emitting diodes mounted at the edge of PCBs driven by simple reliable logic buffers and whose on or off states give a visible indication of normal circuit condition or existence of a fault. Additionally, some equipments still provide metering facilities where quantitative values are important. The more sophisticated arrangements use the microprocessor. On powering up the radar, the computer runs through a logical series of checks, testing its own sub-system and all other sub-systems, devices and signals. Failure routines held in memory throw up a diagnostic message on the screen if a fail condition is met with, for example *MOD PWR DWN*. The diagnostic message *OK* or *SYSTEM PASSED* indicates a satisfactory test run. Other designs enable the operator to call up specified test routines during normal radar operation without interfering with its navigating capability. Indications of power level, performance monitoring pulse length, etc., can be displayed as a menu in a small area of the PPI or on a separate indicator, or covertly monitored. Any divergence from acceptable norms throws up a diagnostic message and can activate an audible alarm. Facilities vary with equipment sophistication but generally work on the fail-safe principle so that sections of the radar related to automatic plotting may fail whilst not affecting the integrity of the digitized display. That too may fail without compromising the analogue display facility. Most modern displays incorporate means to generate geometric test patterns to assist in the checking of PPI accuracy and picture alignment after major replacements, such as the tube.

8.5.4 Automatic radar plotting aids; introduction

Automatic radar plotting aids have developed to meet an increasing demand for improved collision avoidance at sea. During the period of that development new technologies have emerged in the electronics industries to provide reliable, low cost computer-controlled automatic tracking radars capable of operation in a demanding shipboard environment. An ARPA comprises a PPI type display capable of simultaneously providing continuously updated navigational information on a number of targets. The targets may be manually selected or, according to the operator's requirements, acquired automatically. Information extracted by the tracking computer concerning the target motion is used to generate vectors and marker symbols giving a rapid visual assessment of traffic movement. In addition alphanumeric readout is provided on any tracked target regarding range and bearing of the target, evaluated true course and speed of the target, the predicted target range at the closest point of approach (CPA) and the predicted time to CPA (TCPA). Since the task of plotting on the PPI is no longer a manual chore, the radar observer is more effectively able to keep a vigilant radar watch on many targets of interest, and may quickly evaluate potentially hazardous conditions. This ability reduces the possibility of collision. The ARPA also allows the navigator to evaluate the consequences of an own-ship manoeuvre by simulation of its effect on the displayed target information. Provision is made to warn the navigator of hazardous target movements both visually and audibly and also of lost targets which may be of importance.

The main contribution of an ARPA is to present a great deal of traffic information in palatable form, and the system requires the same degree of skill in operation and interpretation as does the 'conventional' radar display in order to optimize performance. This is only obtained by applying the well established rules for adjustment of operational controls and carrying out frequent system checks during use.

Manufacturers of ARPAs have complied with the IMO ARPA plotting requirements by adopting various individual design approaches. Most designs provide facilities in excess of those requirements; indeed each year produces some new additions as the microprocessor power is expanded. In the following text a generalized approach is used, since the tracking systems very closely follow well established methods adapted from military and civil aviation radars. Section 18.5.5 gives a brief overview of the Marconi ARPA to outline one design solution to the problem of information interchange between man and machine.

18.5.4.1 ARPA sub-systems

The block schematic diagram of Figure 18.5.18 illustrates a typical ARPA display system. This is broadly divided into three sub-systems each under the control of an individual microprocessor and each capable of transmitting and receiving data from the others. Processor *A* is recognizable as the digital video display described in section 18.5.3 and is capable of providing relative unstabilized and stabilized displays together with the functions outlined in those sections. Additionally this sub-system has a true-motion capability with some input from processor *B* sub-system. Means are provided to present a raw radar signal display. Failure of ARPA sections does not affect the integrity of the radar display which, in type tested radars, meets the required radar performance specifications.

Processor *B* sub-system incorporates the target extraction and target tracking hardware, deriving video input from *A* and processing directives concerning manual target acquisition, guard zones, range in use, etc., from *C*. Digital information relating to targets is passed to *C*, which generates appropriate marker symbols, vectors and alphanumeric messages. In Figure 18.5.18 alphanumerics are sent to a separate display but may equally well appear at the PPI. If the PPI is used for messages, radar specifications are met if the characters are not less than 3mm in height, appear in a rectangle not greater than 70mm by 40mm in a position under operator control via a spring-loaded switch and do not interfere with the presentation of radar or ARPA data. Audible alarms are output from sub-system *B* and visual (flashing symbol) alarms are superimposed on the PPI.

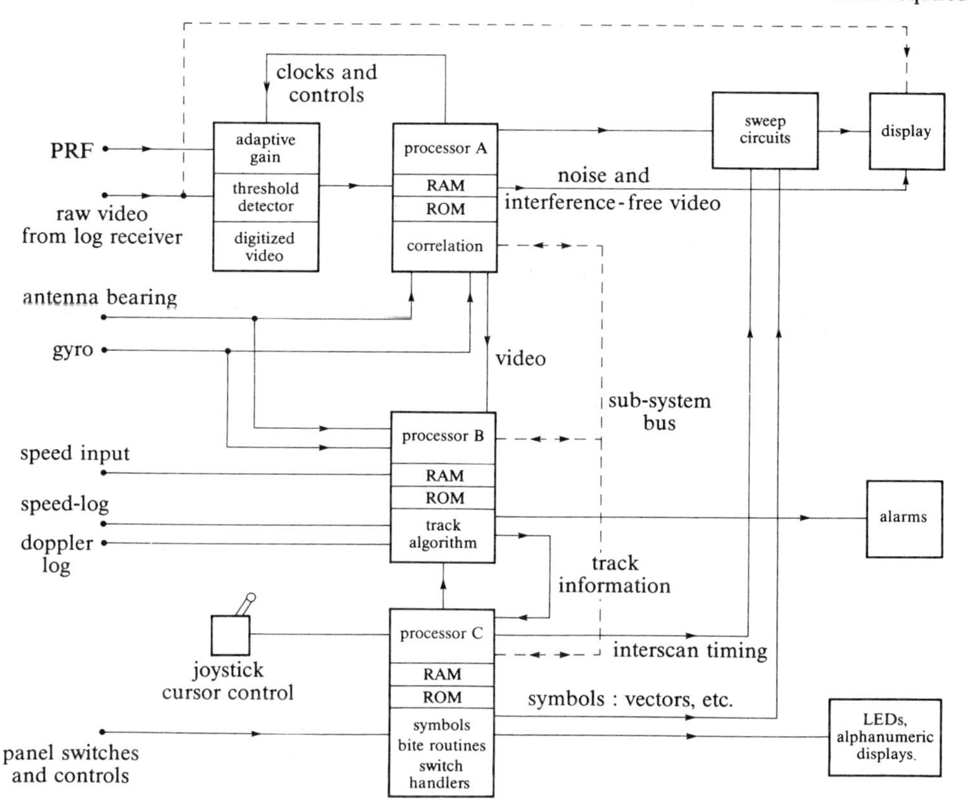

Figure 18.5.18 Typical ARPA system arrangement.

Sub-system *C* also has a supervisory role interpreting switch and push-button actions at the ARPA console and may, in some ARPAs, interact with the user by displaying messages or asking for specific input. In the later case, a separate video monitor is often used (section 18.5.5.1).

18.5.4.2 Tracking principles

Radar signals are conditioned and digitized as described in previous sections of the text. Targets of interest may be manually selected by positioning a cursor over the target and depressing a target acquisition button. ARPAs offer either a manual or an automatic mode of acquisition, the default condition being manual for acquisition and cancellation. In the manual acquisition mode a typical system will allow up to 20 targets to be acquired and displayed with vectors. Alternatively, the operator can construct a specific search area around own-ship within which targets can be automatically tracked. The tracker may be storing data on fifty or so targets but will probably display about 20 of these with vectors. In addition, tracked targets which violate specified CPA and TCPA values will cause a collision warning.

The tracking computer measures the coordinates of an extracted target, the available data being range and bearing, the latter being referred to true north by the gyro input. A track-while-scan process takes place when successive scans over an area containing those coordinates are used to update range and bearing information. After a few initial scans the computer determines the target path and predicts its future course from the sampled data at each pass.

Figure 18.5.19 shows the basic sampling and tracking process. The range is quantized into equal time and range intervals according to the desired resolution. As the antenna beam sweeps over the target the extractor samples *n* pulses where *n* is the anticipated minimum number of returns from a target. One commonly used detector examines the last number of samples from within the identified range interval and assumes the existence of a valid target if *m* out of *n* of these cross the detector threshold. Since the target must lie within the *n* pulses and these have a spatial dimension, the centre of the group is computed (sometimes known as its 'centre of gravity') and from this the range and bearing estimate is made.

It follows that a good CFAR receiver is required as, in the majority of marine situations there will be other targets present as well as clutter. The tracking detector outlined is known as a 'moving window detector' and upon establishing a new target track the computer algorithm sets up a search region enveloping a number of contiguous range intervals or cells which is itself known as the 'tracking window' or 'tracking gate'. The computer expects the target, normally relatively slow moving, to lie somewhere within this window on each subsequent scan. As the target is successively detected the algorithm is arranged to reduce the window size and move it to a predicted detection zone in anticipation of the target's presence. At the same time the motion of the target relative to own-ship motion must be calculated with a fair degree of accuracy. As successive predictions continue to return the *m* out of *n* hits, the computer's confidence is increased in its estimate of target speed and direction and it shrinks the window to a minimum compatible with the detector sampling criteria. Making the window as small as possible is a fundamental design requirement to avoid where

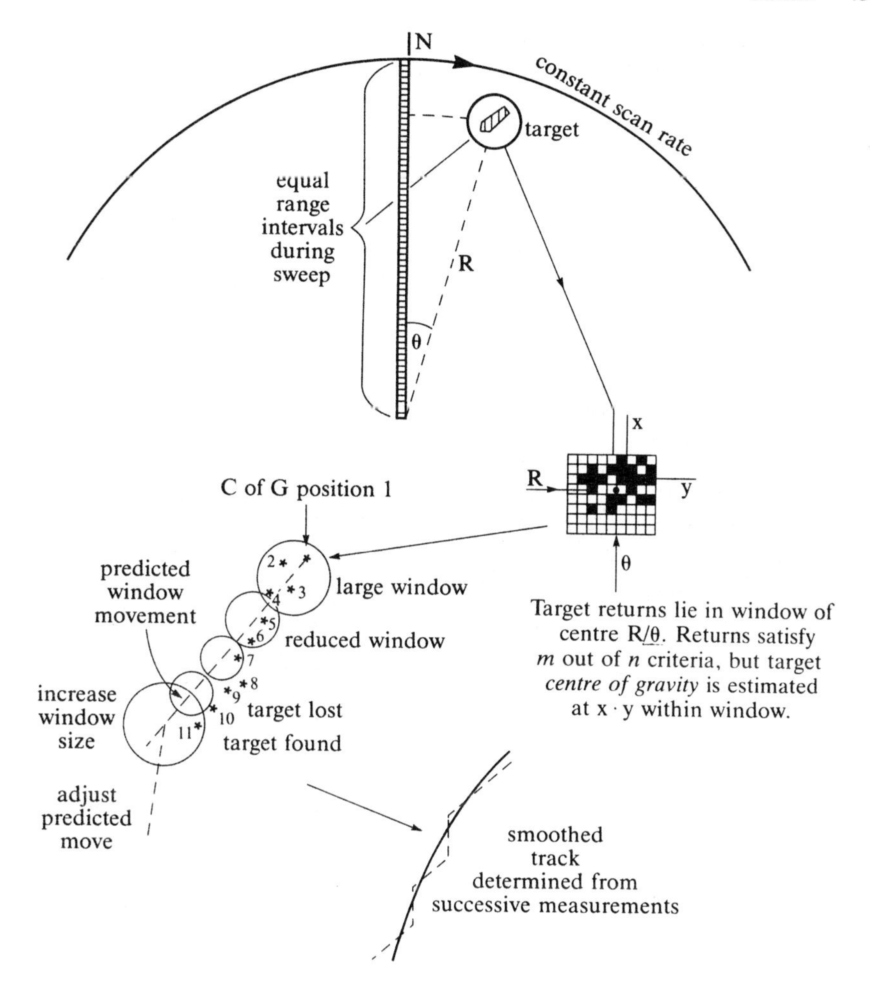

Figure 18.5.19 Basic principle of quantizing and smoothing track.

possible, more than one target within the window. Such occurrences can lead to *target swop* which is explained later.

18.5.4.3 Target vectors; track smoothing

These initial tracking estimates are undertaken over the first several scans and are invisible to the operator. When first manually acquired the tracker will enclose the selected target with a distinctive symbol, commonly a circle. It will continue to extract data on successive scans, at the same time tracking any other selected targets as they are scanned. After a couple of scans the tracker indicates acquisition, usually by shrinking the circle size, and continues to track covertly. When tracking confidence is established data are sent to the symbol generation circuits and a *vector* appears extending from the tracked target. Its direction indicates target bearing and the vector length indicates target speed. The option is usually provided to vary the vector timescale which must be clearly displayed. It is also possible to have true or relative vector presentation, the selected mode must be clearly indicated to the user. Examples of acquired targets and vectors are shown in Figure 18.5.20.

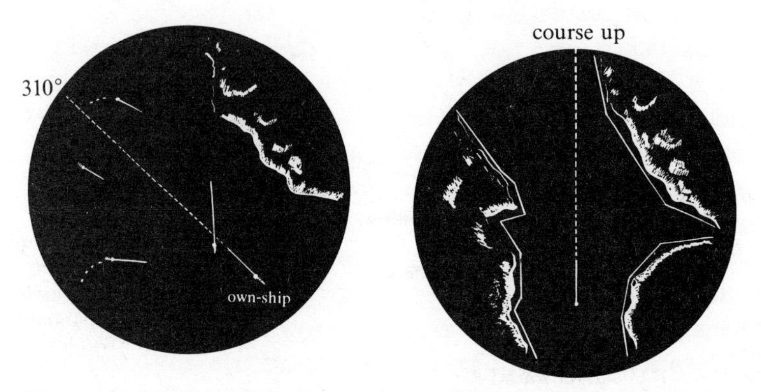

Figure 18.5.20 (a) Simplified PPI display showing true vectors, range 6 nm. Vector length 12 mins. Dotted trails are history plots showing course change; dots are 3 mins apart. (b) Mapping principle. Maps are aligned with a distinct reference point.

Evidently, the accuracy with which the track can be established also determines the moving window size and most algorithms continue to improve the track and vector accuracy over the succeeding 45 to 50 scans. Slow-moving targets are accommodated in smaller track gates than faster targets whose window area is calculated on the estimated maximum acceleration in turning. Such estimates are of a statistical nature.

Once a reliable track has been determined there is a need to minimize false or spurious pointing of vectors. Range quantization and radar errors are present and *filtering* of the measured target velocity is carried out to produce a *smoothed* estimate of target velocity. The filtering or smoothing action is a compromise between a need to follow the target when it manoeuvres and the need for accurate measurement. A widely-used method generates a smoothed estimate of the target's present position and velocity and a prediction of these based on past detection data. The track filter is a so-called non-linear adaptive α–β tracker; α is known as the *position smoothing parameter*, β is the *velocity smoothing parameter*. Digital methods are used, one being to determine the smoothed position by multiplying the difference between predicted and measured positions at the *nth* scan by α and adding the result to the predicted value. Smoothed velocity is derived by multiplying the difference by β/T where T is the time interval between observations, and adding the result to the smoothed velocity calculated at the previous scan. The values of α and β are derived from statistical knowledge of system measurement errors and prediction errors.

If the tracker does not receive any returns from a target or the *m* out of *n* criteria are not met during one scan, it will continue to predict a track based on the last data. At the same time the window is enlarged on the assumption that during the next scan it may envelop the missed target. This iteration continues until the target is either relocated or determined to be a *lost target*. In the meantime, smoothed predicted tracking continues.

A re-established target will have tracking confidence renewed over successive scans as its track window is minimized. A lost target has its track terminated and displayed vector removed. If 15 scans have been used to establish a track, 25 consecutive misses will be a suitable criterion to terminate the track. ARPAs must indicate lost targets and usually accomplish this by flashing the target marker symbol and issuing an audible alarm. Others give the visual alarm on a separate VDU.

18.5.4.4 Guard zones; out of range targets

The operator may set up an area surrounding own-ship clearly delineated on the PPI and usually by manipulation of the range marker strobe and EBL controls which have a dual role under computer control. A simple *guard ring* will alert the processor to any target track which transits the guard zone. Audible and visual alarms occur to draw the observer's attention.

Guard zones can be roughly contoured to coast lines to avoid persistent false alarms. The degree of sophistication of this facility varies considerably between manufacturers. Alarms are also triggered if a tracked target is predicted to approach to within a minimum range and time of own-ship, as determined by the user.

Typically, automatic cancelling of targets will occur if the range exceeds 20 nm or if a persistently bad echo is returned and the target is not violating CPA or TCPA rules or if the target is behind and moving away from own-ship at a safe range.

18.5.4.5 Target swop

When two targets fall within the tracking window of one of those targets it is possible for the tracker to transfer the tracking data from the tracked target to the interloper. The situation is exacerbated if traffic density is high by having a range gate large enough to admit two such targets and by unpredictable vessel movement when within the gate. With a small gate and two tracked targets which maintain course and speed the predicted track usually holds good and swop is avoided. If a tracked and untracked target fall within the same gate and the tracked vessel alters its course, the predicted move of the window may place it over the untracked vessel. The original target of importance ceases to be tracked. Some ARPA designs have adopted a design approach placing the gate size under operator control together with the facility to create a restricted navigation zone ahead of own-ship. Manual acquisition only is allowed under these conditions, generally found in narrow congested waterways. The tracker will still monitor targets manoeuvring outside the zone and issue alarms if targets are closing in time and range.

18.5.4.6 Ground lock

ARPAs are able to acquire a fixed prominent target and use it as a ground stabilising reference point in the tracker. Caution should be exercised when using this facility since a sea stabilised ARPA showing true vectors is calculating those from target motion through the water and includes the effects of any tidal influences on target and own-ship. Ground lock can lead to misinterpretation of tracked vessel vectors.

18.5.4.7 Maps and simulated targets

Mapping facilities comprise a simple graphics display composed of straight-line sections whose 'start' and 'finish' coordinates can be entered into the map memory using the joystick control. The map elements are locked to a point of the PPI shown by a distinctive symbol. The number of available elements to form a map varies between manufacturers and is of the order of 15 to 20. Maps are used to show coastlines, shoal areas, routes and waymarks and can be saved in memory. A battery back-up power supply to the memory preserves its contents when radar supplies are removed, the battery being recharged on powering up the set. Storage of maps under power-down conditions is about one month but varies according to the battery capacity. Stored maps can be recalled and displayed when required, and typically about 15 maps can be stored in this way.

Simulated target echoes can be generated in the ARPA by appropriating parts of the video extractor circuit. These are shown on the PPI with a distinctive symbol attached and the ARPA is able to track these targets. A number of simulated targets may be generated with known course and speed and are used to check vector and numerical readout accuracy. A larger number of separately generated targets whose initial positions, speeds and courses are under operator control may also be generated. Manipulation of such targets varies with the type of ARPA, the main purpose being to provide operator-controlled simulation of situations for training and exercise purposes.

18.5.4.8 Trial manoeuvre

The ARPA is mandatorily required to be able to simulate the effect on all tracked targets, of a proposed change in own-ship course and/or speed, without interrupting the updating of target information. The facility permits evaluation of the likely consequences of the manoeuvre which is planned in advance by the observer and its effect shown in true or relative vector display. Manoeuvring characteristics of own-ship are accounted for in the tracker calculations and video presentation, these factors having been entered into the computer during equipment trials. Some displays show a 'dynamic' trial manoeuvre which displays an accelerated true-motion display of own-ship and tracked targets, taking into account own-ship characteristics and predicted target motion. Typically, a 30 minute prediction takes one minute of display time, holding the 'trial manoeuvre' switch closed. At the end of that time the display 'holds' the 30 minute manoeuvring situation. Regulations require that the switch be arranged so that the display cannot accidentally be left in the trial manoeuvre condition.

18.5.4.9 History plot

A facility is provided which displays a distinctive symbol to indicate the past track of a target as a series of discrete steps. Small circular symbols are commonly shown indicating say, four to six immediate past locii of target track. The history plot is a useful adjunct to indicate changes of course made by tracked targets.

18.5.4.10 Land echo suppression

If the near and far edges of returned echoes are measured in time this may be converted into distance. The tracking detector can be arranged to determine this value for all returns together with the angle of arc subtended by the return. After adjustment for range the computer determines whether the echo size exceeds in angular width and range that echo which might be expected from a large vessel. If so, the echo is assumed to be from a coastline. It is possible to average the coastal contour from several scans, determined as the near edge of these echoes and arrange for the contour to be superimposed on the displayed picture as a dotted line or some other convenient configuration. Automatic acquisition and tracking of smaller targets beyond the synthetic coastline can then be suppressed.

18.5.5 A modern ARPA

The objective of this text is to deal with the principles of automatic radar plotting aids not with particular systems. It will be instructive to consider a modern example of an ARPA and take a brief look at the major features which it affords. A good example is that of the Marconi ARPA shown in Figure 18.5.21. The operational console comprises the indicator and a compact video display unit (VDU); printed circuit boards carry the essential electronics and are held in racks in the cabinet space beneath the VDU. Circuits appertaining to the power supplies, IF and deflection amplifiers are housed in the cabinet section beneath the indicator.

Figure 18.5.22 illustrates the indicator operational panel controls which, with a few exceptions, have purposes assigned to them common to most radar displays. These are dealt with in section 18.4. The synthetic video brilliance control (symbol brill) is not found on conventional displays. To the right of the console (Figure 18.5.22) is the video display unit, the face of the tube being angled to permit the user standing at the indicator easily to read information from the VDU.

ARPA controls have been minimized in an elegant manner by use of a keypad forming a column of ten soft switches and a joystick control used to steer an electronic cursor over the surface of the PPI and which carries the manual target acquisition button. Software held in ROM permits user interaction with the equipment via the VDU, using a displayed 'menu' of multiple functions for each key

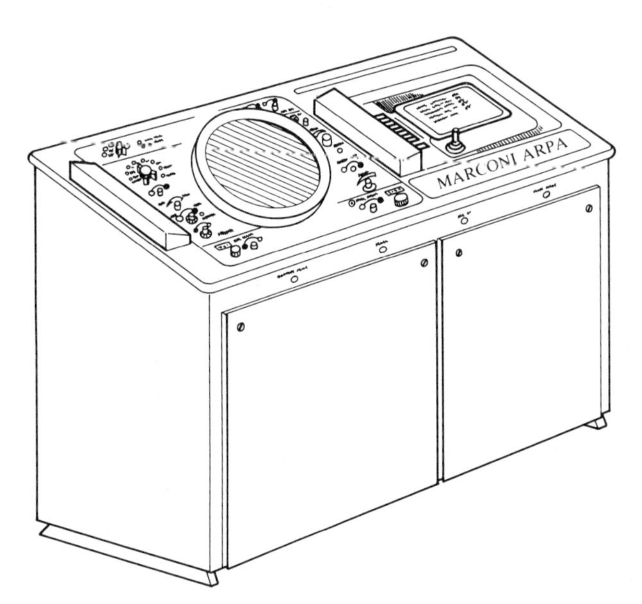

Figure 18.5.21 The Marconi ARPA illustrating the convenient operational control, display and VDU layout of a modern ARPA. (Redrawn with permission from the Marconi International Marine Company Limited.)

numbered one to eight. Default to the master index or menu can be obtained by depressing the master index key from which position the novice user can be guided by the interactive screen messages, to set up and use the ARPA. As stated in earlier chapters the ARPA does not supplant the experience of the navigator and skill is required to interpret data provided by any ARPA.

Targets may manually be acquired by manipulating the joystick to place a target cursor or marker in the form of a cross, over the desired target and then pressing the 'acquire' button on the joystick. Automatic target acquisition can also be selected. Individual targets may be cancelled by similarly positioning the marker and pressing the CANCEL key; a facility exists for cancellation of all targets except that which may be used for target lock.

In common with the generalized ARPA outlined earlier in this chapter, the Marconi ARPA has provision for compass stabilized north up, course up, unstabilized ship's head up and true-motion presentation. In the latter case sea or ground stabilization is available. Unlike the true-motion unit described in section 18.5.1.4 the ARPA can provide 'target lock' to a stationary target giving an automatically ground stabilized display. Other true-motion facilities are conventional.

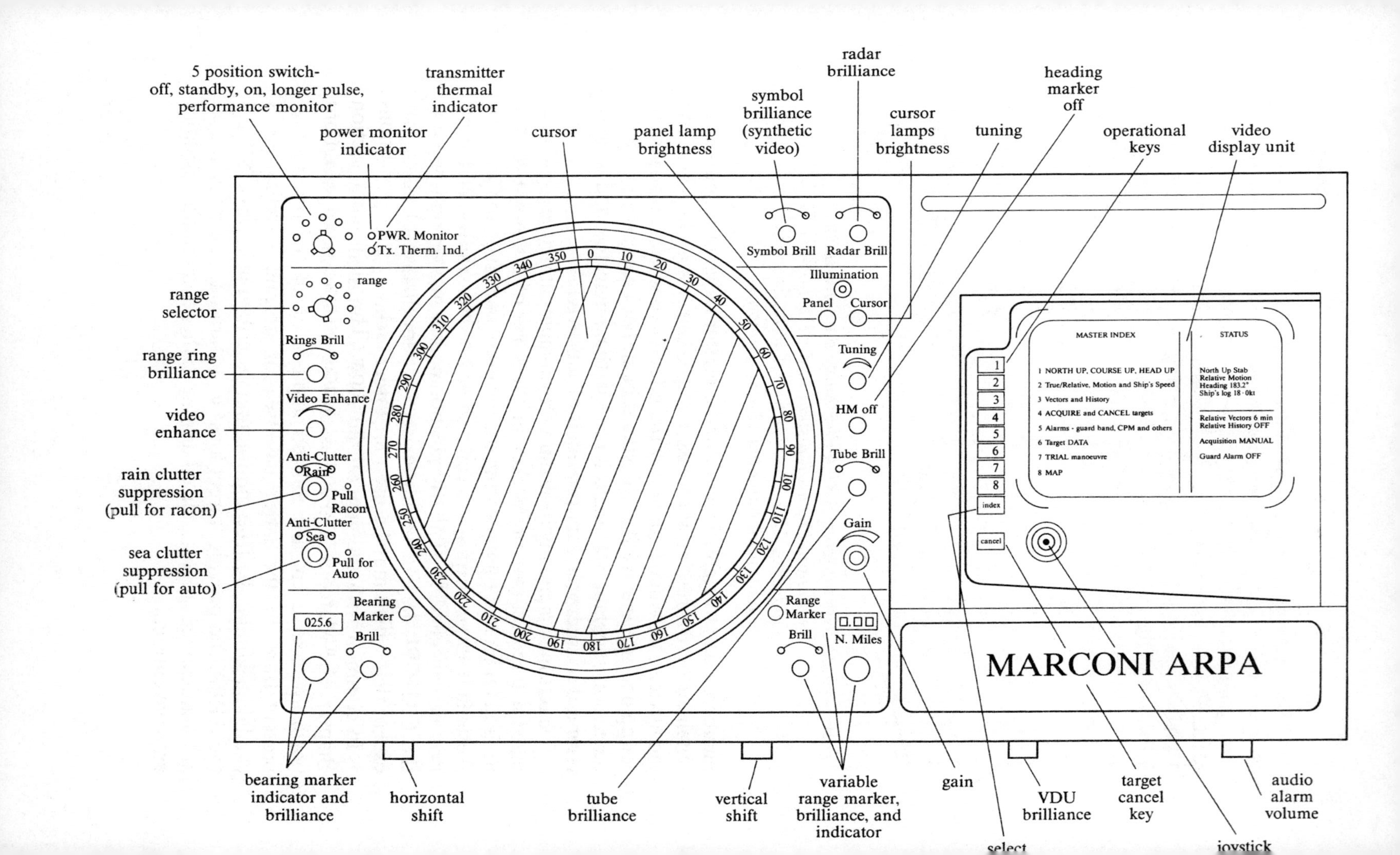

5 position switch-
off, standby, on, longer pulse,
performance monitor
transmitter
thermal
indicator
power monitor
indicator
cursor
panel lamp
brightness
symbol
brilliance
(synthetic
video)
radar
brilliance
cursor
lamps
brightness
tuning
heading
marker
off
operational
keys
video
display unit
PWR. Monitor
Tx. Therm. Ind.
range
Symbol Brill
Radar Brill
Illumination
Panel
Cursor
range
selector
Rings Brill
range ring
brilliance
Video Enhance
video
enhance
Anti-Clutter
Rain
Pull
Racon
rain clutter
suppression
(pull for racon)
Anti-Clutter
Sea
Pull for
Auto
sea clutter
suppression
(pull for auto)
Tuning
HM off
Tube Brill
Gain
MASTER INDEX
1 NORTH UP, COURSE UP, HEAD UP
2 True/Relative, Motion and Ship's Speed
3 Vectors and History
4 ACQUIRE and CANCEL targets
5 Alarms - guard band, CPM and others
6 Target DATA
7 TRIAL manoeuvre
8 MAP
STATUS
North Up Stab
Relative Motion
Heading 183.2°
Ship's log 18·0kt
Relative Vectors 6 min
Relative History OFF
Acquisition MANUAL
Guard Alarm OFF
index
cancel
Bearing
Marker
025.6
Brill
Range
Marker
Brill
N. Miles
MARCONI ARPA
bearing marker
indicator and
brilliance
horizontal
shift
tube
brilliance
vertical
shift
variable
range marker,
brilliance, and
indicator
gain
VDU
brilliance
target
cancel
key
audio
alarm
volume
select
joystick

18.5.5.1 Built-in test and some VDU routines

At 'switch-on' the equipment runs through a self-diagnosis routine executing progressive tests on the microcomputers, VDU, PPI and tracking sub systems and on successful completion returns to the VDU the message shown in Figure 18.5.23. Failure of a test is indicated in the diagnostic message which points to the general area at fault. Failure of the computing section may still permit the display to be used as a basic radar indicator.

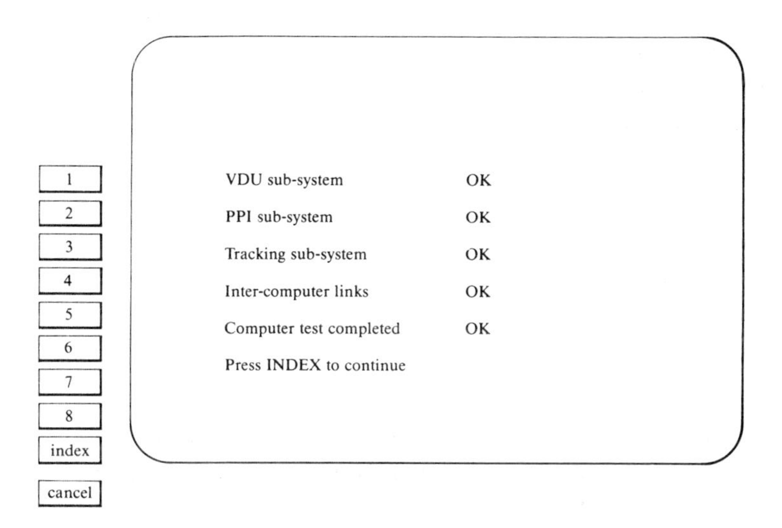

Figure 18.5.23 Diagnostic messages returned by VDU following successful self-test. (With permission from the Marconi International Marine Company Limited.)

At this point pressure on the INDEX switch causes an audible beep and the VDU returns the menu shown in Figure 18.5.24. The VDU screen represents a page of information divided into three sections. Numbered text on the left of the screen indicates the function appropriate to each soft key for the displayed mode. The mid-vertical strip is reserved to display visual alarm conditions appropriate to any numbered function. The right-hand section displays information concerning the 'status' or operational conditions of own-ship and any selected radar target.

Figure 18.5.22 (Opposite) Details of console controls display and video display unit of the Marconi ARPA. (Redrawn with permission of the Marconi International Marine Company Limited.)

It will not be possible to deal with all the facilities offered by the Marconi ARPA since the permutation of keypad combinations is very large; reference to the manufacturer's operational manual is recommended. Some examples of the messages generated at the VDU will serve to show the relative ease of use of the ARPA.

MASTER INDEX

1 NORTH UP, COURSE UP, HEAD UP
2 TRUE/RELATIVE MOTION and ship's speed
3 VECTORS and HISTORY
4 ACQUIRE and CANCEL targets
5 ALARMS - guard band, CPA and others
6 Target DATA
7 TRIAL manoeuvre
8 MAP

STATUS

North Up Stab
Relative Motion
Heading 183.2°
Ship's Log 15.0kt
Relative Vectors 6 min
Relative History OFF
CPA Alarm OFF
TCPA Alarm OFF
Acquisition MANUAL
Guard Alarm OFF

1
2
3
4
5
6
7
8
index
cancel

Figure 18.5.24 The Master Index. (With permission from the Marconi International Marine Company Limited.)

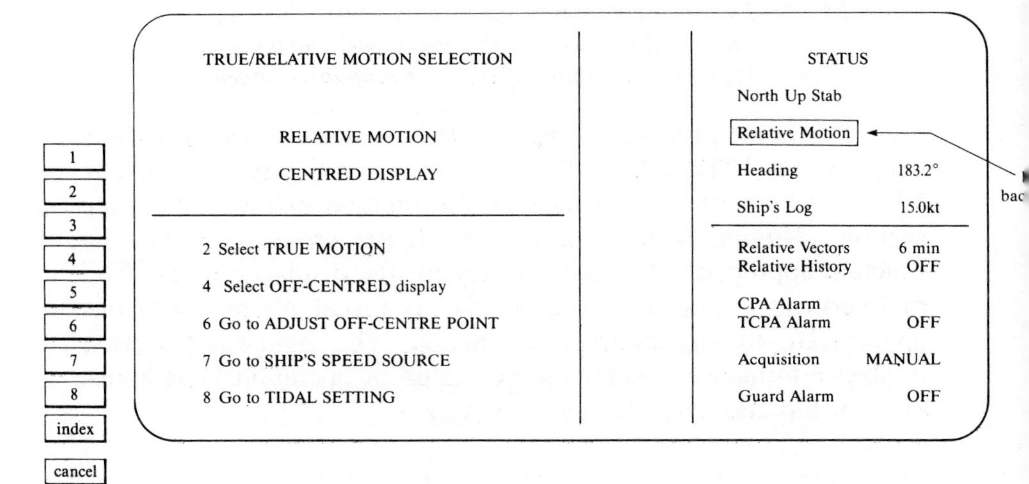

Figure 18.5.25 Options presented as result of pressing key 2 on master index. (With permission from the Marconi International Marine Company Limited.)

One of the ARPA ranges (lying in the 1.5–24 nm range) will have been selected and compass repeater alignment carried out. From the master index display (Figure 18.5.24) switch 2 has been pressed to display the page of information in Figure 18.5.25. From switch-on the dcfault condition is to present a relative motion, centred display. The status function indicates the condition in inverse video text (green characters on a light rectangular background). Heading is 183.2° in this example. The menu now presented in Figure 18.5.26 uses the same soft keypad as before but offers several different key functions. Pressing appropriate keys allows the user to select the options offered, in each case the VDU shows interactive text which guides the user in the manipulation of the ARPA. Figure 18.5.26 shows an example of such text called up by pressing switch 4 in the master index mode; here, specific instructions are given concerning the use of the joystick and acquire button. Other options concerning guard rings and lines, automatic target acquisition and cancellation of all targets are presented to the user. The text of Figure 18.5.26 is self explanatory.

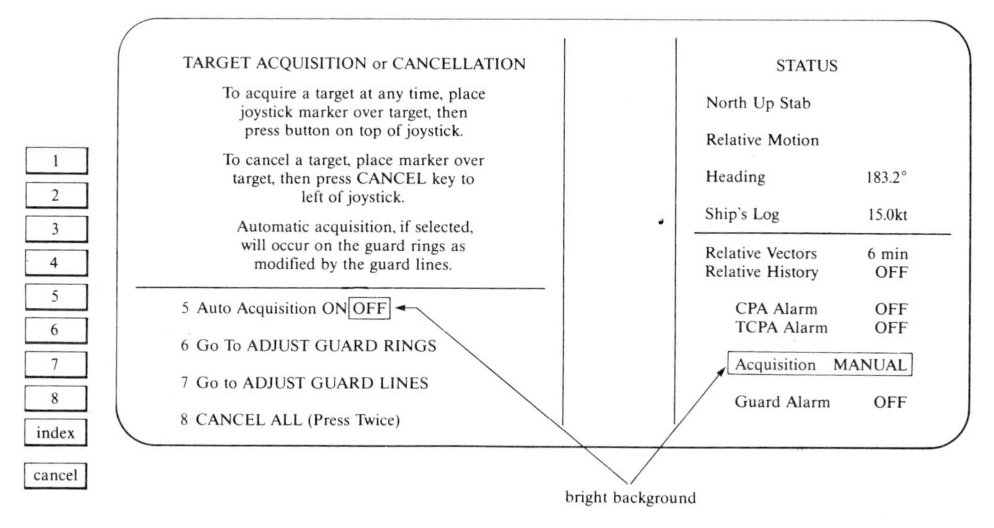

Figure 18.5.26 An example of operator tutoring text provided by Marconi ARPA. (With permission from the Marconi International Marine Company Limited.)

18.5.5.2 Synthetic video symbols and their application

Synthetic video symbols created by the Marconi ARPA are simple and have been minimized to provide a clear and uncluttered plotting aid. A manually acquired target is initially marked on the PPI by a

6mm diameter circle which, after a few antenna revolutions, shrinks to a 4mm circle. From this circle a 'track ahead' vector appears as a line whose direction and length indicate the course and speed of the tracked target. The vector length indicates the distance over which the target will travel in a time selected by the user (between one and sixty minutes). A tracked target may be selected for presentation of its data which then appears on the VDU which displays details concerning its bearing, speed, course, range CPA, TCPA, bow crossing range and bow crossing time. Such targets are marked by two concentric cicles (4mm and 6mm) together with the track ahead vector, thus signalling to the user that the VDU data relates to that target. A target selected for target lock permits automatic ground stabilization and, in addition to the true-motion application previously mentioned, it may be used to lock a map display. Target-lock targets are surrounded by a square of side 6mm.

Tracked target positions can be stored and displayed to provide the recent history of a target's movements and history plots are shown as a bright circular spot. The four most recent positions are shown at selected intervals extending backward from the present target position which is itself shown by its circle and vector. Whether true or relative history plot is shown depends on the mode selected for the track ahead vector presentation. Guard band limits are indicated by guard rings and guard lines, shown on the display as dotted lines. The rings are centred on own-ship. A target entering the guard band will normally trigger the guard band alarm, this being a mandatory requirement, and displays a 10mm diameter flashing circle to indicate the point of entry. Guard lines are manipulated by joystick and electronic bearing marker control to modify the circular shape of the guard ring and hence the guard band so that it does not, for instance, extend over land to give a continuous alarm condition. Examples of other alarms are: lost target alarm (target no longer being tracked), CPA alarm, TCPA alarm.

18.5.5.3 Trial manoeuvre

The user may set up trial manoeuvre by changing the parameters of course and speed of own-ship, together with a time for their initiation. The situation resulting can be displayed in either true or relative mode to assess the possible navigational consequences of such a manoeuvre.

18.5.5.4 General

ARPA is operational on 1.5, 3, 6, 12 and 24 nm. A maximum of 25 targets can be tracked manually. Auto acquisition (if selected) does not operate if 20 or more targets are already being tracked. Automatic acquisition will occur on a guard band. When a target is detected the ARPA tracks it 'invisibly' to ascertain whether the target is closing in range. If so, the ARPA displays the target's track marker and a vector. Vector lengths are common to all tracked targets depending upon user selection at the VDU control.

18.5.5.5 Ship characteristics; sources of system error

When own-ship manoeuvres, the path which it follows is deduced by the tracker and the information is used to display the situation. Deviations between calculated and true track result in the error being apparent as a manoeuvre of all tracked targets. Figure 18.5.27(a) illustrates the effect of own-ship making a turn to starboard, the true track being a tighter turn than the calculated track. Targets on the starboard side are apparently turned toward own-ship. It follows that targets to port appear to turn away from own-ship.

Such errors occur because of the ship dynamics. The gyro and log input provide a basis for deadreckoning of own-ship track and do not show the true course through the water. Any turning manoeuvre will produce a sideways slewing motion which will vary with ship's speed, its condition of loading and its size. At installation the ARPA and devices with which it is interfaced such as the speed-log, gyro and antenna bearing system are carefully adjusted. In addition, the ship's characteristics are input, usually by means of wired links or switches. The turning error described is commonly countered by means of a compass delay time constant. The algorithm devised for the tracker contains a model of the own-ship's predicted track for various ship sizes. The rate at which the ship loses speed during a turn, the time it takes for a ship to reduce its speed by a given value, the distance by which the ship runs on after the helm is altered, the radius of turn are parameters used in the model to calculate the track as accurately as possible. An ARPA's trial manoeuvre facility will make its predictions based, in part, on the ship's handling characteristics. Bearing errors can appear due to gyro and bearing transmission backlash and due to resolution errors in bearing quantization. Displayed target vectors may appear to point erratically, an effect generally made more obvious for targets at the longer ranges.

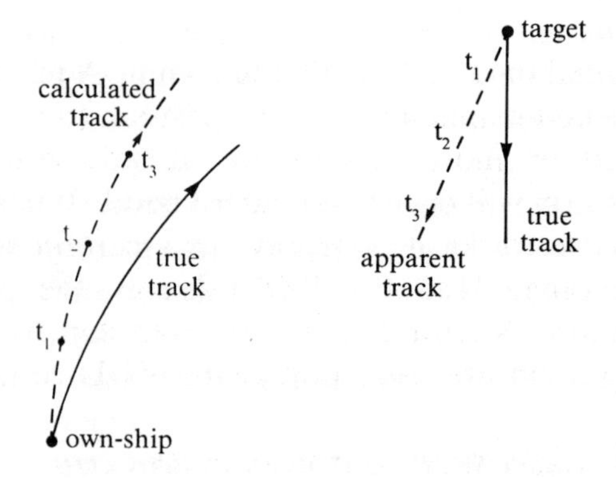

Figure 18.5.27 (a) Effect of error between true and calculated track during own-ship manœuvre.

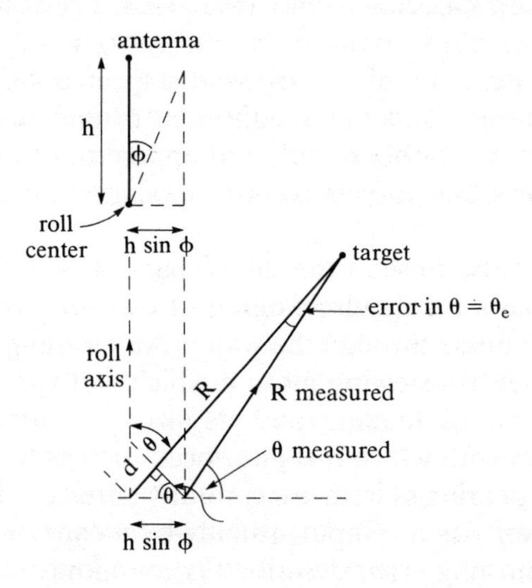

Figure 18.5.27 (b) Indicating range and bearing error due to ship roll. ϕ. $R >> h \sin \phi \cos \phi$; θ_e = bearing error, d = range error, h = antenna height above roll centre.

Range and bearing errors appear because the radar is mounted on an unstable platform. These are not system errors but depend on such factors as the antenna height, roll and pitch angles. Figure 18.5.27(b) shows how a range and bearing error can be introduced due to ship's roll. Figure 18.5.19 indicates that a target track is computed from a series of discrete quantified values of range and bearing. These are separated in time by an amount determined by the detection and recognition parameters. The tracking algorithm produces a smoothed track from these values, the final track accuracy being dependent on the 'filtering' operations executed on the available quantized values.

18.5.5.6 Master–slave displays; interswitch units

It is common practice to have more than one display viewing point in a ship's radar system. Both displays will conform to regulatory requirements. One of the displays is designated the *master* and it is from this point that the observer can adjust tuning of the receiver, select pulse length, place the system in standby, operate performance monitor controls, etc. The slave display is provided with the same video and syncs as the master but cannot effect adjustment of the controls outlined. This arrangement is necessary to prevent confusion arising at either display by a conflict of operator control settings.

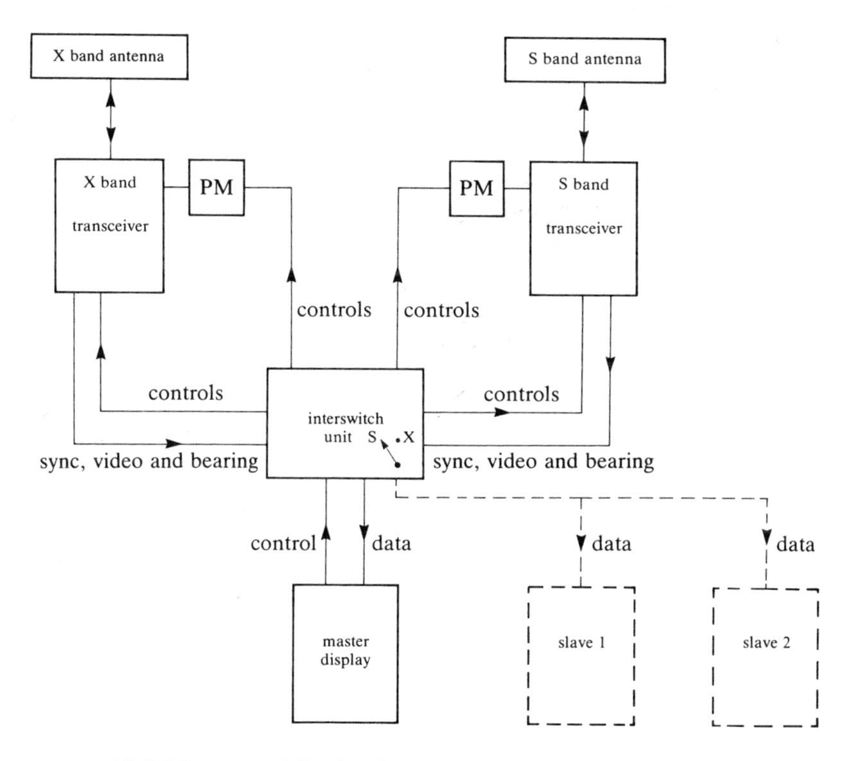

Figure 18.5.28 Typical dual radar arrangement.

The increased use of dual radar systems using three-centimetre and ten-centimetre transceivers and installation of two X band sets has led to the development of complex interswitching units. Figure 18.5.28 shows a block diagram of a dual arrangement. This unit permits one display to be switched between the two transceivers, and would also be used to route performance and power monitor leads, synchro or resolver signals, etc., and as an interconnecting point for slave displays. Where two X band transceivers are installed the unit can control choice of antenna for either of the transceivers. Signals are re-routed by waveguide switching.

18.6 Secondary radar and passive reflectors

The term *secondary radar* embraces various types of small low-power radar beacons sited at navigationally important positions such as landfalls, lighthouses, lightships and essential pilotage marks. Their aim is to provide some means of identification on a ship's radar display together with range and bearing information. Most of these devices are grouped under the generic name of *racon*, from RAdar beaCON, and although producing similar responses on radar screens there exist various operational differences between various types of racon. Essentially a racon is a small radar transceiver which upon receiving pulsed signals from a ship's primary radar is triggered to *respond* automatically, transmitting a unique signal to that radar in its vicinity. The response signal power reaching the radar receiver is much higher than the echo power normally received from the marked site and provides the navigator with unambiguous targeting. Low-lying land, sloping mudbanks and shoals are sometimes difficult to identify at long range without a radar beacon as reference.

The *Admiralty List of Radio Signals, Volume 2* provides details of the majority of radar beacons worldwide. To date, there are about 700 beacons of all types, about 85 operate in both the 3cm and 10cm band, the majority of those being in Swedish waters (53). Of the remainder, 32 are of a type known as the ramark, with a scattering of a few *F Racons* or *fast sweep racons* which have largely been experimental. The ramarks are used in Japanese waters and are gradually being displaced by the more modern type of racon.

18.6.1 The ramark

Ramarks comprise only a small minority of the radar beacons. They act independently of the ship's primary radar and contain transmitters only, sending pulses according to a predetermined time schedule. During transmission the frequency of radiation is swept through the band 9320 to 9500 MHz and back to ensure that all radar receivers within range will receive a beacon signal. Figure 18.6.1 illustrates the frequency sweeping principle. It becomes apparent that the effect seen at the display of a particular radar will depend on the receiver overall bandwidth during reception and the rate at which the beacon transmission frequency is swept. A relatively narrow band receiver in conjunction with a rapid sweep will produce a series of intermittent radial dots or dashes to be displayed. On the other hand a broad band receiver and a slow sweep combine to produce the effect of a bright line extending from the scan origin on the display, its thickness being a function of the beacon sweep time and receiver bandwidth.

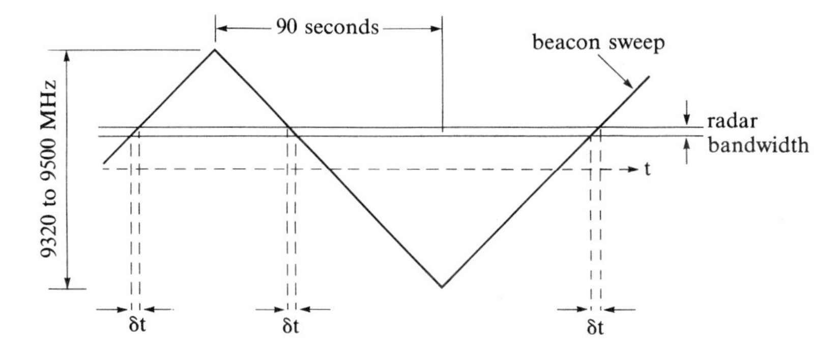

Figure 18.6.1 Slow sweeping principle. Depending on position in band and bandwidth of receiver (usually a function of selected transmit pulse width), the receiver may detect and display response during δt.

Ramarks are useful for bearing measurement only, their main purpose being to identify landfalls. One of the main disadvantages of the device is that all radar displays within range of the ramark will display its signal irrespective of individual requirement. At long range this usually poses no problem, but if a vessel is navigating within close range the ramark output can severely mask low-level signal returns from small vessels. The clutter appearing at the display can be particularly severe at very close range where antenna sidelobes can, during reception, produce on the PPI total obliteration of target paint through a wide arc near the picture centre. Use of the rain clutter or differentiator circuit can reduce the ramark's paint but there is a danger also of subduing desired echoes. Figure 18.6.2 illustrates a typical ramark paint.

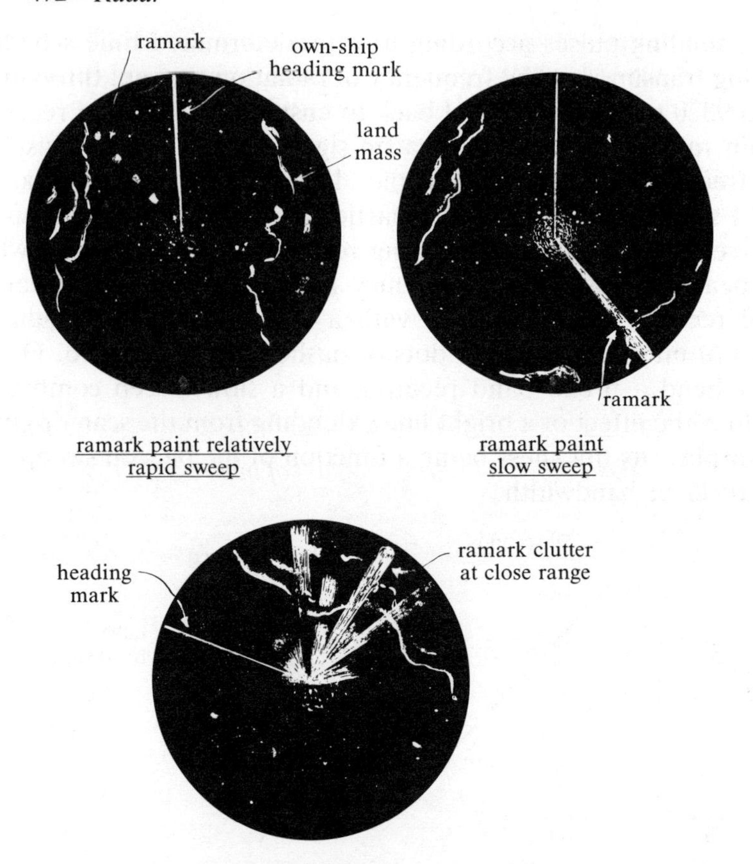

Figure 18.6.2 The lower display shows the effect of remark clutter at close range. Essential targets may be masked.

18.6.2 Racons

Several different types of racon exist or are in a state of development or experimental assessment. The vast majority of those at present in use are classified as *non-selectable* and output a response when they detect the presence of a radar's pulses irrespective of the navigator's requirements. The main advantage of the arrangement is that special modifications do not have to be made to the ship radar.

18.6.2.1 Slow sweep racons

These form the bulk of radar beacons worldwide at the present date, although international consensus seeks to update them over the next decade or so. The block diagram of Figure 18.6.3 shows the basic

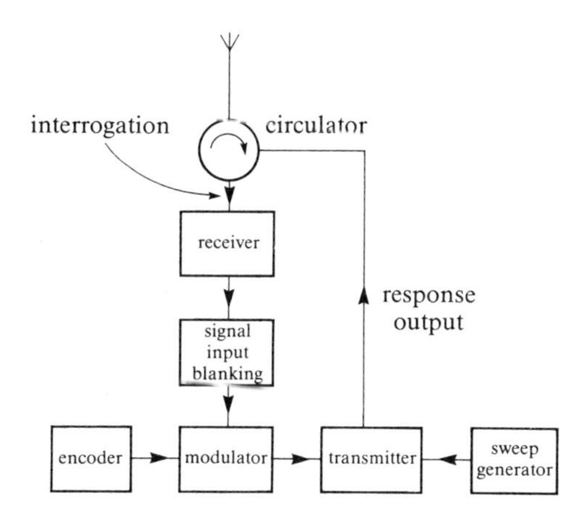

Figure 18.6.3 Basic block schematic diagram of interrogated radar beacon (racon).

principle embodied in the beacon operation. A broad band receiver, either X or S band (9320–9500 MHz or 2900–3100 MHz), triggers an on-board beacon transmitter when a pulse is received from a radar within the appropriate band. The transmitter sends in the same band a morse-encoded signal on the frequency to which it is fortuitously tuned when the radar pulse interrogates the beacon. The transmitter master oscillator is swept at a relatively slow rate over the band at between 60 and 150 seconds per sweep and a displayed response from the beacon usually appears on any radar display within range for about two or three scans during the beacon sweep time. Evidently the slow sweep beacon presentation at a radar display is not continuous and depends upon receiver bandwidth

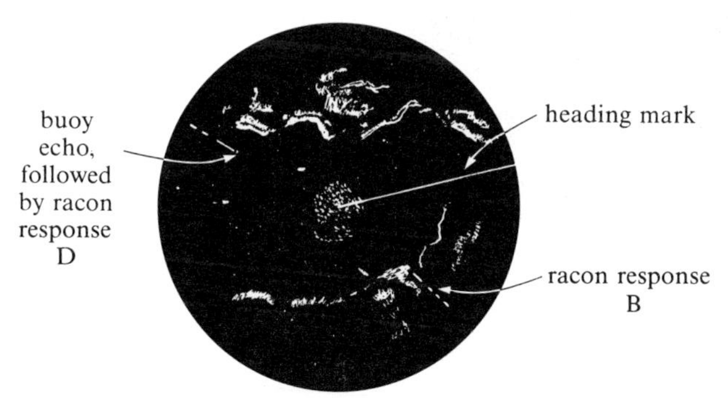

Figure 18.6.4 Two racon responses, coded B and D. Due to system delays in beacons the range displayed is slightly greater than true range. For beacon D, the actual site of the beacon is displayed. At these ranges between 6 and 20 n.m the range disparity is negligible. At very close range in good weather the beacon structure is visible.

and beacon sweep rate to a great extent. Figure 18.6.1 shows this relationship. Figure 18.6.4 depicts a coded response on a display.

18.6.2.2 Sidelobe effects

Marine radar antennas produce spurious sidelobes. When a primary radar is within one or two miles of a racon the sidelobe power level reaching the racon receiver may be sufficient to trigger the racon. A response is then returned to the radar via the sidelobe and is displayed at a false bearing. The effect can cause excessive clutter in the form of *spoking* at the PPI centre. Modern racons can be fitted with optional side lobe suppression, depending on the beacon service. In principle the suppression circuit determines the strongest interrogations and then inhibits weaker interrogations at or near the same frequency assuming them to be from sidelobes, for about four seconds. The arrangements could, of course, inhibit response to a distant radar operating on the same primary frequency. Sidelobe suppression is applied alternately to signals in each half of the appropriate frequency range in a time-shared fashion using an on-board electronic timer at the beacon, which prevents permanent deprivation of the beacon access for those radars. Figure 18.6.5 illustrates the effects of spoking.

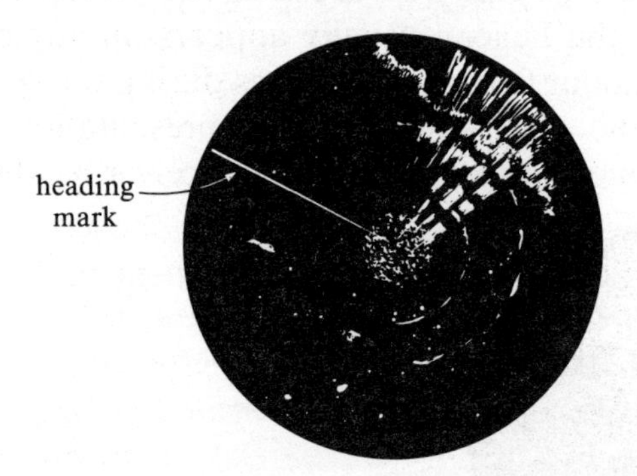

Figure 18.6.5 Effect of slow sweep racon interference on PPI.

18.6.2.3 Fast sweep racons

Few of this type exist. Most are experimental and operate in a manner similar to the slow sweep type, the main difference being the sweep rate. When the beacon is interrogated its response transmitter frequency is swept through the S or X band in about 1.2 to 3.6 μs. For X band operation this represents a maximum rate of

150 MHz per micro second, and the number of sweeps from one interrogation lies between 8 and 40 depending on the beacon service thus producing a response on every radar picture scan. Note that with such a fast sweep a distant primary radar transmitting a long pulse and having a possible 5 MHz bandwidth will receive beacon response for only 0.033 μs, tending to produce faint response. Because of the high number of sweeps per interrogation a number of dots appear on a radial line extending approximately from the range of the beacon to the tube periphery. Successive interrogations produce similar fast responses and the paint is integrated at the screen. The number of morse-type characters which can be encoded is severely limited by the high sweep speed.

18.6.2.4 Frequency agile radar beacons

One of the most recent developments which is gaining wide acceptance is the so-called frequency agile racon. Further operational details are given in section 18.6.6. Beacon response is almost immediate, returning a coded signal on the frequency of the interrogating radar. This is displayed as a morse group extending from the beacon site toward the tube edge, as shown in Figure 18.6.4 and a display appears in each antenna scan. Due to the continuous beacon display other target returns can be hidden beneath the beacon paint and the beacon is equipped to be muted for varying periods depending on the service requirement. Sidelobe suppression is also provided.

18.6.2.5 Stepped sweep racon

This is a type of beacon which has found limited use, mainly in the United States of America, in the 3cm band. It is principally derived from the slow sweep concept. The band 9320 to 9500 MHz is sub-divided into four equal bands each of 45 MHz width. Transmitter sweeping occurs within one band until interrogated. The beacon responds on the frequency to which it happens to be tuned and then performs a discrete 45 MHz step to the next highest band. Here it continues sweeping until a further interrogation initiates the next step. The transmitter cycles to the top of the radar band in this manner, stepping to the bottom in order to begin the cycle once more, in accordance with interrogating pulses.

For a single interrogating radar, beacon response will occur in each of the four sub-bands during an antenna scan. At the radar display the beacon response occurs approximately every 22 seconds, the sweep time of the beacon for each sub-band.

18.6.3 *User selectable radar beacons*

A category of beacons which are at present largely experimental are the user selectable generation. The simplest of these in principle is the fixed-frequency beacon. When interrogated, the beacon responds on a fixed frequency at the edge of the radar band. Two sub-bands have been allocated for their use, 2900 to 2920 MHz and 9300 to 9320 MHz. The receiver at the primary radar has its local oscillator and receiver tuning switched to racon mode to accommodate the frequency off-set. Normal radar echoes are eliminated and the fixed frequency racon's response appears at the display. One of the system's disadvantages is the special adaptation of the receiver circuits and additional cost entailed. Few radars offer this facility as standard (see Marconi ARPA, section 18.5). Interscan switching which will alternately display echoes/racons ensures that no echo masking occurs under racon echoes. Fixed frequency racons have not been accepted on a large scale.

18.6.3.1 Off-set frequency agile radar beacon

A beacon which is basically similar to the frequency agile device. During response the beacon, having measured the interrogation frequency, then returns a signal automatically off-set by a fixed amount – typically 50 MHz above the received frequency. The primary radar has its receiver switched to that off-set value when in beacon mode. The frequency agile beacon of Figure 18.6.9 can provide the facility.

18.6.3.2 Time off-set frequency agile radar beacon

Another beacon being assessed uses basic frequency agile techniques with known time delay introduced into the response. When interrogated the beacon delays its response by up to a few hundred microseconds. The primary radar thus receives the beacon response during a time when relatively weak echoes are returned from beyond about 30 nm. One objective is to minimize cluttering the PPI, particularly at close ranges; targets which may be masked are at distant ranges.

If the delay at the beacon is known the same delay can be introduced at the radar display during beacon reception to trigger the radial scan and display the beacon response at its correct scaled range. Obviously some modification in ship-borne equipment is necessary to realize the full potential of the system which is only broadly outlined here. Development is continuing.

18.6.3.3 Interrogated time off-set frequency agile racon

Some interest is now being shown in the ITOFAR which is a derivative of the time off-set frequency agile arrangement. At interrogation the ITOFAR produces normal frequency agile response at the frequency of the primary radar transmission. In addition, however, the radar operator can select operational modes offered by the ITOFAR by transmitting one of a selection of PRFs. Recognition of a specific PRF may, for instance, put the beacon into time off-set frequency agile mode. Other specific PRFs switch it to transmit positional information, identification, etc. The PRFs suggested are not commonly used in radar sets. It is seen that the sophistication of the racon is now beginning to blur its role as a pure range and bearing marker; moving more toward the definition of a *transponder*.

18.6.3.4 Ship-borne transponders; principles

A transponder is electronically very like a racon. However, it responds automatically only upon reception of a recognized coded interrogation. Transponders are then able to send encoded data. Responses are decoded and can be displayed on a PPI or other display. It is envisaged that transponders will be used for automatic exchange of information between ships to improve collision avoidance and for search and rescue purposes. At the Sixteenth CCIR Plenary Assembly held at Dubrovnik it was determined that 'transponders should be used to meet the following purposes':

(i) identification of certain classes of ships (ship-to-ship)
(ii) identification of ships for the purposes of shore surveillance
(iii) search and rescue operations
(iv) identification of individual ships and data transfer
(v) establishing position for hydrographic purposes.

Development of transponder systems to meet these requirements is at present being carried out. In the USA the system is known as the MRIT, or Maritime Radar Interrogator-Transponder. In the USSR a system nearing completion is the SIT, or Ship-borne Interrogator Transponder, whilst in Japan small boat and life-raft transponders have been developed. UK manufacturers are developing an interrogator-transponder system known as MIDAR, Microwave Identification Data Automatic Response.

18.6.4 *Passive reflectors*

18.6.4.1 Target returns

Section 18.2.4.8 indicates that the signal echo power returned to a radar receiver depends, among other things, upon the target aspect. The aspect presented to an incident electromagnetic wavefront is for many targets a variable factor. A changing aspect produces wide variations in return and an effect of 'glint' or scintillation is often produced at the display. Targets having high reflectivity (or reflection coefficient) produce best signal reflection but this energy does not necessarily return along a path back to the radar antenna. So, although high reflectivity may occur, the *attitude* of the reflecting surface relative to the receiving antenna at the time of return is of vital importance. Figure 18.6.6 illustrates that smooth, highly reflective surfaces which are perpendicular to the line of propagation form the best reflectors. Steel has a reflection coefficient of about 0.95. Surfaces which slope or are spherical tend to return poor echoes; they have small radar echo areas even though physically large compared to the wavelength of the propagation.

The reflection process is complex but depends upon the conductivity of the reflecting surface wherein the incident wave induces

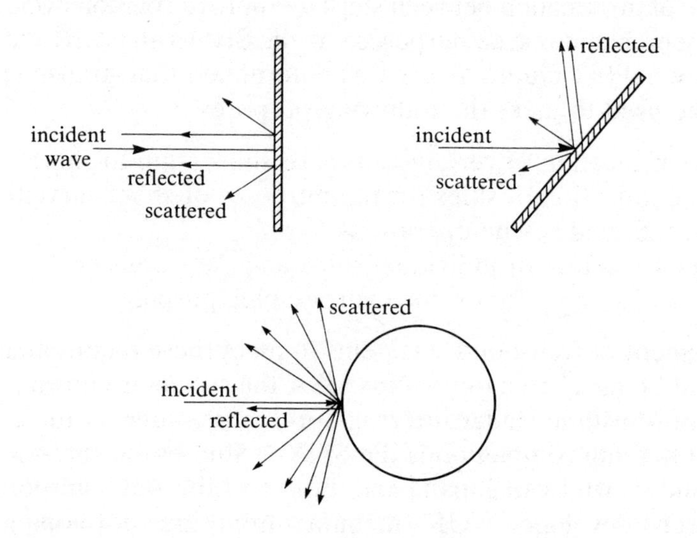

Figure 18.6.6 Simple optical illustrations showing how a vertical surface returns more energy than sloping or spherical surfaces. Surface aberrations always produce some reflected power incident upon the radar antenna. Reflectivity is also a function of conductivity of the reflecting surface.

RF EMFs to produce currents of a phase which cause re-radiation from the surface. Sea water reflects well, ice less so. Glassfibre and wooden boats are notoriously poor reflectors. Targets which, in general, lie along the horizon, in the general direction of the wave's polarization also produce an integrating effect as they are scanned and return good echoes.

18.6.4.2 Corner reflector

Targets which are navigationally of importance, such as buoys, small wooden and glassfibre boats, need to enhance their echo return so that radar observers can easily identify them as clear, consistent and unambiguous targets. A *radar reflector* is used for that purpose and is known as a *passive reflector* since it contains no active responder. The majority of these are fabricated from steel and embody the principle of incorporating three mutually perpendicular reflecting faces. Primary radar energy irradiating such a structure as in Figure 18.6.7 will always return a proportion of the energy falling on the effective collecting area. The level of return depends on the aspect of that surface. In practice, a *cluster* of such corner reflectors is constructed and various arrangements have been devised, all with the idea of returning a strong echo despite possibly violent motion of the mounting structure. A very common arrangement is the *octahedral* cluster shown in Figure 18.6.7.

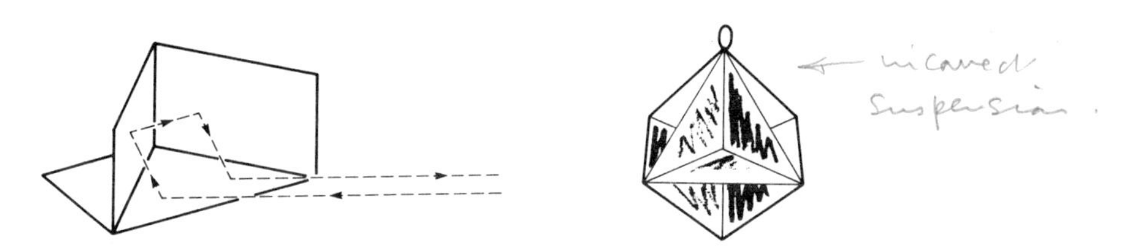

Figure 18.6.7 Principle of corner reflector (left) and octahedral cluster (right).

18.6.4.3 Luneberg lens reflector

At microwave frequencies the index of refraction of materials is proportional to the square root of the material's permittivity. Materials can be manufactured with accurately reproducible values of permittivity fairly cheaply. The Luneberg lens is spherical and is constructed having a variable index of refraction along its radius, from a minimum of one at its surface to $\sqrt{2}$ at the centre. This produces a lens whose property is to focus a plane wave incident at its surface onto a point diametrically opposite to the line of

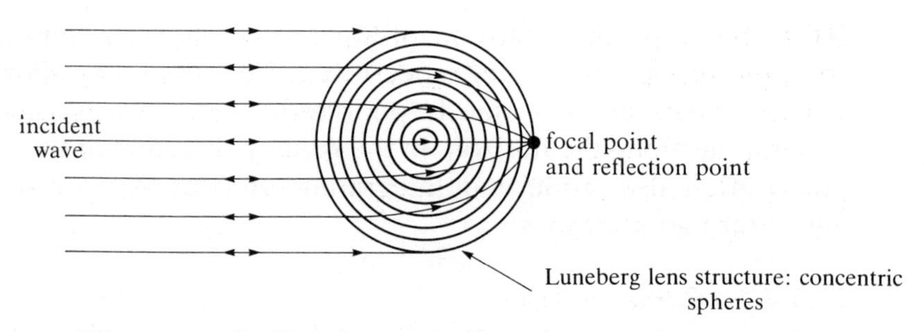

Figure 18.6.8 Principle of Luneberg Lens.

propagation. Figure 18.6.8 shows the principle. Conversely, a reciprocal action will occur where energy reflected from that point emerges at the opposite side of the sphere as a plane wave. It follows that the lens presents a large electrical plane reflecting surface, *regardless of the angle from which it is irradiated*.

The Luneberg lens arrangement therefore overcomes some of the main objections associated with other passive reflectors mounted on moving platforms. Practical lenses comprise several concentric shells having different permittivities and of low dielectric loss, a 304mm diameter lens having an effective radar cross-section of about 35 square metres.

18.6.5 *Principle of a frequency agile beacon*

The block diagram of Figure 18.6.9 shows the main units of a modern frequency agile radar beacon and is based on one British design.

Interrogation pulses from a ship-borne radar transmitter pass through a circulator which provides isolation from the transponder section and directs the pulses to a balanced mixer. Sensitivity is adjusted by an attenuator whose main purpose is to protect the mixer from overload. A Gunn diode acts as local oscillator to produce intermediate frequency outputs between 50 and 230 MHz dependent on the interrogation carrier frequency. The mixer is succeeded by a wide band IF amplifier and a series of logarithmic amplifiers. This arrangement prevents receiver paralysis.

The IF output drives a frequency discriminator via an amplitude limiter and also drives an envelope detector. A radar pulse reaching the beacon is converted to an IF pulse and demodulated to produce a rectangular pulse of width equal to that of the interrogating radar

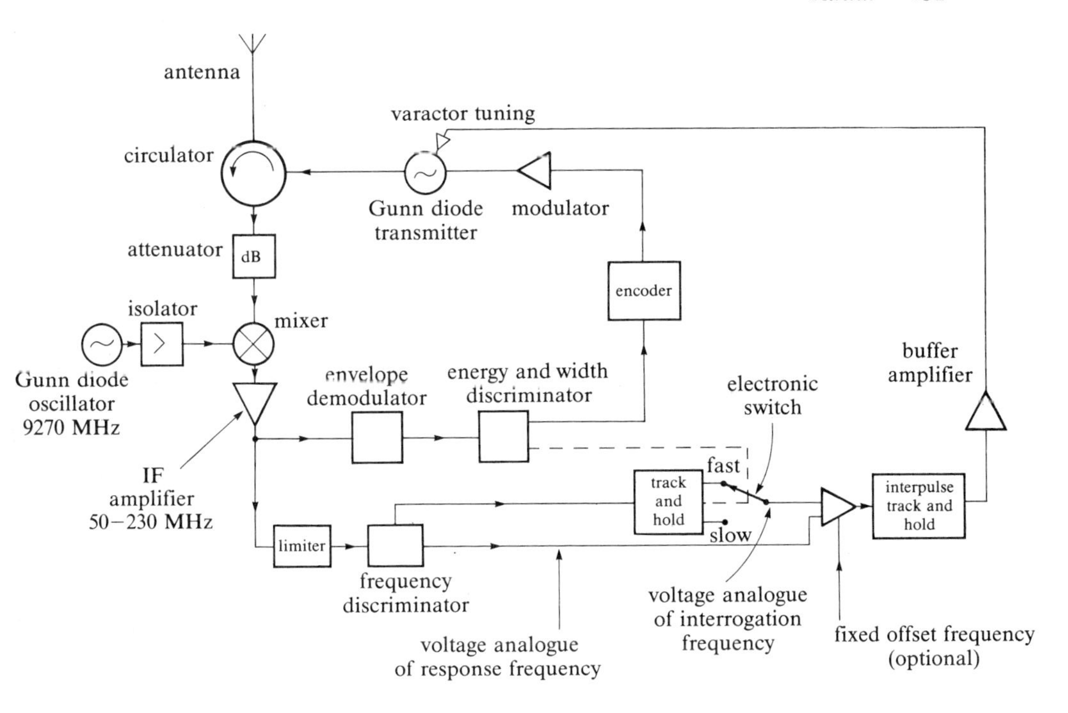

Figure 18.6.9 Simplified block diagram of X band beacon. S band version differs in microwave sections.

and having an amplitude proportional to the received signal amplitude. If the energy content of the pulse (width and height) are adequate, the encoder circuit is triggered. Received signals not meeting preset criteria are discriminated against and not permitted to generate a response. Such signals are generally weak in power and of short pulse duration received from distant radars using short display ranges, or from sidelobes of the main antenna at closer ranges. The arrangement helps to minimize unnecessary interference.

Preset switches allow a unique identification code to be generated by the encoder from a sixteen-bit word; bit lengths are adjustable in 0.25 μs increments. Unaccented morse coded letter pairs are chosen preferably with a leading dash to aid recognition at the distant display. Overall length of beacon response is limited to about 45 μs of which the marks should occupy about 25 μs maximum: operation then occurs within the recommended duty cycle of the transmitter. An inhibit is imposed on the encoder following the transmission to ensure that response echoes (from waves, etc.) do not retrigger the beacon.

Transmitted response powers of the order of 200 mW or 1W are available depending on siting and application using Gunn diode structures in X band beacons. Tuning of the transmitter is by varactor diode and slewing onto the desired response frequency occurs within 2.5 μs of interrogation inception. S band beacons use a master oscillator – power amplifier configuration, tuning the oscillator and pulsing the p.a. Minimum transmitter power is about 750 mW.

18.6.5.1 Frequency control of response

Frequency agile beacons automatically and swiftly adjust their responding frequency to that of the interrogating radar, an action which ensures that response signals received by that radar will lie within its receiver bandwidth. No signal degradation should then result at the display, nor is it necessary to sweep the responder output through the radar band as in the slow sweep beacon of section 18.6.2.1.

After limiting, the IF signal is fed to a frequency discriminator which produces a pulse output proportional to the frequency and independent of IF amplitude. This is the voltage analogue of frequency received and is simultaneously applied to two track and hold circuits. One of these has a very fast tracking capability, the other is as much as twenty times slower. Immediately a radar pulse is detected the fast track and hold circuit is latched thus memorizing the analogue of the interrogation frequency. The circuit can hold information from pulses as short as 0.05 μs. If the pulse width exceeds 0.3 μs, the slow track and hold circuit is initiated which stores a voltage analogue of the average frequency within the longer received pulse. If the slow track and hold circuit has been triggered an electronic switch preferably uses that circuit output as reference to determine the transmitted beacon response frequency since the circuit has more accurately averaged the received frequency information, otherwise the fast output is taken. The resultant tuning voltage is held in the interpulse track and hold circuit (Figure 18.6.9); when the transmitter fires, the varactor tuning is adjusted via a buffer amplifier.

The energy which enters the receiver section during response time exceeds most interrogation signal levels and the discriminator now outputs a voltage analogue of response frequency. The response frequency is compared to the memorized preferred value and any error is amplified and applied to the transmitter varactor tuning feedback circuit. This causes swift adjustment of the transmitter frequency to equal the interrogation frequency. If the fixed off-set frequency mode (FOF) is in use then a fixed off-set voltage is applied to the error amplifier so deviating the response frequency from the interrogation frequency by about 50 MHz. The off-set unit is not shown in Figure 18.6.9. The interpulse track and hold circuit is arranged to memorize the tuning voltage between response and consecutive interrogations to reduce unnecessary frequency adjustments when responding to one radar.

18.6.5.2 Power supplies and structure

Since radar beacons must operate in hostile environments for long unattended periods the structure must be particularly rugged and of high precision. Power is generally provided from batteries and typically the type of beacon described will consume about 8–10W in operation. The apparatus operates from a wide range of d.c. voltages, the regulated system supply being derived using inverter techniques. In addition an electronic timer is available which can be programmed to shut down the beacon for prescribed periods during which the power consumption is considerably reduced to about 100 mW. Figure 18.6.10 illustrates the outward appearance of the beacon, all circuitry and the antenna being enclosed in a weatherproof cylindrical container, transparent to microwaves and shaped to shed snow and dust and to prevent birds from perching. The power supply is external to this structure. Familiar waveguide technology is used in the RF section whilst integrated circuit techniques are widely applied elsewhere. Additional available facilities not shown in Figure 18.6.9 include sidelobe suppression (SLS) and intermittent beacon operation.

18.6.5.3 Circular polarization

In almost all cases horizontally polarized transmissions are used in the civil marine radar bands. However an occasional need arises for vertically polarized radars to interrogate and receive a compatible response from beacons which service a mainly horizontally polarized demand. In such cases an antenna carrying a circular polarizer for X band will convert either horizontal or vertically polarized interrogations into circular polarized format. The horizontal component can then be received by the antenna. A reciprocal action occurs during transmission but a 3dB loss must be supported in either direction. There is a significant number of vertically polarized S band marine radars. Beacons operating in this band use a circularly polarized antenna. Twin-band beacons capable

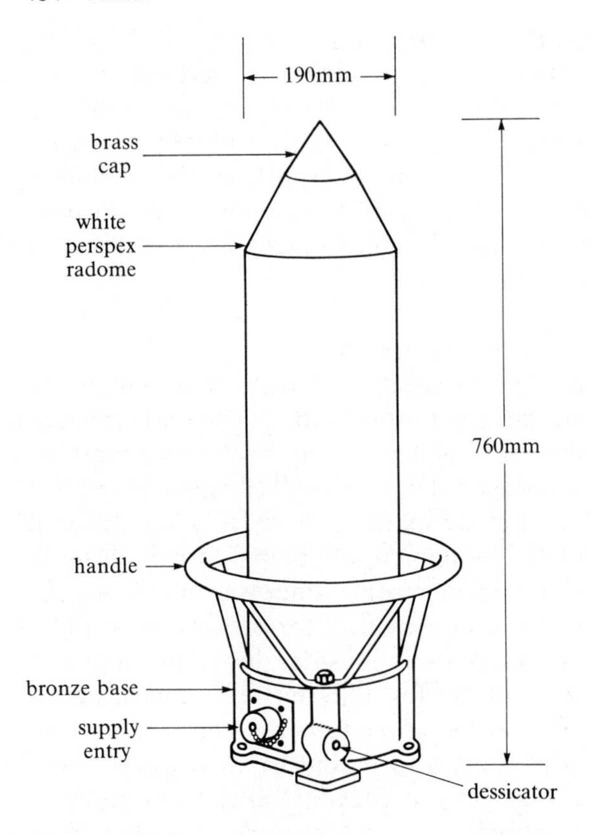

Figure 18.6.10 External appearance and dimensions of a typical radar beacon.

of operating in the S and X bands use a common coder, logic circuitry, sidelobe suppression and power supply section but otherwise operate as two independent racons.

18.6.5.4 PPI clutter due to racon response

It is possible to receive racon responses not initiated by own-ship's radar. In such instances the response may be displayed at any point on the PPI and range information is lost. In addition, desirable echo paints can be obscured by the clutter created. This occurrence is not very common and does not normally pose a serious navigation problem. A more alarming situation can arise where at reduced range the beacon response is displayed as a large proportion of the display radius. In these situations the signal strength of interrogation is sufficient to trigger racons on sidelobes and ship reflections. The display can become cluttered as indicated by Figure 18.6.5. If the signals are differentiated the racon display is lost, except for the leading edge of the response, but so unfortunately are minimal echoes which lie beneath the clutter. Some modern radars which use video processing methods apply a constant amount of signal differentiation (section 18.5) rendering the racon response useless. To implement the racon facility such circuits need temporarily to be disabled. It will be appreciated from the various PPI diagrams that racon response can very often be obscured by land echoes and the devices are best sited where a clear view of their response is obtained. The ideal is not always achieved since the lie of the land beyond the racon structure depends on the relative position of the interrogating vessel.

A great deal of development has occurred in racon circuitry in the past decade due mainly to the availability of improved solid state devices which has resulted in improved response, effective sidelobe suppression and reduction of racon clutter.

18.7 Recent advances in civil marine radar systems

8.7.1 Colour displays; digital scan conversion

Sections 18.4 and 18.5 have outlined the advances made in the last decade or so in microprocessors, solid state memories and high-speed logic devices which have made possible the application of very powerful digital video processing and computing techniques in marine radar sets – mainly in the display. These have been aimed at improving the PPI information presentation and to provide automated target tracking and surveillance of many targets simultaneously. A task which would fatigue a team of human observers.

In keeping with advances in solid state technology and colour CRT development, Racal-Decca Marine Limited have developed a colour digital scan conversion display designed to meet international type-approval requirements. The display monitor uses a 20 inch medium resolution shadow-mask colour tube (section 18.4.5.2) and manual selection is provided for bright daylight viewing or a more subdued mode for night viewing, both in full colour. The method of target and synthetic video presentation has been engineered to maximize information presented to the radar observer whilst minimizing viewing fatigue. Additionally the unit offers some simple but powerful relative motion target plotting and guard zone facilities.

A conventional non-coherent pulsed marine radar transmitter with logarithmic receiver provides heading mark, PRF trigger, antenna azimuth bearing pulses and raw video signals which are input to the display control module for digital processing and handling. Figure 18.7.4 shows familiar panel controls bearing standard IMO symbols and providing manual adjustment to gain, rain and sea anti-clutter, receiver tuning and display brilliance. Automatic anti-clutter is provided, disabling manual control and suppressing clutter by an adaptive process (section 18.5.3).

18.7.2 *Principle of DSC*

The block diagram of Figure 18.7.1 shows a simplified version of the bright-track DSC display unit. Analogue video signals are applied as the raw radar input and quantized in synchronism with the timing of their related sweep trigger into short rectangular pulses of high or low level. All echoes are adjusted either manually or automatically to display above the noise threshold to minimize displayed system noise. Scan-to-scan, two-out-of-two correlation techniques performed on stored radar information virtually eliminate interference from other radars. Digitized video is then clocked at a high rate in real time into a video memory having 252 locations, the clock rate being arranged for the range in use to effectively divide the PPI picture radius (sweep) into 240 equal intervals. On lower ranges 80 intervals are generated. The contents of this temporary storage (known as the retimed-video memory) are used in conjunction with the control circuit to store echo coordinates at unique addresses in dynamic ram.

18.7.3 *Polar to rectangular conversion; the display raster*

A 625 line television raster scan is used at the colour display monitor similar to a domestic television receiver. Not all the lines are displayed and the field scan is 60 Hz interlaced 2:1 to produce 30 interlaced pictures per second and a flicker-free picture. Odd and even fields display different information, effectively doubling the picture resolution. Inputs are the red, green and blue video and mixed syncs. Picture brightness is controlled from the operator

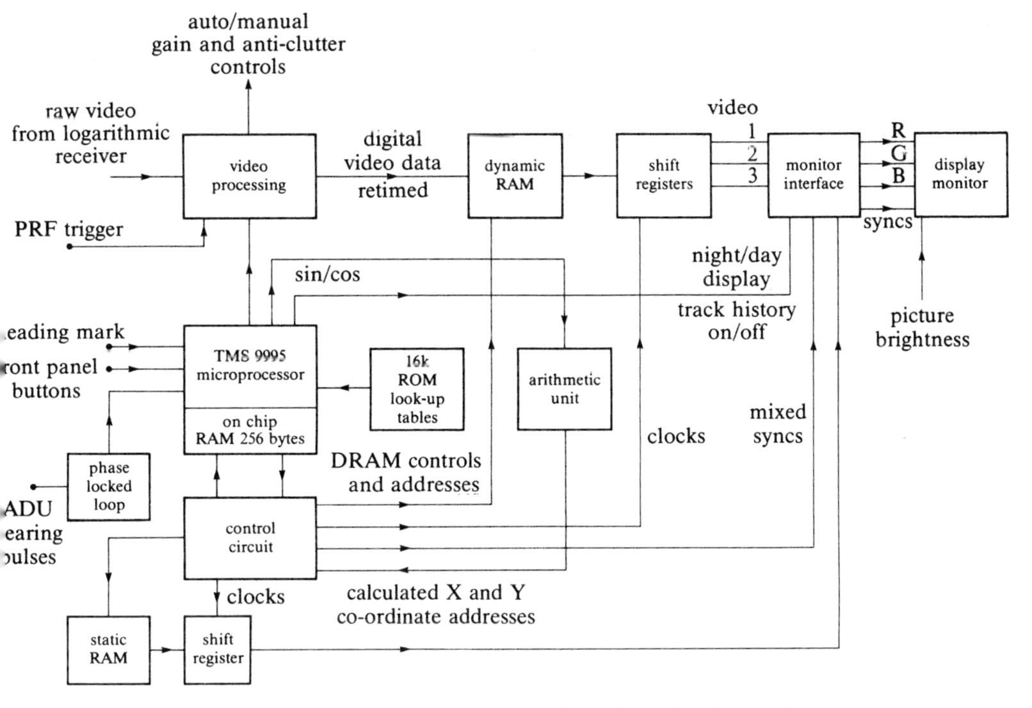

Figure 18.7.1 Basic block diagram of DSC colour display unit. (With permission from Racal Marine Radar Limited.)

control panel. Since the tube is a low persistence device the picture information must be refreshed at a high rate by operating at a line frequency of 18.75 kHz.

Information is displayed using digital video graphics methods. Target information is gathered in polar form (range and azimuth bearing) whereas the television display shows information in rectangular coordinates, hence a conversion from polar to cartesian form is made in the microprocessor-controlled section of the unit. A bearing pulse is received from the ADU for each degree of rotation of the antenna; this is converted for processor purposes via a phase locked loop to ten pulses per degree, providing 3600 pulses per radar picture counted from the heading mark. A TMS 9995 microprocessor carrying 256 bytes of on-chip RAM acts in a supervisory capacity also polling the front panel controls and appropriately updating displayed synthetic video.

Initializing the system bootstraps the cold-start program into the processor RAM from firmware residing in an associated 16K of ROM; the heading mark is found and the processor, acting on interrupt from a bearing pulse, uses look-up tables stored in the ROM to output corresponding sin-cos values to the arithmetic unit (see also section 18.5). The arithmetic unit calculates X and Y coordinate addresses corresponding to the memory map in dynamic RAM which determines, according to the control unit action, the displayed position of an echo on the television display. In this manner each bearing pulse successively arranges the display coordinates to form a rotating radial line and simulate the familiar PPI presentation on the TV display. The dynamic RAM is used only to generate the radar echo image, essential timing and addressing is provided by the control circuit.

18.7.4 Memory mapping; synthetic video

Digitized video information is shown on the display as small discrete areas or *pixels* (picture cells). Figure 18.7.2 illustrates pixel arrangement in a hypothetical case. The total screen area is divided into 576 vertical by 768 horizontal cells, depicted in Figure 18.7.3. Each has an X, Y coordinate and is therefore uniquely addressable. The position of each pixel is therefore *mapped*, i.e. has coincidence with a unique address in an area of static RAM. A cell can be addressed under program control and brightened up by applying appropriate control to the CRT beam. In the case of colour pixels

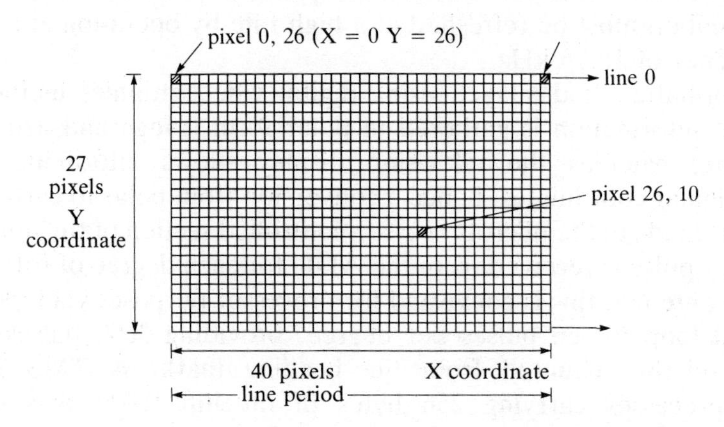

Figure 18.7.2 Hypothetical arrangement of 1080 pixels in 27 rows by 40 columns. Pixel data can be stored in 1080 static or dynamic RAM 'bits'. A 1 in a bit position which coincides at read-out time with a pixel indicates that the pixel is to be brightened.

the principle is similar, acting on three beams and the triad of primary colour phosphor dots.

The static RAM operates under microprocessor control, being used to display synthetic video. This comprises the bearing scale, heading mark, EBL, VRM, range rings, index lines and the 'tote' which is information displayed in the four corners of the display concerning EBL bearing, VRM setting, range in use, pulse length, state of display, etc. Figures 18.7.3 and 18.7.4 illustrate. The static RAM does not store radar information. A second memory, the previously mentioned dynamic RAM, maps the displayed picture storing all digitized video representing the received picture information. The mapped area comprises a square of 512 pixels centred on the screen; a PPI circle is formed of those pixels within a 252 pixel radius so that a radial line is formed of 252 equal sections corresponding to one sweep and not all of the 512 by 512 pixels are used. The dynamic RAM is accessed directly by the addressing action of the control unit which multiplexes the various read and write cycles, generates the master clock, and effects time sharing between the static and dynamic RAM so that a steady composite image is formed of radar information and synthetic video. It also generates various other timing cycles.

The dynamic RAM stores data concerning any radar pixel as a three-bit code thus giving eight possible *status codes* per pixel. Each time a pixel is addressed in the radar circle its status is first examined to determine the status attributed to that pixel on the previous scan. A decision is then made which determines the current status of that pixel, in this way data stored during previous scans is correlated with present scan data and the nature of its displayed image is decided. For example, if the code 000 represents an empty pixel in 'daylight mode' and no current echo arises there, the decision is made to fill that pixel with a light blue colour (daylight background). The code 001 encountered might indicate that the pixel in the previous scan was loaded with an uncorrelated echo; if the current scan again shows an echo that pixel is promoted to status of a correlated echo by storing, say, code 010 and displaying at the pixel the colour for a correlated target (yellow/amber). On the other hand, the code may be 001 and the current scan shows no correlated echo hence the decision is made to demote the dynamic RAM data to 'afterglow' colour (black) and store 011 to represent the afterglow status. It is thus possible to present as a daylight display the targets in normal yellow/amber, a blue background and recent track history in black. Uncorrelated single sweep echoes are shown in low intensity brown

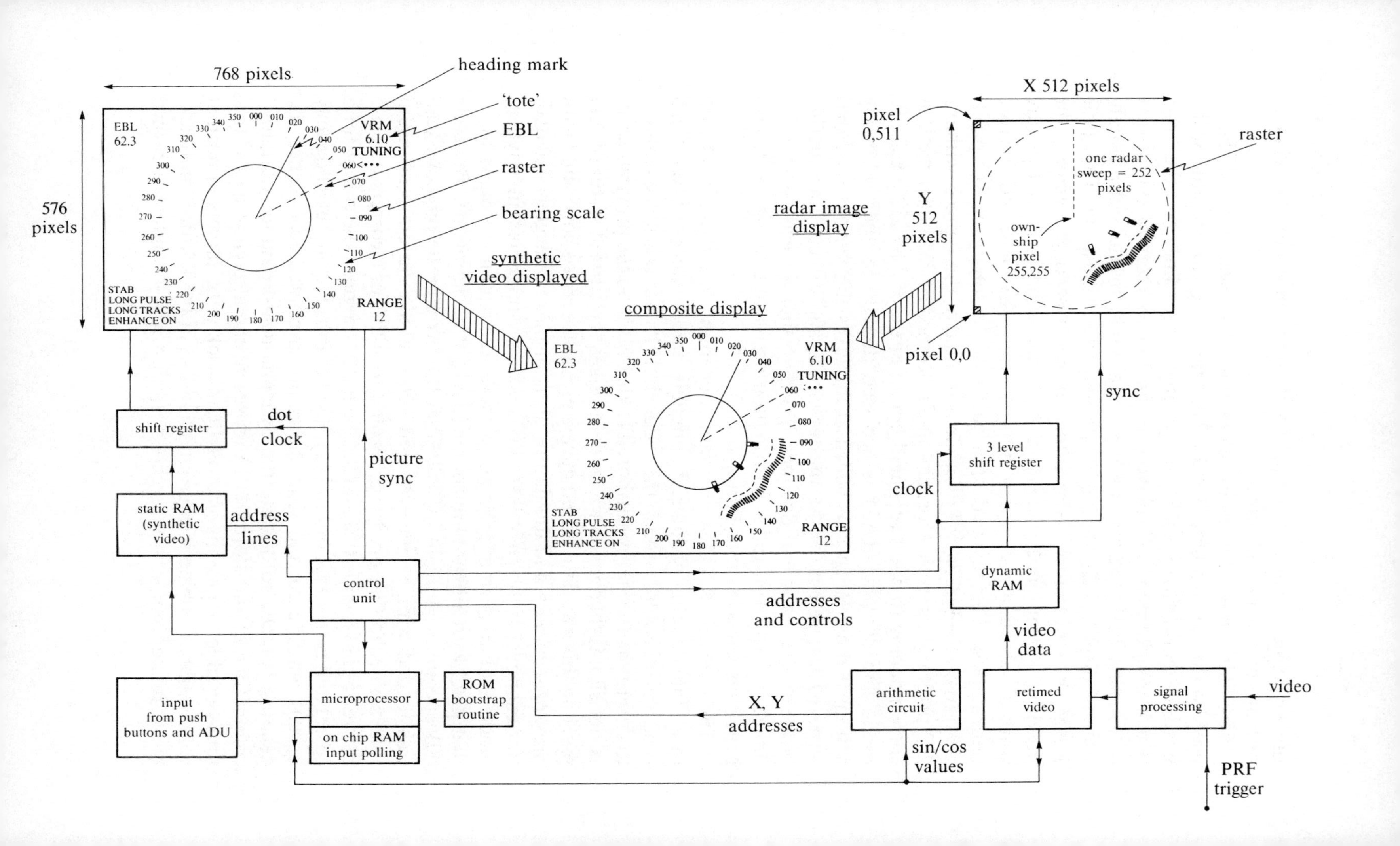

768 pixels
576 pixels
heading mark
'tote'
EBL
raster
bearing scale
EBL 62.3
VRM 6.10
TUNING
STAB
LONG PULSE
LONG TRACKS
ENHANCE ON
RANGE 12
synthetic video displayed
composite display
radar image display
X 512 pixels
Y 512 pixels
pixel 0,511
pixel 0,0
raster
one radar sweep = 252 pixels
own-ship pixel 255,255
shift register
dot clock
picture sync
static RAM (synthetic video)
address lines
control unit
input from push buttons and ADU
microprocessor
on chip RAM input polling
ROM bootstrap routine
addresses and controls
X, Y addresses
3 level shift register
clock
sync
dynamic RAM
video data
arithmetic circuit
sin/cos values
retimed video
signal processing
video
PRF trigger

to indicate where first 'hits' were recorded. Synthetic video is shown in white in the daylight mode.

A night viewing mode is also available under operator control in which case the background becomes black, with blue track history and green synthetic video (see Figure 18.7.4). Track history is plotted by operation of the TRACKS push-button to select 'long' or short tracks. In daylight mode the short track history simply provides a short black 'afterglow' to indicate relative motion of targets. Long tracks provide a three minute display on all ranges and facilitate RM plotting.

18.7.5 *Enhance, index, guard facilities*

Many of the panel push buttons have a multiple function requiring one push for one function and a second push (or move) for the alternate function. A single push of the enhance control initiates the 'integrated video' mode (correlation of echoes); a first scan target appears in brown and a second and subsequent confirmatory scan promotes the displayed echo to yellow/amber. A second push activates the 'echo stretch' facility on ranges 6 nm and above (see section 18.5 for echo stretching techniques). Echoes beyond approximately 3 nm and of minimum length are elongated to improve contrast of desired echoes at sea.

A maximum of two index lines (hatched) can be displayed, positioned by manipulation of the VRM and EBL, the latter control revolves the chosen index line about a small cross which indicates the intersection of EBL and VRM. The same control permits the second index line to be set up as required. A sustained push on the control erases all index lines.

Selecting the guard option will produce a guard zone at half range value either in sectors or for 360 degree coverage. Multiple operation of the guard push button and the EBL, together with messages shown on the screen allow selection of the start and finish of the guard sector. Targets moving into the guard zone activate an audible alarm.

Figure 18.7.3 (Opposite) Basic DSC units indicating the separate functions of static RAM for synthetic video and dynamic RAM for radar video. The two memory mapped areas are effectively time division multiplexed by the control unit to produce the composite display (with permission from Racal Marine Radar Limited).

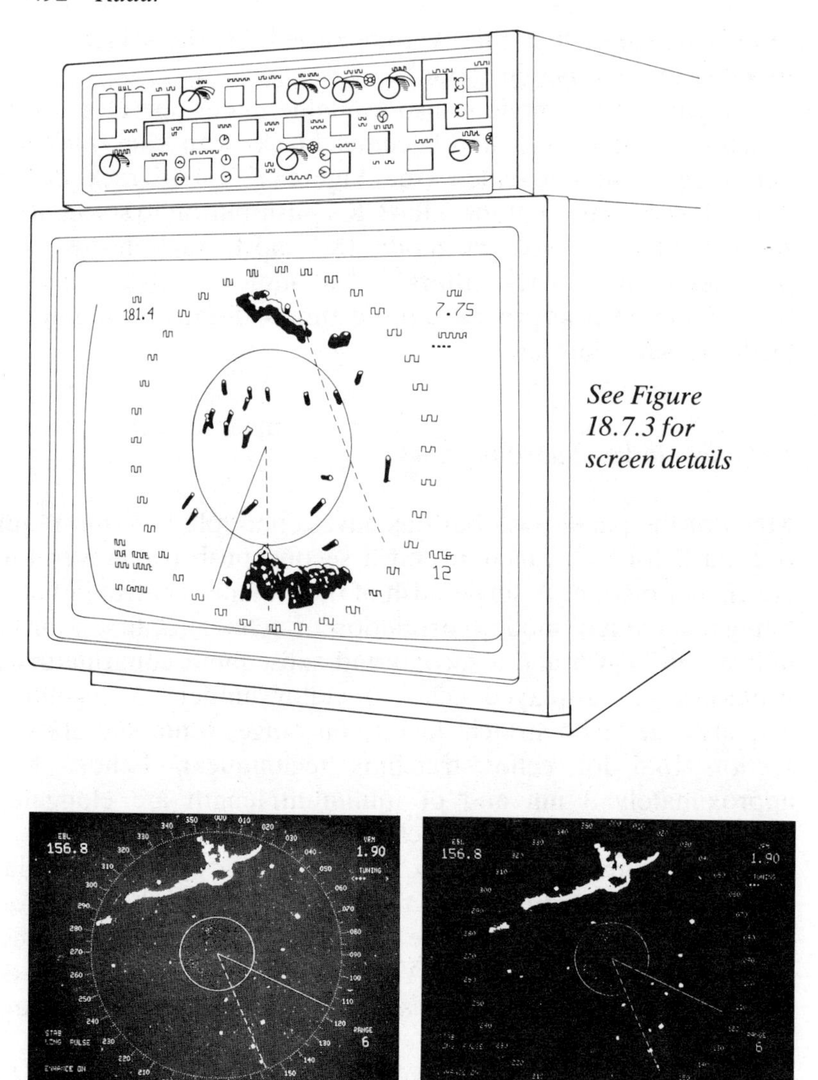

choice of day or night colour coding

Figure 18.7.4 *Typical displays of a DSC colour display unit*
Below left: daylight colour codes – blue background, yellow/amber targets, black track history, white synthetic video.
Below right: night colour codes – black background, blue track history, green synthetic video.
(With permission from Racal Marine Radar Limited.)

18.7.5.1 Target trails

The correlation and status checking which occurs for each cell in a radar sweep is evidently a dynamic process in which pixels are being promoted to firm echoes and due to subsequent relative motion are being demoted to afterglow and finally to background status, a process which occurs sometime after a cell was first promoted. In order to avoid an unrealistic afterglow tail of fixed length attached to such targets the final demotion from afterglow to background colour is achieved by operating a pseudo-random algorithm and using the unsynchronized TV cycle time for afterglow erasure. This ensures that the observed trail of a moving target breaks up randomly to produce a more realistic conventional PPI effect.

Appendix A

Consider two power amplifiers having power gains A_1 and A_2. The first amplifier has a power input of 1.0 μW and an output of 1.0 mW. The second amplifier is connected in cascade with the first and produces a 1.0W output from the 1.0 mW signal. Figure A.1 shows the arrangement.

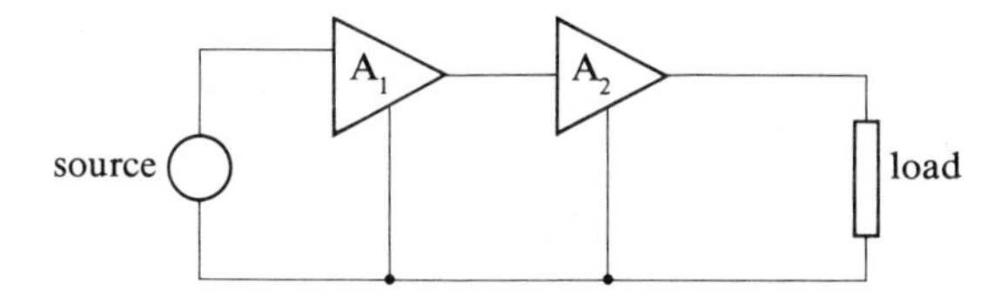

Figure A.1

The first amplifier can be said to have an output/input power ratio of 1000:1 or a *per-unit* power gain of 1000. The second amplifier also has an output/input power ratio of 1000:1. The total power gain from input at amplifier 1 to output of amplifier 2 is 1 000 000:1.

Expressing power ratios in this manner produces unwieldy values, being either very large or exceedingly small and a more convenient method is to express the power gain as a common logarithm. The overall gain can then be determined by adding the logarithms. If the per-unit ratio is needed the antilog can be found.

The unit chosen is the BEL and two powers differ by D Bels when:

$$D = \log_{10} \frac{P_1}{P_2} \text{ Bels}$$

The Bel is a large unit for most electronics applications and the DECIBEL, equal to one-tenth of a Bel, is used hence:

$$D = 10 \log_{10} \frac{P_1}{P_2} \text{ decibels (dB)}$$

Our single example then becomes:

$$D_1 = 10 \log_{10} \frac{1000}{1} = 30\text{dB}$$

D_2 is also 30 dB, thus the overall power gain is:

$$D_{overall} = (30 + 30) = 60\text{dB}$$

If necessary to revert to per-unit values:

$$60\text{dB} = 6 \text{ Bel}$$

$$6 = \log_{10} \frac{P_{load}}{P_{input}}$$

The number whose logarithm is 6, is of course 1 000 000.

A.1 *Losses*

In some circuits the signal path introduces losses where the power input may be greater than the power output. Suppose in our example that the power output P_2 is less than the power input P_1.

For example $P_2 = 1$ mW and $P_1 = 1$W:

$$D = 10 \log_{10} \frac{P_2}{P_1} = 10 \log_{10} \frac{1/10^3}{1} \text{ dB}$$

$= -30$dB, the minus sign indicating power loss.

A.2 *Absolute values*

In many instances a result is desired in absolute values. It is common practice to provide a reference level to which the result can be referred. For example, the statement ‘the power gain is 30 dBm’, means that the output is related to a reference of 1 milliwatt. Any reference can conveniently be chosen. For example:

$$3\text{dBm} = 10 \log_{10} \frac{P_2}{P_1} \text{ (referred to 1 mW)}$$

hence:

$$\log_{10} \frac{P_2}{P_1} = 0.3$$

giving an antilog of 2.0.

The result tells us that there has been a doubling of power. For an absolute value referred to 1 mW it can be seen that $P_2 = 2$ mW. Similarly:

$$-3\text{dB} = 10 \log_{10} \frac{P_2}{P_1} = 0.5$$

and when referred to 1 mW, produces 0.5 mW, indicating a loss of half the power input.
Since power ratios are involved the calculation $P = V^2/R$ or I^2R watts may be used.

$$D = 10 \log_{10} \frac{\dfrac{(V_{out})^2}{R_{out}}}{\dfrac{(V_{in})^2}{R_{in}}} \text{ dB}$$

$$= 20 \log_{10} \frac{V_{out}}{V_{in}} + 10 \log_{10} \frac{R_{in}}{R_{out}} \text{ dB}$$

Evidently it is necessary to know the values of resistance and exceptionally where $R_1 = R_2$ we can write:

$$D = 20 \log_{10} \frac{(V_{out})}{(V_{in})}$$

Unfortunately, many engineers assume $R_1 = R_2$ and loosely use the decibel for arbitrary voltage and current ratios. It is misleading and bad practice.

A.3 Normalizing

When plotting response curves of circuits or radiation patterns of antennas and in many other similar circumstances it is convenient to select an easily identifiable point in the plot and measure power gain or loss relative to that point. The resultant measurements are said to be normalized to that level which is commonly given the value of unity. As an example consider the simple radiation pattern shown in

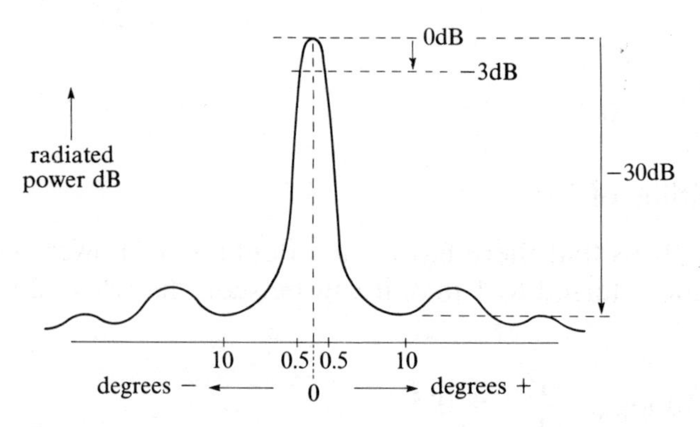

Figure A.2 Application of dB log unit to a simple radiation pattern.

Figure A.2. Here the maximum relative output power point is chosen as the normalized level and given the label '0dB'. All other values measured relative to that point will (in this case) be below 0dB. Hence at a point plus or minus 10 degrees from the main radiation axis, we find that the radiated power is −30dB, or 30dB below the normalized level.

Power measured at these relative angles is, in per-unit terms only 1/1000 of the power at the main axis. If absolute values are applied so that the normalized level does, in fact, represent, say, 10W; at plus or minus 10 degrees we would expect to find 0.01W radiated power (at a constant measuring range).

Appendix B

B.1 Obtaining the basic radar equation

Consider an isotropic radiator isolated in space and supplied with RF power P_0 watts. A spherical wavefront will appear with the radiator at its centre. At range R metres the power density at the surface of the sphere will be:

$$\text{power density} = \frac{P_o}{4\pi R^2}\ \text{Wm}^{-2}$$

Figure B.1 depicts the condition.
A target situated at distance R metres has an effective echo area of A_t m^2 and is assumed to radiate equally well in all directions. The

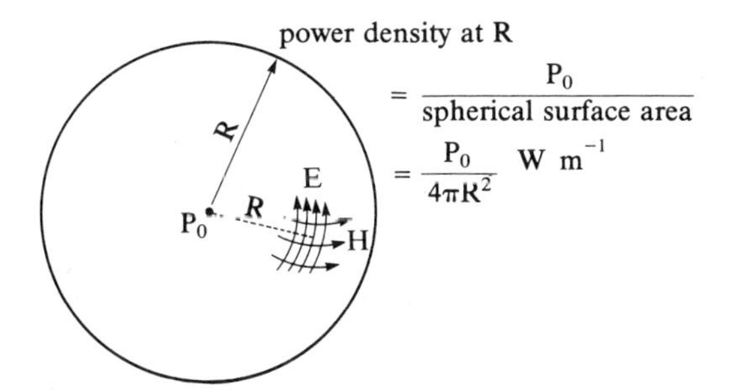

Figure B.1 Spherical wavefront due to ideal isotropic radiator.

echo power will thus form a spherical wavefront and the echo power density reaching the antenna would be:

$$\text{echo power density} = \frac{P_o.A_t}{4\pi R^2.4\pi R^2}\ \text{Wm}^{-2}$$

The radiating antenna has a power gain, however, of G_a and the target power density is G_a times greater than the isotropic value, hence:

$$\text{echo power density} = \frac{G_a.P_o.A_t}{(4\pi)^2R^4}\ \text{Wm}^{-2}$$

At the antenna the received echo power density is 'collected' by the antenna having an effective area A_e. Hence the received power P_r is given by:

$$P_r = \frac{G_a.P_o.A_t.A_e}{(4\pi)^2.R^4}\ \text{watts}$$

At maximum detection range, the value P_r will have its minimum detectable value $P_{r\ min}$. Hence for $P_{r\ min}$ we have:

$$P_{r\ min} = \frac{G_a.P_o.A_t.A_e}{(4\pi)^2.R^4_{max}}$$

and R_{max} is given by:

$$R_{max} = \sqrt[4]{\frac{G_a.P_o.A_t.A_e}{(4\pi)^2P_{r\ min}}}\ \text{metres} \qquad [\mathbf{18.20}]$$

Expression [**18.20**] is a basic form of the range equation used in section 2. It is sometimes shown in other forms and caution should be exercised in its interpretation as free space conditions have been assumed which cannot be attained in practice; the equation is considerably affected by earth's proximity.

B.2 *Optical range between antenna and target*

Assuming that the earth is a spheroid of radius 6370km, and neglecting diffraction or refraction effects, Figure B.2 shows that the range at which the target at height h_2 is just visible from the antenna at height h_1 is $r_1 + r_2$.

By Pythagoras:

$R^2 + r_1^2 = (R + h_1)^2$ and h_1^2 can be ignored compared to R_2.

Hence $r_1 = \sqrt{2Rh_1}$

and $r_2 = \sqrt{2Rh_2}$

thus range is $= \sqrt{2R}\,(\sqrt{h_1} + \sqrt{h_2})$ **[18.21]**

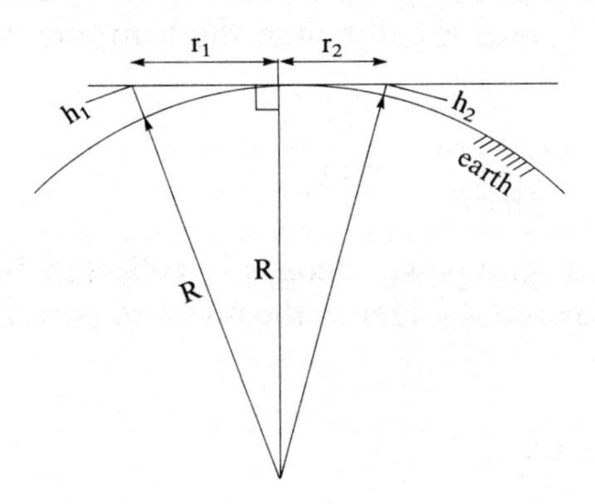

Figure B.2

B.3 *Earth radius factor*

In practice, some diffraction occurs at the earth's surface together with some refraction in the atmosphere. A 'standard' atmosphere is adopted which has a refractive index slightly greater than unity at sea level. The total effect is to extend line-of-sight path as if it were passing over an earth having a greater radius than the nominal 6370km. The factor by which the earth radius is increased is called the *earth radius factor* and has a value 4/3, whereupon expression [**18.21**] becomes:

$$\text{range} = 4/3(\sqrt{2R}\,(\sqrt{h_1} + \sqrt{h_2}) \qquad [18.22]$$

which gives a close approximation for theoretical consideration of range.

Appendix C

C.1 Waveguide principle

Whatever frequency value is involved, an electromagnetic wave comprises an alternating electric field acting at right angles to an alternating magnetic field both being perpendicular to the line of propagation. Such waves are able to be guided from one point to another by any surface which is bounded by a medium having a totally different conductivity or permittivity to that of the surface. Best results are obtained when the EM wave is guided by a good conductor such as copper which itself bounded by a good dielectric. In principle the simplest guide comprises two parallel conducting lines suspended in dry air, the arrangement being known as a *transmission line*. There are structural difficulties and unacceptable inefficiencies in such an arrangement for marine radar usage and EM energy is guided to and from the antenna either by means of a *coaxial cable* or a hollow copper tube. Figure C.1 illustrates the basic coaxial line structure comprising a solid uniform centre conductor surrounded and supported by a low-loss dielectric and having an outer conducting surface concentric with the inner conductor.

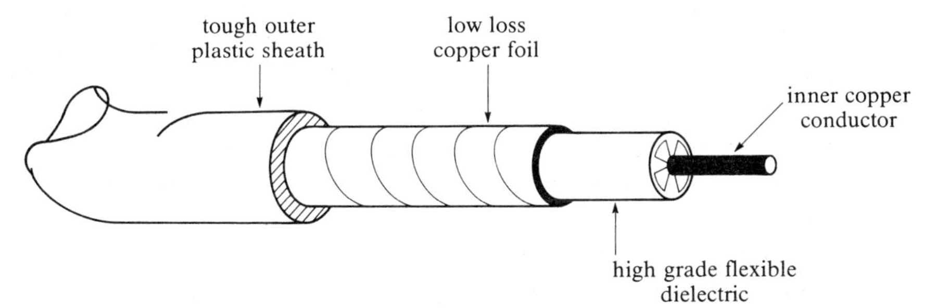

Figure C.1 Typical high grade flexible coaxial cable structure.

The use of such cable is prevalent in ten-centimetre radars where the losses incurred are acceptable. Except for very short lengths, the use of coaxial cable in three-centimetre radars at the magnetron frequency produces high signal loss and the hollow waveguide tube is invariably used at the higher frequencies. Many installations use tubular waveguide for both three-centimetre and ten-centimetre equipments; choice of guide in the latter case being dictated mainly by cost and relative ease of installation. By common usage the term *waveguide* implies a hollow tube but modern radars also use assemblies of *stripline*, commonly known as *micro-strip* which occur in the duplexing RF head section and are used as guides due to their ease of manufacture (see Figure C.16).

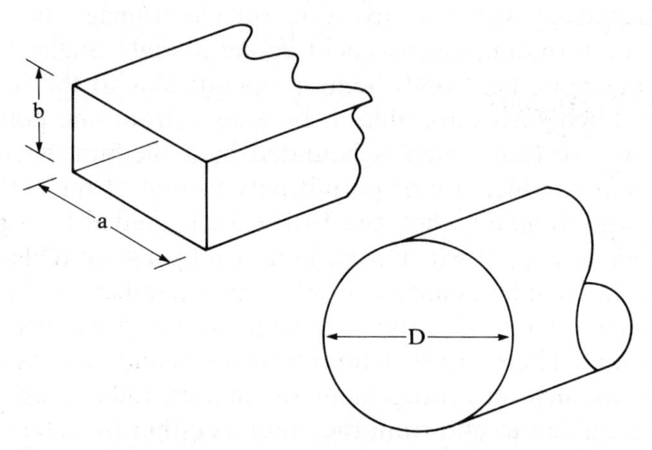

Figure C.2 Rectangular and circular waveguide sections. Lowest propagation frequencies correspond to a wavelength λ in air. Dimension 'a' ≥ λ/2, dimension D approximates to λ/2.

Waveguide shapes are numerous; those most often found in marine radar installations are of rectangular or circular cross-section as shown in Figure C.2. A flexible elliptical waveguide is also used, having the advantage of being provided in one continuous length without intervening joints. Because waveguides and coaxial cables totally screen the propagated fields there are no radiation losses. Dimensions of a waveguide are related to the propagation frequency and power handling capability. Referring to Figure C.2 the dimension *a* is approximately half the wavelength of the lowest frequency to be propagated, dimension *b* is approximately one-half that of *a*, but is determined mainly by the power handling requirement of the guide.

For a rectangular waveguide the operating pattern of the fields within the guide is the *dominant* mode of operation. This is a

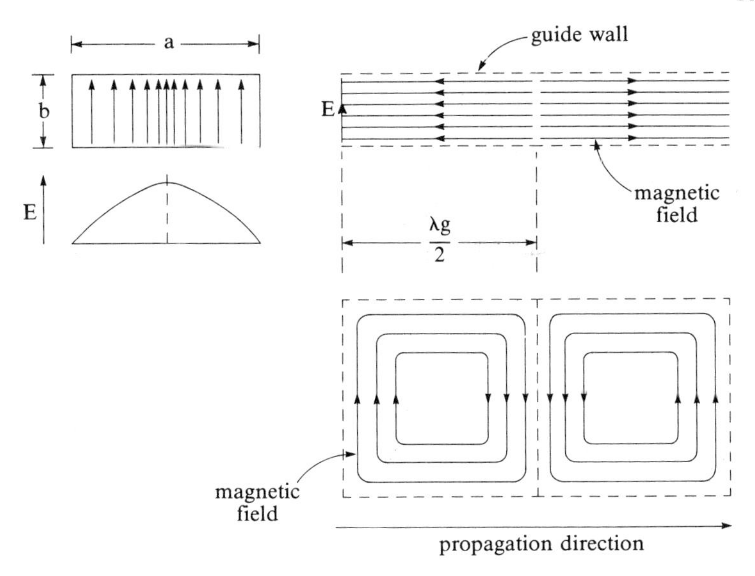

Figure C.3 A basic diagram showing E field acting across dimension 'b' of the guide. Magnetic H field acts along guide. The TE_{10} mode is shown.

complex electromagnetic field pattern produced when the RF energy travels down the guide from the point of excitation in a series of reflections. The electric field acts across the narrow *b* dimension whilst the accompanying magnetic field acts perpendicular to the electric field. In a rectangular waveguide the field pattern will be the TE_{10} mode, implying that only the electric field is transverse to the direction of propagation (TE), the subscript number denoting by the first digit that one-half wave *E* field pattern acts along the *a* dimension and by the second digit indicates that there are zero half waves of the electric field along the *b* dimension. A conceptual diagram of the field pattern for TE_{10} mode is shown in Figure C.3. If the guide is correctly matched at the source (magnetron) and the antenna, the travelling wave pattern represents RF energy conveyed along the guide from source to load and contained within the guide boundaries. The alternating fields so produced induce alternating currents to flow in the inner surface or skin of the guide; a simple idea of those current paths is given in Figure C.4. An EM wave may be launched into the guide by one of three methods, probe coupling, loop coupling and aperture coupling each being represented by Figures C.5 (a), (b), (c).

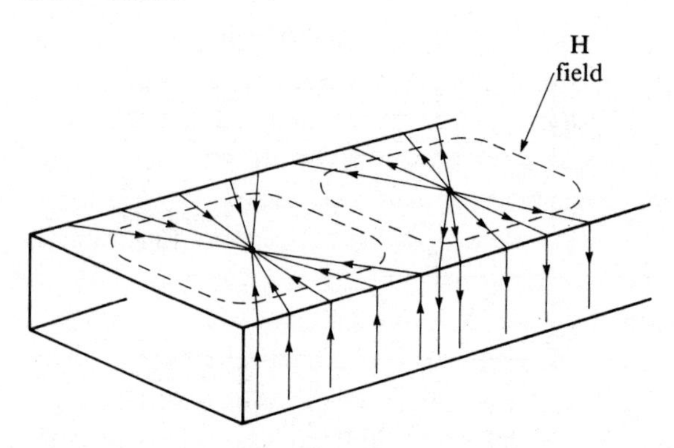

Figure C.4 Basic idea of RF current flow in rectangular waveguide wall.

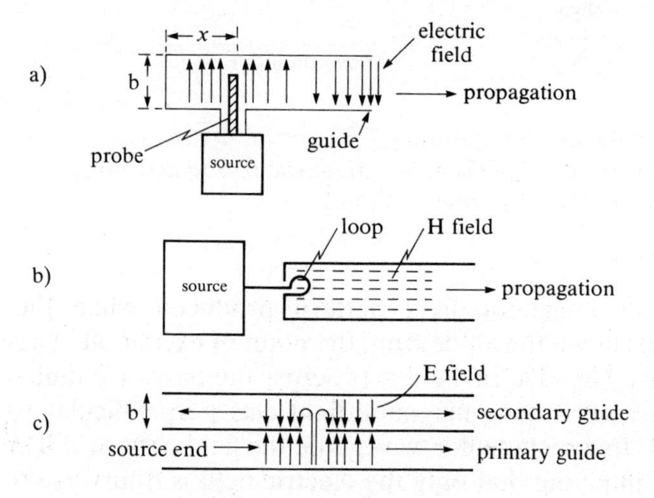

Figure C.5 (a) Probe launching EM field, x is any odd number of quarter wavelengths
(b) Loop coupling via the magnetic field component
(c) E field excitation of secondary guide via an aperture from the E field of primary guide which is connected to the RF source.

C.2 Arrayed antenna principle

C.2.1 The dipole

It can be demonstrated that a section of two-wire transmission line connected to a source of radio frequency power at one end and open circuited at the other end will exhibit a *standing wave* pattern of current and voltage due to the interaction of electromagnetic waves travelling from the source to the open end and reflected electromagnetic waves travelling back from the open end *back towards* the source. The alternating standing wave pattern will always have a voltage maxima (voltage antinode), and a current minima (current node) at the open end, regardless of the line length. The standing waves of voltage and current will adopt a particular pattern on the line, according to the wavelength of the source. Figure C.6 indicates that a section of line which is exactly $\lambda/4$ in length for that particular wavelength, will produce an antinode of current and a node of voltage at the source. The ratio of the magnitude of V to the magnitude of I anywhere along the section of line will give the magnitude of impedance. Thus at the open end the impedance is infinite and at the source it is zero. In a practical situation, since energy is lost, these represent very high and very low impedances.

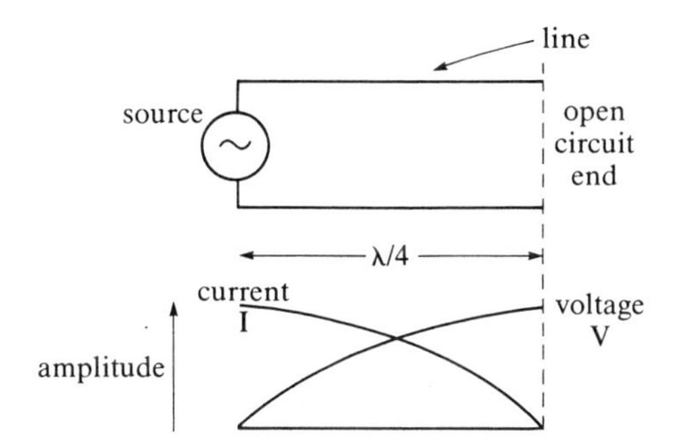

Figure C.6 A section of line open circuited and $\lambda/2$ in length resonates and produces a standing wave pattern shown.

The section of line is said to be *resonant* since it exhibits a standing wave pattern and at the frequencies considered it would tend to radiate RF energy. However, in the state represented by Figure C.6, insufficient radiation would take place due to the proximity of the two lines. If the lines are rearranged, as in Figure C.7, the resonant line becomes an *open oscillatory* circuit having the standing wave voltage and current distribution shown and is now a $\lambda/2$ antenna or dipole.

The dipole is characterized by its point of feed having a low impedance and is therefore fed from a current source into two equal quarter wavelength halves. Figure C.8 shows the section through a very short dipole's radiation pattern. The pattern is seen to be

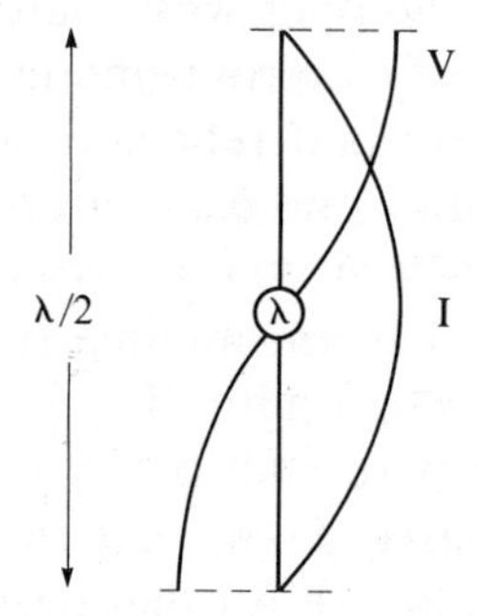

Figure C.7 Standing wave pattern on a centre fed λ/2 antenna.

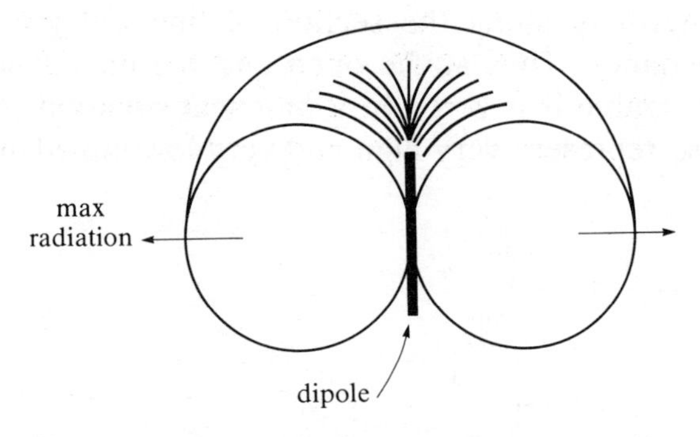

Figure C.8 Vertical cross-section through vertical dipole's ideal radiation pattern.

directive. A dipole radiates best in a direction perpendicular to the length of the antenna and does not radiate off its ends.

C.2.2 The dipole array principle

A number of short dipoles can be arranged so that their centres lie along the same line and the elements of the dipoles are in the same plane. If the spacing between the dipoles is made one-half of a wavelength, a highly directional radiation pattern is produced, as shown in Figure C.9, providing all antennas are fed in phase with each other. Very little radiation takes place in the direction *A,A*. Since each dipole radiates identically and in phase, dipole *a* radiating along *A,A* towards dipole *b* will experience a cancelling field from *b* because of the half wave spacing. Similar reasoning applies to other dipoles in the array. In a direction perpendicular to the plane in which the dipoles lie, the various fields produce an interaction which results in a narrow beam and enhanced gain along B,B. In principle, the greater the number of dipoles the narrower will be the beamwidth. If the dipoles lie along a vertical plane the vertical beamwidth will be very wide and the horizontal beamwidth narrow. The unwanted 'backward' radiation can be reflected forward to enhance the power gain in the desired direction by placing a reflecting conductive plane one quarter of a wavelength behind the array. Such an array will radiate an electromagnetic wavefront with the electric flux lying parallel to the length of the dipoles. Hence vertical dipoles radiate vertically polarized wavefronts.

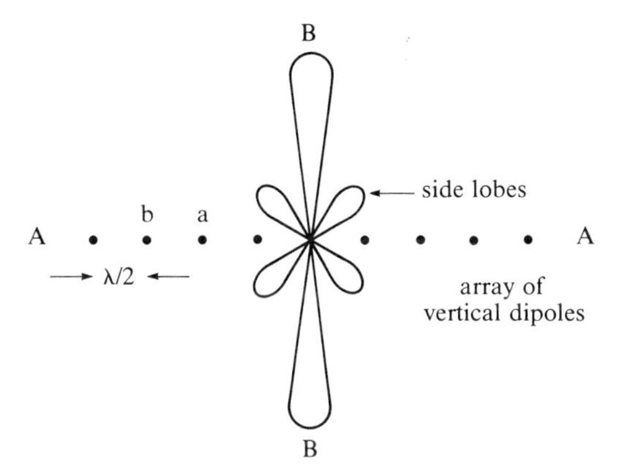

Figure C.9 Beaming effect in the horizontal plane of a dipole array.

C.2. Arrayed antenna principle

C.2.3 Slot antenna principle

A simple slot (rectangular aperture) cut in the surface of a conducting plane exhibits properties of inductance and capacitance, the space across the opposite sides of the slot being an elemental capacitor, the conductive strip surrounding the slot being an elemental single turn inductor. Figure C.10 illustrates the idea. The figure also shows that the slot may be energized from an RF source at its centre. Such excitation sets up alternating electric fields across the slot, being maximum at the centre and minimum at the ends. Radio frequency currents also flow around the perimeter of the slot. If the slot is arranged to have a length which is one-half wavelength of the source, it will behave as a resonant half wave dipole in a manner similar to the dipole described in section C.2.1. There are significant differences between the two arrangements; a vertical slot antenna has its electric field acting across the slot and will therefore radiate horizontally polarized EM waves. The slot has a similar standing wave pattern to the wire dipole but the current and voltage nodes and antinodes are interchanged. The radiation patterns are similar but the slot is a relatively high impedance load compared to the wire dipole.

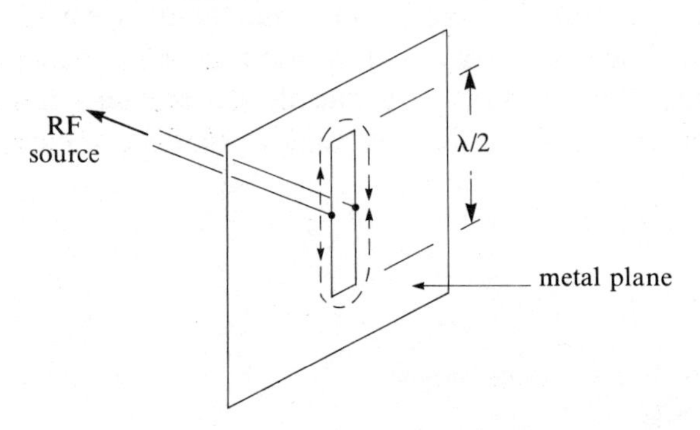

Figure C.10 A λ/2 slot cut in a metal plane and fed from an RF source at correct frequency will radiate. Dotted lines around the slot show paths of current flow.

C.2.4 The Gunn diode

A Gunn diode is a gallium arsenide semiconductor device capable of operating in the super high frequency band. It operates on the principles of negative resistance which is a phenomenon occurring in the device bulk material when the diode has a low voltage applied.

A high field domain travels from the cathode of the material and falls to zero at the anode as another domain is formed at the cathode. The sudden cessation of the domain creates a current pulse. The transit time of the domains through the bulk material of the diode can be arranged in manufacture such that current pulses occur at microwave frequencies. In order to extract microwave power from the diode (about 5 mW) it is placed in a resonant cavity which must be tuned at or close to the Gunn diode frequency, energy being loop or probe coupled out of the cavity. Such devices and cavity structures are commonly used in modern radars where the cavity may be electronically tuned by placing a variable-capacitance diode effectively across the cavity in the direction of the *E* field. The bias upon the diode and hence its capacitance affects tuning of the cavity, this being controlled by the receiver tuning line from the display console. The Gunn diode requires a low voltage regulated supply, is compact, reliable, cheap and has long life compared to the vacuum tube re-entrant klystron which it has supplanted as the local oscillator in the radar receiver. Perhaps its most important operating characteristic is its low noise performance when compared with other microwave generators. This enhances the signal to noise ratio of the receiver overall (see Figure C.11).

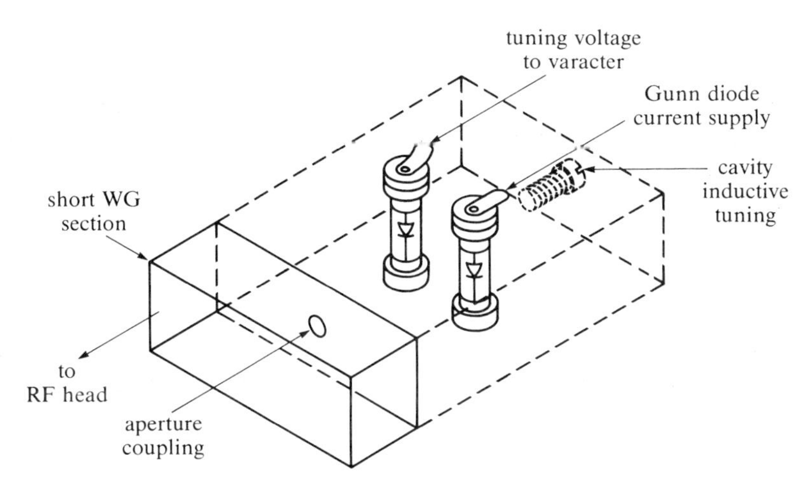

Figure C.11 Principle of local oscillator waveguide assembly. Usually an adjustable attenuator screw is placed to obstruct the aperture.

C.2.5 Circulator principle

The basic structure of a circulator is shown in Figure C.18. It comprises an arrangement of three micro-strip arms which meet at a central junction. These are fabricated on a substrate contiguous with a ferrite material through which a permanent magnetic field acts. Energy launched into *A* emerges only at *B*, energy launched into *B* emerges only at *C*, energy launched into *C* would emerge at *A*. The device is useful for coupling transmitter energy to antenna and echo energy to receiver whilst providing the necessary isolation of the receiver port from the transmitter. The structure of Figure C.18 would be suitable for low transmitter power.

Ferrites are compounds of various oxides of metals. These act as dielectrics due to the compound structure and are able to propagate electromagnetic waves. However, when an external permanent magnetic field is applied, the electron spin axes are aligned and they exhibit a gyroscopic effect. At a particular resonant frequency determined by the ferrite and magnetic field, an alternating electromagnetic field causes a large precession of the spin axes and energy from the a.c. field is dissipated as heat in the ferrite. The substance finds applications in attenuators, isolators and a special adaptation of the isolator, the circulator. The isolator comprises a piece of ferrite in a d.c. magnetic field transverse to the direction of electromagnetic propagation. When propagation occurs in one direction it causes the resonance loss by aiding precession; propagation in the opposite direction opposes precession and there is little loss. About 30dB of isolation is provided in one direction. The circulator shown operates on this principle, with RF energy at the resonant value being forced to flow or circulate in one direction.

C.2.6 Optical encoder disc principle

One form of a frequently used device to derive feedback from rotating shafts is shown in Figure C.12, which depicts a two-channel

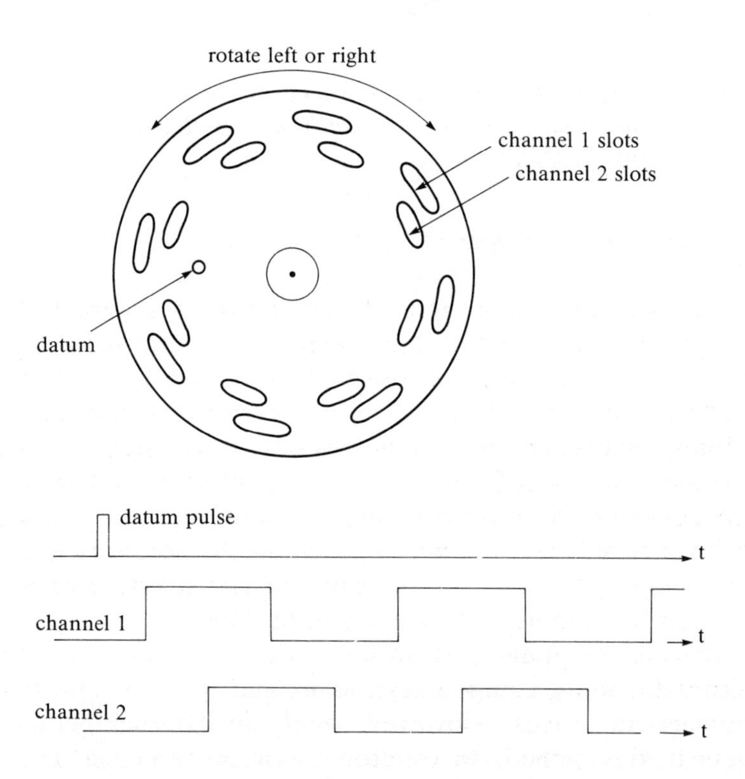

Figure C.12 Principle of two-channel optical encoder disc.

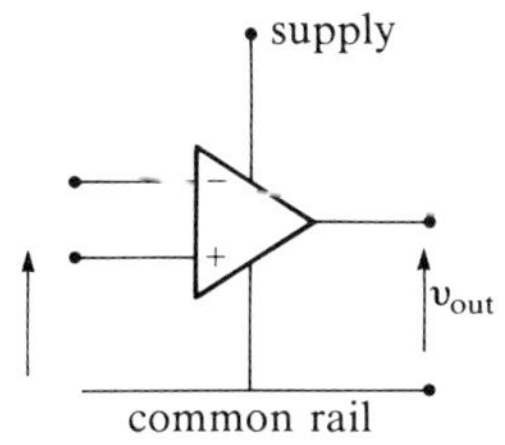

(a) Symbol for differential operational amplifier. Inputs applied between common rail and −VE input terminal are operated on and appear inverted at the output. Inputs at +VE terminal are not inverted.

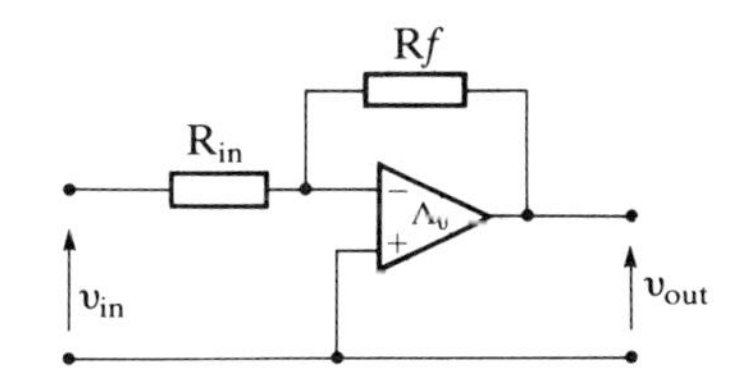

(b) Basic inverting amplifier arrangement.

$$v_{out} = -A_v v_{in}; \quad A_v = \frac{Rf}{R_{in}}$$

$$v_{out} = v_{in} \times \frac{Rf}{R_{in}}$$

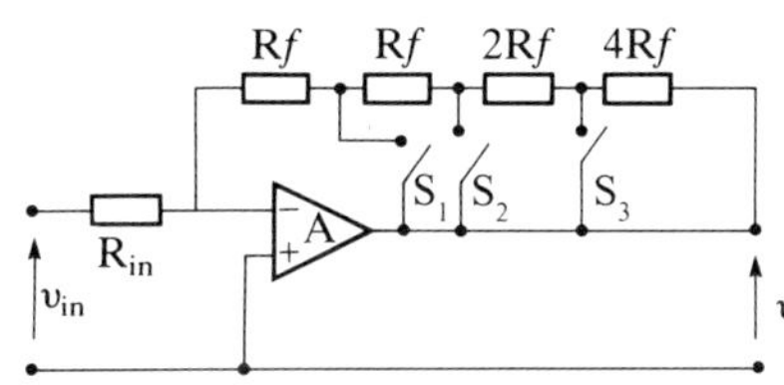

(c) Simple 4-range switched gain, speed scaling arrangement. S_1 to S_3 are range selector switches. When open as shown the voltage gain is $-\frac{8\,Rf}{R_{in}}$ maximum, used on shortest range. Progressivley closing S_3, S_2, S_1 alters gain in ratios of 4, 2, 1 Rf/R_{in}.

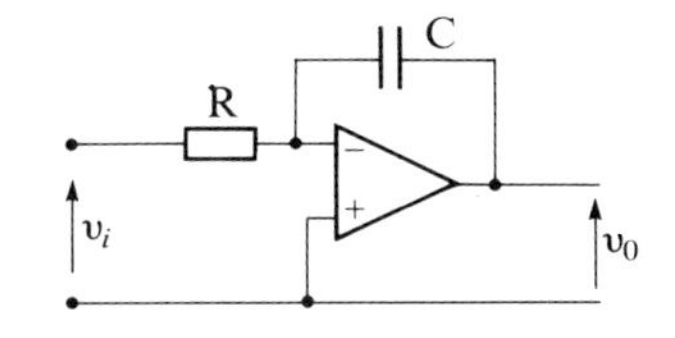

(d) Basic integrator circuit.

$$v_0\,(t) = -\frac{1}{CR}\int v_i\,(t)dt + v_0(0)$$

Output voltage is proportional to the time integral of the input voltage, plus an initial condition $v_0(0)$. The latter is the constant of integration and shows that at time t = 0 the output may be non-zero since the capacitor may carry some charge from a previous action.

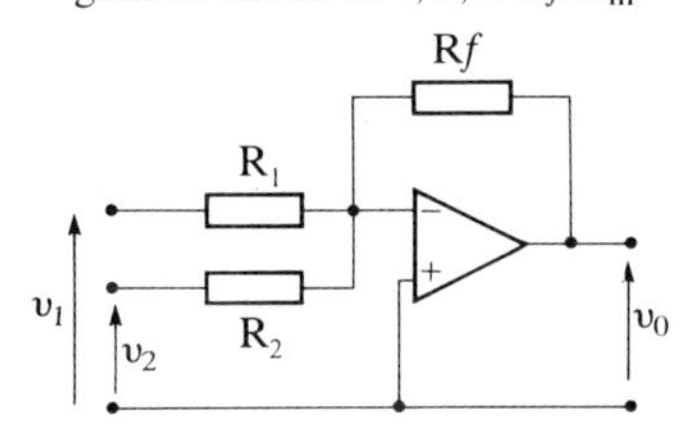

(e) principle of summing amplifier.

$$v_0 = -\left(\left(v_1\frac{Rf}{R_1} + v_2\frac{Rf}{R_2}\right)\right.$$

Output is the sum of individual scaled inputs. Several inputs can be summed in this manner. The Integrator (d) can also have summed inputs, the output being the sum of the separate integrals.

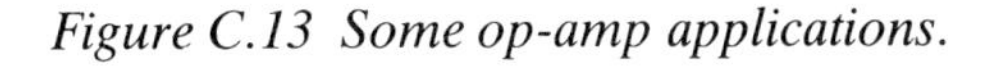

Figure C.13 Some op-amp applications.

optically encoded disc. These are substituted in many applications for resolvers and synchros of the analogue type, typically used for antenna and gyro-repeater positional information. The encoder shown comprises a series of apertures in a disc, through which a

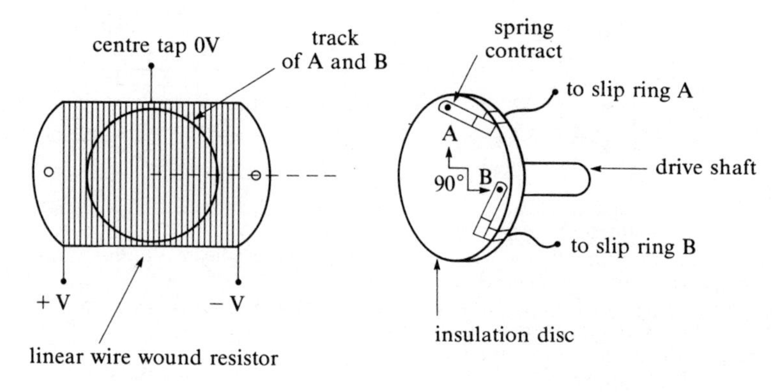

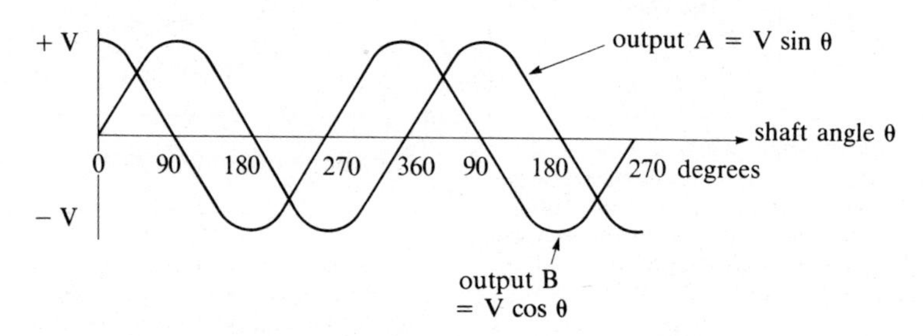

Figure C.14 Above: basic detail of flat card potentiometer. Contacts A and B bear on the winding at points on a circle subtending 90 ° at the centre. Lower: output from sliding contacts A and B.

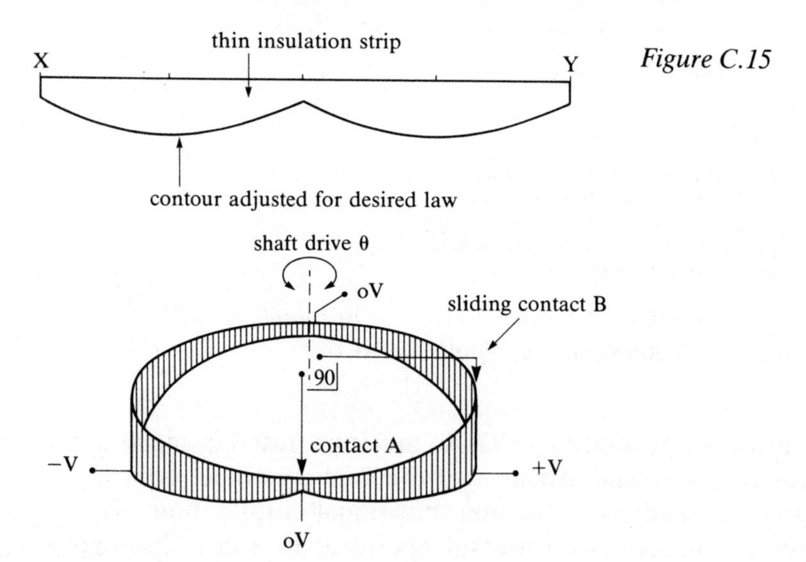

Figure C.15

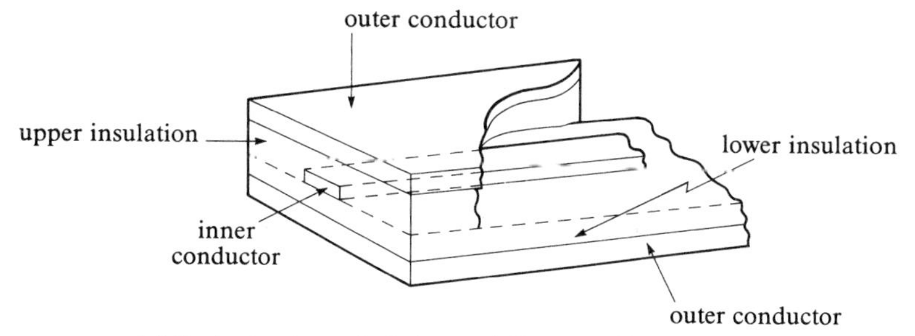

(a) Basic structure of tri-plate stripline. The inner conductor is printed on lower insulation. TEM propagation occurs.

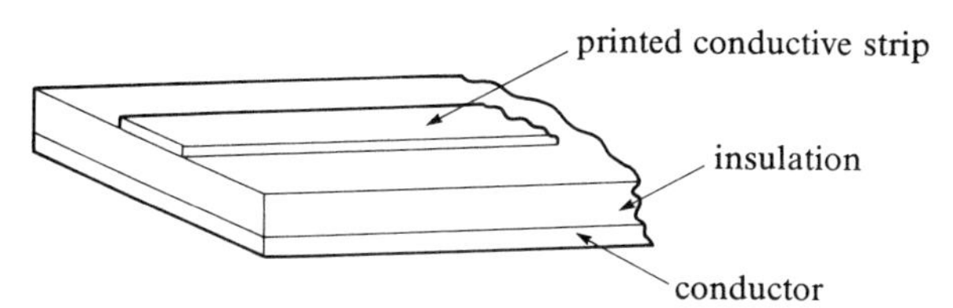

(b) Basic microstrip structure. Does not propagate true TEM mode and has greater loss than tri-plate construction. However, it is cheap to manufacture. Microwave integrated circuits are easily integrated with the line if the substrate used is a semiconductor. Relatively complex circuits, like the hybrid ring used in mixers, can easily be fabricated.

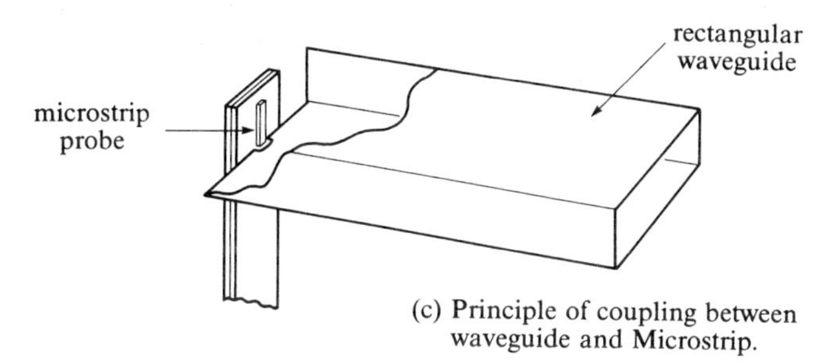

(c) Principle of coupling between waveguide and Microstrip.

Figure C.16.

Figure C.15 (Opposite below) Principle of contoured card potentiometer. Points X and Y of the card are joined to produce a circular former. On this is wound the fine linear resistance wire. Sliding contacts A and B bear on the wire as shown and maintain a constant angular displacement of 90°. Output signals A and B are similar to those of Figure C.14.

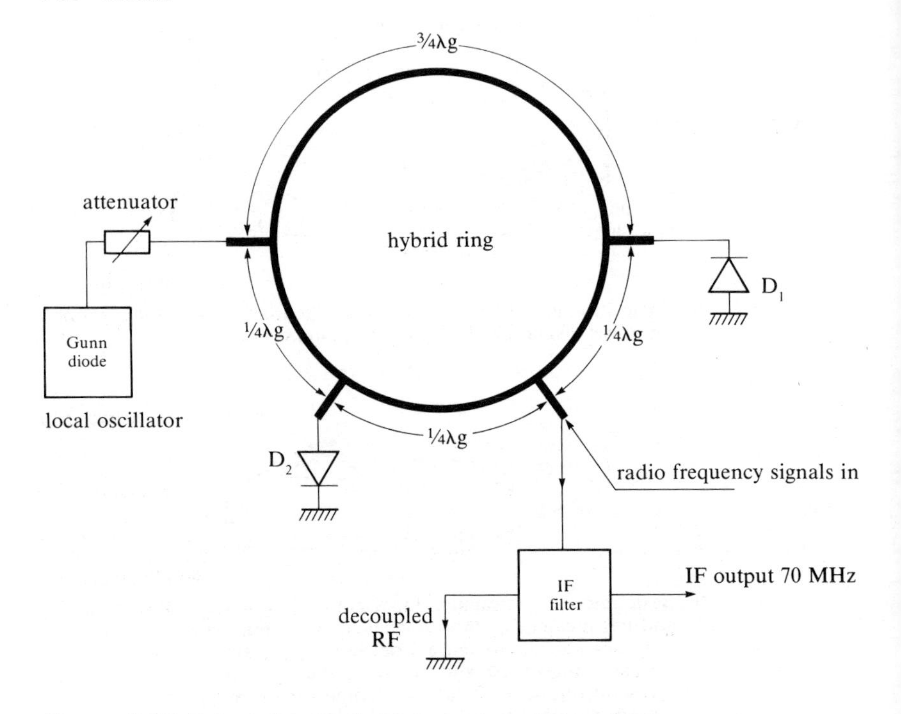

Figure C.17 Principle of hybrid ring mixer. D1 and D2 comprise balanced crystal detectors. Echo signals arrive at the local oscillator branch in anti-phase producing no effect there. The distances which the signal has to travel to D1 and D2 cause in-phase components. In a similar way, energy from the local oscillator arrives at the signal input port in anti-phase so that no energy is radiated. Local oscillator energy at the balanced crystals is in anti-phase and so is local oscillator noise. At the 70 MH IF the stripline is a simple conductor and the rectified signals add; local oscillator noise subtracts. The description neglects slight differences between λg and λLO.

light source is passed to fall on photo sensors in fixed positions on the opposing side of the rotating disc.

Modern encoders can output 2048 pulses on each channel per revolution of the disc. An alternative is to arrange the channels to provide unique digital codes for a small increment in rotation angle thus providing recognition of absolute angle values.

The diagram of Figure C.12 shows a typical two-channel arrangement generating two rectangular phases 1 and 2. These are

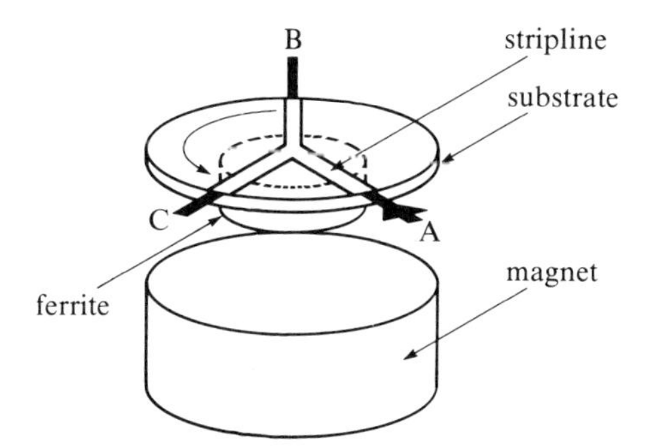

Figure C.18 Basic structure of stripline circulator.

arranged to be 90 degrees out of phase. A synchronizing pulse occurs once in each revolution to provide datum. A pulse count in either direction from that point represents angular displacement whilst the direction of shaft rotation, clockwise or anticlockwise can be determined by examining the edge of one phase relative to the level of the other.

ography

Admiralty List of Radio Signals, volume 2, Hydrographer of the Navy, 1985.

Documents of Sixteenth CCIR Plenary Assembly, Dubrovnik 1986
8/1058-E 10 January 1986: *Technical characteristics for search and rescue transponders*
8/1063-E 31 January 1986: *Technical parameters of radar beacons (racons) fixed frequency radar beacons*
8/1067-E: *Frequency requirements for ship-borne transponders.*

Long, M.W., *Radar Reflectivity of Land and Sea*, Lexington Books.

Marine Automatic Radar Plotting Aid (ARPA) Performance Specification, HMSO, 1981.

Marine Radar Performance Specification, HMSO, 1982.

Miller, G.M., *Modern Electronic Communication*, second edition, Prentice Hall.

Skolnik, M.I., *Introduction to Radar Systems*, second edition, McGraw Hill, Kogakusha Ltd.

Skolnik, M.I., *Radar Handbook*, McGraw Hill.

19 Marine communication systems

R. S. Linford

Introduction

The development of radio communication has undoubtedly had a major impact on the safety and profitability of maritime enterprises. It has also provided the mariner with rapid communication facilities and on-board entertainment.

With advances in technology has come a steady improvement in the quality, reliability and choice of communication services available. This trend continues with the introduction of high quality satellite communication services, satellite distress beacons and automatic information services, such as Navtex.

The aim of this section is to highlight and illustrate the operational features and limitations of modern marine communication systems, so that those concerned with their operation or management may gain an increased awareness of equipment performance and some insight into the underlying engineering. To this end we start by reviewing some fundamental communication engineering concepts. It is hoped that the reader unfamiliar with such work will find the effort of studying these chapters well worth while.

19.1.1 The communication system

We can represent all communication systems by the simple model shown in Figure 19.1. The essential difference between each system being the type of information that it can send and the channel by which it is sent. It is therefore important to understand the nature and characteristics of both signals and channels in order that we can fully appreciate their implications for equipment design and system performance. In the case of marine communication, the channel will inevitably be defined by one of the various radio propagation mechanisms in common use.

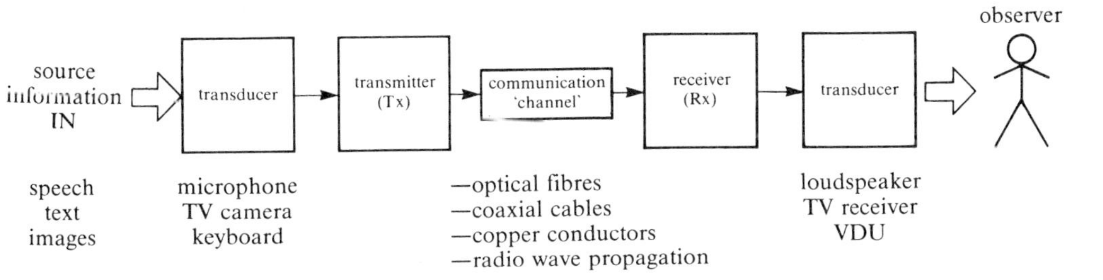

Figure 19.1 The basic communication system model.

19.1.2 *Signal-to-noise ratio*

A fundamental requirement of any practical communication system is that it shall deliver source information to a distant observer with an acceptable degree of degradation. For telephony or visual systems that level of acceptable voice or picture degradation is established via subjective testing of representative groups of users.

Such tests are often elaborate, but basically involve the introduction of measured amounts of noise and/or distortion to the information bearing signal, resulting in degrees of degradation as perceived by the observers. Clearly, there will be a level of degradation which is judged just acceptable for the service envisaged. This will correspond to measured signal and noise power levels which are usually expressed by their ratio in dB, the signal-to-noise ratio (SNR dB).

$$\text{SNR} = 10_{\log}\,(\text{signal power/noise power})\ \text{dB} \qquad \textbf{[19.1]}$$

Systems designed to carry coded machine information in digital form, such as data or telex, also suffer from the effects of noise. Here subjective judgements can be replaced by specifying the bit error rate. This is defined as the probability that a bit will be received in error. Once specified, the signal-to-noise ratio required to ensure this performance can then be established by trial or calculation.

The minimum permissible signal-to-noise ratio is therefore vitally important as it allows us to reflect the user requirements into the domain of the electrical system. Signal-to-noise requirements for marine communication are, of course, well established and influence heavily the technical specifications for system equipment such as transmitters, receivers and antennas.

In order to understand how our information bearing signals are conveyed over a communication link we must first examine and characterize the different types of signal and noise that we may encounter.

19.2 Signals

The most common forms of information that we may wish to transmit over a marine communications link are speech, music, text, data and charts. All of these are converted into equivalent electrical signals via some form of transducer such as a microphone, camera tube or telex keyboard, and are known as baseband signals.

The key signal characteristic of interest to us is its bandwidth. We will see later that the bandwidth occupied by a signal will influence the destination signal-to-noise ratio.

The bandwidth of a signal is set by the range of frequency components it contains. For example, the bulk of the energy contained within a human voice will be distributed over the range of frequencies extending from 400 Hz to 3.4 kHz. Figure 19.2a shows us this distribution, or frequency spectrum as it is more usually known. Figure 19.2b, on the other hand, represents the same signal as its amplitude varies with time, this being called the signal waveform.

Note that frequencies outside the above range only add character to the voice and their loss is considered acceptable for commercial speech communication. We can therefore say that a commercial grade speech signal occupies a bandwidth of approximately 3 kHz.

If we now consider the waveform and frequency spectrum of a standard baseband television picture, as shown in Figure 19.3 we notice that they are quite different in shape and form to those of speech. It should also be noticed that such a signal will occupy a bandwidth of some 5 MHz; vastly more than a speech signal.

Finally, we must ask the question: 'What will be the bandwidth of a digital data signal?' Well, it turns out that the faster the data is sent, the greater the bandwidth of the signal. This can be seen by looking at a typical digital signal feeding a marine telex system (Figure 19.4) and a high-speed data signal that might feed a satellite telemetry system (Figure 19.5.)

In both cases we see that the bulk of the signal energy resides within the first lobe of the spectrum. However, if we look a little closer we see that the telex signal only occupies a bandwidth of approximately 100 Hz, whereas the high-speed data requires 100 000 times more bandwidth or 10 MHz.

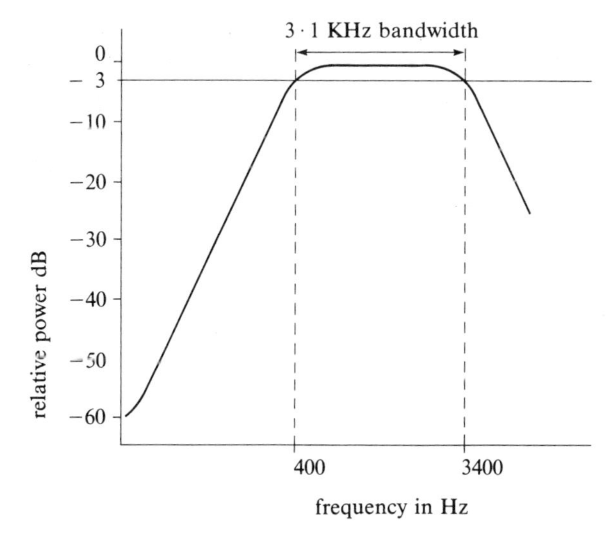

Figure 19.2a Frequency spectrum of human voice.

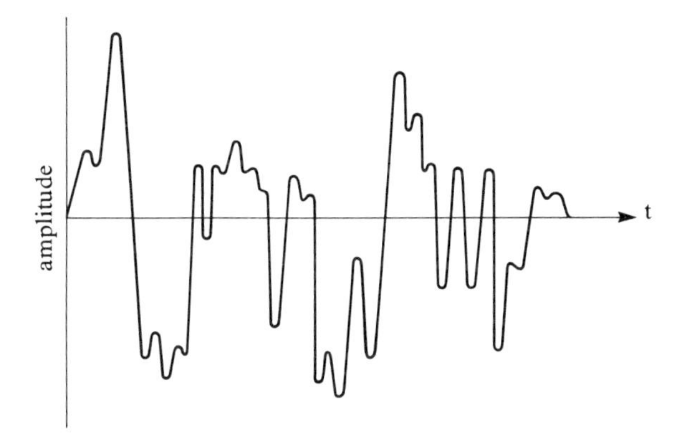

Figure 19.2b Typical speech waveform.

An important relationship between a digital signal and its bandwidth emerges, i.e. that the bandwidth of the signal is dictated by the narrowness of the data bits, i.e.:

$$\text{BW approx} = 1/T \text{ Hz.} \qquad \textbf{[19.2]}$$

where T sec. is the width of a data bit.

You may be wondering about the secondary spectral lobes of the digital signal. They are characteristic of the very sharp edges of the data bits. In practice digital signals are often pre-shaped such that their edges are less sharp and hence these spectral sidelobes reduce to insignificance. However as a consequence of this the main lobe width will be somewhat increased. This is shown in Figure 19.6.

Having established the notion of signal bandwidth it is useful to list the baseband bandwidths of signals commonly encountered in marine communications (see Table 19.1).

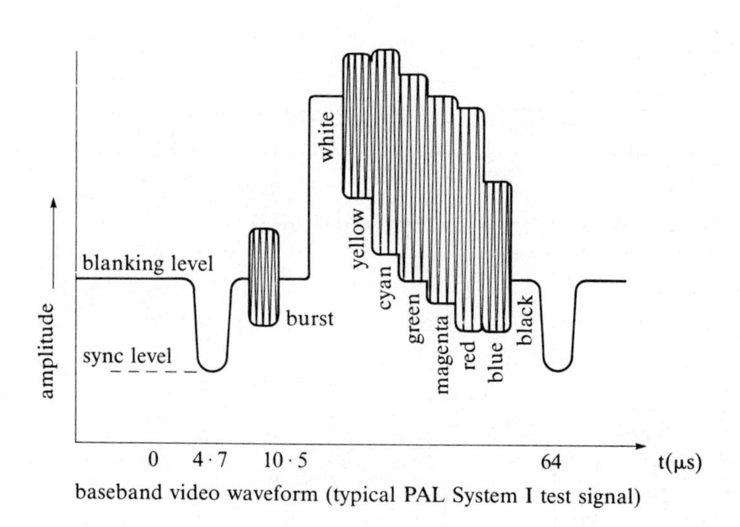

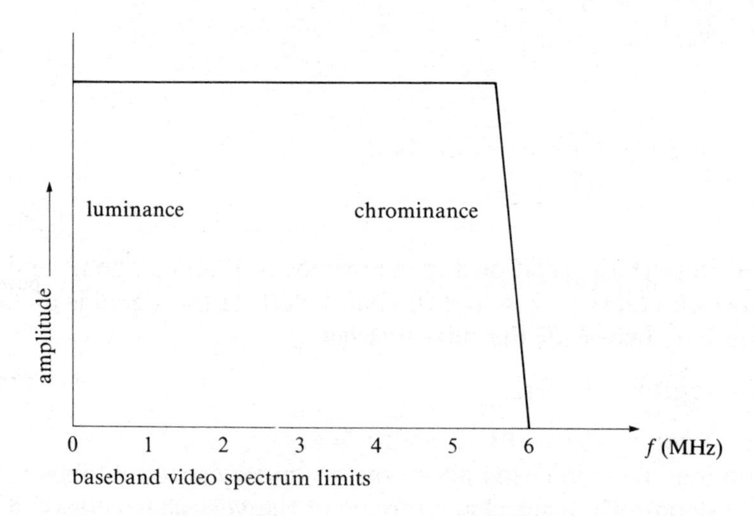

Figure 19.3 Baseband television waveform and spectrum.

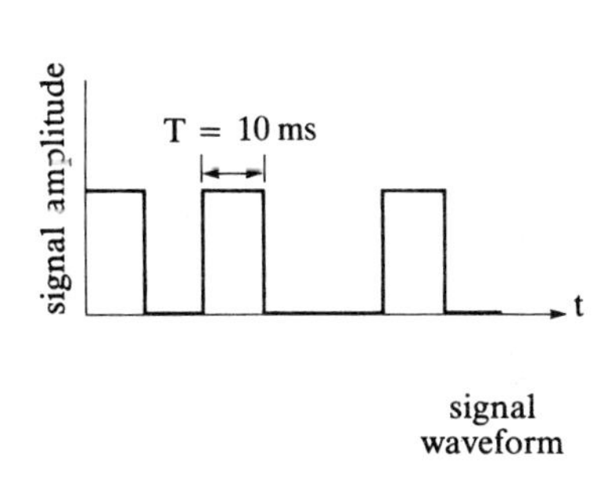

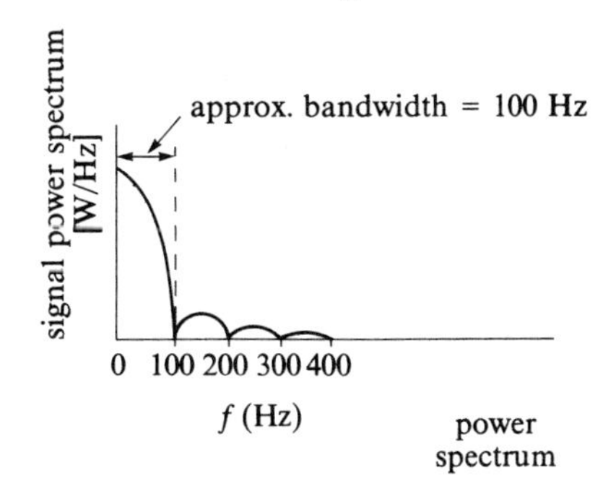

Figure 19.4 Typical marine telex signal (baseband).

Table 19.1 Typical baseband signal bandwidths.

Signal type	Bandwidth
Telegraphy (hand-keyed)	up to 100 Hz
Telex and Navtex	100 Hz
Telephony (commercial grade)	3 kHz
Music (FM broadcast quality)	15 kHz
Facsimile (weather charts, etc.)	2 kHz
625 line colour television (video)	5.5 MHz
High-speed data	above 1 MHz

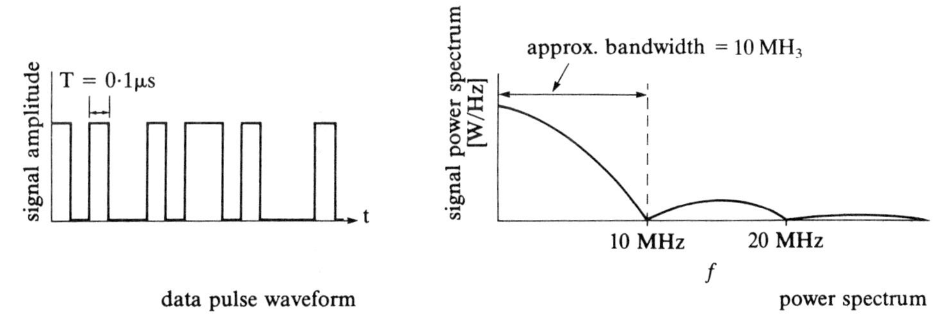

Figure 19.5 Typical high-speed data signal.

19.2.1 *System and channel bandwidths*

So far we have considered various baseband signals and examined their bandwidths. We must now review the implications of signal bandwidth on the communication system.

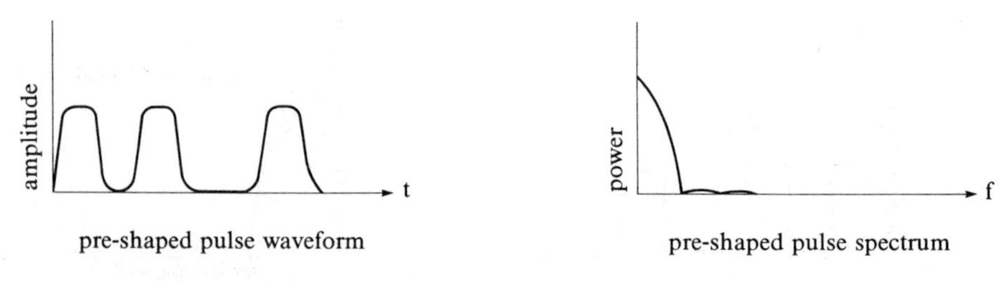

Figure 19.6 Pre-shaped data pulses.

Intuitively, we might suppose that a communication system must be able to transmit all the frequency components within a signal. In other words, the system must have a bandwidth at least as wide as that of the signal to be transmitted. In general this is true, but we will discuss one important exception, that of the minimum bandwidth required to transmit a digital signal of the form shown in Figure 19.7. We can, in fact, reduce the system bandwidth to a point where the output resembles a sinewave and yet still differentiate between '1' and '0' bits. This reduced bandwidth is theoretically sufficient for

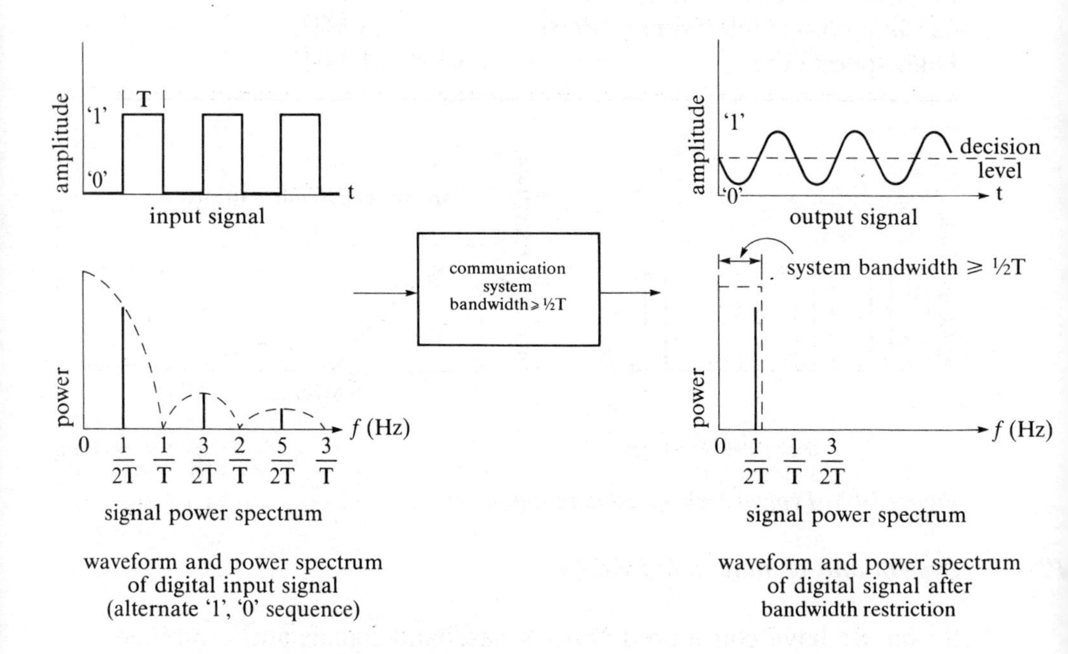

Figure 19.7 Minimum system bandwith for baseband digital signal transmission.

the transmission of the original information and equals half the bit rate, i.e.:

$$\text{system BW} = 1/2T\ \text{Hz}. \qquad \textbf{[19.3]}$$

In practice however, somewhat wider system bandwidths are required.

Returning to the general question of system bandwidth, we find that it is the communication channel itself that sets the ultimate bandwidth limit. This restricts the type of service that can be used over a given channel. We could not, for instance, provide a conventional television service over a high frequency (HF) communication channel as there would be insufficient channel bandwidth to do so. Even if this were possible, it would preclude thousands of narrow band users from the resource. Table 19.2 lists typical channels and their maximum bandwidths.

Table 19.2 Typical communication channels and their maximum usable bandwidths.

Channel	Maximum usable bandwidth
Copper wire	50 kHz
Coaxial cable	200 MHz
Waveguide	200 MHz
Optical fibre	1 GHz
Low frequency radio	6 kHz
Medium frequency radio	10 kHz
High frequency radio	10 kHz
Very high frequency radio	1 MHz
Ultra high frequency radio	10 MHz
Super high frequency radio	100 MHz

In practice, the signals that we transmit by radio only occupy a small fraction of the total available channel bandwidth. We therefore need to match the equipment bandwidth to that of the signal. In this way we minimize the amount of noise that reaches the system output and hence maximize the destination signal-to-noise ratio.

19.2.2 *Noise bandwidth*

There are many sources and types of noise encountered in communication systems and these will be reviewed in section 19.6. Generally, they have extremely wide frequency spectra, which can often be considered to be flat over bandwidths typical of the signal encountered. Figure 19.8 shows the power spectrum and typical waveform of thermal noise, which is present in all communication systems. The spectrum shows how the noise power is distributed with frequency and is often called the noise power spectral density or NPSD.

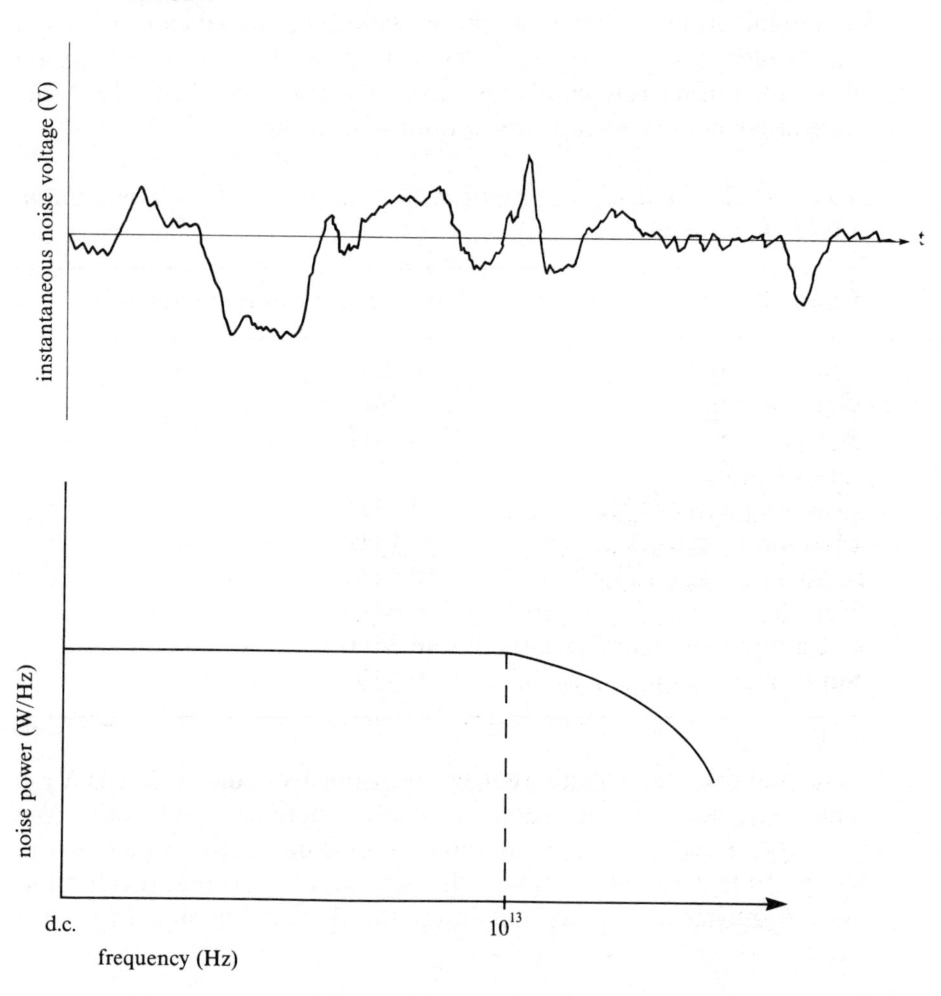

Figure 19.8 Above: typical waveform of thermal noise. Below: noise power spectrum of thermal noise.

We see that the NPSD of thermal noise is featureless and has equal energy at all frequencies up to about 10 000 GHz. For this reason it is often called 'white noise'. When such noise 'enters' a communication system we need to know how much of it appears at the output. This noise power can then be compared with the output signal power and thence the output signal-to-noise ratio obtained. The system will, of course, only allow noise through at the frequencies within its passband. It is therefore necessary to estimate the noise power at the output of a communication system by integrating the noise over the system bandwidth. This is very straightforward if we make the assumption that the system bandwidth is rectangular, as shown in Figure 19.9.

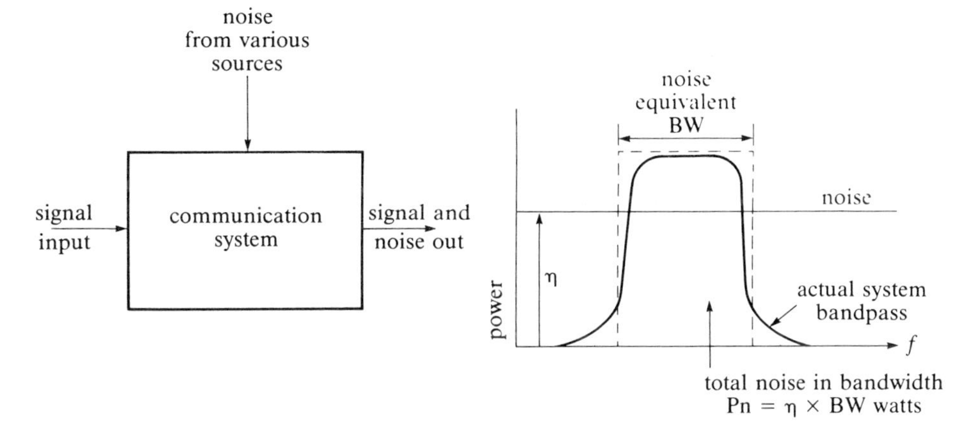

Figure 19.9 Noise power observed at the output of a system with noise equivalent bandwidth BW when subjected to white noise of power density η [watts/Hz].

The noise power P_n at the output of a system that is subjected to white noise of power spectral density η is therefore:

$$P_n = \eta \times \text{BW watts} \qquad [19.4]$$

where nita is expressed in watts per Hertz and bandwidth (BW) in Hz.

19.3 Modulation

The signals so far considered have been direct electrical equivalents of the source information, often called baseband signals. In order to transmit these signals over a radio channel we must convert them to a frequency and from such that they can energize a suitable antenna and propagate efficiently, to a distant receiver. This conversion process is known as modulation and results in a new signal; the radio frequency or RF signal.

Modulation is achieved by varying the amplitude, frequency or phase of a radio frequency oscillation, in sympathy with a baseband signal. This RF oscillation is more usually known as the carrier.

Here we will review the modulation methods commonly used in marine communication systems. The method used for a particular system will be dictated by the baseband signal type, signal-to-noise requirements and the characteristics of the radio channel to be used.

A standard designation system for radio emissions has been adopted by the ITU and is summarized in Appendix 19.1. Throughout the remaining text these designations will be shown in triangular brackets.

19.3.1 *Amplitude modulation*

We start with the simplest form of amplitude modulation, on-off keying or CW as it is often called. Here, the carrier amplitude changes from zero to its maximum value in sympathy with some baseband code, traditionally morse code <A1A>. This is shown in Figure 19.10.

Being a form of digital signal, we now know that the width of the resultant signal spectrum will depend upon the narrowness, and hence speed, of the modulating code. Notice, however, that the modulation process has not only frequency translated the baseband signal to the carrier frequency (fc), but has also doubled the required bandwidth by adding a spectral image.

A typical 25 word/minute baseband morse signal requiring a bandwidth of approximately 50 Hz, will therefore require at least 100 Hz of RF bandwidth.

A natural extension of the amplitude modulation process is to impress more complex baseband signals, such as speech, onto the carrier (See Figure 19.11).

Inspection of the modulated carrier waveform shows that its envelope or amplitude varies with the baseband information to produce the modulated RF signal. The depth of modulation is governed by the modulation index (M) of the system, which is usually expressed as a percentage and is defined by:

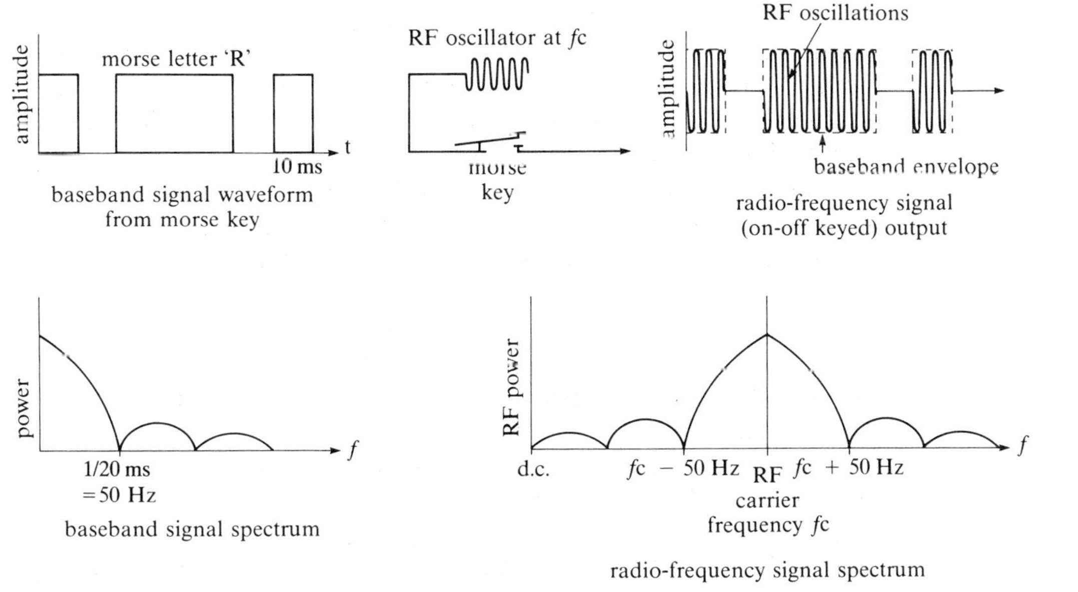

Figure 19.10 Basic on–off keyed modulation system.

$$M = (a - b)/(a + b) \times 100\% \qquad \textbf{[19.5]}$$

where a is the peak-to-peak amplitude of the envelope of the modulated waveform and b is its trough-to-trough value. M must not exceed 100 per cent otherwise signal distortion will result, but in the interest of efficiency it should be as close to this figure as practical i.e. 80–90 per cent.

The frequency spectrum consists of a component at *fc* (the carrier), plus two mirrored replicas of the baseband signal, known as the upper and lower sidebands. This form of amplitude modulation is therefore known as double sideband full carrier modulation, or AM, which will have the designation <A3E> when used for single channel analogue telephony. The advantage of this form of modulation is that it is very simple to demodulate at the receiver. All that is required is a circuit that will follow the variations in carrier amplitude and eliminate the carrier oscillations. Such a circuit is known as an envelope detector.

There are two major drawbacks in the use of <A3E> for commercial marine communications. The first is that all the information is effectively duplicated in the second sideband, which means that the emission wastes valuable spectrum which could have been used by other stations. The second problem is that the carrier component and the redundant sideband waste power that could have been used to increase the information power of the wanted sideband. In short, double sideband full carrier modulation is simple, but power and bandwidth inefficient. With the ever growing demand for radio services and the increasing congestion of the radio waves, this kind of modulation has been disallowed for marine communication, but remains in use for domestic AM radio broadcasting.

19.3.1.1 Single sideband suppressed carrier modulation (SSB)

As the heading implies, this form of modulation eliminates the redundant sideband and significantly reduces the power of the carrier, typically −26dB. In this way the inefficiencies of conventional AM are reduced.

The power spectrum of a single channel telephony signal, using SSB <J3E>, is illustrated in Figure 19.12. and shows the translated baseband spectrum and suppressed carrier position. In this case, the

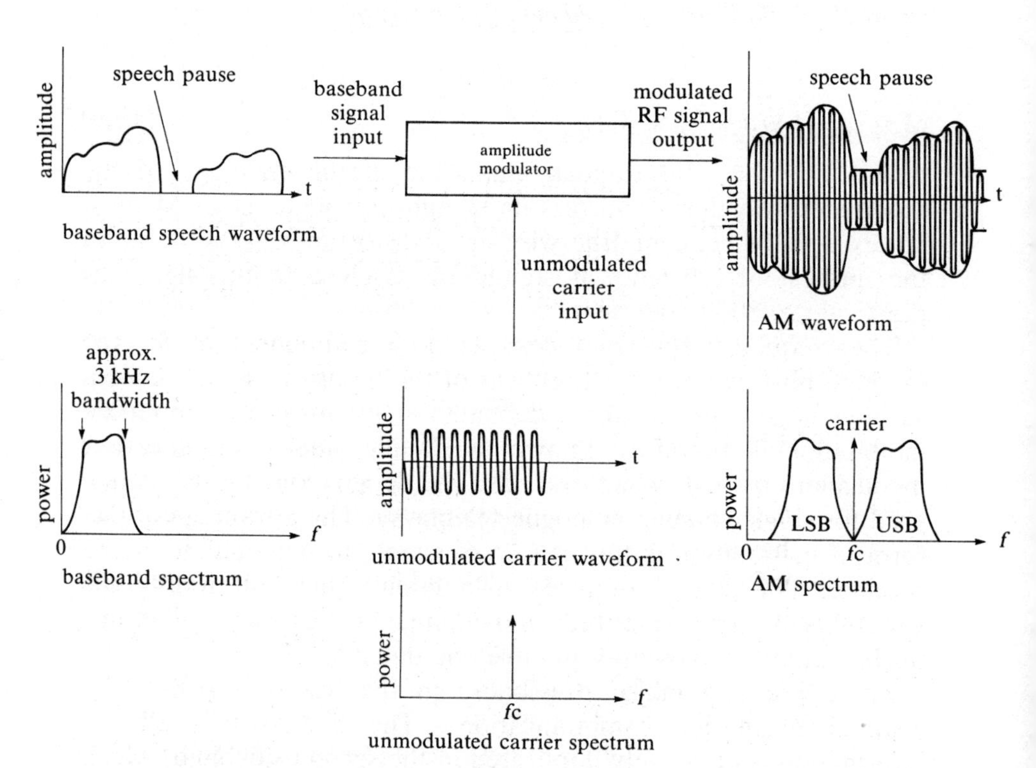

Figure 19.11 Basic amplitude modulation.

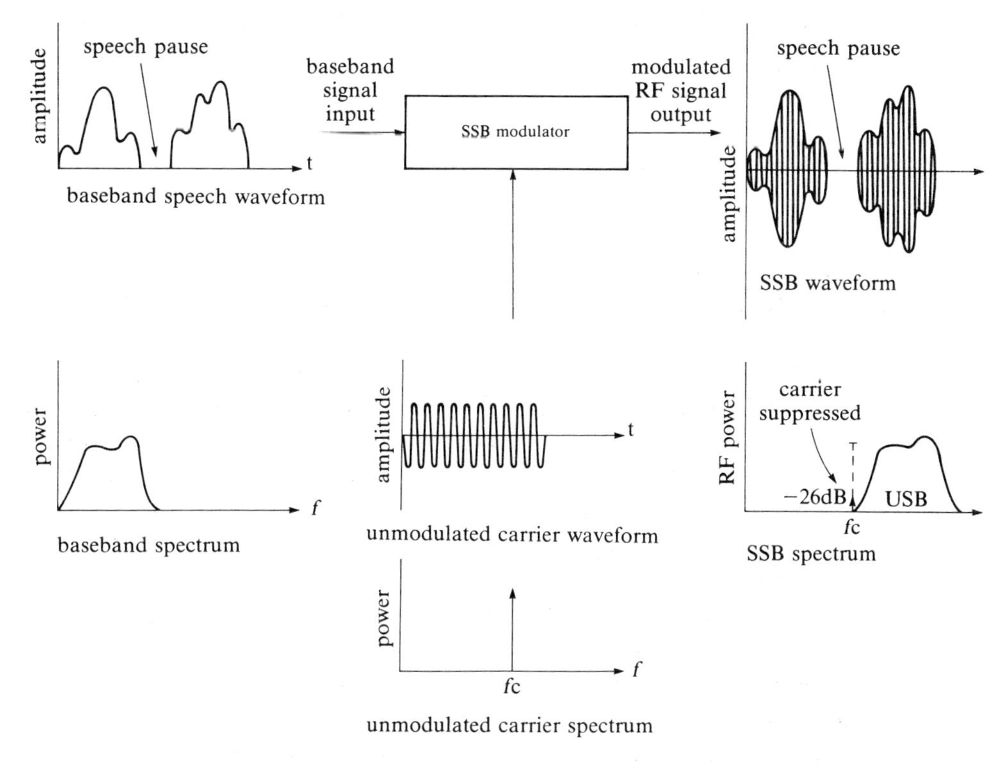

Figure 19.12 Single sideband modulation.

upper sideband has been selected for transmission, but equally the lower sideband may have been used. Notice that the RF spectrum is now asymmetric about the carrier position, unlike AM. This apparently simple modification has a major influence upon the nature of the transmitted signal, which has components both in phase with the carrier and in quadrature with it (at 90 degrees). The quadrature component being a modified version of the baseband signal, its Hilbert transform. As a result the shape of the resultant waveform is less obvious than that of an AM wave, but clearly illustrates that the RF signal only exists during periods of speech as the carrier has been suppressed.

As the carrier and redundant sideband have been eliminated, all the transmitted power can be used for the information bearing sideband. For conventional speech signals this can yield a 13dB mean power advantage over AM, i.e. a 100W SSB transmitter is as effective as a 2 kW AM transmitter.

There is of course, a price to pay, in the form of more complex transmitters and receivers. Despite this, single sideband (as it is more usually called) is now well established for both telephony and telex transmission.

19.3.2 *Frequency modulation (FM)*

The amplitude modulation techniques considered in section 19.3.1 occupy no more than twice the baseband signal bandwidth. They are therefore well suited for use in systems where bandwidth is either restricted or at a premium.

In contrast, frequency modulated signals may require transmission bandwidths ranging from twice to many times the baseband signal bandwidth. This apparent drawback is mitigated by a destination signal-to-noise advantage over that obtainable with amplitude modulation. This means that we can improve the quality of the service by reducing the background noise, without increasing transmitter power. This improvement is gained at the cost of increased transmission bandwidth and inferior performance under weak signal conditions (threshold effect).

Before evaluating this improvement we must first examine some of the major features of a typical frequency modulated signal, in this case a single channel analogue speech modulated transmission <F3E>. Figure 19.13 shows the waveform and spectrum of a radio frequency carrier, frequency modulated by a 3 kHz tone. This

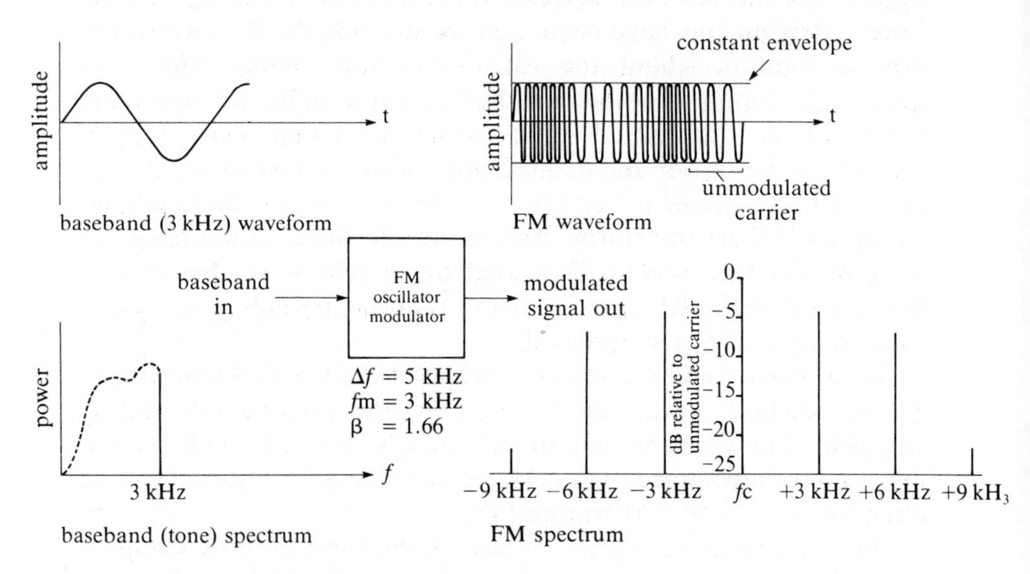

Figure 19.13 Frequency modulation.

represents the highest permissible component of a commercial grade speech signal.

Notice that the instantaneous frequency of the carrier is varied in accordance with the 3 kHz baseband signal amplitude. The amount by which the carrier frequency departs from its unmodulated value depends upon the amplitude of the baseband signal and the equipment design. This is known as the system deviation, the peak value of which is given the symbol $\triangle f$ Hz. For marine communication <F3E> this is set to 5 kHz.

The ratio of the peak frequency deviation to the highest modulation frequency FM is known as the modulation index (β) i.e.:

$$\beta = \triangle f / FM \qquad \textbf{[19.6]}$$

This is analogous to the modulation index of an AM signal, but unlike AM it is usually greater than unity. In this case β = 1.66. Further inspection of the <F3E> wave reveals that it has a constant zero information bearing envelope. This means that all the information is held in the zero crossing positions of the carrier, rather than its amplitude.

Turning to the spectrum we see that it bears little resemblance to the baseband signal, unlike amplitude modulation. It consists of a series of side frequencies spaced at the baseband signal frequency (3 kHz). If the deviation of the system shown in Figure 19.13 were doubled (10 kHz), then the spectrum would widen and β would equal 3.33. This can be seen in Figure 19.14.

If, on the other hand, we keep $\triangle f$ constant at 5 kHz and reduce FM to 400 Hz (the lowest permissible commercial speech frequency),

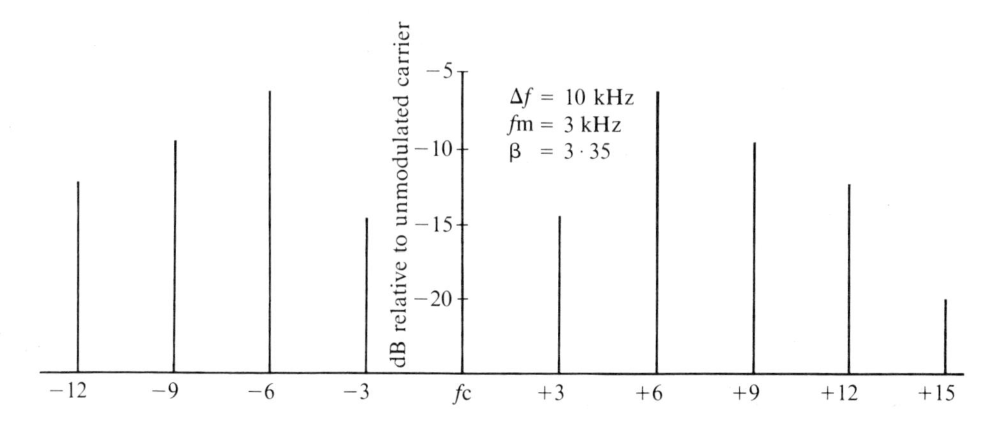

Figure 19.14 Single tone FM spectrum (fm = 3KHz, β = 3·33).

then the bandwidth reduces despite an increase in significant side frequencies and β to 12.5. This is shown in Figure 19.15. The bandwidth of the signal therefore depends upon the deviation of the system and the highest modulating frequency.

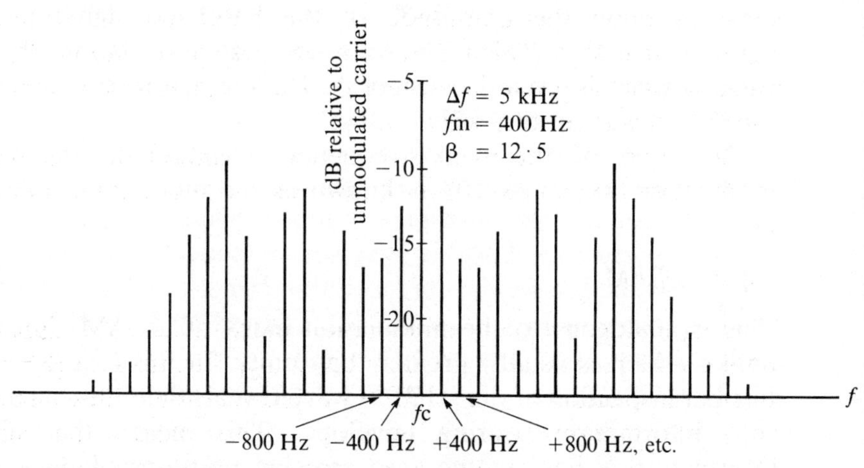

Figure 19.15 Single tone FM spectrum (FM = 400 Hz, β = 12·5).

Theoretically, a frequency modulated signal has an infinite bandwidth, but in practice we can restrict it to include only side frequency components that are significant, i.e. equal to or greater than 1 per cent of the unmodulated carrier amplitude. A useful expression, based upon this assumption, known as Carson's rule, allows us to estimate the practical transmission bandwidth (BT):

$$\mathrm{BT} = 2\,(\Delta f + \mathrm{FM})\ \mathrm{Hz} \qquad \textbf{[19.7]}$$

If we apply this rule to the 3 kHz signal, with the modulator set to $\Delta f = 5$ kHz, then the bandwidth necessary to transmit the signal is 16 kHz. Over five times the baseband bandwidth!

So, what about the signal-to-noise improvement for this bandwidth expansion? We find that the FM improvement (after demodulation) is proportional to β^2:

$$\text{FM Improvement (I)} = (\beta^2) \qquad \textbf{[19.8]}$$

For our single 3 kHz sinusoidal tone, this becomes:

$$\mathrm{I} = 3/2\,\beta^2 \qquad \textbf{[19.9]}$$

or $20\log(\beta)\mathrm{dB} + 3.5\mathrm{dB}$

In this case, 7.9dB, which is equivalent to increasing the transmitter power by a factor of 6.2. You may have noticed that the FM improvement for a 400 Hz signal with the same peak frequency deviation will be 25.4dB, equivalent to 347 times the transmitter power!

We conclude, therefore, that low frequency components of the baseband signal enjoy a greater noise quieting than the higher frequencies. This improvement inequality may be reduced by artificially increasing the amplitude of the high frequencies in order to increase their deviation and hence increase their noise improvement. This is known as pre-emphasis. Clearly, this distortion of the signal must be corrected after demodulation to yield a noise equalized replica of the original signal. This is achieved by a de-emphasis network after the receiver demodulator. Figure 19.16 shows how the received or predetection signal-to-noise ratio (often called carrier-to-noise ratio, CNR) is improved upon after frequency demodulation as a function of β.

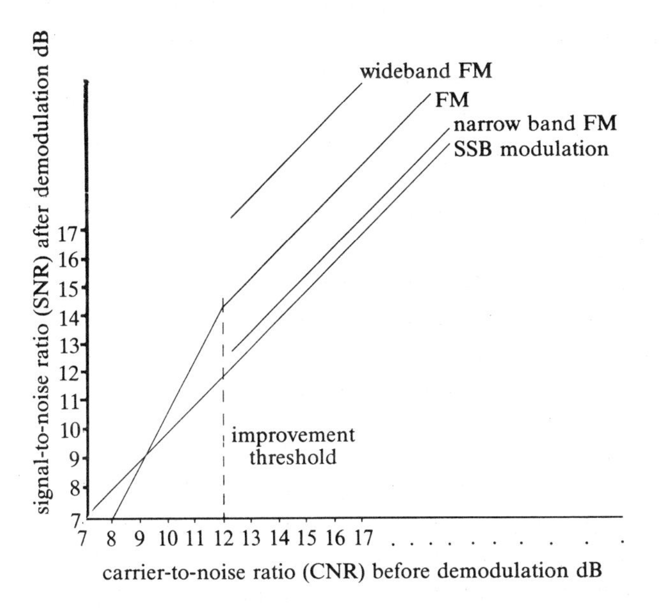

Figure 19.16 Carrier-to-noise vs signal-to-noise for FM.

Systems with wide deviations clearly yield significant improvements for the increased transmission bandwidth. These are often called wideband FM systems.

We notice however, that a CNR threshold exists below which the destination SNR crashes, i.e. the system fails. This effect does not occur in amplitude modulated systems which degrade more gracefully.

A great deal of research and development effort has gone into the design of low threshold demodulators for use in satellite communication receivers where low CNRs are often encountered. FM is used for maritime VHF radio telephony as well as satellite communication telephony. For these services signal levels are reasonably predictable and signal threshold can be avoided in the service area. The result is a high quality (low noise) telephony service that is only suitable for VHF and above due to the increased bandwidth requirements.

19.3.2.1 Frequency shift keying (FSK)

The FM signals so far considered have been based on analogue modulation. When a carrier is frequency modulated by a digital signal the result is known as frequency shift keying, or FSK. In the simplest case the carrier hops between two discrete frequencies representing a '1' or '0'.

FSK is often the preferred method of modulation for maritime data communications, due mainly to its immunity from amplitude interference and equipment simplicity. Radio telex, Navtex and some facsimile transmissions use a narrow band form of FSK known as audio FSK <FIB and FIC>. Here a conventional single sideband transmitter is modulated by two audio tones. Let us consider, for example, the standard radio telex and Navtex signal, format <FIB> (see Figure 19.17).

The input code, at 100 baud, switches 85 Hz either side of a nominal audio centre frequency of 1700 Hz. This signal then amplitude modulates a single sideband transmitter which translates the signal up to the required radiation frequency. The frequency deviation (Δf) is therefore 85 Hz and the highest baseband signal frequency (FM) is 100/2 = 50 Hz. This occurs when the data alternates from '1' to '0'. Therefore $\beta = 85/50$ or 1.7. From equation [**19.7**] we obtain a minimum transmission bandwidth of 270 Hz for a telex signal, very narrow when compared with speech frequency bandwidths. Note that a slightly more conservative figure of 304 Hz is usually assumed for equipment design.

19.3.2.2 Phase modulation (PM and PSK)

As frequency can be considered as the rate-of-change of phase angle we find that phase and frequency modulation are very similar. Indeed both systems involve the changing of the carrier phase angle in order to achieve modulation.

Analogue phase modulation is not used in maritime communica-

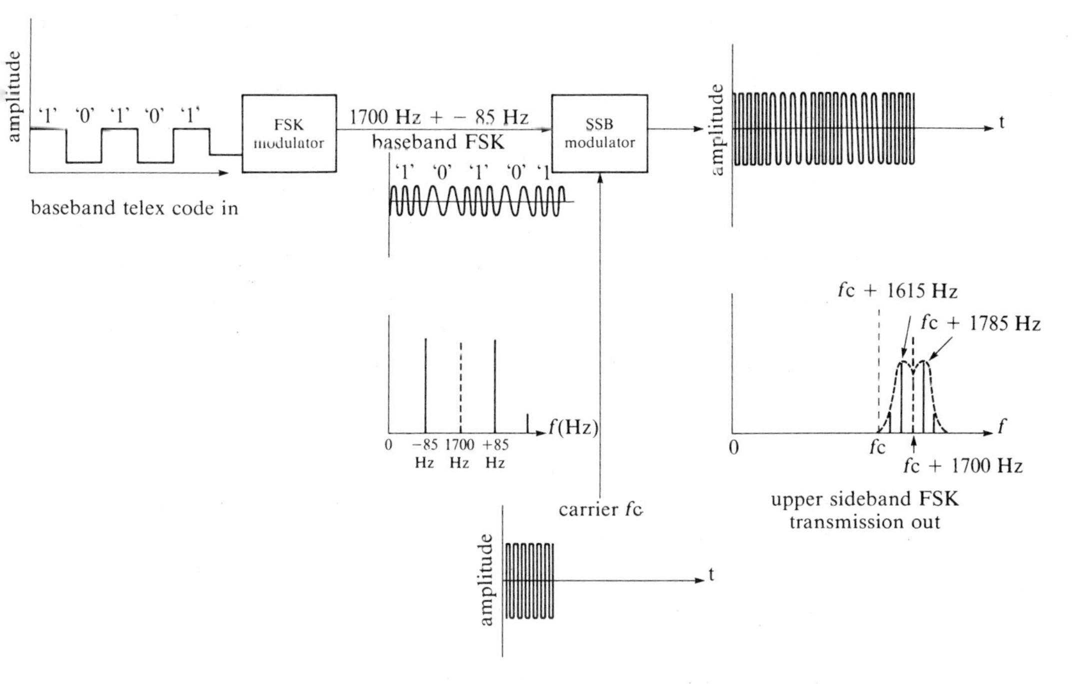

Figure 19.17 Generation of a typical telex/navtex (FIB and FIC) frequency shift transmission.

tions as it offers no advantage over frequency modulation and requires more complex equipment.

Digital phase shift keying is, however, highly efficient and is used extensively for digital speech transmission and telemetry for satellite systems. Figure 19.18 shows a two-level (binary) phase shift keyed signal waveform (BPSK). Notice that the baseband data reverse the phase of the carrier at each transition, i.e. there is a 180 degree difference between phases for a two-bit system. Multi-level PSK is also used in which 4, 8 or 16 phase positions define the code number.

19.3.3 *Pulse code modulation (PCM)*

PCM is an increasingly important class of modulation which takes baseband analogue signals, and converts them into equivalent digital signals that can be transmitted via FSK or PSK. As with all digital modulation schemes, PCM has the advantage that occasional signal regeneration (the digital equivalent of amplification) allows infinite transmission without further degradation (see Figure 19.19). This contrasts with analogue modulated signals that degrade continually with distance despite amplification. PCM is used increasingly for terestrial telecommunications where digital telephone exchanges are employed.

The conversion of speech to PCM is a two-part process. First, the baseband signal is sampled as shown in Figure 19.20. This results in an intermediate signal known as a pulse amplitude modulated signal. This rather unlikely looking process will not result in any distortion of the signal on demodulation provided that the signal is sampled at greater than twice the highest frequency component of the signal:

$$\text{or } 1/T \geqslant 2f\text{max} \qquad \textbf{[19.10]}$$

This is known as Nyquist's sampling theorem and is an axiom of communication theory.

The second process converts the sampled amplitudes into digital words, as shown in Figure 19.21 and is known as quantization and digitization. The number of bits used to form the digital word describing the sampled amplitude will dictate the quantization resolution. Clearly this process will introduce some distortion, due to quantization step size, but it has been found that eight bits

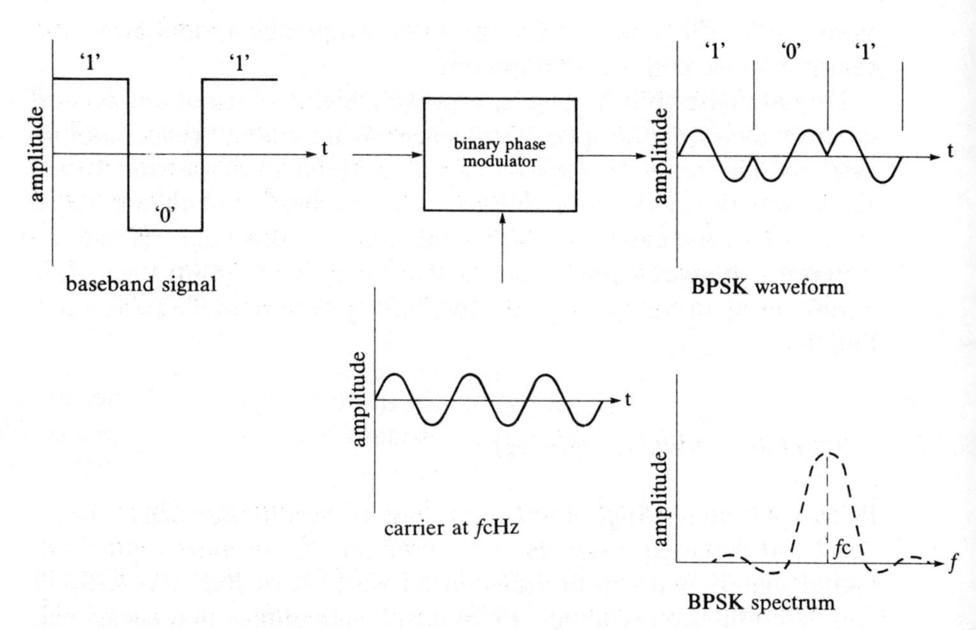

Figure 19.18 Binary phase shift keying (BPSK).

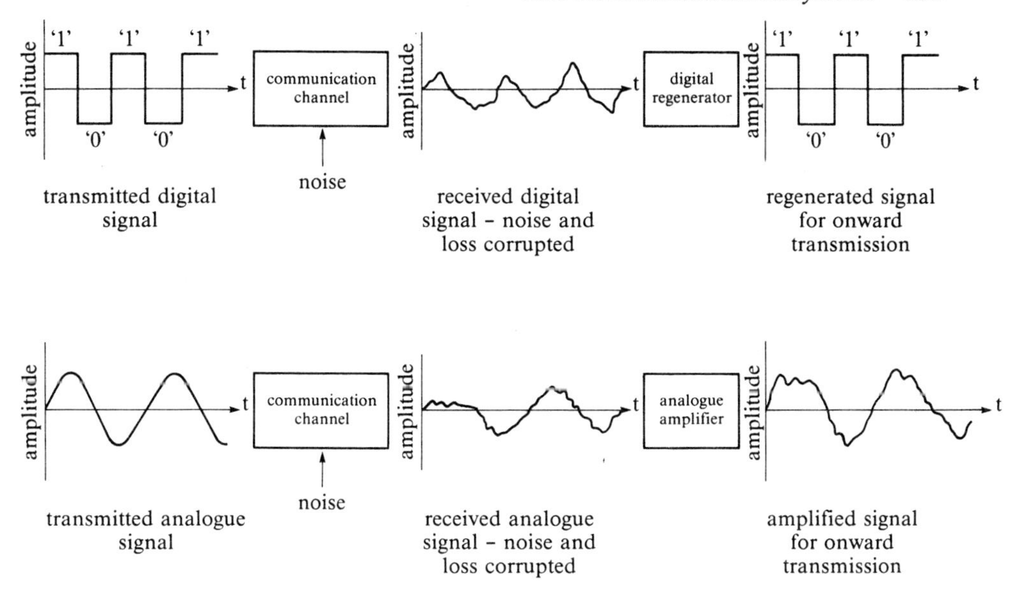

Figure 19.19 Regeneration and amplification of digital and analogue signals.

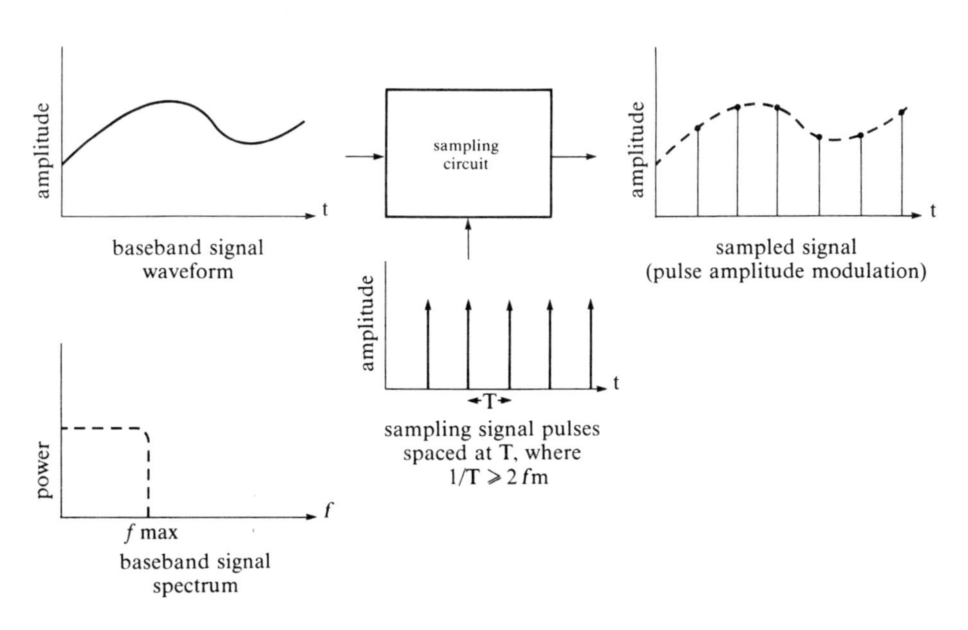

Figure 19.20 Signal sampling.

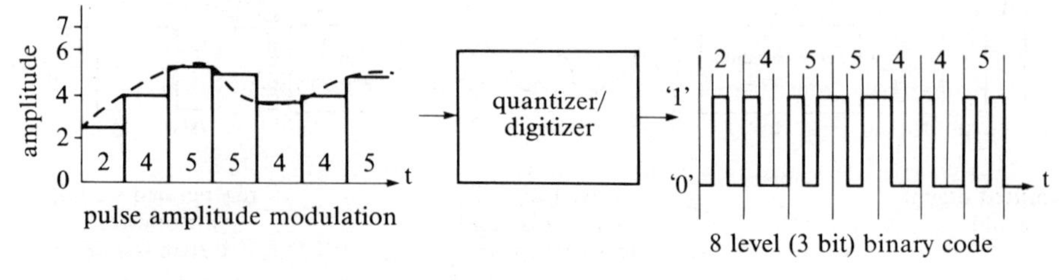

quantization level	binary code
0	000
1	001
2	010
3	011
4	100
5	101
6	110
7	111

Figure 19.21 Quantization.

provide sufficient resolution for good quality telephony. On reception, the PCM signal goes through a digital to analogue conversion process which reconstructs the original baseband signal, albeit with some distortion due to the quantization process (see Figure 19.22).

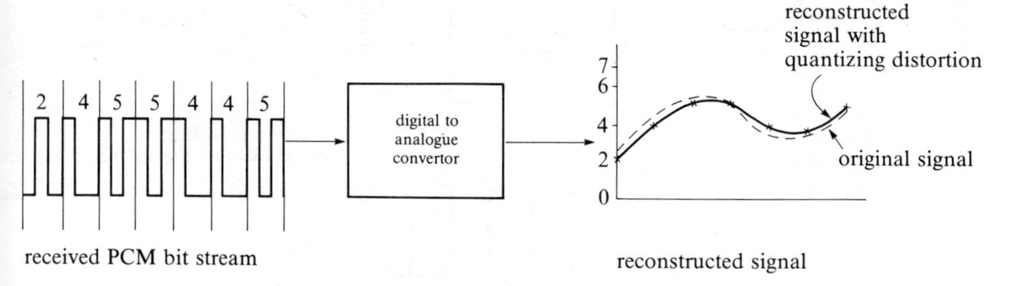

Figure 19.22 The demodulation of a PCM signal.

19.3.4 Signal multiplexing

Signal multiplexing is used extensively in telecommunications an satellite communications in order to allow a system to transmit, effectively, more than one signal at a time.

There are several methods of multiplexing, the two most common being frequency and time division. An example of frequency division multiplexing (FDM) can be seen in Figure 19.23.

We see that a number of independent baseband signals are frequency translated such that they each occupy a different slot within a band. The ensemble may then be treated as a single signal and transmitted over the transmission system. FDM is employed in maritime satellite communication systems for multi-channel telephony transmission.

Time division multiplexing (TDM), on the other hand, allocates the entire communication system to each baseband signal for a short period of time. The result is that all the baseband signals appear to be transmitted simultaneously. TDM is used for marine satellite telex and system control functions. A representation of a TDM system is shown in Figure 19.24.

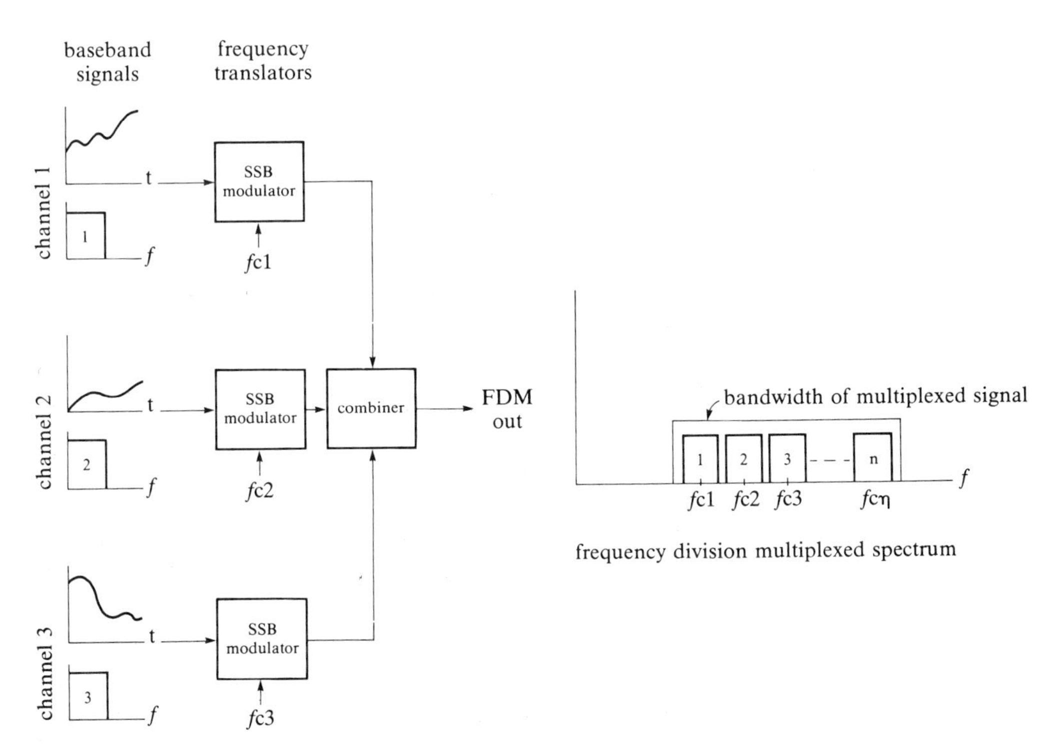

Figure 19.23 Frequency division multiplexing (FDM).

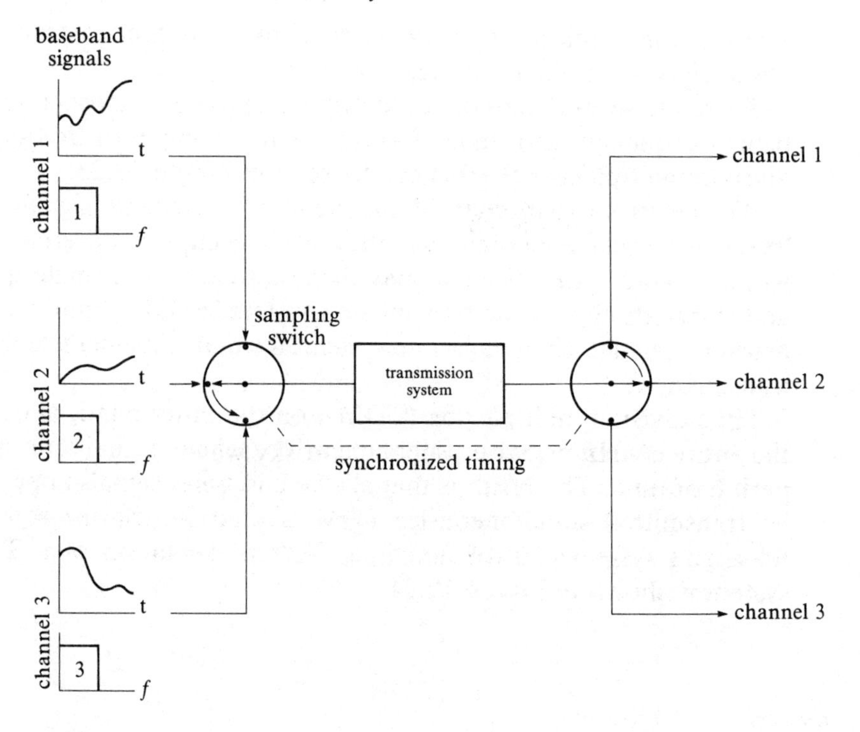

Figure 19.24 Basic principle of time division multiplexing.

19.4 Principles of wave propagation

In order to design, operate or assess any communication system we must first understand the capabilities and limitations of the channel over which we are to pass our information. In the case of radio communication, one or more of the many electromagnetic wave propagation mechanisms will act as the transmission channel. These are highly frequency dependent and are broadly categorized in Table 19.3.

Other channels found in communication systems, such as optical fibres, coaxial cables and telephone wires, have very well-defined characteristics that do not change with time. Radio signals may, however, be modified or degraded by alterations to the channel resulting from changes in sunlight, weather conditions, sunspot activity or path terrain, to mention but a few. With so many, largely unpredictable, factors affecting radio wave propagation we are forced to rely on simple models refined by observations and experience in order to predict the performance of a radio communication system.

Before looking at the various mechanisms of radio wave propagation, we must first review some basic concepts relating to radiation, antennas and electromagnetic wave behaviour.

Table 19.3 Radio propagation channels.

Frequency	Band	Propagation mechanism
3–30 kHz	(VLF)	Earth-ionosphere waveguide
30–300 kHz	(LF)	Surface wave
300 kHz–3 MHz	(MF)	Surface and sky wave
3–30 MHz	(HF)	Sky wave
30–300 MHz	(VHF)	Space wave
300 MHz–3 GHz	(UHF)	Space and scatter wave
3–30 GHz	(SHF)	Space wave
30–300 GHz	(EHF)	Space wave

9.4.1 *Electromagnetic waves*

An electromagnetic wave may be considered as an oscillating electric, field component (E), at right angles to an oscillating magnetic field (H). The direction of propagation of the wave being at right angles to the E and H components as shown in Figure 19.25.

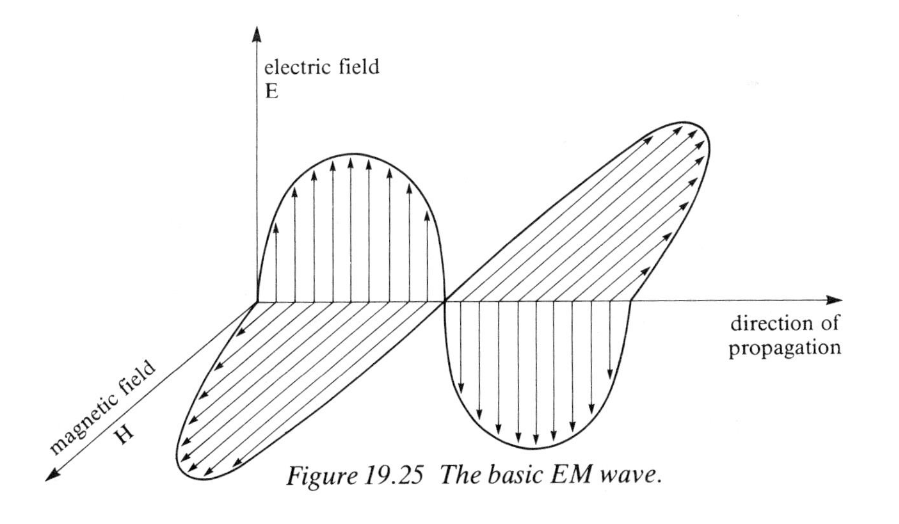

Figure 19.25 The basic EM wave.

By convention, the direction of the E field establishes the polarization of the wave. When the E field is in the vertical plane the wave is classified as vertically polarized. Similarly, when the E field is horizontal the wave is said to be horizontally polarized. Other rotating polarizations are sometimes employed, including right- and left-hand circular polarization (RCP and LCP). The polarization of a wave does not affect its velocity of propagation but can influence it in other ways, as we will see later.

19.4.2 Propagation in free space

The simplest radio communication scenario is that of a transmitted wave emanating from a point source antenna and propagating through free space (see Figure 19.26).

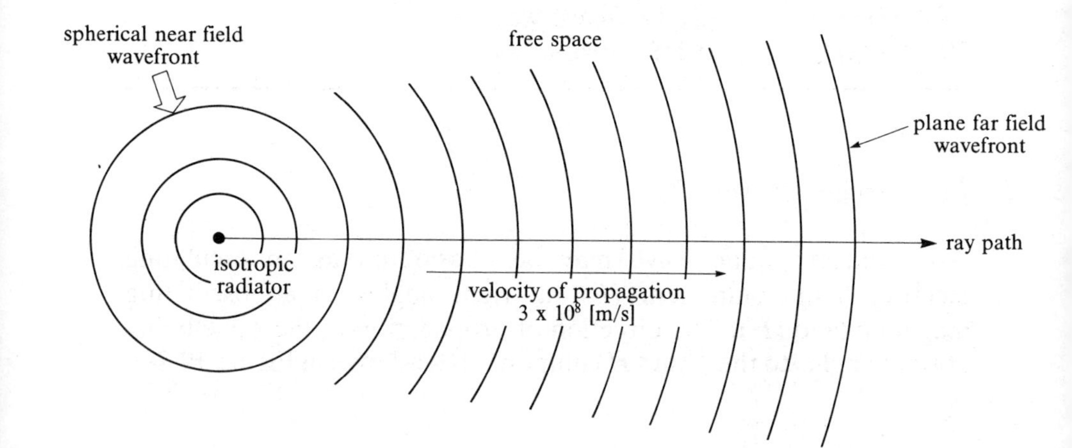

Figure 19.26 Propagation from a point source through free space.

Point source antennas are known as isotropic radiators and radiate uniformly in all directions to produce a spherical radiation pattern. This expands at a rate of 300 000 km/s, the speed of light. After 1 μs the spherical wave front would therefore have a radius of 300m. At such distances the wave can usually be considered plane and simple ray optics may be used to describe the wave's path.

As the wave expands, so its power is distributed over an ever increasing area. The power density, $Pd(\mathrm{W/m})^2$, of the wave, measured at a distance d metres from the transmitter will be:

$$Pd = Pi/4\pi d^2 \text{ [watts/metre}^2\text{]} \qquad \textbf{[19.11]}$$

where *Pi* is the effective isotropic radiated power (EIRP).

This power density, or flux as it is sometimes called, could be converted to an equivalent field strength, expressed in volts/metre, by using the relationship, $p = E^2/R$. But we need to know the value of *R*, the resistance of free space. It turns out that *R* for free space is equal to $\sqrt{\mu/\epsilon}$ or 120 π/Ω (see later).

The field strength, or intensity, *E* (V/m) is therefore given by:

$$E = \sqrt{P_D R}$$

$$E = \sqrt{\frac{120\pi Pi}{4\pi d^2}} = \frac{\sqrt{30Pi}}{d} \quad \text{[volts/metre]} \qquad [19.12]$$

Field strength is often expressed in dB relative to $1\mu V$, i.e.:

$$E\text{dB} = 20\log_{10}\left(\frac{E}{(1 \times 10^{-6})}\right) \text{dB } \mu\text{V/m} \qquad [19.13]$$

Equation [**19.12**] clearly shows that the field strength of a wave, propagated into free space, will fall as the inverse of the distance. This loss in field strength as a function of distance is common to all radio wave propagation and provides a foundation upon which we can build a model of a practical communication link which will have additional losses.

9.4.3 *Antennas*

It should be appreciated that an isotropic radiator has no physical reality, but merely serves as a useful reference against which to judge the effectiveness of practical antennas. All practical antennas will exhibit some directional properties; that is, they radiate more power in some directions at the expense of less in others. The directivity gain (*Gti*) is the ratio of the actual power density along the main axis of radiation of the antenna to that which would be produced by an isotropic antenna at the same distance and fed with the same power. The power density measured at some distance *d* from a practical antenna of gain *Gti*, fed by a transmitter of output power *Pt*, will be:

$$Pd = Gti \times Pt/4\pi d^2 \text{ (w/m}^2\text{)} \qquad [19.14]$$

An example of a simple, practical antenna is a half-wave dipole. Figure 19.27 illustrates the radiation pattern or polar diagram of such an antenna. We see that the field strength resulting from a half-

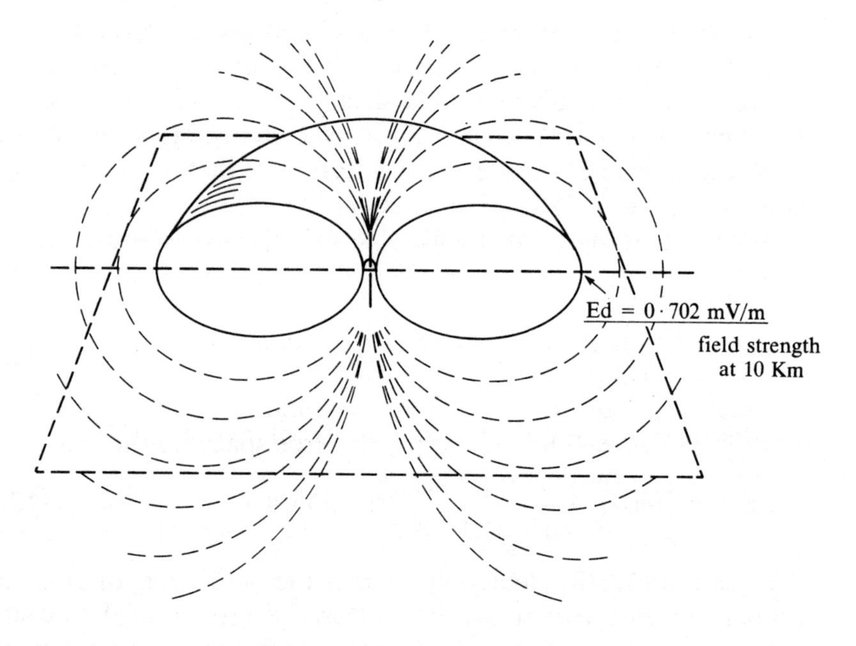

Figure 19.27 Half-wave dipole field pattern.

wave dipole, radiating one watt into free space and measured at a distance of 10km, is 702 μV/m. An isotropic antenna, radiating a similar power, would only produce a field intensity of 548 μV/m at the same distance. The ratio of the two field amplitudes is therefore 1.28, or 1.64 when expressed as a power ratio. This results in a power gain for the dipole relative to that of an isotropic radiator of 2.15dBi. Other practical antenna gains are listed in Table 19.4.

Table 19.4 Antenna gains (where D is the dish diameter and k *is an efficiency factor, usually between 0.5 and 0.8 for dishes which are very large compared to a wavelength).*

Antenna type	Gain over isotropic
Short monopole ($1<\lambda/8$)	G = 3 or 4.77dB
Parabolic dish	$G = (\pi^2 K D^2/\lambda^2)$ or $20\log_{10}[(\pi\sqrt{k}D/\lambda)]$ dBi **[19.15]**

So far we have not referred directly to receiving antennas where signal collecting ability is important. It turns out that the properties of receiving and transmitting antennas are reciprocal, i.e. the directional gain of an antenna is exactly the same whether used for transmission or reception.

The effective signal collecting area (*Aeff*) or aperture of an antenna is related to its gain (G) via the following ratio:

$$Aeff/G = \lambda^2/4\pi \quad [\mathrm{m}^2] \qquad \textbf{[19.16]}$$

A half-wave dipole will therefore have an effective area of $0.13\lambda^2$.

We can now relate the actual power (*Pr*) entering the receiver to the power density about the antenna via:

$$Pr = Pd \times Aeff\,(\mathrm{W}) \qquad \textbf{[19.17]}$$

Having knowledge of the transmitting and receiving antenna gains enables us conveniently to express the free space loss in terms of the attenuation between two antennas:

If $Pr = Pd \times Aeff \quad (\mathrm{W})$

then substituting for *Pd* from equation [**19.14**] and for *Aeff* from equation [**19.16**] gives:

$$Pr = \frac{PtGti}{4\pi d^2} Aeff = \frac{PtGti}{4\pi d^2} \cdot \frac{Gri\,\lambda^2}{4\pi}$$

$$\text{or } \frac{Pr}{Pt} = GtiGri \left(\frac{\lambda}{4\pi d}\right)^2$$

For isotropic antennas $Gti = Gri = 1$ and therefore the free space loss factor between isotropic radiators becomes:

$$\mathrm{FSL} = (\lambda/4\pi \mathrm{d})^2$$

Generally, we express this loss as a power loss in dB given by:

$$\mathrm{FSL} = 20 \log_{10} \left(\frac{4\pi \mathrm{d}}{\lambda}\right) \qquad \textbf{[19.18]}$$

9.4.4 *Propagation through the atmosphere*

In practice, radio waves do not propagate in free space, but through various layers of the earth's atmosphere whose characteristics may influence the wave's progress. These layers are broadly categorized in Figure 19.28.

The lower atmosphere in which we live is known as the troposphere. In this region the temperature reduces as the altitude increases until the tropopause is reached and the stratosphere

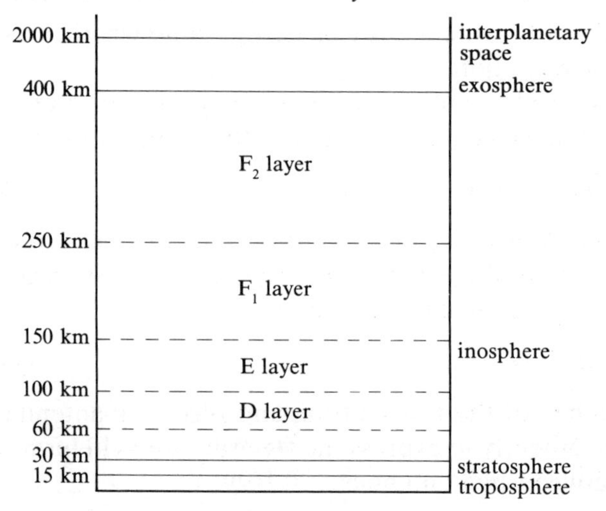

Figure 19.28 The earth's atmospheric layers.

begins. At still higher altitudes a multi-layered region of ionized gas, the ionosphere, gives way to the exosphere and interplanetary space. Radio waves are not only influenced by the media through which they pass but also by the terrain over which they travel. This may include water, soil, rock or vegetation.

We can model the atmosphere and terrain, and thence predict their influence upon a radio wave, by using the following three parameters:

i Conductivity (σ):the ability of a material to pass electrical energy. Being the reciprocal of resistance, its unit is the siemens/metre.
ii Permeability (μ):the measure of the superiority of a material over a vacuum to pass magnetic lines of force. Its unit is the henry per metre.
iii Dielectric constant or permittivity (ϵ):the amount of electrostatic energy that can be stored by any particular medium.

It turns out that the velocity of propagation of an electromagnetic wave is related to μ and ϵ by:

$$V = 1/\sqrt{\mu\epsilon} \text{ metres/sec} \qquad \textbf{[19.19]}$$

The values of μ and ϵ for dry air are essentially the same as for a vacuum or free space, i.e.:

$$\mu_0 = 1.257 \times 10^{-6} \text{ H/m and } \epsilon_o = 8.854 \times 10^{-12} \text{ F/m}$$

yielding a velocity of approximately 3×10^{-8} m/s, or 300,000 km/s (the speed of light). We have already noted in section 19.4.2 that the resistance of free space is given by:

$$R = \sqrt{\frac{\mu_0}{\epsilon_0}} = 120\pi \text{ohms}$$

9.4.5 *Refraction*

When an electromagnetic wave passes from one medium to another its velocity will change. If the density of the new medium is less than that of the surrounding medium, the speed of the penetrating portion of the wave will increase. This will cause the wave as a whole to bend in a downward directon from its original trajectory. Similarly, if the new medium were more dense, then the wave would be refracted in an upward direction. This is illustrated in Figure 19.29.

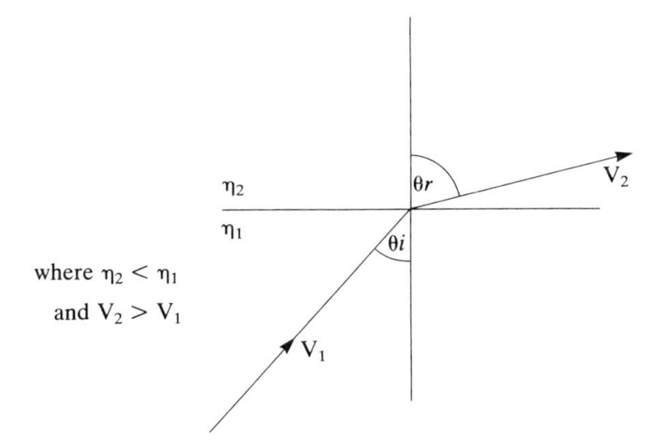

Figure 19.29 Refraction.

According to Snell's law, the ratio of the sin of the angle of incidence (θ_i) to the sin of the angle of refraction (θ_r) is the same as the ratio of the respective wave velocities, v_1 and v_2. That is:

$$\eta = \sin\theta_r/\sin\theta_i = v_1/v_2 \quad [19.20]$$

where η is called the refractive index of the second medium relative to the first. This is an important system parameter, as we will see later. Of course, if the first medium were free space, a perfect vacuum, then η becomes the absolute refractive index η_o.

In practice, radio waves will propagate through several media of gradually differing refractive index prior to reaching their destination. Such changes in refractive index will cause the waves to be bent or refracted. This refraction may have a very significant effect upon radio communications. In the case of HF communication, the wave may be refracted so much by the ionized layers of the upper atmosphere (ionosphere) that the wave returns to earth at some distant point, thereby enabling long range over-the-horizon communication.

At VHF abnormal changes in refractive index of the lower atmosphere (troposphere) may lead to interference between stations using the same frequency but located out of normal range of one another.

19.4.6 Reflection

An electromagnetic wave may be completely or partially reflected by a terrain feature, rain cloud or some other surface. We can represent the ratio of the incident and reflected waves by the reflection coefficient ρ.

Where the incident and reflected energies are the same, then the surface is a perfect reflector with $\rho = 1$. But if $\rho<1$ then not all the incident energy is reflected, and the residue is either absorbed by the medium or passed through as a refracted wave. Multiple reflections form the basic mechanism for communication via tropospheric scatter, but are in general a source of interference to most communication systems, as we will see later.

19.4.7 Diffraction

So far we have no mechanism that allows a wave to bend around obstacles such as the earth's curvature, mountain ranges or high buildings, yet we know that at certain frequencies communication around such obstacles is quite routine.

Hygen postulated that a wave front may be considered to be made up of an infinite number of isotropic radiators, as shown in Figure 19.30. He further proposed that a wave does not reflect from a single point but radiates from the entire surface of an obstacle in its path. As a result of this the wave is apparently bent or diffracted around objects as it grazes their surfaces.

Diffraction is also dependent upon the properties of the medium

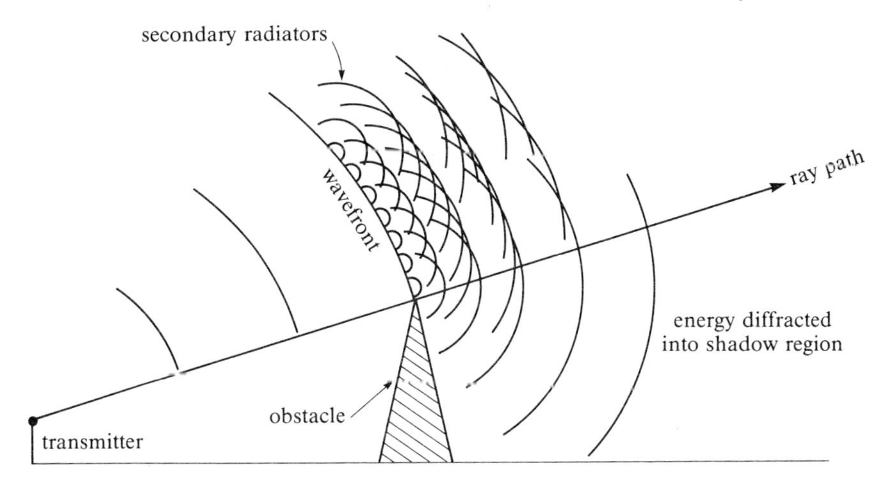

Figure 19.30 Wave diffraction.

over which the wave propagates. The earth is an imperfect dielectric and will have finite conductivity. The electric field component of the surface wave travelling over the earth will thus induce a horizontal component into it, giving rise to indirect currents and a resultant loss of energy from the wave. This continuous flow of energy from the wave downward into the earth effectively tilts the wavefront and assists the diffraction bending of the wave. The energy dissipated in the ground represents an attenuation of the wave which varies with the conductivity and permittivity of the ground below the transmission path.

19.5 Radio propagation in practice

Having reviewed some of the basic physical laws that influence the propagation of radio waves, we will now examine the various propagation mechanisms that are used for practical communication. Referring to Figure 19.31 we see that the most obvious propagation path is through the troposphere via the direct component of the space wave. Here both transmitter (*TX*) and receiver (*RX*) are in sight of one another. A ground reflected component of the space wave will also be present, leading to possible interference with the direct wave. Depending upon the path geometry and the coefficient of reflection of the ground, the effects of such interference can range from minor wave enhancement to total cancellation.

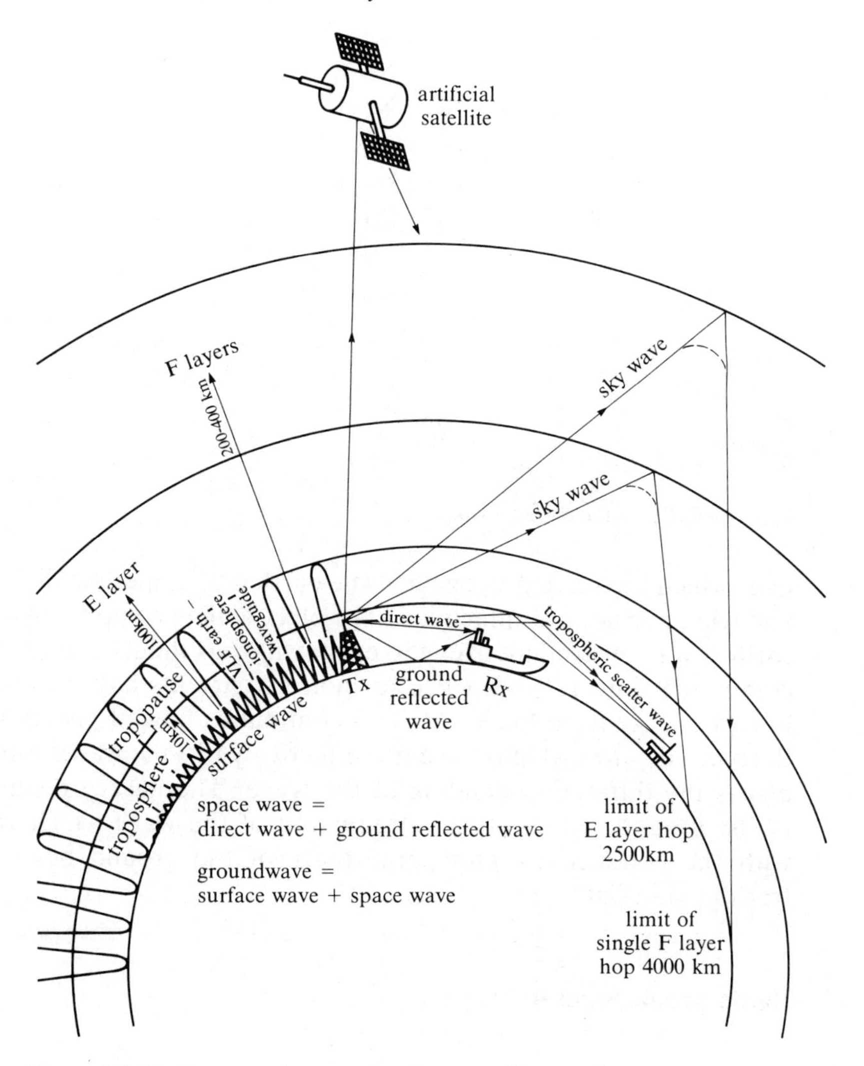

Figure 19.31 Propagation mechanisms used in marine communication systems.

Communication via the space wave will, in any case, be restricted in range to little more than the optical horizon. This will be defined by the earth's curvature, other obstacles, tropospheric refraction and antenna heights.

When over-the-horizon communication is required, other transmitted wave components may become important. These may include:

- surface or diffracted waves

- ionosphere returned sky waves
- artificial satellite returned waves
- earth ionosphere guided waves
- tropospheric scattered waves.

The existence and propagation efficiency of each wave is strongly frequency dependent. For this reason, we will examine them in the context of their significance in the various frequency bands.

19.5.1 *Low frequency (LF) propagation (30–300 kHz)*

At low frequencies, over-the-horizon communication is primarily via the surface or ground diffracted wave, as the sky wave component is severely attenuated by the lower layers of the ionosphere. The strength of the surface wave at a distant receiver will depend upon the ratio of the wavelength to the radius of the earth. In addition, diffractive bending and signal losses are influenced by the permittivity and conductivity of the ground over which the wave passes. Sea water for instance has a very high conductivity (σ = 5 S/m) and relative permittivity ($\epsilon_r = 80$), leading to low loss and long range propagation. Conversely, dry soil, typically has a σ = 1/1000th S/m and $\epsilon_r = 4$, resulting in increased ground absorption and reduced range. Surface wave ground losses increase with frequency and largely depend upon σ at low frequencies and ϵ_r at VHF.

We have then three main factors affecting LF surface wave propagation range:

1 free space loss
2 loss proportional to frequency due to diffraction process
3 ground absorption losses.

These effects are illustrated in Figure 19.32 which shows how LF field strength, due to a 1 kW transmitter using a short vertical monopole antenna on the ground, falls with distance and frequency for dry land and sea paths. Notice that the difference between propagation over sea and land is negligible at the low frequencies but increases rapidly with frequency. We also note from Figure 19.32 that all the curves are initially asymptotic to the theoretical free space loss curve $1/d$, but ultimately fall short, due to diffraction and terrain losses. Long range LF transmitters tend to radiate high powers, typically between 1 kW and 1 MW, in order to provide an

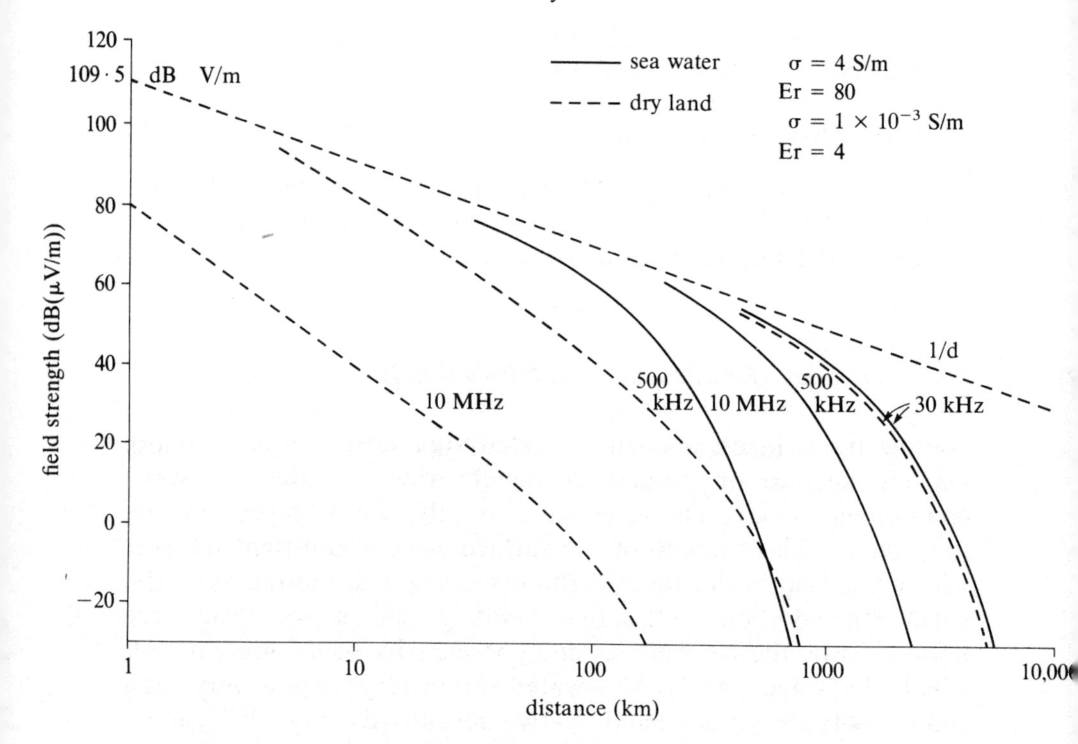

Figure 19.32 Groundwave propagation curves for land and sea.

adequate signal-to-noise margin against the high levels of atmospheric noise that prevail at these frequencies. In addition antennas are necessarily small compared with a wavelength, and very inefficient. In order to utilize successfully the surface wave, these antennas must be located as close as possible to the ground.

The LF surface wave channel clearly has a relatively narrow bandwidth by virtue of its frequency range and loss characteristics, but it is very stable with time, being largely independent of daylight; this makes it suitable for the transmission of navigation system and time standard signals over distances of up to 1500km. Systems using LF communication are listed in Table 19.5.

19.5.2 *High frequency (HF) propagation 3–30 MHz*

At high frequencies surface wave coverage is severely restricted, as the curves of Figure 19.32 would indicate. However, high frequencies are primarily used for medium to global range communications via ionospheric reflections, utilizing the sky wave component.

Table 19.5 LF systems.

System	Frequency	Bandwidth	Range
Loran C	100 kHz	20 kHz	1000 nm
Decca	85–128 kHz		240 nm
RDF	200–500 kHz	2 kHz	70 nm
AM radio	150–285 kHz	8 kHz	1000 nm
Telegraphy	30–150 kHz	300 Hz	1000 nm
Standard time	100 kHz band	300 Hz	1500 nm

In general, ionospherically returned sky wave signals are less stable than those of the groundwave. Their strength and existence will depend greatly upon the condition of the ionosphere, which varies with time of day, sunspot number, season to season and area to area. These variations are recorded by monitoring stations around the world and correlated with previous observations. This information is used to produce ionospheric prediction publications, which enable the system user to select the optimum frequency for transmission over a given route.

19.5.2.1 The ionosphere

A large part of the sun's energy reaches the earth's atmosphere in the form of ultra violet radiation. This can lead to the liberation of free electrons from the gas molecules of the atmosphere and the creation of an ionized region. The amount of ionization will depend upon the strength of the radiation and its wavelength.

The ionization process will be offset by recombinations of ions and electrons to reform the original gas atoms. The rate of recombination and hence the ionization density will depend upon the number of molecules available in a particular region. In the lower atmosphere the number of molecules is great so that the recombination rate is very high, but in the upper atmosphere, where the density is low, this rate is much slower.

It is clear that there will be regions within the atmosphere where the ionization density will reach a maximum, this being due to an equilibrium between creation and annihilation of ions. The necessary conditions for this to occur are illustrated in Figure 19.33.

There are in fact several layers in which ionization densities reach a maximum. These are designated by letters D, E and F in order of height and are shown in Figure 19.34. The altitude and density of

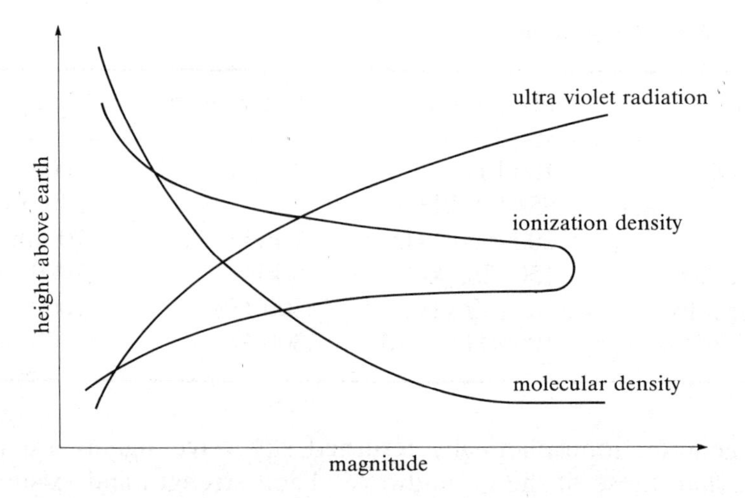

Figure 19.33 Formation of a single ionized layer.

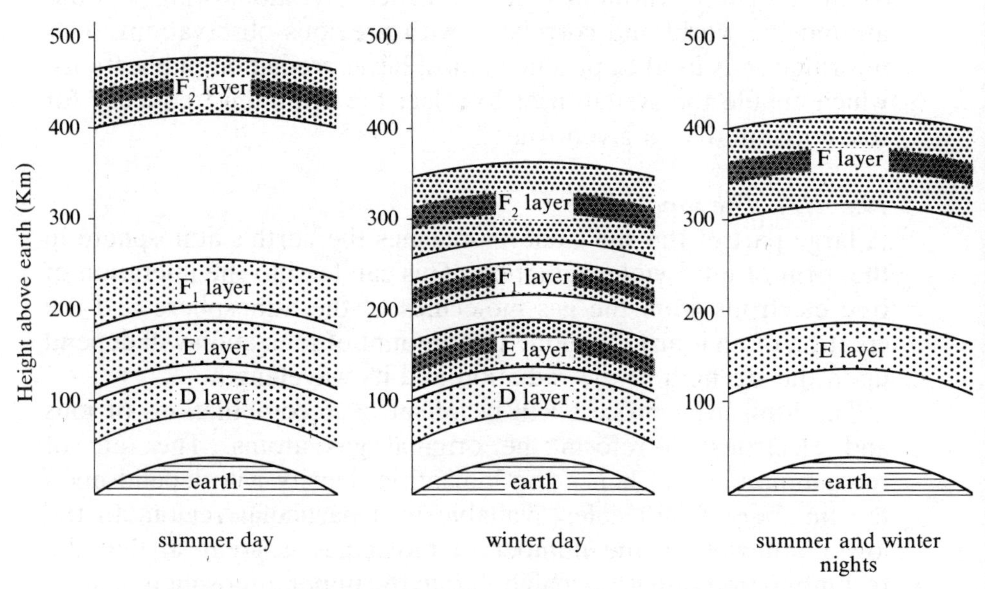

Figure 19.34 The layers of the ionosphere.

these layers depends upon many factors including time of day, latitude, season and the phase of the eleven-year sunspot cycle.

The highest layer is the F layer which exists between 150km and 500km above the earth's surface. During daylight hours the F layer splits in two to form the F1 and F2 layers; the F1 being relatively stable at around 200km. The F2 layer, on the other hand, undergoes significant variation in both altitude and density; being strongly dependent upon the solar heating of the upper atmosphere.

Below the F layer, the E layer at between 100 and 150km, remains substantially constant during daylight hours, but weakens considerably at night. Like the F region, its density is a maximum during summer months.

The D region is a rather indistinct layer that only exists during daylight hours. It extends rather diffusely from about 50km to 100km.

The ionosphere may affect the progress of a radio wave which passes through it in several ways, the two most significant being:

1 Beam bending – due to the changes in radio refractive index of the ionized layers. At frequencies below a critical value this bending may be sufficient to return the wave to earth; the basis of long range HF communication.
2 Wave attenuation – the electric field of a radio wave will set the free electrons of the ionosphere in motion. When these energized electrons collide with gas molecules, an interchange of energy will occur. In this way energy is absorbed from the wave and signal attenuation results. Absorption is therefore greatest in regions of high molecular density such as the D layer and to a lesser extent the E layer. As the D layer only exists during daylight hours, it follows that attenuation is greatest at that time. Wave attenuation is found to be proportional to $1/f^2$, and the higher the frequency, the lower the loss.

19.5.2.2 Ionospheric propagation

At HF the ionosphere may be treated as a dielectric with a continuously variable refractive index η. This index will depend on the ion density N, frequency f and is related by:

$$\eta = \sqrt{1 - \frac{81N}{f^2}} \quad [\mathbf{19.21}]$$

Typical ionization density variations with height are shown in Figure 19.35.

Notice that the value of η will always be less than unity and decreases as N increases. An incident radio wave entering from a non-ionized region ($\eta = 1$) will therefore experience a decreasing refractive index as the ionization density increases with height. If we regard the wave as a ray, and approximate an ionospheric layer by a

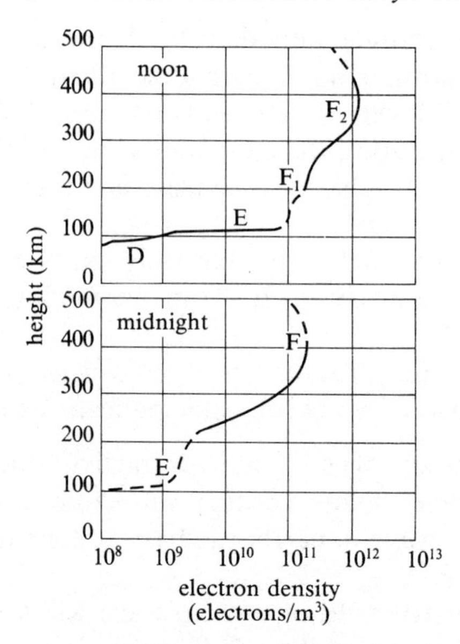

Figure 19.35 Ionization density vs height.

series of thin strips of constant refractive index, each strip having a greater electron density than the one beneath it, then successive refraction at the strip boundaries will lead to ray bending, as shown in Figure 19.36. If the layer is sufficiently thick, refraction will continue until the angle of refraction reaches 90 degrees, when the

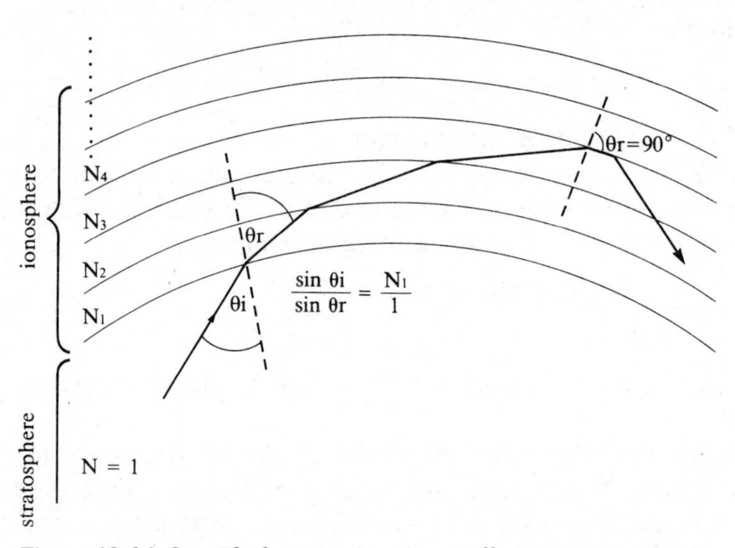

Figure 19.36 Stratified approximation to illustrate ionospheric refraction.

ray will have reached its highest point and will then start its downward journey back to earth. Under these conditions the ray appears to have been reflected from a plane surface within the ionosphere, the altitude of which is known as the virtual height of reflection (see Figure 19.37).

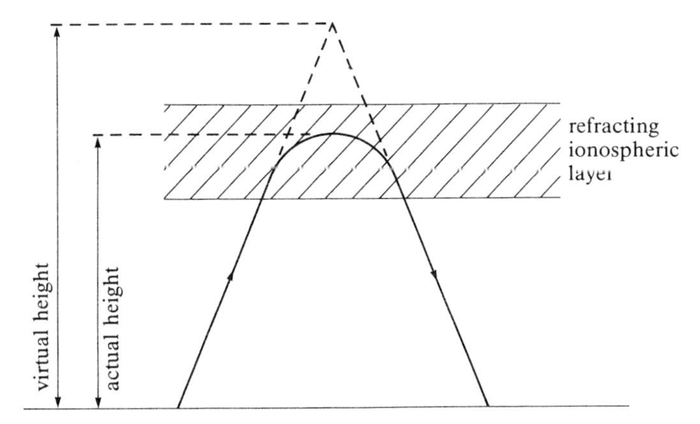

Figure 19.37 Virtual height of ionospheric reflection.

Total internal refraction, or effective ionospheric reflection, not only depends upon the thickness and density of the layer, but also on the ray's launch angle. Below a critical angle of incidence, refraction may be insufficient to return the signal to earth and the wave will escape into space, as shown in Figure 19.38. Still assuming a ray-like wave, we may apply Snell's law (equation [**19.20**]) to find the critical angle of incidence that will result in a refraction angle of 90° and ensure reflection, i.e.:

$$\frac{\sin \theta_i}{\sin \theta_r} = \frac{\eta}{1} \text{ as } \theta_r = 90^\circ$$

$$\text{therefore } \sin \theta_i = \sqrt{1 - \frac{81N}{f^2}} \text{ or } \theta i = \text{Sin}^{-1} \sqrt{1 - \frac{81N}{f^2}} \qquad [\mathbf{19.22}]$$

The critical angle is a function of frequency, and for any given ionization density maximum (Nm), the frequency at which the critical angle reaches zero is known as the critical frequency f_c. This being the maximum frequency which will be reflected at vertical incidence, i.e. $\theta i = 0$:

$$\text{therefore } f_c = 9\sqrt{Nm} \text{ (Hz) as } \sin \theta_i = 0 \qquad [\mathbf{19.23}]$$

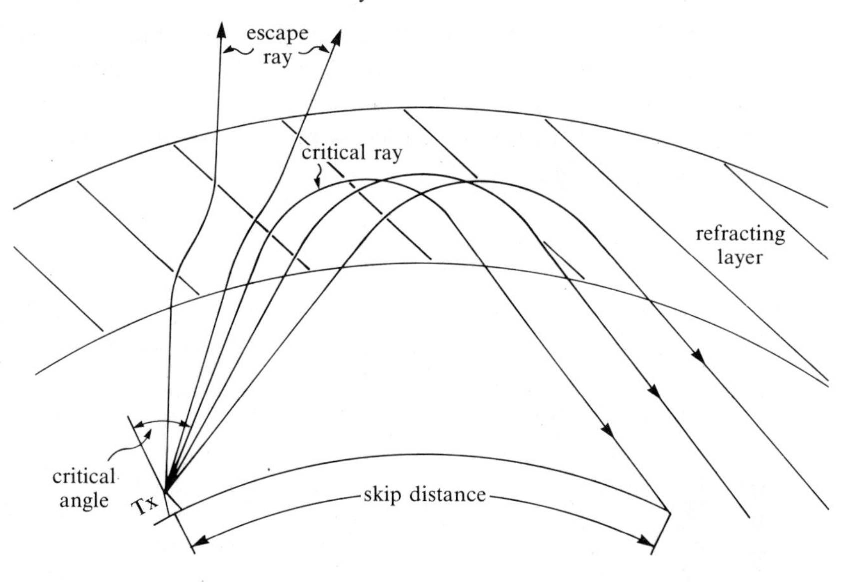

Figure 19.38 Escape ray and skip distance.

For a given angle of incidence, the maximum frequency at which reflection takes place is called the maximum usable frequency (MUF) and is related to the critical frequency by:

$$\text{MUF} = fc.\ \text{Sec}\ \theta_i\ (\text{Hz}) \qquad \textbf{[19.24]}$$

This is the maximum communication frequency for reflection from a particular layer of critical frequency *fc*.

At the MUF the wave path is critical and the transmission distance between two points on the earth's surface is the minimum distance at which the sky wave returns to earth. This is called the skip distance. More accurately, it refers to the first skip between transmitter and receiver, since it is possible for a second and even third skip to take place.

Maximum transmission distances for typical E and F layer reflections are plotted against antenna beam elevation in Figure 19.39. In practice, elevation angles near to zero are avoided, as the wave will suffer high ground loss. The lowest usable angle is generally accepted as five degrees. The largest angle of incidence obtained with *F* layer reflection is of the order of 74 degrees, where the MUF is 3.6 *fc*.

Typical daily MUF predictions for various transmission distances are shown in Figure 19.40. These are based upon monthly averages and do not account for daily variations of up to 15 per cent. For this

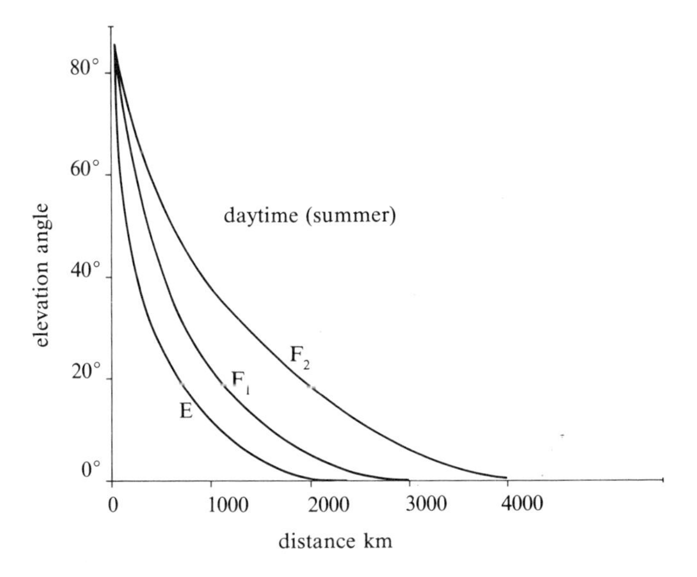

Figure 19.39 Transmission distance vs. antenna elevation for E, F_1 and F_2 layers.

reason, it is usual to work at a frequency somewhat below the MUF, in order to ensure that the transmission path is maintained for the duration of the communication. This is known as the optimum working frequency (OWF) and is 85 per cent of the MUF. It is desirable to work as close to the OWF as possible, as ionospheric absorption is proportional to $1/f^2$.

19.5.2.3 Ionospheric variations and disturbances

We have seen that the monthly average density and make up of the ionosphere, and thence the critical frequency, varies in a fairly smooth and predictable manner with the time of day and the season of the year. It is also found that these quantities vary in synchronism with the eleven-year sunspot cycle, as shown in Figure 19.41. The ultra violet radiation incident upon the ionosphere is dependent upon the sunspot cycle, which in turn governs the critical frequency. Note that the critical frequencies are considerably higher during a year near a sunspot maximum than during a period near a sunspot minimum.

In addition to the regular ionospheric variations, there are also irregular disturbances that are often unpredictable in both occurrence and duration. Many of the causes of these disturbances are not well understood, and include plasma instability within the ionosphere, and changes in solar activity, the latter being of major importance.

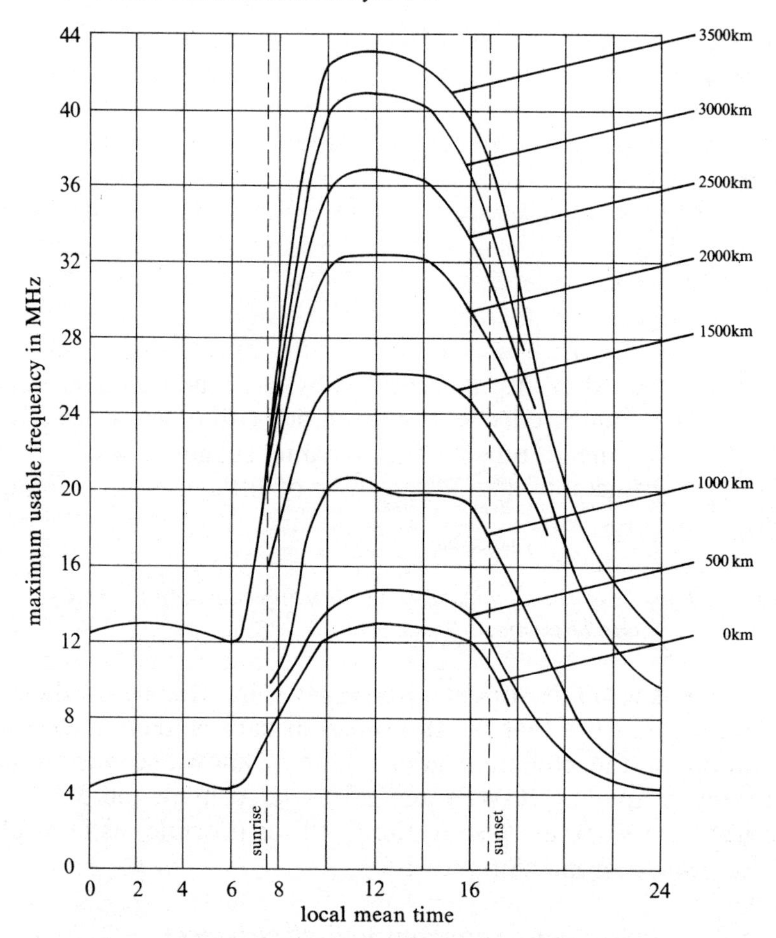

Figure 19.40 Typical diurnal variation of MUF for various transmission distances.

Sudden bright eruptions on the sun can produce a large increase in the ionizing radiation that reaches the D layer. This can result in total absorption of all sky waves above about one MHz and a complete radio fade out lasting from a few minutes to an hour or more. Such sudden ionospheric disturbances (SIDs) used to be called Dellinger fade-outs. Their instant of occurrence is unpredictable, but they are associated with periods of high sunspot activity. SIDs do not occur at night and are most intense at low latitudes.

Another disturbance, the so-called ionospheric storm, causes a breakup of the normal stratification of the ionosphere. This can lead to highly erratic propagation with rapid flutter, fading and abnormally low signal strengths. These storms are caused by streams of solar particles that are thought to emanate from sunspot regions or solar flares. Ionospheric storms usually last for several days and affect both sunlight and darkened regions of the earth. Paths that pass near the geomagnetic poles have been found to be most severely affected as are the higher HF frequencies.

An occasional abnormality of a different type occurs when cloud-like layers of exceptionally high ionization drift through the E layer. The resulting sporadic E layer, as it is called, can extend the E layer critical frequency up to 10 or even 20 MHz, with the result that signals intended for F layer reflection may be reflected prematurely.

VHF signals may also be reflected by a sporadic E layer, leading to possible interference between stations that are normally out of range of one another. As its name suggests, sporadic E is highly unpredictable and is therefore of little use for commercial communication, except where sophisticated real-time ionospheric soundings or channel evaluation systems are employed.

19.5.2.4 Signal fading and bandwidth limitations

A major drawback with HF propagation is the tendency for received signals to undergo wide fluctuations in level. These are known as

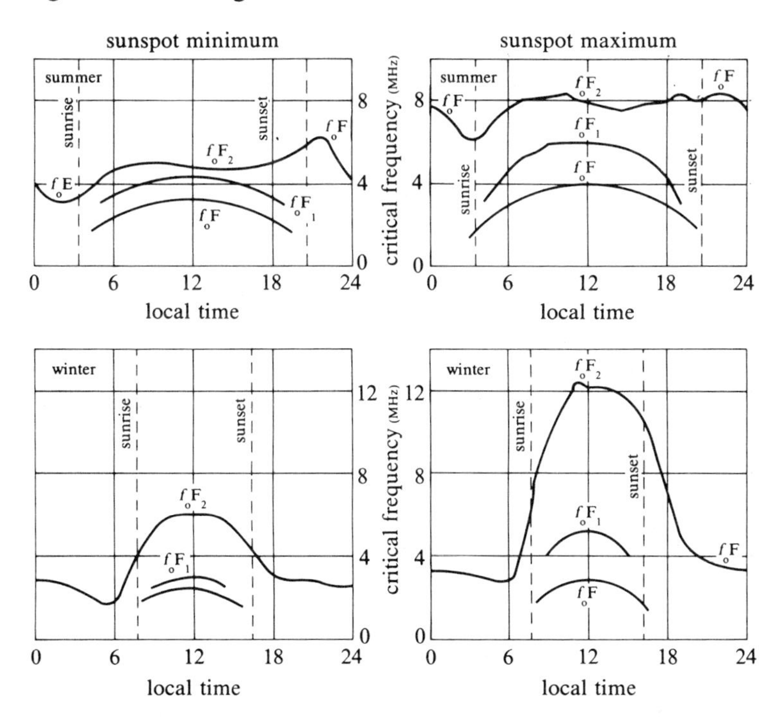

Figure 19.41 Typical diurnal variations of critical frequency.

signal fades and may be characterized by three parameters: depth, rate and duration.

There are a number of reasons why fading may occur. We have already discussed two important causes, those due to the natural cycles of the ionosphere and those due to unexpected disturbances. In the first case, fading will be relatively slow and predictable, and may be partially compensated for by providing additional signal margin to ensure that received power does not fall below the required level. This may be achieved by using increased transmitter power or antenna gain. However fades of greater than 10dB would require the selection of different transmission frequencies.

In the second case, the fade characteristics are largely unpredictable and may be so severe that communication is completely interrupted. Such effects may last from a few minutes to several hours or even days and may only be partially mitigated by frequency changes.

Short term rapid fading may also be experienced, particularly when the higher layers of the ionosphere are used for reflection. These are due to short-term fluctuations in the layer and can lead to fade rates of between 10 and 20 fades per minute. Variations in depth can range from a few dB to 30dB and can generally be reduced by the automatic gain control circuits of the receiver (AGC).

An important type of fading, known as interference or selective fading, is due to the propagation of waves over several paths between transmitter and receiver. These waves will be displaced in time (out-of-phase) due to differing path geometries and may combine to produce signal enhancement or cancellation or some intermediate level. Unlike other forms of fading, interference fading is highly frequency sensitive and may be restricted to individual frequencies or small bands of frequencies a few hundred hertz wide. This means that different frequencies within a signal bandwidth may suffer different amplitude and phase changes. Selective fading, therefore, sets a fundamental limit on usable HF bandwidth, and can lead to severe distortion of analogue modulated transmissions and the mutilation of digital signals.

19.5.3 Medium frequency propagation (MF) 300 kHz to 3 Mhz

Medium frequency communication draws upon the main modes of propagation associated with HF and LF transmission; these being ionospheric reflection of sky waves and surface diffraction of groundwaves.

During the hours of daylight, communication will be via the groundwave, as the sky wave will be almost completely absorbed in the D layer. The surface wave component of the groundwave will suffer greater ground loss than that experienced at LF, and these losses will be greatest over land and least over the sea. The daylight range of MF signals is therefore restricted to about 100km over land and 500km over sea paths.

After sunset the D layer will disappear and the E layer will weaken, leading to a significant reduction in sky wave attenuation and a subsequent increase in the strength of the ionospherically reflected wave; typically 20–30db.

Ionospheric reflections are from the E layer at an altitude of about 100km, although returns from the somewhat higher F layer are possible on short paths, at the higher frequencies in the MF band. Thus we have both ground and surface waves coexisting at night. At short distances the direct and surface waves will predominate and provide a stable primary coverage signal. At long range the reflected sky wave will provide a much less stable secondary coverage with rather pronounced fading, due to the temporal variations in density of the E layer. At intermediate distances the mean strength of the two waves will be comparable and very severe cancellation fading may occur. This area is known as the night fading zone.

The centre portion of the medium frequency band is allocated for local broadcast transmissions, 535 kHz to 1.6 MHz.

Maritime mobile telegraphy, telex, Navtex and distress services use the lower part of the band, while telephony and telex are used at the higher frequencies (see Table 19.6). These services enjoy stable day-time ground wave coverage, with little co-channel interference and moderate noise levels. However, they may suffer night-time interference from distant stations and severe sky wave interference fading, especially when the listener is near the limit of groundwave coverage.

19.5.4 Very low frequency propagation (VLF) 3 kHz to 30 kHz

For short to medium range communication using very low frequencies, the surface wave is the dominant propagation mechanism. As we observed with LF communication, the field strength of the received

Table 19.6 MF systems.

Service	Frequency	Bandwidth	Range
Maritime mobile	405–535 kHz		
Telegraphy		100–200 Hz	1200km
Distress (telegraphy)		100–300 Hz	1200km
Selective calling		4 kHz	750km
Navtex		300 Hz	1200km
Sound broadcast	535 kHz–1.6 MHz	8 kHz	
Land paths	550 kHz		50km
Sea paths	556 kHz		240km
Land path	1.6 MHz		20km
Sea path	1.6 MHz		224km
Maritime mobile	1.625–3.8 MHz		
Telephony		6 kHz	600km
Distress (telephony)		6 kHz	600km
Telex		300 Hz	1000km
Selective calling		4 kHz	600km

wave will initially reduce as the inverse of the distance. At greater distances, however, ground losses and the earth's curvature lead to much higher reduction rates. At very long distances, in excess of 3000km, communication is still possible via repeated earth ionosphere reflections.

At these frequencies the change in density of the ionosphere, from zero to a maximum, occurs over a distance that is small compared to a wavelength. Consequently, the ionosphere appears as an abrupt discontinuity and acts like a near perfect reflector. As the earth ionosphere interval becomes comparable with a wavelength, it is no longer useful to employ ray optic notions to describe such propagation.

A more realistic method of describing VLF propagation is to consider the wave as propagating along a waveguide, one wall of the guide being the ionosphere and the other the earth. When this waveguide is excited by a radiating antenna, many waveguide modes will initially exist, giving rise to a relatively complex modal pattern. High order modes are rapidly attenuated and so beyond about 500km only the first- and second-order modes need be considered. These remaining modes are illustrated in Figure 19.42.

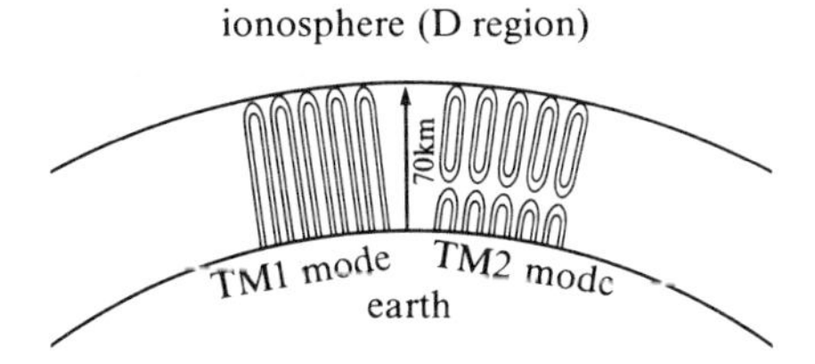

Figure 19.42 VLF propagation within the earth ionosophere waveguide, illustrating the two primary modes: TM1 and TM2.

VLF signals are highly stable and are therefore suitable for long range radio navigation systems and standard frequency emissions. At these frequencies, a small proportion of the wave will penetrate the surface of the sea and may be used for submarine communication. Unfortunately signal losses and atmospheric noise levels are very high, which means that extremely narrow receiver bandwidths must be used, typically 10 Hz. This, of course, means that speech communication between submarines is quite impossible and only very low speed morse and data can be conveyed. Signal attenuation, in sea water, at 10 kHz over a distance of 25 metres will be approximately 87dB. This means that communication is restricted to a few tens of kilometres even when high transmission powers are used and the submarine antenna is close to the surface (5–15m). Usually, however, VLF is used to broadcast to submarines with very high power and large shore station antennas. In this world-wide coverage is achieved. Systems using VLF are listed in Table 19.7.

Table 19.7 VLF systems.

System	Frequency	Bandwidth	Range
Omega	10–13 kHz	< 10 Hz	world-wide
Submarine comms	10 kHz	10 Hz	20km

19.5.5 *VHF to SHF propagation 30 MHz to 30 GHz*

At frequencies above about 30 Mhz, the ionosphere is not normally capable of reflecting sky waves back to earth, as the ionization density is insufficient for total internal refraction to occur. Such waves will pass through the ionosphere, escaping into space, and may be utilized for satellite communication. Waves travelling close to the earth's surface will be severely attenuated at these frequencies due to the ground losses associated with the diffraction process. For this reason, it is important that antennas are elevated above the ground by a few wavelengths, in order to ensure efficient space wave propagation.

It becomes apparent that terrestrial communication at VHF and above relies upon the space wave component, which will propagate through the troposphere.

Under normal conditions, communication range will be limited to a little more than line-of-sight, but localized, refractive index fluctuations in the troposphere can scatter radio energy, which may be received at distances of up to 1000km. This form of communication is known as tropospheric scatter and is most efficient at frequencies between one and five GHz.

Radio waves passing through the troposphere are not only influenced by its refractive index, but also by obstacles and reflecting surfaces that may exist along the route. Generally these will tend to disrupt communication either by restricting range or by causing wave cancellation and hence fading. Under certain circumstances the presence of a sharp obstacle in the waves' path may actually improve communication range beyond the normal radio horizon. This phenomenon is known as obstacle gain and is due to wave diffraction.

19.5.5.1 Tropospheric refraction

The troposphere is the lower part of the atmosphere which extends from ground level to an altitude of about 9km at the poles and 17km at the equator.

The refractive index (η) of the air within the troposphere will be very close to unity (1.00035) with a similarly valued dielectric constant and a near zero conductivity. Even very slight changes in η can have a significant influence upon wave propagation. These changes depend to a great extent upon the prevailing weather conditions which may be characterized by temperature, pressure and humidity.

These factors influence the refractive index according to:

$$\eta = \left[\frac{77.6P}{T \times 10^6} + \frac{0.373e}{T^2}\right] + 1 \qquad \textbf{[19.25]}$$

where T is the absolute temperature in degrees kelvin, P is the atmospheric pressure in millibars and e is the water vapour pressure in millibars.

Clearly, the refractive index will change with time, location and altitude and will be linked to weather conditions. At first sight it would appear that wave behaviour in such a variable medium would be almost impossible to predict, but we find that changes in refractive index with height are fairly predictable.

Under normal conditions, the temperature of the troposphere will fall with altitude, at a rate of approximately 6.5°C per km, until the tropopause is reached, where it remains substantially constant at about −50°C. This situation may be used as the basis for a much simplified model of the troposphere, known as the standard atmosphere model, which assumes a constant gradient of refractive index with height, given by:

$$\eta = 1 + 289 \times 10^{-6} e^{-0.136h} \qquad \textbf{[19.26]}$$

where the height h is in km above sea level (ASL). Figure 19.43 shows this linear variation of refractive index with height.

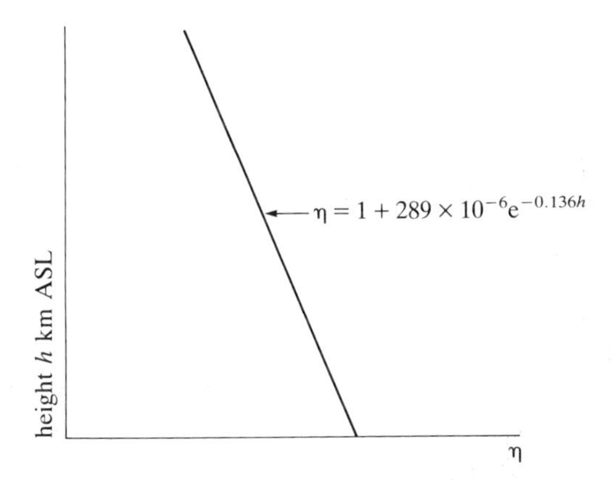

Figure 19.43 Standard atmosphere refractive index profile.

As η is continually changing with height, so the velocity of propagation of a wave will vary. If we assume a ray optic model for wave propagation (Figure 19.44) we can see that a ray emanating horizontally from an antenna and crossing the curved earth will experience a decreasing η and hence an increase in velocity, which will cause beam bending. This bending leads to a slight increase in range, over the optical horizon. For the standard atmosphere this increase is 4/3. Under these conditions the effective radio horizon, d km, may be obtained from:

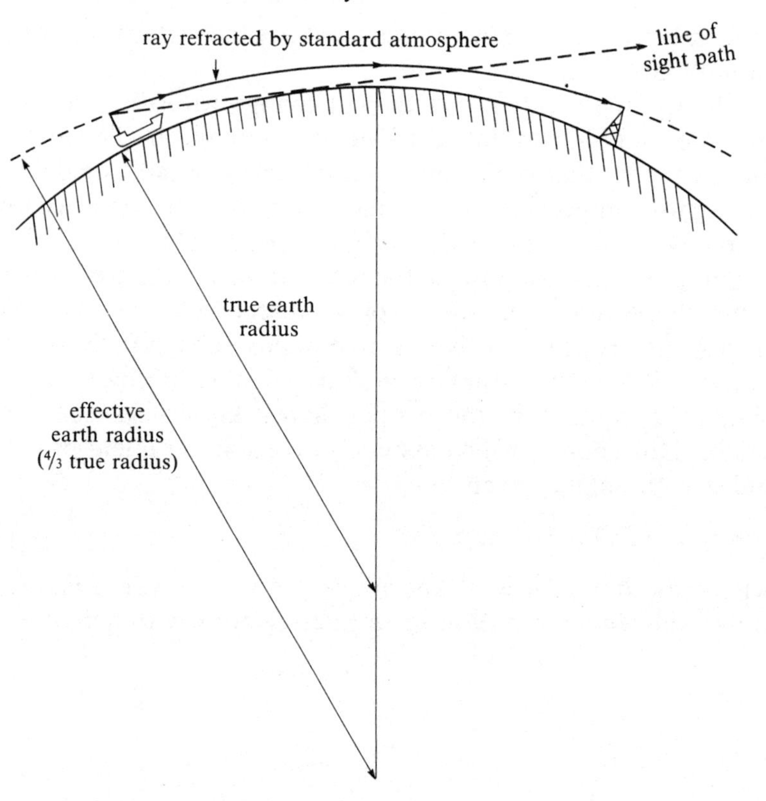

Figure 19.44 Radio refraction in standard atmosphere.

$$d = 4.13\sqrt{h}\ (\mathrm{km}) \qquad [19.27]$$

where h is the antenna height in metres ASL.

Significant departures from standard conditions may occur, which can lead to super-standard and sub-standard refraction, as shown in Figure 19.45.

For the purposes of terrestrial marine communication, refraction provides a small over-the-horizon capability, which is seldom disrupted by such anomalies. Conversely, navigation, satellite and point-to-point systems may have their performances severely degraded.

19.5.5.2 Reflections and multipath effects

The space wave consists not only of a direct wave, but also of a ground reflected wave whose magnitude and phase will depend upon the path geometry and the reflection coefficient of the terrain. For long distances and small antenna heights, the grazing angle is

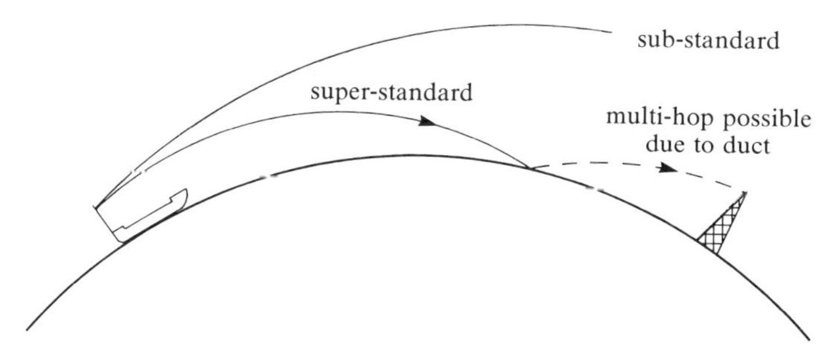

Figure 19.45 Sub- and super-standard refraction.

small and the reflection coefficient ρ tends toward a value of −1, i.e. a wave will be reflected with little loss and will experience a 180 degree phase change. In addition, the phase of the reflected wave at the receiver relative to that of the direct wave will depend upon the path length differences between the two waves and the frequency. If, for example, the net phase difference between the two waves is some n 360 degrees, where n is a whole number, then wave addition and hence signal enhancement will occur. Should the net phase difference be n 180 degrees, then wave subtraction may cause signal fading, known as interference or multipath fading.

As a result of these effects, the received signal will vary about the free space value, as the distance between transmitter and receiver increases. These variations are shown in Figure 19.46.

Once the radio horizon is approached, the two ray path distances will converge and the waves will begin to cancel, leading to a cessation of space wave communication and the onset of diffraction losses. Beyond this horizon, propagation by diffraction begins to dominate, albeit with a much greater rate of attenuation. Reflections may occur from one or more surfaces, but strong single reflections will usually cause the deepest fades, as the sum of many smaller reflections is unlikely to approach the magnitude of the direct wave. Reflections from calm water and metallic objects, such as ships and aeroplanes, may be particularly troublesome and cause deep flutter-type fading.

19.5.5.3 Diffraction effects

Communication beyond the radio horizon is possible via wave diffraction and is most efficient when the wave path is blocked by a relatively sharp obstruction, such as a mountain, building or, in certain circumstances, a cliff formation.

Figure 19.47 shows how the path loss experienced by a

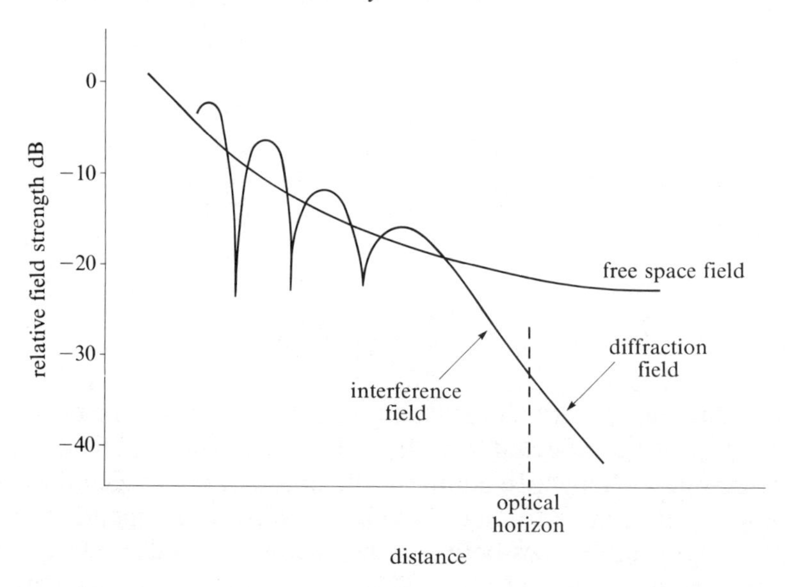

Figure 19.46 Field produced by interference between direct and reflected waves.

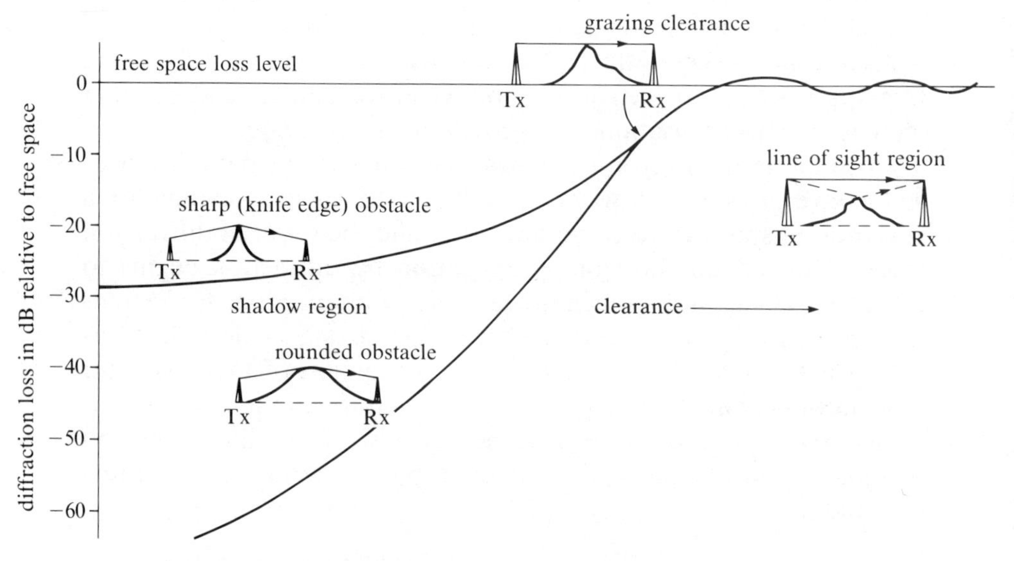

Figure 19.47 Diffraction loss vs path clearance.

communication link which encounters a single obstacle, varies as a function of path clearance. When the clearance is positive, i.e. the wave passes well above the object, then only slight variations in attenuation about the free space value will be experienced, due to ground reflections. As clearance reduces, however, the wave begins to graze the obstacle and once again diffraction losses set in and the rate of attenuation increases. Once path clearance becomes negative, i.e. the receiver is in the shadow of the obstacle, propagation is via the diffraction field alone, whose strength depends upon the sharpness of the feature, i.e. diffraction around a smooth object is much less effective than that achieved across a knife edge.

It should be appreciated that these effects are frequency dependent, as the diffraction loss across a given structure will increase with increasing frequency.

In terms of marine communication, the diffraction field may increase penetration into cluttered port areas or enable extended range working over sea paths. Wide band communication systems for oil rigs often utilize the diffraction field for distances up to about 80km, albeit with high power transmitters and high gain antennas. Diffraction losses, over and above those for free space propagation, may restrict satellite communication at low antenna elevation angles.

19.5.5.4 Tropospheric scatter

As we have seen in the previous section, over-the-horizon communication may be achieved by utilizing the diffraction field, which extends beyond the line-of-sight. The rate of field decay with distance is, however, much higher than for free space propagation and only modest range extensions are obtainable.

At still longer ranges, i.e. once the transmission loss has reached some tens of dB over the LOS value, the rate of attenuation reduces appreciably, to about 0.1dB/km. This is due to the signal contribution from tropospheric scatter which tends to become dominant once the diffraction field becomes very weak. As a result tropospheric scatter communication is possible over distances of up to 1000km. Many theories exist for the presence of this field but no satisfactory explanation prevails. System planning is based upon previous experience, statistical projections and extensive trial programmes.

Severe signal fading and high path losses mean that such links require high transmitter powers, high gain (fixed) antennas, sensitive receivers and usually some form of diversity arrangement. Despite these requirements, troposcatter remains the only practical wideband reliable ground-based system that is capable of achieving over-the-horizon communication.

19.5.5.5 Fading and atmospheric absorption

As we have seen, fading due to single and multiple reflections can occur. This is often rapid, particularly if the transmitter, receiver or reflector is moving fast. Fades due to single reflections are generally much more severe than those due to the combined effect of many secondary reflections, and may cause breaks in communication.

Long-term slow fading may occur as the make up of the troposphere changes due to temperature, humidity and pressure variations. Such fading is generally most pronounced at sunrise and sunset during summer months, due to the high water content of the atmosphere which increases beam bending.

At frequencies above 3 GHz, a wavelength becomes comparable to the size of a water droplet and energy absorption will occur. This may lead to significant signal attenuation during heavy rain, particularly over long paths. Figure 19.48 shows attenuation in dB/km that will occur for differing precipitation rates. Above 15

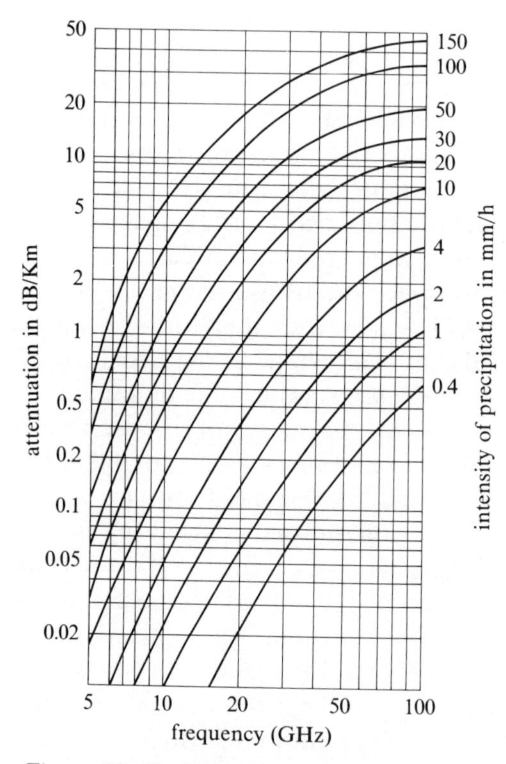

Figure 19.48 Attenuation due to precipitation. (CCIR XIIIth Plenary Assembly, Vol. V, Report 233–3, Geneva, 1974.)

GHz, attenuation due to oxygen and water vapour absorption will also limit line-of-sight communication range.

19.5.5.6 Earth–space propagation

Due to the presence of the ionosphere, signals below about 30 MHz are either absorbed or reflected back to earth and are therefore unable to pass into space. This sets a lower frequency limit for earth to space communication.

Many of the factors which influence terrestrial communication also affect satellite-based systems. These become particularly important when the satellite appears close to the horizon and the earth station antenna elevation is low. This situation often occurs when communicating with geosynchronous satellites from high earth latitudes, or when non-geosynchronous satellites appear near the horizon.

Under these conditions, signals will pass through a significant portion of the troposphere in much the same way as terrestrial system signals. As a result, refractive beam bending, terrain reflections, diffraction losses and atmospheric absorption may all be significant. In addition, the wave must pass through the ionosphere and may be subjected to absorption, delay and changes of polarization.

As free space path losses are high and satellite power is generally limited, such communication systems are often working close to their limit, and the high losses associated with low elevation working become unacceptable. Generally, the lowest usable earth station antenna elevation with a clear horizon is approximately five degrees, but this may have to be increased if diffraction losses occur due to obstacles such as mountains, cliffs and buildings.

When operating well above the horizon, propagation problems are much reduced and ionospheric influences become insignificant, particularly at GHz frequencies. Here attenuation due to precipitation and atmospheric gases becomes the major propagation limitation. As this attenuation varies with frequency, certain portions of the spectrum are better suited for earth-space communication than others. These are shown in Figure 19.49 and are often called space windows, the most obvious and heavily used of which lies between 100 MHz and 20 GHz.

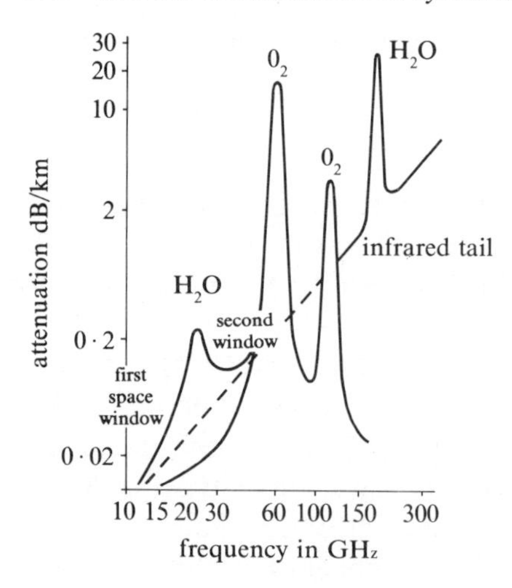

Figure 19.49 Atmospheric loss.

19.6 Noise

Introduction

There are many sources of noise that may degrade a radio signal. Some are of natural origin, others are man-made and some are due to equipment imperfections and inadequacies. It is convenient to organize these into one of two broad categories: internal or equipment noise and external noise.

Internal noise consists primarily of thermal noise, which pervades all systems, plus the combined contribution of other circuit noise such as shot, partition and flicker noise. For the purpose of noise power evaluation and comparison, these latter sources are generally represented by the equipment noise figure, which adds directly to the thermal noise 'floor' to define the total internal noise contribution.

All equipment will contribute noise to the system, but in general the noise associated with receiving equipment is usually of prime importance, as received signals are usually very small and of comparable amplitude to the noise. Systems are often equipment noise limited at frequencies above about 300 MHz and external noise limited below. Of the external noise sources, atmospheric noise places the most severe limitations on VLF and LF communication, whereas galactic and man-made noise is usually significant in the high HF and VHF bands. The relative significance of these noise sources can be seen in Figure 19.50.

Interference from other radio services, whilst not being true noise, can often cause reception failure particularly in the highly congested HF bands. Another impairment to signal reception, which is related to interference, is distortion due to equipment overload. This can cause intermodulation noise which may degrade multi-channel systems and limit HF communication.

19.6.1 *Internal noise sources*

19.6.1.1 Thermal noise

Thermal or Johnson noise, as it is sometimes called, is caused by the random motion of electrons within a conductor, which results from the application of thermal energy. This noise will appear across any conductor with resistance and its power is given by:

$$\text{Pnth} = KTB \text{ (watts)} \qquad [19.28]$$

where T is the absolute temperature of the resistance in degrees kelvin, B is the bandwidth in Hz over which the power is observed, and K is Boltzmann's constant $= 1.374 \times 10^{-23}$ joules/° kelvin. Such noise may also be expressed in terms of a mean square voltage or current appearing across or passing through a conductor of resistance R ohms:

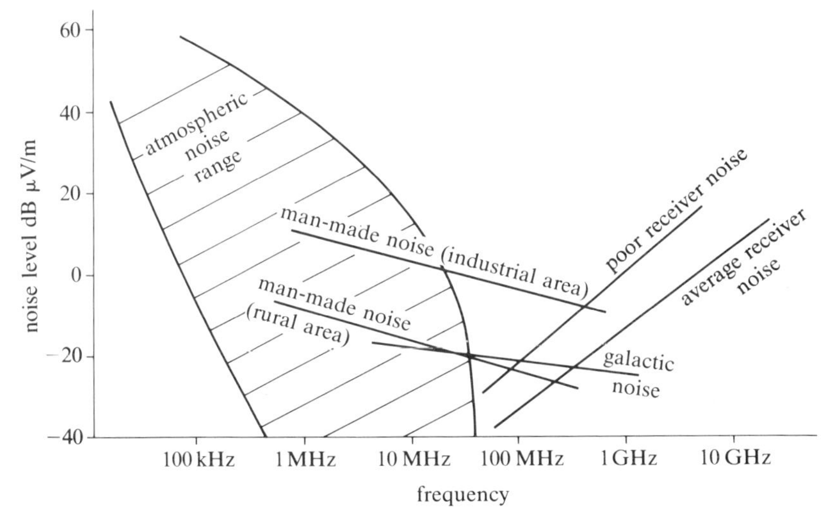

Figure 19.50 Major noise sources affecting radio communication.

$$\overline{vn^2} = 4KTBR(volts) \quad [19.29]$$

$$\overline{in^2} = 4KTB/R(amperes) \quad [19.30]$$

Thermal noise has a virtually flat power spectrum and, for all practical purposes, may be considered to be white (see again Figure 19.8). It is present in all systems and often forms a limit to achievable signal-to-noise ratios. For this reason it is used as a reference against which the noise contribution from other sources may be compared.

19.6.1.2 Antenna noise

An antenna appears to a receiver input as an impedance whose real component consists of its ohmic resistance and its radiation resistance. The temperature of the ohmic component will be the same as the physical or ambient temperature of the antenna and for terrestrial systems this is usually taken as 290°K. The temperature of the radiation resistance will be the same as the medium from which the antenna receives radiation. For antennas that look into space this is difficult to determine and a highly variable quantity but for ground-based systems 290°K is often assumed. In this case the thermal noise power at the receiver input due to the antenna resistance may be found by applying equation **[19.28]** directly. Under these conditions it is often more convenient to express the noise power at the receiver input in dB relative to 1 mW by:

$$Pn_{th} = K290/1\text{ mw} + 10\log(B)$$

$$\text{or} = -174 + 10\log(B)\text{ dBm} \quad [19.31]$$

As we can see, thermal noise is quite unavoidable in practical systems and may only be reduced, for a given bandwidth, by a reduction in temperature. Very often, however, it is external noise or the noise introduced by the receiving equipment itself that limits the receiver's sensitivity or ability to detect small signals.

19.6.1.3 Equipment noise factor

Noise, over and above thermal noise, will be introduced by all receiving equipment due to the electronic processes within.

A major source of such noise is due to electron flow within devices and is known as shot noise. Each electron carries a discrete amount of charge which produces a small current pulse. The sum of all such pulses produces an average current flow plus a small mean-square noise current or shot noise. This may be expressed by:

$$\overline{in^2} = 2eBI \qquad [19.32]$$

where I is the direct current, e is the charge on an electron = 1.6×10^{-19} coulomb. Another random process known as partition noise occurs whenever currents divide between one or more paths. Fluctuations in division between the paths lead to noise with similar characteristics to white noise.

A form of noise which is particularly prevalent in semiconductor devices is known as flicker noise. This is essentially a low frequency effect and as a result it is sometimes referred to as $1/f$ noise. It results from fluctuations in the carrier densities within the semiconductor. Such noise will not only degrade low frequency circuit performance, but may also produce noise side frequencies within a few kHz either side of an oscillator output.

Clearly there are many devices and circuits within equipment and whilst it would be impossible to account for the noise contribution from each individual element, we can assess the net effect of all such impairments by invoking the concept of noise figure or factor.

The noise factor of a network is defined as the ratio of the noise at the output of that network to that which would be present at the output of a perfect, noiseless network having the same gain, response, temperature and ideal terminations.

Noise factor is usually expressed in dB and is known as the noise figure (NF), although both terms are frequently interchanged. The noise figure may be added directly to the thermal noise $10\log_{10}$ (KTB) to yield the total noise associated with the equipment and its termination (in the case of a receiver this will be the antenna). Expressed in dB relative to 1 mW and assuming a temperature of 290°K, the noise power due to thermal and equipment noise will be:

$$\mathrm{Pn} = -174 + 10\log_{10}(B) + NF \text{ dBm} \qquad [19.33]$$

This is often referred to as the equipment noise floor.

Receiving equipment will be made up of many cascaded circuits with differing gains, losses and noise factors, all of which will contribute to the overall noise factors (see Figure 19.51). A simple analysis of cascaded networks with gains $G1$, $G2$, $G3$ . . . and noise factors $NF1$, $NF2$, $NF3$ . . . yields the important expression for the overall noise figure:

$$NFo = NF1 + (NF2-1)/G1 + (NF3-1)/G1G2 + \dots \qquad [19.34]$$

NB gains and noise factor must be in ratio form not in dB for this expression.

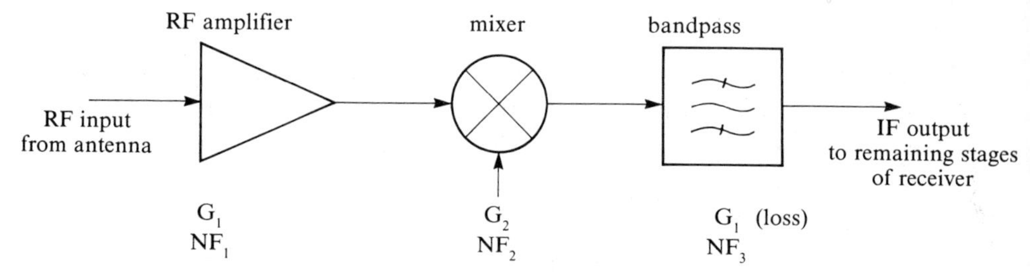

Figure 19.51 Simplified receiver front-end.

We can see that the overall noise factor will depend mainly upon the early stages of the receiver, as the contribution from later stages is diluted by the gain of the early stages. Very often the first stage is an RF amplifier with gain sufficiently high to enable the noise factor contribution from the second and subsequent stages to be ignored.

If we require a receiver to have a very low noise factor we must concentrate on designing a low noise first stage or front-end. Low noise front-end amplifiers are often referred to as LNAs. Typical noise figures currently achievable are listed in Table 19.8.

A low noise factor receiver is only required when the overall system is internal or equipment noise limited, as is usually the case at frequencies above about 300 MHz. Below this frequency, however, external noise sources may predominate and expensive low noise equipment is not needed.

Table 19.8 Typical front-end amplifier noise factors.

Frequency	Device	Gain	Noise factor
100 MHz	bipolar	25dB	0.5dB
up to 4 GHz	bipolar	8dB	3.0dB
30 MHz	mosfet	30dB	0.5dB
200 MHz	mosfet	20dB	0.7dB
1 GHz	gasfet	20dB	0.3dB
6 GHz	gasfet	15dB	0.4dB
12 GHz	gasfet	10dB	<1.5dB
20 GHz	gasfet	8dB	2.5dB

19.6.1.4 Noise temperature

An alternative way of expressing noise is to use the concept of equivalent noise temperature (*Te*). Here, the basic relationship for thermal noise $Pn = KTB$ is borrowed and rearranged, as shown in equation [**19.35**]:

$$Te = Pn/KB\ (°K) \qquad [19.35]$$

A given noise power Pn observed over some bandwidth B will therefore have an effective or equivalent noise temperature Te. This representation is very convenient when very low noise factors are encountered. Noise factor and noise temperature are related by:

$$NF = 1 + Te/Ts \qquad [19.36]$$

where Ts is the actual temperature of the equipment or circuit, often taken as 290°K.

19.6.2 *External noise sources*

19.6.2.1 Atmospheric noise

Atmospheric noise results from the cumulative effect of the many atmospheric disturbances, such as thunderstorms, that occur around the world. There are tens of thousands of such storms a year resulting in an estimated 200 lightning discharges per second, with currents ranging from 10 to 100 kA and an average discharge period of about 10 μs. These discharges appear as a random sequence of multi-gigawatt pulses which spread energy over a very wide portion of the electromagnetic spectrum.

The energy that results from lightning strikes will be subject to the same laws of propagation as communication signals, but the enormous powers will also enable modes of propagation that would be too lossy for communication purposes. The noise energy associated with a single discharge will, therefore, travel via a large number of paths with differing time delays giving rise to multiple receptions. The net result of many such discharges is a continuous noise spectrum whose amplitude varies slowly with frequency (see Figure 19.50).

Atmospheric noise energy peaks at about 10 kHz and tails off very rapidly above 30 MHz, as it is not reflected by the ionosphere, and passes on into space. For this reason atmospheric noise is negligible above VHF but forms a major limitation to communication at VLF and LF.

Thunderstorm activity is concentrated around the main storm belt of the world, which spans the tropical or equatorial regions, and includes central Africa, Indonesia, New Guinea, the Philippines and

Central America. Such activity is seasonal with marked differences between summer and winter. Normal diurnal, seasonal and sunspot variations which effect the ionosphere also influence atmospheric noise levels. As a result of all these variations, noise levels vary markedly with location, time, season and frequency.

An estimate of the likely atmospheric noise level at a particular time, season, frequency and location may be obtained from the contours of an atmospheric noise atlas such as that published in CCIR Report 322, *World distribution and characteristics of atmospheric radio noise*. Figure 19.52 shows a typical chart valid from 0000 to 0400 (local time) during winter months. The contours show mean noise factor *Fam* in dB above thermal noise at 1 MHz. Having obtained *Fam* for our receiver location we can transfer it to Figure 19.53 where it will define a particular curve. The mean noise figure for the frequency of interest may then be obtained. For example the atmospheric noise factor for the North Sea area is 60dB at 1 MHz. If we wish to estimate the noise factor for this location at 10 kHz we find that it is some 154dB above thermal noise, which represents a colossal amount of noise. At 30 MHz, however, this factor has fallen below zero in which case atmospheric noise may be neglected.

Over practical communication bandwidths, atmospheric noise looks very similar in character to white noise and may be treated as such. The exception to this occurs during local thunderstorms, in which case the noise will be much more impulsive or spike-like and will resemble shot noise.

19.6.2.2 Galactic noise

The origins of galactic or cosmic noise are largely unknown, but significant noise does emanate from the sun and a large number of discrete sources distributed throughout the galaxy. Noise from such sources is superimposed upon the general radiation background of the sky which has similar characteristics to white noise, although the received level is a reducing function of frequency (see Figure 19.50). The galactic noise that reaches the earth's surface is band-limited to between 15 MHz and 100 GHz, the lower limit being set by ionospheric absorption and the upper by atmospheric absorption. In practice, however, such noise is negligible above 1–2 GHz when compared to other sources. Galactic noise may be the dominant noise source for communication between 30 MHz and 300 MHz, particularly in areas of low atmospheric and man-made noise such as polar regions.

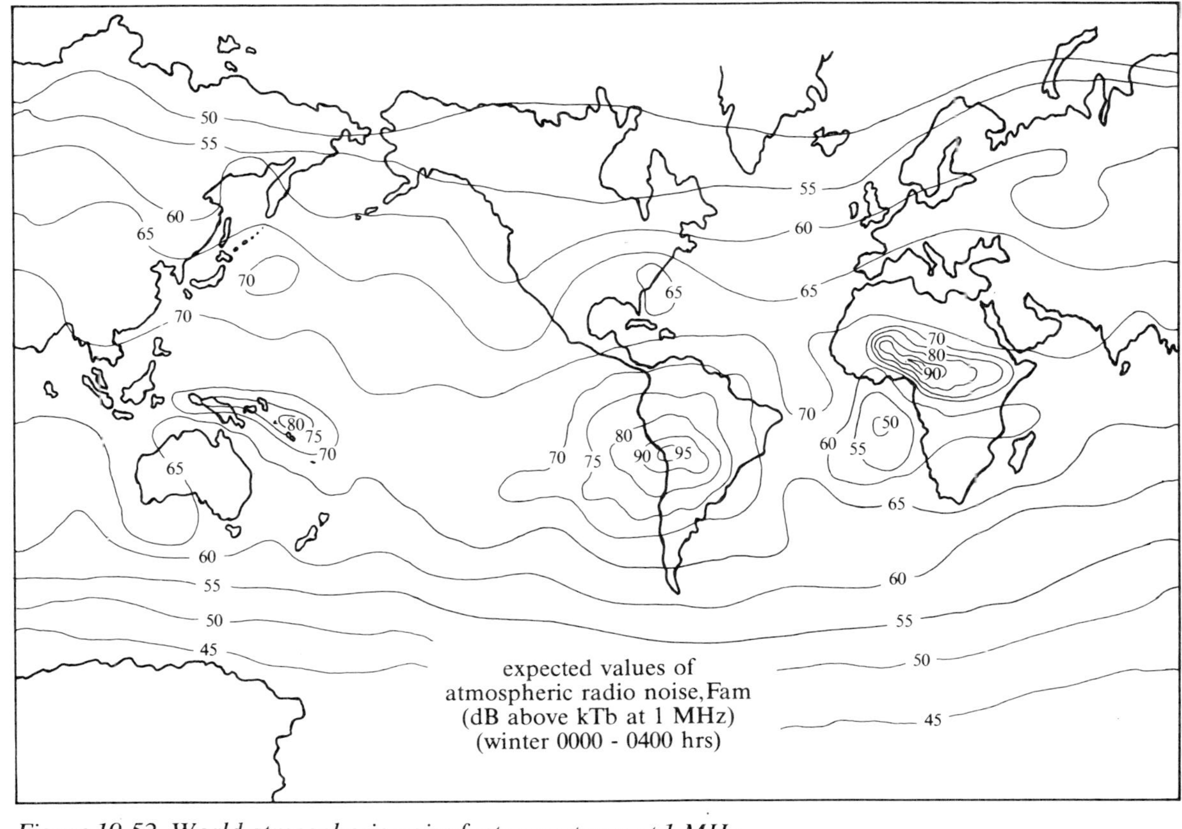

Figure 19.52 World atmospheric noise factor contours at 1 MHz.

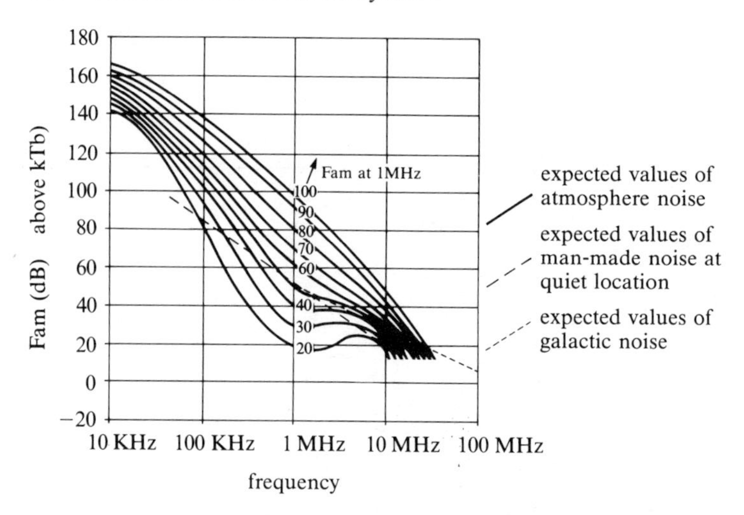

Figure 19.53 Variation of atmospheric radio noise with frequency (winter 0000–0400 hrs).

In certain circumstances, galactic noise may be reduced, and specific hot spots in the sky avoided, by the use of directional antennas.

19.6.2.3 Man-made noise

Man-made noise is very varied in nature but is generally more impulsive than atmospheric and galactic noise. Industrial areas, hospitals, airports, motor cars and high voltage power lines are all common sources of such noise and should be avoided if sensitive receiving equipment is to be installed. Although ships generally operate in areas of zero industrial noise, on-board electric motors, contactors, and power switching semiconductors may become dominant noise sources unless proper screening, filtering and suppression is applied.

19.6.2.4 Interference

Interference from other communication services and on-board electronic equipment falls into one of two categories, out-of-band and in-band. Fortunately, most interference will fall outside the band of frequencies occupied by the wanted signal, in which case their effects can usually be mitigated by filters within the receiver. These will reject the interfering signals and pass the wanted ones. The amount of rejection will depend upon how far removed the interference is from the wanted signal and upon the receiver complexity and cost. The ability of the receiver to reject such adjacent channel interference is described by receiver's selectivity.

In the case of in-band interference, there is often little that can be done to reject the interference, particularly if its amplitude is greater than that of the wanted signal.

9.6.3 *Intermodulation noise*

When a linear system, circuit or component is overloaded, distortion of the signal will occur (this is illustrated in Figure 19.54).

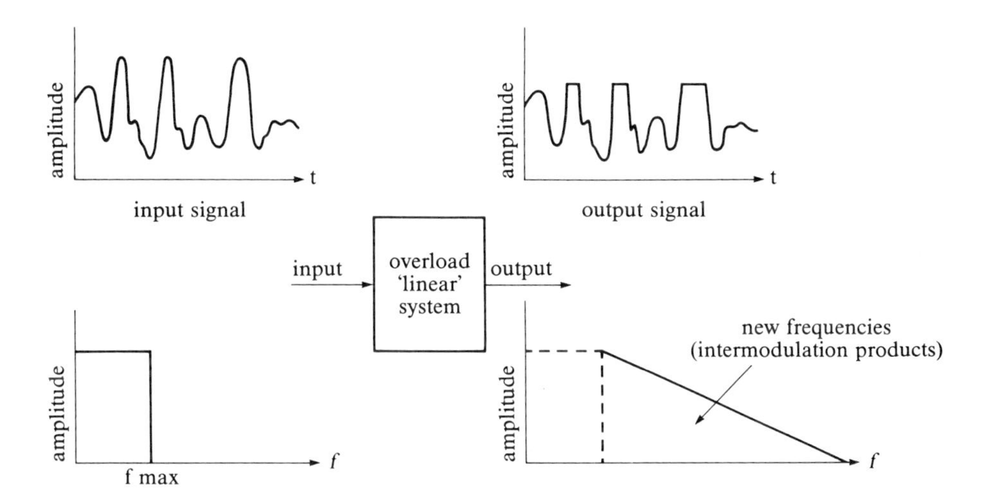

Figure 19.54 Overload distortion.

Viewed in the frequency domain we see that the distortion has produced new frequency components which are related to the original input signals. These new frequencies are called intermodulation products and are designated 2nd, 3rd, 4th, etc, in order.

There are often so many products produced when overload occurs, that they take on a noise-like appearance and extend over large bandwidths. These intermodulation products may cause interference by falling into the wanted signal band or into an adjacent band within an FDM signal. This situation may be avoided by ensuring that linear circuits are not overloaded or driven into saturation. In some situations this may be difficult to avoid; for instance, an HF communication receiver, connected to an efficient antenna, will receive signals from thousands of transmitters across the HF spectrum. Whilst the individual power contribution from each station will be minute, the combined power may be very high. The effect of subjecting the front-end of a receiver to such power, may be to cause overload which will lead to the generation of a vast number of intermodulation products and a subsequent raising of the receiver noise floor. For this reason high quality HF receivers either have pre-selection filters to reduce the total input power or very robust front-ends which are capable of linear power amplification with low noise.

19.6.4 *Overall operating noise factor of a receiving system*

Combining the effects of internal noise sources, external noise sources and equipment losses we can obtain an overall operating noise factor (f) for a particular system:

$$f = (fa - 1) + fc.ft.fr \qquad \textbf{[19.37]}$$

Where

fa = effective antenna noise factor resulting from external noise (assuming a perfect lossless receiving antenna)
fc = noise factor of lossy antenna
ft = noise factor of lossy feeders
fr = noise factor of receiver.

The overall noise factor f is often expressed as a figure, *NF*, where NF = 10log (f).

As an example, consider the system illustrated in Figure 19.55. Here the external noise factor of 20dB (fa = 100) combines with a

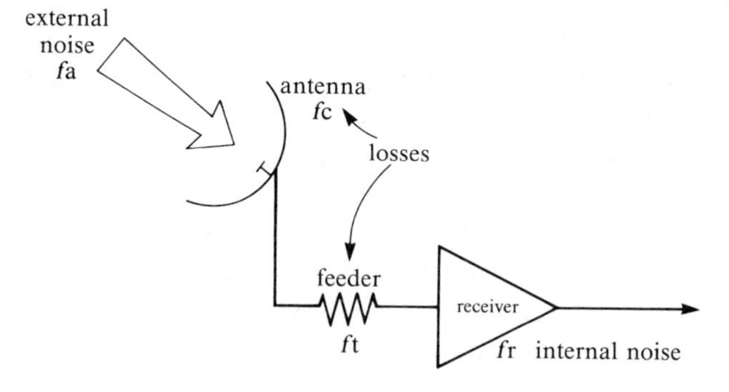

Figure 19.55 Noise contributions to receiving system.

receiver of noise factor 15dB ($fr = 31.62$) and feeder and antenna losses of 3dB each ($fc = ft = 2$) via equation [**19.37**] to yield an overall operating noise factor of:

$$f = (100-1) + (2 \times 2 \times 31.62) = 225.5$$

or, expressed as a noise figure, 10log (225.5) = 23.5dB.

Clearly, in this case, the external noise dominates the overall noise factor but lower losses and a better receiver would make a useful difference. (NB the overall operating noise factor could equally well have been obtained by converting all noise figures to equivalent noise temperatures via equation [**19.36**] summing and converting back to noise factor.)

Appendix 19.1

Classification of emissions

Emissions are classified according to their basic characteristics by a set of symbols:

First symbol – type of modulation of the main carrier:

1 Unmodulated carrier	N
2 Double sideband	A
3 Single sideband reduced carrier	R
4 Single sideband suppressed carrier	J
5 Independent sideband	B
6 Vestigial sideband	C
7 Frequency modulation	F
8 Phase modulation	G
9 Single sideband full carrier	H

Second symbol – nature of signals modulating the main carrier:

1 No modulation	0
2 Single channel containing quantized or digital information without the use of a sub-carrier	1
3 A single channel containing quantized or digital information with a sub-carrier	2
4 A single channel containing analogue information	3

5 Two or more channels containing quantized or digital information	7

Third symbol – type of information to be transmitted:

1 None	N
2 Telegraphy for aural reception	A
3 Telegraphy for automatic reception	B
4 Facsimile	C
5 Data transmission/telemetry	D
6 Telephony	E
7 Television	F

NOTE: This list of classifications has been truncated for brevity. A fourth and fifth symbol also exist giving details of signal type and multiplexing method. Refer Appendix 6 of *ITU Manual For Use by the Maritime Mobile Services*, edition of 1982.

20 Marine communication services

R. S. Linford

Introduction

In chapter 19 we reviewed some of the fundamental concepts that lie behind the various marine communication systems. In this chapter we will examine specific system examples in the context of the services that they offer and their equipment requirements and limitations.

Organizational and operational details have been reduced to a minimum as these tend to cloud the basic system principles and are well documented in the various operational publications.

We begin by reviewing the marine communication requirements which have led to the adoption of specific systems.

20.1 Marine communication overview

One of the first major applications of Marconi's wireless communication system was its use by ships to enhance the safety of life at sea. Since those early days there have been many innovations and improvements in radio communication which have led to a wide range of services. However, distress and safety working remains the primary service requirement for all marine communication systems.

The earliest marine radio systems were based on simple morse telegraphy which, despite the later introduction of radio telephony, has endured as the dominant communication system to date. Morse telegraphy equipment is essentially simple, cheap and reasonably reliable. With a trained operator it can provide a world-wide communication capability even when noise and interference levels are high. In fact, it is only in recent years that automatic terrestrial systems with comparable noise performance have become available at a realistic cost. Such systems include telex over radio (TOR), Navtex, selective calling and facsimile transmission.

Satellite communication systems are now maturing and offer some significant advantages over terrestrial communication, particularly for deep-sea users. The most significant advantage of satellite communication over terrestrial systems is that signals are no longer subject to the vagaries of ionospheric propagation. This has meant that major improvements in service availability, reliability and quality have been achieved. These new services have inevitably lead to a decline in manual telegraphy but this has been slower than predicted, mainly due to economic pressures, legislative inertia and nationalistic opposition. It is now widely felt within the industry that morse telegraphy is unlikely to be replaced totally by any other media until well into the twenty-first century.

In the context of safety of life at sea, however, these new systems will at last enable major changes in marine distress alerting and search and rescue (SAR) philosophy.

The present maritime distress and safety system is based on the principle that adequate maritime safety may be obtained by means of assistance being rendered by other shipping in the vicinity of a distress incident.

The current carriage requirements for vessels subject to the 1974 Safety of Life Convention (SOLAS) are based on manual communication equipment operated by a suitably qualified officer. In the case of vessels over 1600 gross tons this means MF (500 kHz) radio telegraphy equipment operated by a morse-qualified radio officer. For vessels of over 300 tons an IF (2182 kHz) and VHF (156.8 MHz) radio telephony system operated by a trained telephony operator is required.

This means that an inter-ship alerting system with a range of 100–150 nautical miles, reliant upon manual watch-keeping by specially trained personnel, is assumed to be adequate for vessels operating world-wide.

The weaknesses of the old system have, of course, long been recognized and, to a certain extent, awaited new technology to enable radical improvements to take place. During the late 70s the International Maritime Organization (IMO) defined the general requirements for a new maritime distress and safety system and has since been working, in conjunction with other organizations, to formulate and plan a Global Maritime Distress and Safety Service (GMDSS).

This system is based on the introduction of maritime communication satellites together with a system of digital selective calling (DSC) and an automated direct printing system for the transmission of navigational and meteorological warnings and urgent information to ships (Navtex). In addition a system, based on polar orbiting satellites, will be used for distress alerting and position fixing from float-free EPIRBs (emergency position indicating radio beacons).

The functional and operational requirements of GMDSS are as follows:

1 Areas of operation will be used to determine the equipment appropriate for the ship. These are:

Area A1 Within range of shore-based VHF stations
Area A2 Within range of shore-based MF stations (excluding A1 area)
Area A3 Within the coverage area of geostationary maritime communication and satellites (excluding areas A1 and A2)
Area A4 Outside the coverage area of geostationary maritime communication satellites.

2 Existing MF arrangements will be rationalized.

3 A long-range capability using satellites and HF will be provided.

4 Watch-keeping on relevant distress and safety channels will be maintained by automatic means.

5 Means will be provided for the automatic reception of all relevant safety information including meteorological and navigational warnings.

6 Radio telephony, digital selective calling and narrow band direct printing (NBDP) will be used in terrestrial radio systems. Morse telegraphy will not be used in the new system.

The basic philosophy of the GMDSS system is that search and rescue authorities ashore as well as shipping in the vicinity will be alerted to a distress incident. In this way, effective assistance can be provided with a minimum of delay via a properly integrated terrestrial and satellite-based communication network.

20.2 Navtex

Navtex is an international service which provides for the automatic promulgation of meteorological and navigational warnings and urgent messages to vessels sailing within approximately 400km of the coast.

The service is intended for use with simple low cost receivers which have the ability to automatically select and reject messages in order to ensure that only the information requested by the mariner is presented. In this way the system can be used to convey a wide range of information of both local and regional importance without overwhelming the user with unnecessary information.

The Navtex system will fulfill an integral role in the Future Global Maritime Distress and Safety System (FGMDSS) being developed by the International Maritime Organization (IMO). Following a period of five years of trials, development and international debate, the Navtex service was officially declared operational in the Baltic and North Sea areas in 1983. Since that time the service, which will ultimately become world-wide, has expanded to include:

Mediterranean/Aegean and Black Sea
USA east and west coast
Atlantic coast of Spain/Portugal
Canary Islands
Azores
South America
China
Japan

The service is sponsored by the International Maritime Organization (IMO) and is coordinated by the Navtex coordinating panel.

Messages are broadcast mainly from transmitters sited at existing coast radio stations, which time-share the common transmission frequency of 518 kHz. Each message contains a coded preamble, enabling a microprocessor-controlled user/receiver to select or reject messages according to their category, station of origin or serial number.

20.2.1 *The Navtex service*

Figure 20.1 shows the basic organization of the Navtex service. Navigation, meteorological and search and rescue (SAR) coordinators feed source messages to the Navtex coordinator who relays them, usually via telex, over the public switched network, to appropriate coast station for broadcast on 518kHz.

The Navtex coordinator allocates messages to specific transmitters

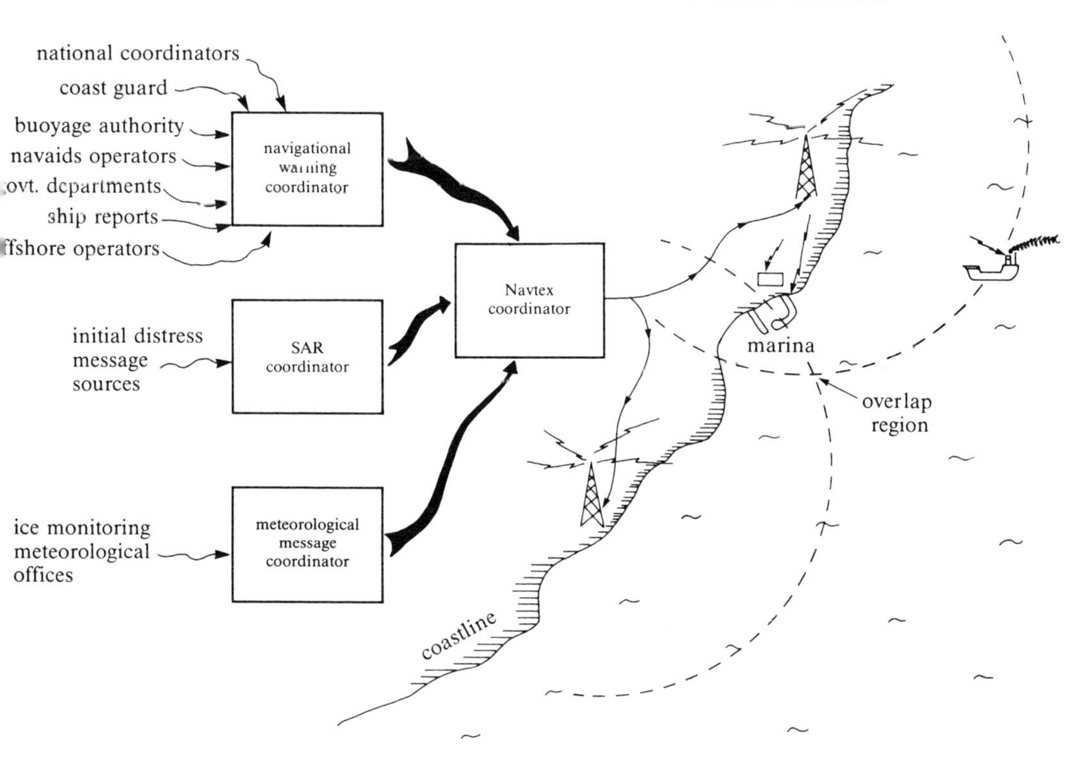

Figure 20.1 Navtex service organization.

according to the information contained in each message and the geographical coverage required. As a result the user may choose to accept messages either from the single transmitter which serves the sea area around his position, or from a number of transmitters as appropriate.

Navareas are serviced by a number of Navtex-equipped transmitters which provide appropriate coverage for that area (see Navarea 1 allocation, Figure 20.2).

As a common transmission frequency is used, mutual interference is avoided by allocating transmission time slots to each of the participating stations. Radiated power is reduced to the minimum necessary for the required sub-area coverage.

Received messages are in English and have a coded preamble consisting of four characters (B_1, B_2, B_3, B_4) which may be inspected by the processor within an automatic receiver and used as a basis for message acceptance. The characters are as follows:

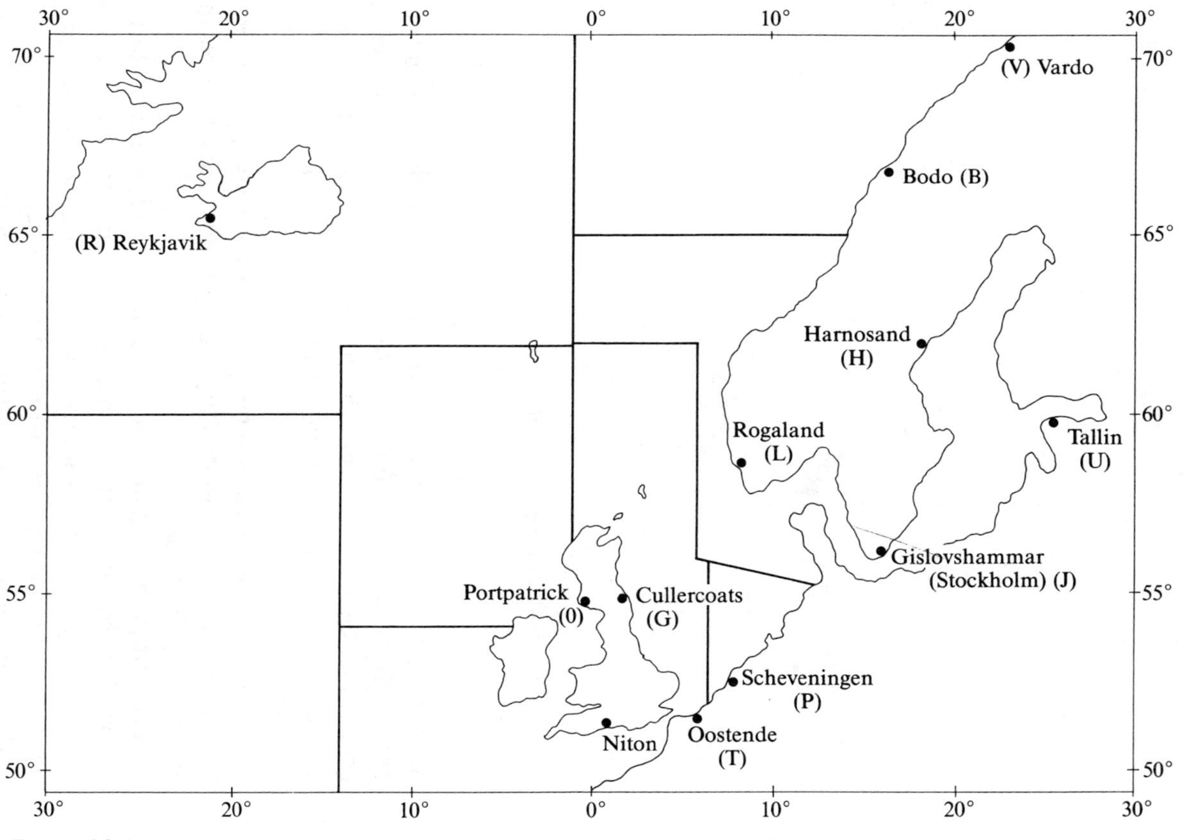

Figure 20.2 Navtex areas within Navarea one.

Transmitter identification (B_1)

A single unique letter is allocated to each transmitter. This enables broadcast identification and message editing on the basis of transmitter service area. For example; Niton radio (S), Rogaland (L), Scheveningen (P), etc.

Subject identification (B_2)

Information is grouped by subject on the Navtex broadcast and each subject is allocated a subject indicator (B_2) as follows:

A = Navigational warnings
B = Meteorological warnings
C = Ice reports
D = Search and rescue information
E = Meteorological forecasts
F = Pilot service messages
G = Decca warnings
H = Loran warnings
I = Omega warnings
J = Satnav warnings
K = Other electronic navaid messages
L = Navigation warnings additional to A
M–Y = For service use or not assigned
Z = No messages in hand

These indicators may be used by the receiver processor to accept or reject messages on a subject basis as pre-selected by the user. Certain essential classes of safety information such as navigational and meteorological warnings and SAR information are non-rejectable to ensure that ships using Navtex always receive the most vital information.

Message numbering (B_3, B_4)

Each message within a subject group is allocated a serial number B_3, B_4, between 01 and 99. On reaching 99 numbering will recommence at 01 but avoid the use of message numbers still in force. The user receiver can store the serial number of each messages received and thereby avoid unnecessary repetitions. Figure 20.3 shows a typical Navtex messages from Cullercoats (G) and Niton (S) – a met. warning number 59 and nav. warnings 59, 64 and 65.

ZCZC GB59
CULLERCOATSRADIO
GALE WARNING.

GALE WARNING FRIDAY 14
NOVEMBER 1615GMT.

VIKING NORTH UTSIRE SOUTH
UTSIRE
SOUTHEASTERLY GALE FORCE 8
CONTINUING.

CROMARTY
SOUTHEASTERLY GALE FORCE 8
EXPECTED SOON.

FORTH TYNE
SOUTHEASTERLY GALE FORCE 8
IMMINENT.

WIGHT
SOUTHERLY SEVERE GALE
FORCE 9 DECREASING GALE
FORCE 8 IMMINENT.

PORTLAND PLYMOUTH BISCAY
LUNDY IRISH SEA
GALES NOW CEASED.

SHANNON
GALES NOW CEASED BUT
SOUTHERLY SEVERE GALE
FORCE 9 EXPECTED LATER.

ROCKALL
SOUTHERLY SEVERE GALE
FORCE 9 EXPECTED LATER.
NNNN

ZCZC SA59
NITONRADIO
NAVAREA ONE 341.
FRANCE NORTH COAST.
APPROACHES TO OUISTREHAM.
CHART BA 1821.
DANGEROUS WRECK LEAST
DEPTH 14 FEET
LOCATED 49–19. 5N 00–09 4W.
NNNN

ZCZC SA64
NITONRADIO
NAVAREA ONE 346
ENGLAND SOUTH COAST.
TORBAY. BRIXHAM
APPROACHES.
CHART BA 26.
DANGEROUS WRECK 50–25. 0N
03–28. 7W.
NNNN

ZCZC SA65
NITONRADIO
NAVAREA ONE 347.
ISLES OF SCILLY.
CHART BA 34.
PENINNIS HEAD LIGHT (A0006)
49–54N
06–18W
NOW FL 20S 36 METRES 20
MILES.
NNNN

Figure 20.3

20.2.2 *The Navtex system*

20.2.2.1 Propagation and noise considerations

The frequency 518 kHz lies within the medium frequency band allocation for maritime wireless telegraphy, selective calling and distress working (405–535 kHz).

As we have seen in section 19.5.3 propagation at these frequencies will be via surface wave diffraction, with negligible sky wave interference at distances of up to 400km. However, beyond this range E layer reflections at night will cause night fading and possible interference with other Navarea transmitter chains. For this reason transmitter power is kept to the minimum required for service coverage and may even be reduced at night to minimize interference.

In section 19.6.2 we found that from VLF to HF, external noise tends to dominate all other radio noise sources. For the Navtex frequency 518 kHz (MF) atmospheric noise levels are very high, particularly in the tropical storm belts.

Figure 20.4 shows a simplified signal and noise level diagram for a typical Navtex broadcast to a vessel within the service range. In this example we see that a Navtex station radiating 100 watts (+50dBm) will deliver a signal to the ship's receiver of −30dBm over a 320km (200 nm) sea water path (assuming the ship's antenna is isotropic and antenna and feeder losses are negligible). This propagation loss is obtained from the ground wave curves of Figure 20.5.

We must now assess the noise power that the receiver will see in order to estimate the received signal-to-noise ratio for the system.

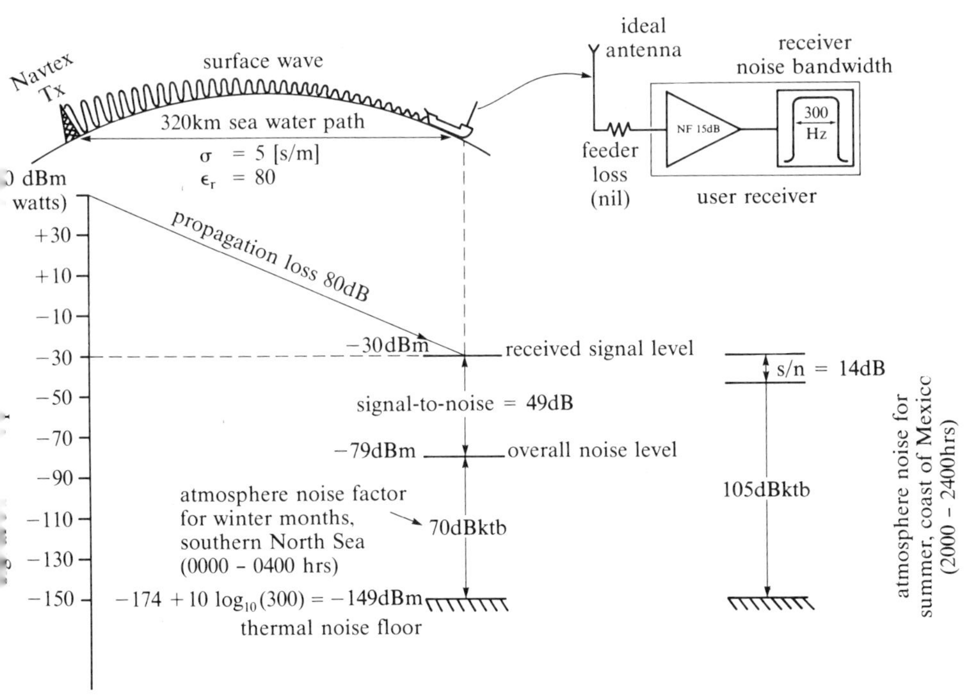

Figure 20.4 Simplified level diagram for typical 518 KHz Navtex path.

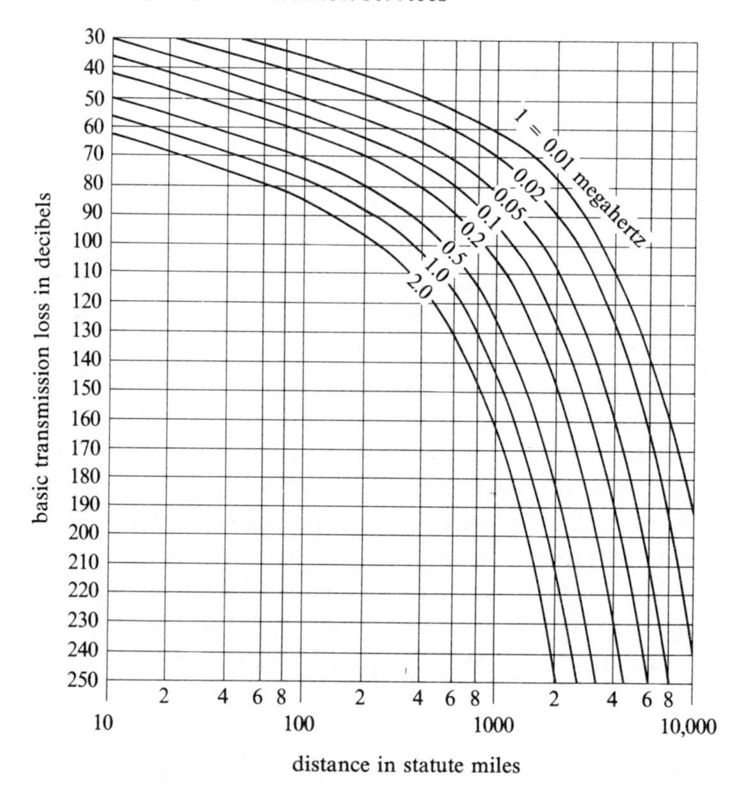

Figure 20.5 Basic transmission loss expected for surface waves propagated over a smooth spherical earth. Over sea water: $\sigma = 5$ mhos/meter, $\epsilon = 80$. Lossless isotropic antenna 30 feet above the surface. Vertical polarization. (Adapted from K. A. Norton, Transmission Loss in Radio Propagation: II, National Bureau of Standards Technical Note 12, Fig. 8; June 1959.)

This should be in excess of 10dB for a usable service, i.e. one with a low-bit error rate.

The basic thermal noise floor, assuming a 300 Hz receiver noise bandwidth (see later), is −149dBm, from equation [**19.31**]. To this must be added the effects of atmospheric noise using a noise factor obtained from the noise map (Figure 19.52). For this example a noise contour for the southern North Sea area (winter 0000–0004 hrs) of +60dBKTb has been chosen. As the contours are based upon observations at 1 Mhz, Figure 19.53 must be used to obtain a factor of approximately 70dBKTb at 518 kHz.

In addition to the effects of atmospheric noise, the noise factor of the receiver may also be considered. This will typically be 10–15dB. However when equation [**19.37**] is used to determine the overall operating noise figure for the system we see that with such a high atmospheric noise contribution the equipment noise effect is totally insignificant.

Adding the atmospheric noise to the thermal noise floor yields an overall received noise level which is approximately 49dB below the signal level. This results in a viable link with some 39dB of signal margin.

If, however, our vessel was operating off the coast of Mexico during the summer (2000–2400 hrs) the atmospheric noise figure could be as high as +105dBKTb, in which case the signal-to-noise ratio from a transmitter at a similar range would only be 14dB, i.e. approaching the threshold of usability.

20.2.2.2 Message coding and error protection

In view of the high noise levels and also interference from other systems, a need arises for a broadcast transmission technique that uses a narrow bandwidth, in order to minimize the received noise power, and a coding system that can be used to detect and correct transmission errors, and thereby improve the character error rate. Such a scheme is said to be forward error correcting (FEC).

In order to service the above need, and for reasons of equipment compatibility, the Navtex system uses one of the standard narrow band transmission techniques employed for telex over radio (TOR) systems. This is the collective B-mode of the direct printing system specified by CCIR recommendations 476 and 540. Prior to transmission, Navtex messages are converted into a continuous stream of binary digits, with each character being represented by a seven-element constant ratio (3:4) code group (see Figure 20.6). This code is shown in Table 20.1. It will be noticed that every character has three zeros and four ones. Such coding enables the receiver processor to identify characters which violate this ratio as incorrect.

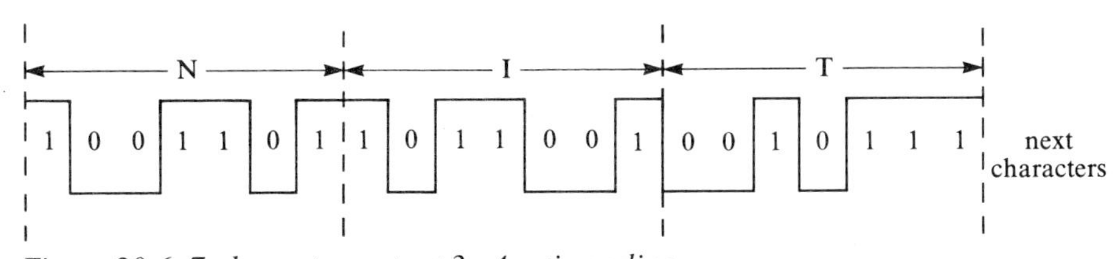

Figure 20.6 7-element constant 3 : 4 ratio coding.

Table 20.1 Seven-unit constant ratio code as used for Navtex and TOR transmission. Also five-unit stop-start code (international telegraph alphabet no. 2).

Letter shift	Figure shift	5 unit	7 unit
A	–	ZZAAA	BBBYYYB
B	?	ZAAZZ	YBYYBBB
C	:	AZZZA	BYBBBYY
D	$	ZAAZA	BBYYBYB
E	3	ZAAAA	YBBYBYB
F	%	ZAZZA	BBYBBYY
G	@	AZAZZ	BYBYBBY
H		AAZAZ	BYYBYBB
I	8	AZZAA	BYBBYYB
J	BEL	ZZAZA	BBBYBYY
K	(	ZZZZA	YBBBBYY
L	)	AZAAZ	BYBYYBB
M	.	AAZZZ	BYYBBBY
N	,	AAZZA	BYYBBYB
O	9	AAAZZ	BYYYBBB
P	0	AZZAZ	BYBBYBY
Q	1	ZZZAZ	YBBBYBY
R	4	AZAZA	BYBYBYB
S	:	ZAZAA	BBYBYYB
T	5	AAAAZ	YYBYBBB
U	7	ZZZAA	YBBBYYB
V	=	AZZZZ	YYBBBBY
W	2	ZZAAZ	BBBYYBY
X	/	ZAZZZ	YBYBBBY
Y	6	ZAZAZ	BBYBYBY
Z	+	ZAAAZ	BBYYYBB
CARR.RET		AAAZA	YYYBBBB
LINE FEED		AZAAA	YYBBYBB
LETTER SHIFT		ZZZZZ	YBYBBYB
FIG.SHIFT		ZZAZZ	YBBYBBY
SPACE		AAZAA	YYBBBYB
NO PERF		AAAAA	YBYBYBB
CONTROL SIG1	CS1		BYBYYBB
CONTROL SIG2	CS2		YBYBYBB
CONTROL SIG3	CS3		BYYBBYB
PHASING	α		BBBBYYY
PHASING	β		BBYYBBY
SIG REP	RQ		YBBYYBB

Where *B* represents the highest emitted frequency and Y the lowest. NB For Navtex and collective *B* mode TOR, B = '1' and Y = '0'.

In addition each character is sent twice, the first transmission (DX) of a specific character is followed by the transmission of four other characters, after which the retransmission (RX) of the first character takes place. This means that a time delay of 280 μs exists between the initial transmission and retransmission of each character (see Figure 20.7). Many noise and interference bursts and short-term propagation fades occur over periods shorter than 280 μs. As a result, this time diversity significantly increases the probability of receiving the correct character, albeit at the cost of repetition.

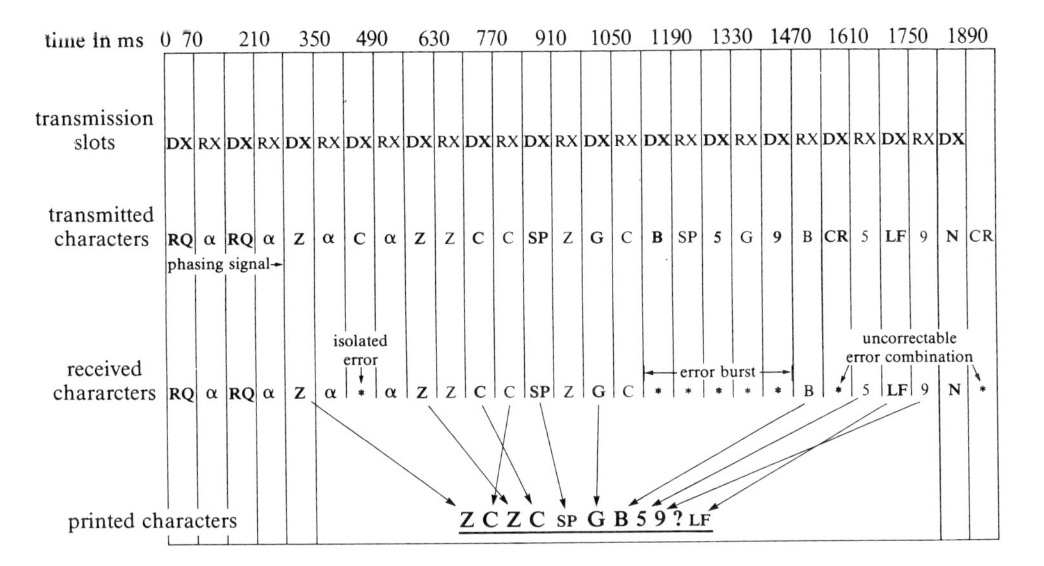

Figure 20.7 Typical Navtex (FEC) transmission reception and print sequence showing correction of single and burst errors.

As the code is received by the user equipment processor each character is inspected to ensure that the 3:4 code is correct. If a received code group violates this ratio the receiver will await the retransmission (RX) and check its ratio. If this is correct the character may then be printed, if not an error character is printed.

Clearly, this form of error protection is fallible since multiple-bit corruptions may be accepted as valid if the correct ratio is maintained. However, in practice the system is very robust and displayed messages contain surprisingly few errors.

20.2.2.3 Message format

The transmission format of each message is identical and contains important features which enable the user equipment to synchronize, process and edit. This format is as follows:

Phasing signal >10 sec	RQαRQαRQα
End of phasing sequence	ZCZC
Single space	
Preamble	$B_1B_2B_3B_4$
Carriage return	
Line-feed	
Message	
End group	NNNN
2 line feeds	
End signal	ααα in DX
or idle signal if more message follows.	

20.2.2.4 Modulation

The Navtex code is transmitted as a continuous data stream at 100 bits/sec via the narrow band audio frequency shift keying <FB1> of a single sideband transmitter. (This form of modulation was discussed in section 19.3.2.1.) The suppressed carrier frequency of the <FB1> transmission is 519 kHz but the energy is actually centred at 518 kHz because the data bits are represented as 517.915 kHz (zero) and 518.085 kHz (one). The user receiver/processor will convert the RF signal into an electrical code suitable for further processing. The minimum necessary transmission bandwidth for such a signal was shown to be 270 Hz, although 300–340 Hz is usually assumed for receiver filters.

20.2.2.5 Synchronization and signal processing

The user equipment processor will be continually processing all received signals and noise in an attempt to detect a valid Navtex transmission.

Its first task is to identify signals that have similar characteristics to a Navtex bit stream, i.e. the tone frequencies and durations ‘look’ correct. Remember that the signal may be partially mutilated by noise or interference. Once the processor is confident that it is looking at a valid stream of data bits (bit synch and timing) it can then set about identifying valid character strings within the continuous data stream and establishing DX and RX positioning (character synch). The synchronization or phasing process is aided by the phasing signal which is sent before the Navtex message (see section 20.2.2.3). This phasing sequence consists of alternate RQ and α signals transmitted for at least 10 seconds, with RQ in the DX slot and α in the RX slot. In addition, four RQ α signals are inserted in the traffic stream every 96 characters to enable stations to phase or re-phase during a message broadcast.

Once synchronized, the equipment must search for the sequence ZCZC in order to identify the all important message preamble which follows. Once identified, the processor can establish whether or not the message is to be printed by the criteria discussed in section 20.2.1. When the processor detects the string NNNN the processor will know that the current message has ended. If that is to be the end of the transmission then three consecutive α signals transmitted in the DX slot will be sent. If, however, a further message is to be sent the phasing sequence will be sent until the next ZCZC occurs.

The overall receiver/processor task is summarized in the flow graph shown in Figure 20.8.

20.2.3 *Equipment requirements*

The general user equipment requirements are for a unit which is fully automatic, save for service pre-selection, display/print control, and self-test initiation. In addition the unit should be compact, inexpensive, reliable and designed for a marine environment. Figure 20.9 shows the basic block diagram of a typical user equipment.

20.2.3.1 Antenna and receiver requirements

At medium frequencies, antennas are very much shorter than a wavelength (600m), and as a result they tend to be inefficient. This inefficiency can be reduced by employing an active antenna which significantly improves the match of the antenna to the receiver input and increases the effective height. A compact structure results which can be easily fitted to the smallest craft.

Signals from the antenna are fed to a receiver which must have adequate sensitivity to detect weak signals yet be able to withstand the high power levels associated with other transmissions on close adjacent channels, i.e. own and nearby ships' MF transmitters. The receiver must also have adequate long-term frequency stability as automatic operation precludes manual tuning. Short term stability must be sufficient for the +/− 85 Hz frequency shift of the modulated signal to be detected.

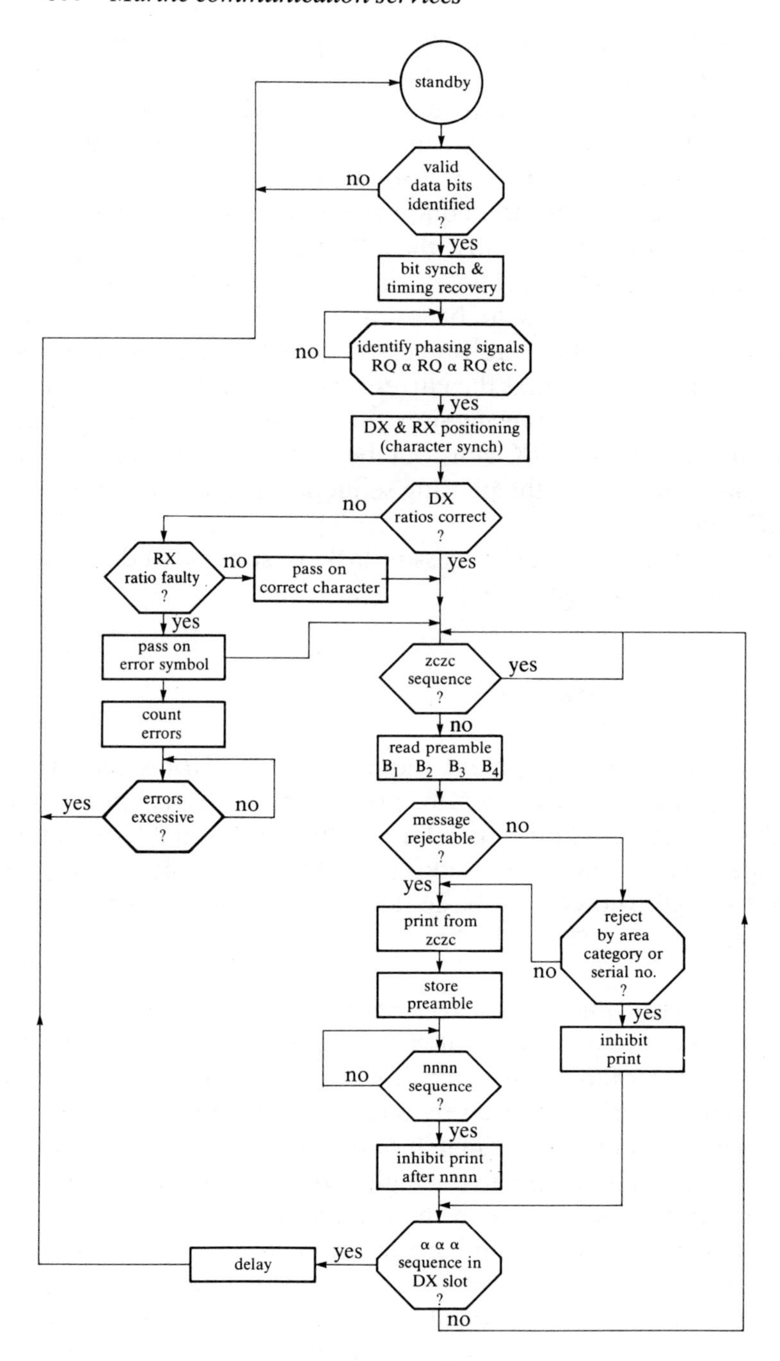

Figure 20.8 Simplified task sequence for Navtex user receiver/processor.

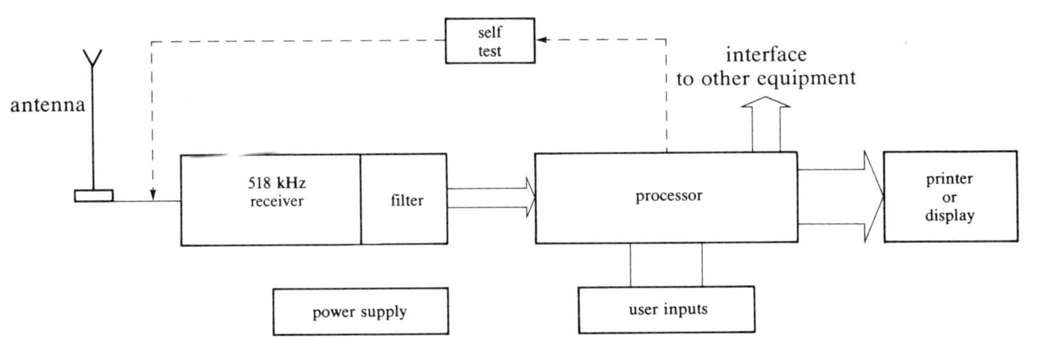

Figure 20.9 Simplified block diagram of Navtex user equipment.

The receiver must have a frequency response that allows the wanted transmission to be de-modulated but rejects noise and interference on adjacent frequencies (adjacent channel rejection). The minimum receiver bandwidth is set by the transmitted Navtex signal spectrum which as we have seen is about 300 Hz. A narrow band steep-sided filter response is therefore required.

20.2.3.2 Processor and display requirements

The processor unit must be able to perform at least the tasks outlined in Figure 20.8 and have a means of storing message preamble information for up to 72 hours. This unit must also accept the user inputs and control and feed the display unit, which may be a printer, VDU or LCD display.

As the equipment is essentially automatic and for use by untrained personnel it is vitally important that an effective and comprehensive self-testing capability is included.

Index